Sociological Footprints

Sociological Footprints

Introductory Readings in Sociology

ELEVENTH EDITION

LEONARD CARGAN
Wright State University

JEANNE H. BALLANTINE
Wright State University

WADSWORTH
CENGAGE Learning

Australia • Brazil • Japan • Korea • Mexico • Singapore • Spain • United Kingdom • United States

WADSWORTH
CENGAGE Learning

**Sociological Footprints:
Introductory Readings in Sociology,
Eleventh Edition**

Leonard Cargan and Jeanne H. Ballantine

Acquisitions Editor: Chris Caldeira

Assistant Editor: Melanie Cregger

Editorial Assistant: Rachael Krapf

Marketing Manager: Kim Russell

Marketing Assistant: Jillian Myers

Marketing Communications Manager:
Martha Pfeiffer

Project Manager, Editorial Production:
Jared Sterzer

Creative Director: Rob Hugel

Art Director: Caryl Gorska

Print Buyer: Paula Vang

Permissions Editor: Timothy Sisler

Production Service: Pre-Press PMG

Copy Editor: Pre-Press PMG

Cover Designer: Medby Pfeiffer

Cover Image: © Alan McEvoy

Compositor: Pre-Press PMG

For product information and technology assistance, contact us at
Cengage Learning Academic Resource Center, 1-800-423-0563

For permission to use material from this text or product, submit all requests online at **www.cengage.com/permissions**. Further permissions questions can be e-mailed to **permissionrequest@cengage.com**.

Library of Congress Control Number: 2008942958

ISBN-13: 978-0-495-60128-9

ISBN-10: 0-495-60128-4

Wadsworth, Cengage Learning
10 Davis Drive
Belmont, CA 94002-3098
USA

Cengage Learning products are represented in Canada by Nelson Education, Ltd.

For your course and learning solutions, visit **academic.cengage.com**

Purchase any of our products at your local college store or at our preferred online store **www.ichapters.com**

Printed in Canada
1 2 3 4 5 6 7 8 9 12 11 10 09

Contents

8 Economics and Politics: Implications for Survival   235

PART IV Inequalities Between Groups   289

9 Stratification: Some Are More Equal Than Others   293

10 Race and Ethnicity: The Problem of Inequality   317

Preface

THE PRIMARY OBJECTIVE OF THIS ANTHOLOGY is to provide a link between theoretical sociology and everyday life by presenting actual samples of both classical and current sociological studies. If students are to grasp the full meaning of sociological terms and topics, they must be able to translate the jargon of sociology into real and useful concepts that are applicable to everyday life. To this end, *Sociological Footprints* presents viewpoints that demonstrate the broad range of sociological applications and the values of sociological research.

Selecting the readings for the eleventh edition involved a number of important steps. As with the previous ten editions, we constantly received feedback from hundreds of students. Feedback was also requested from colleagues who are knowledgeable about the various topics of this anthology. An exhaustive search of the literature was conducted for additional material that was interesting and highly readable, that presented concepts clearly, that represented both recent and classic sociology, and that featured authors of diverse backgrounds. In meeting these criteria we often had to replace popular readings with more comprehensive and up-to-date ones. About one-third of the articles included in this edition are new selections that update all issues and address new concerns. As a final step, we utilized reviewers' comments to make the anthology relevant and useful. In this manner, each edition of *Sociological Footprints* becomes the strongest possible effort in producing a sociologically current, interesting, and highly readable collection.

IN THIS EDITION

The eleventh edition builds on the strengths of early editions but also changes with current research and events. Students will gain from reading selections from classical sociology as well as selections from contemporary societal issues.

The relevance of this collection is seen in the variety of the topics selected and in our efforts to cover the topics found in introductory sociology courses.

A fitting beginning for the material on the sociological perspective are two articles written by the most recognized name in social methods. Earl Babbie. He tells us that the time has come for the recognition of sociology for its contributions to science and for its scientific methodology. These premises of Babbie are confirmed in the third article since it tells us the many things we should know after taking this introductory sociology course (Persell, et al). Mills reaffirms this premise by noting that we will indeed learn these things and more.

The second part of this text explains what it means to be a member of society. To become a member of any society we must be socialized with its norms and values, that is, with the culture of our society—a culture which can be as widely different as the Nacirema and the sacred cow culture in India. We will also be socialized to act in specific ways particular to our biological gender since this is not biologically determined. As Davis notes, none of this can happen if we are isolated from interaction with other members of society. All of the things we learn may not be positive since we could also learn to be violent and to practice gender inequality, but this can also be prevented.

Part III covers the five major institutions in our society. Perhaps the most discussed one of these is that of marriage and family. Despite this title, the first article deals with the large and growing number of adult singles living on this earth—Noah's Ark. Most of us have an idealized image of what marriage and family is supposed to be—an image derived from the good times of the 1950s; but as Skolnick and Skolnick note this institution is constantly changing and change can produce problems.

The two institutions of education and religion have been combined since both are related and are undergoing much criticism—they are, in a sense, institutions in the crossfire. For education, this criticism is seen in the way a child becomes a student and in the fact that education has inequalities and in increasing segregation. For religion the situation is similar but to understand this factor we first have to know what religion is by turning to the first scientific sociologist—Emile Durkheim. Now we are prepared to understand that the different forms of religion can lead to political extremism or even an attempt to retreat from religion.

What many may not realize is the close relationship between the economic and political institutions. All the economic questions dealt with have political repercussions since they will require political efforts to deal with their ramifications—what is the future of work and what welfare should include. At first, that commonly used card, the credit card, would not seem to be politically related, but as Ritzer points out it has both personal troubles and its large use makes it a public issue. Perhaps it is the economic issue that leads to the severe political outlook of groups wanting to kill all the enemies of the system, even a stranger. At this point it is necessary to revisit the social requisites of a democracy.

It would seem that most of the topics in this anthology deal with inequality, yet we have chosen to separate into a part alone only the inequalities pertaining to stratification and ethnicity/race. When it comes to class stratification, it seems that we all want the money and the lifestyle of the upper class but what is true of our democracy—that there is inequality on the basis of class stratification and this is also true on an ethnic and a racial basis.

The final part of this text covers a number of issues dealing with living in a society. It starts fittingly enough with one of the most discussed and misunderstood issues of the day—deviance from the norms of society via crime despite the fact that crime is normal. Despite its normalcy, we live in a culture of fear of it and so all people would like to see crime reduced.

Another important topic is that of health since the baby boomers of the 50s are now reaching retirement age and an age of increasing age- related problems. In short, we are living in a world growing old.

When it comes to the environment, we are increasingly reminded that the world's population is rapidly growing despite cutbacks in the number of children per family and the mass killings by terrorist or war acts. In addition world population are increasingly moving into urban areas.

The final chapter in this anthology reminds us of the fact that we are a society in constant change because of various movements such as the gay/lesbian one, the constant terrorism ones, or those set off by various mega-trends.

As can be seen, this anthology covers a wide range of contemporary sociological topics. A short list shows that it covers the possibility of solving the problem of gender inequality, what is the real story on divorce trends, the inequalities existing in our schools, and political extremism in religion. It also contains an important topic for students preparing for the world of work when there is the possibility of no work being available. Most of us carry one or more credit cards but we consider their impact on the economy and on ourselves, except when payment is due. With terrorist acts almost a daily encounter, one wonders what are the ultimate aims of these killings? At tax time, we are reminded of the growing inequality in our democracy. Another daily item in the news is that of crime. It almost seems to be a normal occurrence. Is it?

In sum, *Sociological Footprints* contains a balance of readings in each major chapter that—according to students, instructors, and reviewers—make this collection a valuable excursion into sociology. Although several of the articles have been condensed, the main theme and ideas of each have not been altered. Digressions, repetitions, and detailed descriptions of quantitative data have been omitted in order to emphasize key points.

FEATURES AND ORGANIZATION

We hope this new edition of *Sociological Footprints* will be as valuable to teachers as it is to students—an intention reflected in the book's organization. First, each major part has an introduction that covers the major themes of the topic area, noting how each reading relates to these themes. Second, each reading is also introduced by a comment about the important points in the reading. Third, we provide questions before each reading to guide the reader toward the important points. Fourth, although anthologies do not usually define concepts used in their readings, before many readings we include a glossary of important terms to give students a basic understanding of special terminology.

SUPPLEMENTS

Instructor's Manual with Test Bank

This instructor resource offers instructors teaching suggestions. The manual also contains a summary analysis of each article in the reader, stating the thesis, findings, and conclusions of each reading. Test items are also provided, including several multiple-choice questions and up to five essay questions for each article.

InfoTrac® College Edition with InfoMarks™

Available as a free option with newly purchased texts, InfoTrac College Edition gives instructors and students four months of free access to an extensive online database of reliable, full-length articles (not just abstracts) from thousands of scholarly and popular publications going back as much as 22 years. Among the journals available 24/7 are *American Journal of Sociology, Social Forces, Social Reasearch*, and *Sociology*. InfoTrac College Edition now also comes with InfoMarks, a tool that allows you to save your research parameters, as well as save links to specific articles. (Available to North American college and university students only; journals are subject to change.)

ACKNOWLEDGMENTS

We wish to thank all those who made this edition of *Sociological Footprints* possible. The reviewers of this edition are Amy Karnehm Willis, North Carolina Wesleyan College; Lynne Moulton, SUNY College at Brockport; Bucky Dann, Southwestern Community College; Craig Reinarman, University of California; Laura Colmenero-Chilberg, Black Hills State University; Jodi Cohen, Bridgewater State College, MA; and Gordon Arnold, Department of Liberal Arts Montserrat College or Art. We also thank reviewers of previous editions: Sarah N. Gatson, Texas A&M University; Kristin Marsh, Mary Washington College; Timothy McGettigan, Ph.D., University of Southern Colorado; Mohsen M. Mobasher, Southern Methodist University; and Carol Ray, San Jose State University; Philip Berg, University of Wisconsin-LaCrosse; Kevin J. Christiano, University of Notre Dame; Rodney B. Coates, University of North Carolina at Charlotte; Thomas F. Courtless, George Washington University; Robert W. Duff, University of Portland; Irene Fiala, Baldwin-Wallace College; Michael Goslin, Tallahassee Community College; Susan F. Greenwood, University of Maine; William J. Miller, Ohio University; Martin Monto, University of Portland; Wilbert Nelson, Phoenix College; Dan J. Pence, Southern Utah University; Ralph Peters, Floyd College; Carol Ray, San

Jose State University; David R. Rudy, Moorehead State University; George Siefert, Daemen College; Eldon E. Snyder, Bowling Green State University; Larry L. Stealey, Charles Stewart Mott Community College; and Jerry Stockdale, University of North Iowa; Debbie A. Storrs, University of Idaho; Donna Trent, Eckerd College; and Assata Zerai, Syracuse University. Heartfelt thanks, also, go to the many students who took the time to give us their opinions, the departmental secretary Susan Schulthesis and aide Jennifer Meininger Deluca, who helped to assemble and type the material, our proofreader, and to all the good people at Wadsworth who aided in the production of this anthology.

To The Student

THE PURPOSE OF THIS ANTHOLOGY IS to introduce you, the beginning student in sociology, to a wide range of sociological perspectives and to demonstrate their relevance to real-life situations. As you apply sociological perspectives to everyday events, you will begin to realize that sociology is more than jargon, more than dry statistics, more than endless terminology to be memorized. It is an exciting and useful field of study. Unfortunately, no textbook can fully describe the many applications of sociology. This anthology should help to fill the gap by supplying classical readings balanced with contemporary readings on current issues and research.

From our experience in teaching introductory sociology, we know some of the problems that anthologies can present to the student: unexplained terms, readings seemingly unrelated to the text, and different emphases from those of the instructor's lectures. Therefore, to enjoy and benefit fully from *Sociological Footprints*, you should take the following steps:

1. Read and study the related textbook chapter and lecture materials. You must be familiar with the concepts and perspectives before you can clearly observe their daily application.

2. Read the introductions to the assigned sections in the anthology. They are designed to summarize the primary themes of the topic area and relate them to specific readings. In fact, the introductions will not only make the readings easier to understand, they will facilitate your application of the readings to other class materials and real-life situations.

3. Use the glossary that precedes the selection before you read each reading. Knowing the terms will make the reading more interesting and understandable.

4. Read each reading thoroughly. Note the problem or issue being discussed, the evidence the author supplies in support of his or her contentions, and

the conclusions drawn from this evidence. Answer the questions posed at the beginning of the piece.

5. Summarize the main ideas of each reading in your own terms, relating them to other material in the course and to your own everyday experiences.

Step 5 is particularly important. Many of the readings address topics of current interest—political issues, population problems, environmental issues, the women's movement, and more. Because these are contemporary problems, you will see related materials in newspapers and magazines and on television. By applying what you have learned from the lectures and this anthology, you should develop a clearer understanding of current issues and of how sociology has aided you in this understanding.

We feel strongly that sociology is a field of study highly relevant to your world and that it can give you a fuller comprehension of day-to-day living. Our aim has been to provide you with a readable, understandable, and enlightening anthology that will convey this relevance.

THE ESSENTIAL WISDOM OF SOCIOLOGY

(Paraphrased from a paper by Earl Babbie from the 1989 ASA annual meeting) "I say this by way of a disclaimer. The essential wisdom of sociology may have twelve or thirteen points, but I'm going to quit at ten."

1. Society has a *sui generis* existence or reality. "You can't fully understand society by understanding individual human beings who comprise it. For example, few people want war, but we have wars all the time."

2. It is possible to study society scientifically. "Society can be more than learned beliefs of 'common sense.' It is actually possible to study society scientifically, just as we study aspects of the physical world."

3. Autopoesis: society creates itself. "Autopoesis (Huberto Maturana's term) might be defined as 'self-creating.' A powerful statement that sociology has to offer is that society is autopoetic: society creates itself."

4. Cultural variations by time and place. "Gaining awareness that differences exist is only the beginning, however . . . Our second task in this regard is to undermine . . . implicit ethnocentrism, offering the possibility of tolerance."

5. Relation of individual and society. "[One might] want to skirt the edge of suggesting that individuals are merely figments of society. Without going quite that far [one might suggest that] individual identity is strongly sociogenetic."

6. System imperatives. "Society is an entity [and] as a system, it has 'needs'."

7. The inherent conservatism of institutions. "The first function of an institution is institutional survival."

8. Determinism. "...We operate with a model that assumes human behavior is determined by forces and factors and circumstances that the individual actors cannot control and/or are unaware of."

9. Paradigms. "...paradigms are ways of looking at life, but not life itself. They focus attention so as to reveal things, but they inevitably conceal other things, rather like microscopes or telescopes, perhaps. They allow us to see things that would otherwise be hidden from us, but they do that at a cost."

10. Sociology is an idea whose time has come. "Finally...on a possibly chauvinist note: All the major problems that face us as a society and as a world are to be found within the territory addressed by sociology. I say this in deliberate contrast to our implicit view that most of our problems will be solved by technology."

Introduction:
Why Study Sociology?

WHAT IS THIS SUBJECT CALLED SOCIOLOGY? What will I learn from studying sociology? Why should I take sociology? What work do sociologists do? How is sociology useful to me or to the world? If I major in sociology, what can I do when I graduate? These are some of the questions that may be in the back of your mind as you approach your study of sociology. Perhaps you are reading this book because you are curious about the subject, or because sociology is a required course, or because you had sociology in high school and wanted to find out more about it, or because your instructor assigned the book and this article. Whatever the reasons, you will find an introduction to the field of sociology in the discussion that follows.

What you read in the next few pages will only begin to answer the questions just posed. As you learn more about sociology, pieces that at first seemed fragmentary will start to come together like pieces in a puzzle. These pages provide the framework into which those pieces can be placed to answer the opening question: Why study sociology?

WHAT IS THIS SUBJECT CALLED SOCIOLOGY?

First questions first: Sociology is the study of people in groups, of people interacting with one another, even of nations interacting during peace or war. Sociologists' interests are sparked when they see two or more people with a common interest talking or working together. They are interested in how groups work and in how nations of the world relate to one another. When two or more people are interacting, sociologists have the tools to study the process. It could be a married couple in conflict or a teacher and students in a classroom situation; it could be individuals interacting in a work group, sports teams on a playing field, or negotiating teams discussing nuclear disarmament.

Sociology shares a common bond with other social sciences. All are concerned with human behavior in society; they share the perspective of the scientific method and some of the same data-collection methods to study their subject matter. Sociology is the broadest of the social sciences; its main concern is with predicting human group behavior.

"That's a lot to be interested in," you may be saying. In fact, most sociologists specialize. No sociologist is likely to be an expert in everything, from studies of a few people or small group interaction (microlevel sociology) to large numbers of people in big groups like organizations or nations (macrolevel sociology). Consider the following examples of sociological specializations:

- determining the factors that lead to marital longevity
- identifying effective teachers by classroom observation
- examining public attitudes about the presidency and its policies
- locating satisfaction and problems in certain jobs

The results of these diverse interests lead sociologists into many different areas. Some sociologists specialize in social psychology, a field that considers such questions as how individuals behave in groups, who leaders are and what types of leaders are effective, why some groups accomplish more than other groups, why individuals usually conform to group expectations, and many other topics involving individuals as functioning members of groups. Another area of specialization is political sociology, which studies political power, voting behavior, bureaucracy, and political behavior of individuals and groups. Anthropology examines the culture of different groups; so does sociology. But the methods of study and primary focus differ. Anthropologists often study pre-literate groups, whereas sociologists focus primarily on modern groups. Another area that concerns sociologists is social history, which emphasizes the use of history to understand social situations. These are only a few examples of the diverse interests of sociologists and how sociology shares its interests with some other social sciences.

WHAT WILL I LEARN FROM STUDYING SOCIOLOGY?

Consider that in some societies premarital sex is not only allowed but expected; in others, premarital sex is cause for banishment and death. Even though sociologists, like everyone else, have personal opinions, the task of the sociologist is not to judge which social attitude is right or wrong but to understand *why* such divergent practices have evolved. We all have opinions. Usually they come from our experiences, common sense, and family teaching. Some opinions are based on stereotypes or prejudices, some on partial information about an issue. Through systematic scientific study, sociologists gain insight into human behavior in groups, insight not possible through common sense alone. They attempt to understand all sides of an issue; they refrain from making judgments on issues of opinion but try instead to deal objectively with human behavior.

Consider the person who is going through the anguish of a divorce. Self-blame or hostility toward the spouse are often reactions to this personal crisis. Sociology can help us move beyond "individual" explanations to consider the social surroundings that influence the situation: economic conditions, disruptions caused by changing sex roles, and pressures on the family to meet the emotional needs of its members. Thus, sociology teaches us to look beyond individual explanations of our problems to group explanations for behavior; this practice broadens our world-view and gives us a better understanding of why events take place.

A typical college sociology program starts with a basic course introducing the general perspective of sociology; sociological terminology and areas of study; how sociologists get their information, that is, their methods; and the ideas, or theories, that lay the foundations for sociological study. Further sociology courses deal in greater depth with the major components of all societies: family, religion, education, politics, and economics. The sociology department may also offer courses on social processes such as social problems, deviance and corrections, stratification, socialization, and change, or in other areas of social life such as medical, community, urban, sports, or minority sociology.

Family sociology, for instance, usually considers the family social life cycle: young people breaking away from their parents' home, forming a home of their own by selecting a spouse through the courtship process, marrying, selecting a career, making parenting decisions, raising a family, having their children leave home, retiring, and moving into old age.

Students who major in sociology generally take courses in *theory*—the basic ideas of the field—and *methods*—how sociologists approach the social world objectively and do their research. Some sociology departments offer practical experiences where students can use their sociological skills in a job setting.

These are a few examples of what you will learn from the study of sociology and how you will learn it. There is much more to the field of sociology than this, however.

WHY SHOULD I TAKE SOCIOLOGY?

Whether you take a number of sociology courses or only one, you will profit in a number of ways. You will gain personal knowledge, new perspectives, skills needed by employers, background training useful in entering other fields, personal growth and development, new perspectives on the world, and a new way of looking at your relationships with others and your place in society. You will gain tolerance for and fascination with the variety of people in the world around you and their cultural systems. You will be able to understand your interactions with your family and friends better; you will be able to watch the news or read the paper with keener perception. You will have an understanding of how to obtain information to answer questions that you or your boss need answered. And the more sociology you take, the more ability you will have to express your thoughts logically, objectively, and coherently.

It is nice to know that the subjects you take in college will have some personal relevance and professional usefulness. Sociology should provide you with a number of "life skills," such as

1. Ability to view the world more objectively
2. Tools to solve problems by designing studies, collecting data, and analyzing results
3. Ability to understand group dynamics
4. Ability to understand and evaluate problems
5. Ability to understand your personal problems in a broader social context

We know from studies that employers value those applicants with the broad training of such fields as sociology because of the skills they provide. The following are skills employers look for, in order of importance:

1. Ability to work with peers
2. Ability to organize thoughts and information
3. Self-motivation
4. Ability to plan effectively
5. Willingness to adapt to the needs of the organization
6. Ability to interact effectively in group situations
7. Self-confidence about job responsibilities
8. Ability to handle pressure
9. Ability to conceptualize problems clearly
10. Effective problem-solving skills
11. Effective leadership skills
12. Ability to listen to others

Although a college graduate in engineering, computer sciences, or business may enter the job market with a higher salary, the sociology liberal arts major is more likely to rise through the managerial and professional ranks to positions of responsibility and high pay. Businesses and organizations value the skills listed here. In today's rapidly altering society, many of us will change jobs or careers several times during a lifetime. Sociological skills can help us adapt to the expectations of new situations.

Because of the knowledge and skills learned in sociology courses, study in this area provides excellent preparation for other undergraduate and graduate fields. From nursing, business, and education to law and medicine, the knowledge of sociology can be applied to a wide variety of group situations. For instance, a current concern of sociologists who study educational settings is what characteristics make schools effective; by singling out certain characteristics, sociologists can make recommendations to improve schools. Teachers and educational administrators profit from this information.

If we are curious about understanding ourselves and our interactions with others and about why our lives take certain directions, sociology can help us understand. For instance, sociologists are interested in how our social-class standing affects how we think, how we dress, how we speak, what our interests are, whom we are likely to marry, what religion (if any) we belong to, and what our "life chances" are, including how long we will live and what we are likely to do in life. Sociologists have even examined how individuals from different social-class backgrounds raise their children, and implications of child-rearing techniques for our lifestyles. Some use physical punishment and others moral chastisement, but the end result is likely to be a perpetuation of the social class into which we are born.

WHAT WORK DO SOCIOLOGISTS DO?

The most obvious answer is that sociologists *teach;* this is primarily at the higher education level, but high school sociology courses are also offered as part of the social science curriculum. There would be nothing to teach if sociologists were not actively engaged in learning about the social world. Their second major activity is to conduct *research* about questions concerning the social world.

Many sociologists work in business organizations, government agencies, and social service agencies. *Practicing sociologists* are engaged in a variety of activities. Some do family counseling with the whole family group; some conduct market research for companies or opinion polls for news or other organizations; some do surveys for the government to determine what people think or need; some work with juvenile delinquents, prison programs and reforms, and police; some predict how population changes will affect schools and communities.

Applied sociologists use their sociological knowledge to help organizations. They assess organizational needs, plan programs to meet those needs, and evaluate the effectiveness of programs. For instance, a community may want to know how many of its elderly citizens need special services to remain at home rather than be moved to nursing institutions. Sociologists assess this need, help plan programs, and evaluate whether programs are meeting the needs they set out to meet.

The position a sociology major ultimately gets depends in part on the degree he or she holds in sociology. The following are some examples of jobs students have gotten with a B.A. or B.S. degree: director of county group home, research assistant, juvenile probation officer, data processing project director, public administration/district manager, public administration/health coordinator, law enforcement, labor relations/personnel, police commander/special investigations, trucking dispatcher, administrator/social worker, counselor, child caseworker, substance abuse therapist, medical social worker, data programming analyst, activities director at senior citizens center, director of student volunteer program, area sales manager, jury verdict research editor, insurance claims adjustor, employment recruiter, tester for civil service, unemployment office manager, child services houseparent, crisis worker volunteer, advertising copywriter, probation officer, travel consultant, recreation therapist, public TV show hostess, adult education coordinator, research and evaluation specialist, neighborhood youth worker.

Sociologists holding an M.A. or Ph.D. degree are more skilled in sociological theory and methods than B.A. degree holders. They are often involved in research, teaching, or clinical work with families and other clients.

HOW IS SOCIOLOGY USEFUL TO ME
AND TO THE WORLD?

Technology is rapidly changing the world. New policies and programs are being implemented in government and private organizations—policies that affect every aspect of our lives. Because sociologists study social processes, they are able to make concrete contributions to the planning of orderly change. Sociological knowledge can also be useful to legislators and courts in making policy decisions. For example, sociologists can assist a juvenile facility to design programs to help young people convicted of crime redirect their energies; how successful such programs are in achieving their goals can be studied by evaluation research.

In summary, sociology is the broadest of the social sciences and, unlike other disciplines, can give us an understanding of the social world. The knowledge and tools of sociology make students of this field valuable in a number of settings, from business to social service to government to education. As you embark on this study, keep in mind that sociology helps us have a deeper understanding of ourselves and our place in the world as well.

Sociology is a study of all people, for all people. To enjoy your encounter with the field and to make the most use of your time in sociology, try to relate the information you read and hear to your own life and relationships with others within the broader context of your social world.

How Will You Spend the 21ˢᵗ Century?

PETER DREIER, OCCIDENTAL COLLEGE

The following article, How Will You Spend the 21st Century?, describes some of the occupations sociology majors have achieved and the many tasks that people with sociology degrees can take on.

...I assume some of the parents in the audience today are like my parents were, 30 years ago, when I told them I was going to major in sociology. They weren't quite sure what sociology was, or whether you could get a job with a degree in sociology.

My father was worried that I might become a *social worker*. My mother was worried that I might become a *socialist*.

Well, let me assure you that your sons and daughters will be able to put their sociology degrees to good use. Some of our nation's most outstanding leaders, today and in the past, majored in sociology.

I'm a sports fan, so I've created a Sociology All-Star Team. The team captain is *Regis Philbin*, who majored in sociology at Notre Dame.... Other sociology majors in the world of entertainment include comedian *Robin Williams*, actor *Dan Aykroyd*, *Paul Shaffer* (the band leader on the David Letterman show), and Oscar-nominated actress *Deborah Winger*. And those of you who grew up in the 50s will remember the singer and TV star *Dinah Shore*, who studied sociology at Vanderbilt. Another sociology major from the world of arts and culture is novelist *Saul Bellow*, winner of the Nobel Prize for literature. The world of sports includes *Alonzo Mourning* (the Miami Heat's All-Star center); *Joe Theisman* (the NFL Hall of Fame quarterback); *Brian Jordan*, the Atlanta Braves' star outfielder; and, from the University of Oregon, *Ahmad Rashad*, the sportscaster and former football star.

Sociology has been a launching pad for people into the world of politics and law. A good example is *Richard Barajas*, chief justice of the Texas Supreme Court, who majored in sociology at Baylor University.

Over the years, quite a few sociology majors have been elected to political office. For example, *Shirley Chisholm*, the first black woman elected to Congress, in 1968, majored in sociology at Brooklyn College. The current Congress includes *Maxine Waters* (the Congresswoman from Los Angeles) and U.S. Senator *Barbara Milkulski* of Maryland.

Quite a few urban majors are on our sociology All-Star team, including *Wellington Webb*, the Democratic mayor of Denver; *Brett Schundler*, the Republican major of Jersey City, New Jersey; and *Annette Strauss*, the former mayor of Dallas.

Source: Reprinted with permission

Who can name a President of the United States who majored in sociology? The answer is *Ronald Reagan*, who has a sociology degree from Eureka College in Illinois.

Sociology is perhaps most well-known as a training ground for social reformers. Whether they go into politics, law, teaching, business, journalism, the arts, urban planning, the clergy, or any other field, they see their professional careers as a means to improve society and help others. Most sociologists, in other words, are *practical idealists*.

I think playwright George Bernard Shaw had the best understanding of sociology. He said: "Some men see things the way they are and ask: why? Others dream things that never were and ask: why not?" One of those practical idealists was *Saul Alinsky*, the founder of community organizing, who studied sociology at the University of Chicago. Another was *Martin Luther King*, who majored in sociology at Morehouse College in Atlanta.

On the list of great Americans who studied sociology, one of my favorites is Frances Perkins. She may not be a well-known name to many of you, but she was one of the most influential social reformers in American history. Frances Perkins was part of the first generation of women to attend college, entering Mt. Holyoke College in 1898. In one of her courses, students were required to visit a factory and do a survey of its working conditions. Perkins visited several textile mills and paper mills. There she saw the dangerous conditions and low pay that workers endured every day. This project opened her eyes to how "the other half lived."

After graduating in 1903, Perkins got involved in social work among poor immigrants and did extensive sociological research about slum housing and unsafe working conditions in laundries, textile mills, and other industries. She was soon recognized as a national expert in the new field of industrial sociology.

The tragedy of the Triangle Fire galvanized New York City's social reform groups. Perkins became the head of a citizens group called the Committee on Safety. Thanks to this group, within a few years, New York State had enacted 36 new laws protecting workers on the job, limiting the hours of women and children, and compensating victims of on-the-job injuries. Perkins continued this kind of social reform work for the rest of her life. In 1932, President Franklin Roosevelt asked her to become the nation's secretary of labor, the first woman ever to hold a cabinet position, where she became the central figure in the New Deal's efforts to improve the lives of America's poor, unemployed, and elderly. These included the passage of the Social Security Act and of the Fair Labor Standards Act, which established the minimum wage and the eight-hour day. This social legislation forever changed the living and working conditions of most Americans. Frances Perkins was in college 100 years ago. Try to imagine yourselves sitting in a commencement ceremony in 1901.

It is the beginning of a new century. What was America like back then? What kind of society were sociology majors like Frances Perkins about to inherit? In 1901, women didn't have the right to vote. Suffragists, who fought to give women that right, were considered radicals and utopians. Few people could look forward to retirement. Most people worked until they were no longer physically able to do so. And when they could no longer work, they often fell into poverty. A hundred years ago, reformers were calling for "social insurance" for the elderly.

In 1901, lynching was a regular occurrence in the South. Lynching kept black people terrorized. The NAACP was founded back then to fight to outlaw lynching and to abolish laws that denied black people the right to vote.

One hundred years ago, conditions in our factories and our urban housing were incredibly dangerous. Many people were regularly killed or seriously injured on the job. Many apartments were constructed so poorly that they were often fire traps, lacking ventilation. Epidemic diseases like TB were widespread because there were no laws dealing with basic sanitation. Back then, sociologists documented these conditions and worked with reformers for basic changes like government regulations regarding minimal safety standards for factories, schools, and apartment buildings as well as for laws outlawing the exploitation of child labor.

One hundred years ago, these and many other ideas, that today we take for granted—laws protecting consumers from unhealthy and unsafe food; laws regulating air pollution from factories and cars; Pell grants to help students pay college tuition; a minimum wage; government health insurance for the elderly and poor—were considered dangerous, or impractical, or even socialistic.

Each of these ideas has improved the day-to-day lives of Americans. Today, Americans enjoy more rights, better working conditions, better living conditions, and more protection from disease in childhood and old age than anyone could have imagined 100 years ago.

Thanks to Frances Perkins and people like her, America is a much better society than it was 100 years ago.

But that doesn't let you off the hook! There are still many problems and much work to do. Like all agents for social change, whether or not they studied sociology in college, Frances Perkins and Martin Luther King understood the basic point of sociology, that is, to look for the connections between people's everyday personal problems and the larger trends in society.

Things that we experience as personal matters— a woman facing domestic violence, or a low-wage worker who cannot afford housing, or middle-class people stuck in daily traffic jams—are really about how our institutions function. Sociologists hold a mirror up to our society and help us see our society *objectively*. One way to do this is by comparing our own society to others. This sometimes makes us uncomfortable—because we take so much about our society for granted. Conditions that *we* may consider "normal," other societies may consider serious problems.

For example, if we compare the U.S. to other advanced industrial countries like Canada, Germany, France, Sweden, Australia, Holland, and Belgium, we find some troubling things:

- The U.S. has the highest per capita income among those countries. At the same time, the U.S. has, by far, the widest gap between the rich and the poor.

- Almost 30 percent of American workers work full-time, year-round, for poverty-level wages.

- The U.S. has the highest overall rate of poverty. More than 33 million Americans live in poverty.

- Over 12 million of these Americans are children. In fact, one out of six American children is poor. They live in slums and trailer parks, eat cold cereal for dinner, share a bed or a cot with their siblings, and sometimes with their parents, and are often one disaster away from becoming homeless.

- Approximately four million American children under age 12 go hungry.

- Only three out of five children eligible for the Head Start program are enrolled because of the lack of funding.

- About seven million students attend school with life-threatening safety code violations.

- The U.S. has the highest infant mortality rate among the major industrial nations.

- One fifth of all children under two are *not immunized* against serious diseases.

- The U.S. is the only one of these nations without universal health insurance. More than 43 million Americans—including 11 million children—have no health insurance.

- Americans spend more hours stuck in traffic jams than people in any of these other countries. This leads to more pollution, more auto accidents and less time spent with families.

- Finally, the U.S. has a much higher proportion of our citizens in prison than any of these societies . . .

. . . What would you like *your* grandchildren to think about how *you* spent the 21st century? . . . No matter what career you pursue, you have choices about how you will live your lives. As citizens, you can sit on the sidelines and merely be *involved* in your society. Or you can decide to become really *committed* to making this a better world.

What's the difference between just being *involved* and really being *committed?* Think about the eggs and bacon that you had for breakfast this morning. The hen was *involved*. But the pig was really *committed!*

Today, there are hundreds of thousands of patriotic Americans committed to making our country live up to its ideals. Some focus on the *environment*, others focus on education, and still others focus on *housing*, or working conditions, or *human rights*, or *global trade*, or *discrimination against women, minorities, and gays and the physically disabled.*

They are asking the same questions that earlier generations of active citizens asked: Why can't our society do a better job of providing equal opportunity, a clean environment, and a decent education for all? They know there are many barriers and obstacles to change, but they want to figure out how to overcome these barriers, and to help build a better society.

So ask yourselves: What are some of the things that *we* take for granted today that need to be changed? What are some ideas for changing things that today might seem "outrageous," but that—25 or 50, or 100 years from now—will be considered common sense?

In fact, your generation has done quite well already. The media stereotypes your generation as being apathetic—but the reality is that a record number of college students today are involved in a wide variety of "community service" activities—such as mentoring young kids in school, volunteering in a homeless shelter, or working in an AIDS hospice.

As a result of this student activism, more than 100 colleges and universities have adopted "anti-sweatshop" codes of conduct for the manufacturers of clothing that bear the names and logos of their institutions.

Positive change *is* possible, but it is *not* inevitable. For about the last decade, America has been holding its breath, trying to decide what kind of country we want to be. I am optimistic that your generation will follow in the footsteps of Frances Perkins and Martin Luther King—not only when you're young, but as a lifelong commitment to positive change.

I know you will *not* be among those who simply "see things the way they are and ask: why?" Instead, you will "dream things that never were and ask: why not?"

PART I

The Discipline of Sociology

Although the term **sociology** *may be familiar, many students are unaware of the areas of study included in the field. Sociology ranges from the study of small groups—perhaps two people—to that of such large entities as corporations and societies. As noted in the following reading, it is the study of interactions within, between, and among groups; and these group interactions encompass all areas of human behavior.*

Leading thinkers in all ages have been concerned about society, human conduct, and the creation of a social order that would bring forth the best man is capable of. But, the study of these problems with the techniques and approaches of science (sociology) is only a little more than a century old.

It was only about 125 years ago that Auguste Comte published his *Cours de philosophie positive*, which first included sociology as one of the scientific disciplines. No course in sociology was available in an American university until 1876 at Yale....Before 1900 all the men who identified themselves as professional sociologists were trained originally in other fields such as history, politics, economics, law, and religion. Today, undergraduate students can obtain training in sociology in almost all American four-year liberal arts colleges and in many agricultural colleges and specialized schools; more than 70 schools offer a doctoral program in sociology and many additional schools offer master's degree programs.

Before World War I the opportunities for employment of men and women with professional training in sociology were largely limited to college teaching and research. Besides teaching and research, sociologists today are engaged in more than 25 different kinds of work in professional schools, in local, state, federal,

Reprinted by permission of the Sociology Department, University of Kentucky.

and private agencies, and in business. They work in the fields of education, medicine, law, theology, corrections, agricultural extension, welfare, population study, community development, health, technological change, and the like. In short, sociologists are working on almost all the problems that concern man in relation to his fellow man and the consequences of this relationship for himself and others.

Chapter 1

The Sociological Perspective

Is It Just Common Sense?

Jeanne Ballantine

In the first reading of this chapter, Babbie notes that few people seem to understand the sociologist's true interests and activities despite the common usage of the term *sociology*. Babbie extends an offer to learn more about the importance of this science. He also notes that to distinguish their conclusions from everyday, common-sense observations or intuitions, sociologists use data-gathering techniques such as surveys via questionnaires and interviews, various observation techniques, laboratory or field experiments, and content analysis of documents and other literary items. *Common sense*, in contrast, is the feeling one has about a situation without scientific, scientific analysis; appearances alone are accepted as the criterion for truth. This being the case, what is true for one person may or may not be true for another. Examples of common-sense truths are those found in such advice columns as "Dear Abby" or "Ann Landers." Other types of common-sense truths are found in proverbs such as "Absence makes the heart grow fonder." How true such sayings are is seen when one realizes that another saying claims, "Out of sight, out of mind." A number of beliefs about human behavior have the patina of truth because they seem logical and have been around for a long time—for example, "Human beings have a natural instinct to mate with the opposite sex"; "A major proportion of those on welfare could work if they wanted to"; "A value found in every society is that of romantic love." Despite their seeming logic, all of these statements have been proven false.

As noted, scientists use various scientific techniques to collect and analyze data. In other words, they use objective, logical, and systematic methods to produce reliable knowledge. This procedure can be clarified by examining the components of the previous sentence in more detail.

The first term claims that the procedure is *objective*; this means that the social scientist is aware of the difference between social behavior and other types of behavior. In short, social scientists must be aware of which activities are actually social facts. Emile Durkheim, who is considered the first scientific sociologist, noted that the rules of the scientific method are an essential part of the discipline of sociology. According to Durkheim, these rules allow us to recognize what is a social fact and thereby help reduce the influence of personal bias on the research effort. Of course, complete freedom from bias is difficult to achieve because we are products of our society and what we "see" is influenced by our observations.

The next two terms used in the scientific procedure are that it must be both logical and systematic.

Logical refers to the arrangement of the facts collected and their interrelations according to the accepted rules of reasoning. Using logic, social scientists construct theories as a guide to possible relationships. A theory states the apparent relationship of observed data to other observed data. For example, Durkheim in his classic study of suicide statistics was able to develop three theories on the causes of suicide: altruistic suicide committed for the good of the group or society; egoistic suicide committed for personal reasons; and anomic suicide committed by those who have been overwhelmed by the rapidity of social change. To reach these conclusions, Durkheim had to be *systematic* in the presentation of his data—that is, consistent in the internal order of the presentation of his materials.

Durkheim used three different theories to predict different types of suicide. Other theories can be much broader in scope and lead to different scenarios for the future. Babbie, in the second article in this chapter, describes several of the broad, overreaching theoretical paradigms that guide social research. As noted, two of the major theories are structural functionalism and conflict theory. Functionalists see social systems remaining pretty much as they are—some things will get better, maybe even disappear, but they will be replaced by other problems. This view, in a sense, is also an optimistic view because it suggests that technological changes may create temporary problems but also bring about solutions to current problems while increasing the wealth of the world on a global basis. Conflict theorists see greed as eventually tearing society apart. These pessimists see an increasing gap between the haves and the have-nots. They predict that this gap will lead to increasing problems for the have-nots and eventual rebellion by them, and on a global scale a devastating war between the rich and poor nations. Do you agree or disagree with these outlooks?

The next term in the scientific procedure is *methods*. This term refers to the techniques that are used to collect data for testing the hypothetical relationships of the variables and that in turn aid in testing the theory. For example, Cargan uses questionnaires to test the variables of being married or being single to see which segment of the population is happier, healthier, and lonelier. Throughout the readings in this anthology, variations of the major science techniques noted previously—surveys using a questionnaire or by interview, different means of observation, field and lab experiments, content analysis of existing materials—will be utilized. An excellent way to become familiar with these methods is to see if you can identify the method used to collect the data in the readings that are in the text when they have not been identified. For example, which method did Miner use in showing the lifestyle of the Nacirema? Which method did Gans use in noting the functions of poverty?

The final term used in the definition of a scientific procedure is *reliable*. Reliable means that the findings produced by the data-collecting methods are dependable because a retest made of the observations will produce similar results. It is this factor that allows social science techniques to gather data that allow predictions regarding human behavior. However, unlike mechanical behavior in the physical sciences, human behavior is subject to people's whims and fads. Therefore, predictions in the social sciences always indicate a reliable range in which the behavior is found. Perhaps the most familiar example of predictable behavior is found in political polls: They always predict a range in which the behavior will be found. For example, one candidate may be said to have a lead of 53% plus or minus 2. This means that the candidate's lead lies in the range of 51% to 55%. The next article in this chapter (by Persell et al.) confirms these ideas by noting how you will better understand human behavior after taking an introductory sociology course.

C. Wright Mills attempts to answer the question of what purpose sociology has beyond that of uncovering facts of human behavior. In the discipline's early days, sociologists believed that there was a rational explanation for all human behavior and that a scientific study of human behavior could lead to the solution of all social problems. Obviously, this prediction did not become true. Ultimately, the goals of sociology evolved to include manifold applications in numerous areas that influence the thinking of all of us.

In his classic essay, Mills reaffirms these ideas when he notes that our perceptions are limited by our family, work, and other social experiences. In short, what we observe is bounded by our particular experiences and, therefore, may lead to biased or common sense beliefs. As a means for overcoming this limited perspective, Mills suggests the use of what he calls the Sociological Imagination, a perspective that allows us to note relationships that exist between our personal experiences and general social issues.

What sociology and the work of the authors in this chapter suggest is that adherence to the scientific method produces reliable knowledge that can be used to prove or disapprove commonsense observations. The sociological approach can also expand our perspective beyond the limitations imposed by time and space, and in so doing help us deal more efficiently with society's needs. Finally, the publication of the results of scientific investigations may lead to the recognition of other truths—truths that may result in benefits to society. As you read this chapter, keep these ideas in mind and apply them to your own common-sense beliefs. In short, challenge your own common-sense beliefs by using your sociological imagination. Take a moment now to attempt to connect your personal experiences with some current social issues.

1

An Idea Whose Time Has Come

EARL BABBIE

The author claims that this is a time of both unprecedented dangers but also of real achievements. To deal with these conditions, we need to learn why people relate to one another sometimes peacefully and sometimes with hostility. This need fits into the role of sociology because it studies interaction and relations among humans, including how humans live together and how rules come into existence. For this reason Babbie claims that sociology is an idea whose time has come.

As you read this selection, consider the following questions as guides:

1. *What do you consider are the dangers now existing in the world and how you would try to resolve them?*
2. *What, if anything, are our politicians doing to resolve these dangers?*
3. *Can you think of any scientific efforts or accomplishments that attempt to deal with these dangers?*
4. *Why do you agree or disagree with Babbie's assertion that sociology is an idea whose time has come?*

There is a more pressing need for sociological insights today than at any time in history....

It is no secret that this generation faces several unprecedented dangers. No sooner has the Cold War seemingly ended than we have come to recognize the danger of localized wars among ethnic groups spilling over into wider, global conflict. This danger is made worse by the possibility that relatively small, impoverished nations could gain access to nuclear weapons, giving terrorists the opportunity to spark an international conflagration. The more we learn about the prospects of a "nuclear winter"—the likely result of the first truly nuclear war—the more evident it is that there would

be no real survivors of such a large-scale nuclear exchange.

If, on the other hand, we escape the threat of nuclear extinction, there is a real possibility that we will overpopulate and pollute the planet beyond its carrying capacity. As a single indicator of this problem, some 13 to 18 million people die of starvation around the world every year, three-fourths of them children. Approximately one-fifth of all humans on the planet go to bed hungry every night.

Add to this such persistent problems as crime, inflation, unemployment, prejudice, totalitarianism, and national debts, and you have sufficient grounds

The Sociological Spirit, Belmont: Wadsworth, 1994.

for understanding the ancient Chinese curse: "May you live in interesting times."

These are unquestionably interesting times. But the picture is not completely gloomy. These are also the times of great achievements in space: humans landing on the moon, a remote-controlled craft landing on Mars, and others photographing the more distant planets. These are the times when human beings, working cooperatively around the globe, eradicated smallpox—a scourge throughout history. These are the times of an awakening of awareness and commitment to ending world hunger. And the breakup of the former USSR, along with its domination of Eastern Europe, is regarded by most as a positive development....

If it were possible to make comparable lists of the positive and negative aspects of life today, my hunch is that they might be about equal in length. At any rate, both lists would be long ones, indicating that you and I and our fellow human beings face both trying challenges and promising opportunities ahead of us.

Assuming that we'd agree in favoring peace over war, prosperity over hunger and poverty, and so forth, the question you should be asking yourself is: what determines how things turn out? Specifically, what would it take for peace to triumph over the continuing specter of wars large and small? We'd all like the answer to that question.

I suggest there is a prior question you should ask. That is: where should we look for the answer to how peace can prevail over war? Before asking what the answer is, we need to ask where the answer is likely to be found. I suggest that up until now we have tended to look for the answer to how peace can prevail over war within the domain of military technology. Most simply, we have tried to create weapons that would preserve the peace....

The point of this discussion on war and peace is to suggest that what we need to know to establish peace around the world is not likely to arise from military technology. If such an answer is to be found at all, we must look elsewhere: in the study of why people relate to one another as they do—sometimes peacefully, sometimes hostilely. This, as we'll see, lies in the domain of sociology.

THE DOMAIN OF SOCIOLOGY

Sociology involves the study of human beings. More specifically, it is the study of interactions and relations *among* human beings. Whereas psychology is the study of what goes on inside individuals, sociology addresses what goes on between them. Sociology addresses simple, face-to-face interactions such as conversations, dating behavior, and students asking a professor to delay the term paper deadline. Equally, sociology is the study of formal organizations, the functioning of whole societies, and even relations among societies.

Sociology is the study of how human beings live together—in both the good times and the bad. It is no more a matter of how we cooperate and get along than of how we compete and conflict. Both are fundamental aspects of our living together and, hence, of sociology.

You might find it useful to view sociology as the study of our *rules for living together*. Let's take a minute to look at that.

To begin, let's consider some of the things that individuals need or want out of life: food, shelter, companionship, security, satisfaction—the list could go on and on. My purpose in considering such a list is to have us see that the things you and I need or want out of life create endless possibilities for conflict and struggle. When food is scarce, for example, I can only satisfy my need at your expense. Even in the case of companionship—where both people get what they want—you and I may fight over a particular companion.

The upshot of all this is that human beings do not seem to be constructed in a way that ensures cooperation. Bees and ants, by contrast, just seem to be wired that way. As a consequence, human

1. Hiram Maxim as quoted in Martin Hellman, *A New Way of Thinking*, Palo Alto, CA: Beyond War, 1985, p. 4.

2. Orville Wright as quoted in Martin Hellman, *A New Way of Thinking*, Palo Alto, CA: Beyond War, 1985, p. 4.

beings *create rules* to establish order in the face of chaos. Sometimes we agree on the rules voluntarily, and other times some people impose the rules on everyone else. In part, sociology is the study of how rules come into existence.

Sociology is also the study of how rules are *organized* and *perpetuated*. It would be worth taking a minute to reflect on the extent and complexity of the rules by which you and I live. There is a rule, for example, that Americans must pay taxes to the government. But it doesn't end there. The rule for paying taxes has been elaborated on by a great many more specific rules indicating how much, when, and to whom taxes are to be paid. In recent years, the index to the IRS tax code has run more than 1,000 pages long, which should give you some idea of the complexity of that set of rules. The much-touted tax simplification of 1986 was 1,855 pages long.

The rules governing our lives are not all legal ones. There are rules about shaking hands when you meet someone, rules about knives and forks at dinner, rules about how long to wear your hair, and rules about what to wear to class, to the symphony, and to mud-wrestling. There are rules of grammar, rules of good grooming, and rules of efficient computer programming.

Many of the rules we've been considering were here long before you and I showed up, and many will still be here after we've left. Moreover, I doubt that you have the experience of having taken part in creating any of the rules I've listed. Nobody asked you to vote on the rules of grammar, for example. But in a critical way, you *did* vote on those rules: you voted by obeying them.

Consider the rule about not going naked in public. Even though you don't recall being asked what you thought about that one, there was a public referendum on that issue this morning—and you voted in favor of clothes. So did I. If this seems silly, by the way, realize that there are other societies in which people voted to accept a different rule this morning.

Sometime today you are likely to be asked to vote on a set of rules about eating. Some of the possibilities are eating spaghetti with a knife, pouring soup on your dessert, and throwing your food against the wall. Let's see how you vote.

The persistence of our rules is largely a function of one generation teaching them to the next generation. We speak of *socialization* as the process of learning the rules, and it becomes apparent that we are all socializing each other all the time through the use of positive and negative *sanctions*—rewards and punishments.

All the rules we've been discussing are fundamentally *arbitrary*—that is, different rules would work just as well. Although Americans have a rule that cars must be driven on the right side of the road, other societies (e.g., England, Japan) manage equally well with people driving on the left side.

Once we've established a rule, however, we tend to add weight to it. We act as though it were better than the other possibilities, that it somehow represents an eternal and universal *truth*. Sociologists often use the term *reification* in reference to the pretense that things are real when they are not, and we often *reify* the rules of society. We make the right side of the road the *right* side for cars to drive on, and we think the British and Japanese strange for not knowing that.

The rules of society take their strongest hold when they are *internalized* by individuals—taken inside ourselves and made our own. Imagine this situation, which you may have actually experienced. It is three o'clock in the morning, and you are driving along a street leading out of a small town. There is no traffic on the street in any direction as you come to a red light at an intersection. You can see there are no cars coming for a mile in every direction. There is no one around. What do you do? There's a good chance that you will sit and wait for the light to change. If someone questioned you about it, you might say, "It just wouldn't feel right."

In the event you would drive through the light and generally regard yourself as above reification and internalization, think again. You have reified and internalized countless rules. How would you feel about having live ants and cockroaches for dinner tonight, for example? Are you willing to give them a try? How do you feel about murder, rape, and child abuse? Are you pretty casual about them, or do you feel they are *really wrong?* If you think about it, you'll find you feel pretty strongly about a lot of our rules.

All of this notwithstanding, sociology is also the study of how we *break* the rules. Some people use bad grammar and pour their soup on the floor, not to mention drive too fast, steal, fix prices, commit murder, and everything in between. Although this may seem like the study of "bad people," beware.

First, the rules of society are so extensive and complex that no one can possibly keep them all. For example, there is probably a street near you that is posted with a twenty-five-mile-per-hour speed limit; that's certainly a rule. And yet if you drive twenty-five miles an hour on that street during rush hour, you may discover you're breaking another rule. Your clue will be the honking horns and shaking fists.

Beyond the impossibility of obeying all the rules, you and I might agree that some of them ought to be broken. Consider the rule, in force a few years ago, that black people had to sit in the back of the bus in some parts of America. The people who finally broke that rule are considered heroes today. By the same token, you might disagree with rules that women can't fix carburetors, that men shouldn't cry, or that professors always know better than students.

The study of how people break the rules is closely related to the study of how the rules *change* over time. Although we are always living in a sea of rules, and many seem to last forever, it is also true that the rules of society are always in a process of change. Rules pertaining to hemlines, hair length, and political views operate a little like yo-yos. Others seem to change in only one direction.

Sociology, then, is an examination of the rules that govern our living together: what they are, how they arise, and how they change. Sociology, however, is a special approach to the rules of social life. As we'll see, there are other approaches.

A SCIENCE OF SOCIETY

...Sociology is a science of social life. Like other sciences, sociology has a *logical/empirical* basis. This means that, to be accepted, assertions must (1) make sense and (2) correspond to the facts. In this sense,

sociology can be characterized by a current buzz-word: *critical thinking*. The simple fact is that most of us, most of the time, are uncritical in our thinking. Much of the time we simply believe what we read or hear. Or, when we disagree, we do so on the basis of ideological points of view and prejudices that are not very well thought out.

1. Hiram Maxim as quoted in Martin Hellman, *A New Way of Thinking*, Palo Alto, CA: Beyond War, 1985, p. 4.
2. Orville Wright as quoted in Martin Hellman, *A New Way of Thinking*, Palo Alto, CA: Beyond War, 1985, p. 4.

Suppose you were talking with a friend about the value of going to college. Your friend disagrees: "College is a waste of time. You should get a head start in the job market instead. Most of today's millionaires never went to college, and there are plenty of college graduates pumping gas or out of work altogether." That's the kind of thing people sometimes say, and it can be convincing—especially if it's said with conviction. But does it stand up to logical and empirical testing?

Logically, it doesn't seem to make much sense, since a college education would seem to give a person access to high-paying occupations not open to people with less education....

There isn't any scientific support for the assertion that education is a worthless financial investment—even though there are some individual exceptions to the rule.

It's important to recognize that human beings generally have opinions about everything...but it needs saying here that the people you deal with every day have a tendency to express opinions about the way things are—and what they say isn't always so. Consequently, you need to protect yourself from false information. That's what critical thinking is all about, and sociology provides some powerful critical-thinking tools.

I hope these few examples will indicate that sociology is not just something you might study in college and never think about again. Sociology deals with powerful issues that determine the quality of your life. Understanding sociology can empower you to be a more effective participant in the social affairs around you whether you are a conscious player or not. Marriage, employment, prejudice, crime, and politics are only a few of the areas of social life that can be importantly affected by your ability to engage in sociological reasoning.

Let's look at the twin foundations of critical thinking and of science in more detail—seeing how they apply to sociology in particular....

SOCIOLOGICAL QUESTIONS AND ANSWERS

The common image of science is of scientists finding the answers to questions. I will conclude this chapter with a somewhat different view of science and of sociology in particular. Science is sometimes better at raising questions than at finding answers to them. It would be useful for you to regard science as an ongoing inquiry, recognizing that questions initiate new avenues of inquiry, while answers close them down.

Science makes an especially powerful contribution when it calls into question those things that "everyone knows." Everyone used to "know," for example, that blacks were inferior to whites and that women were inferior to men. As you'll see, sociological points of view very often raise questions about things everyone else thought had already been answered. Even when we discover new and seemingly better answers about something, it's important to hold them as tentative.

There is a particular *recursive* quality in human life that makes anything we know tentative. Whenever we learn something about ourselves, what we've learned may bring about changes—even to the extent of making what we learned no longer accurate.

Suppose we studied employment opportunities across the country, for example. When our study was complete, we listed the ten cities with the most jobs available. As soon as our findings became widely known, of course, a lot of unemployed people would move to those cities, and soon those cities would not have as many jobs available as before. This is the same thing that happens when a newspaper columnist identifies a local restaurant that has great food, low prices, and no waiting. The "no waiting" part probably won't last another twenty-four hours, and the other two characteristics may disappear, too. In sociology, anything we learn can change things, so no knowledge can be counted on to remain true. Thus, we need to keep asking questions.

Even more fundamentally, sociology deals with a number of questions that will never be answered fully. Who am I? What is a human being? Are we more the result of our genes or of our environment? Is order possible without restricting freedom?... and it is unlikely that they will ever be completely answered. As we'll see, however, it can be very useful to keep asking them anyway.

I point all this out to give you an appropriate context for your own inquiry into sociology. Although there are some facts about sociology that are worth learning, it is more important for you to learn to use sociology for your own ongoing critical thinking. If you were studying brain surgery or medieval history, you might not have an opportunity to use what you learned in your day-to-day life. Sociology is very different. Every day you will wake up into a sociological laboratory with a massive experiment under way. We're all subjects in the experiment, and you now have the opportunity to join the researchers.

2

The Practice of Social Research

EARL BABBIE

The second article in the introductory section explains the major theories used in sociology. These theories are the foundation for a scientific approach to the study of social life and together with research methods provide the tools for scientific investigation.

As you read, ask yourself the following questions:

1. *What are the different paradigms used by sociologists in their examination of social data?*
2. *Which of the paradigms noted by Babbie seems most revealing of social behavior? Why?*
3. *What is the relationship between the prior article on doing research and this article?*

GLOSSARY **Theories** A set of logically interrelated statements that attempt to explain or predict social events. **Paradigm** An example, model, or pattern.

INTRODUCTION

There are restaurants in the United States fond of conducting political polls among their diners whenever an election is in the offing. Some take these polls very seriously because of their uncanny history of predicting winners. Some movie theaters have achieved similar success by offering popcorn in bags picturing either donkeys or elephants. Years ago, granaries in the Midwest offered farmers a chance to indicate their political preferences through the bags of grain they selected.

Such idiosyncratic ways of determining trends, though interesting, all follow the same pattern over time: They work for a while, and then they fail. Moreover, we can't predict when or why they will fail.

These unusual polling techniques point to a significant shortcoming of "research findings" based only on the observation of patterns. Unless we can offer logical explanations for such patterns, the regularities we've observed may be mere flukes, chance occurrences. If you flip coins long enough, you'll get ten heads in a row. Scientists might adapt a street expression to describe this situation: "Patterns happen."

Logical explanations are what theories seek to provide. Theories function three ways in research. First, they prevent our being taken in by flukes. If we can't explain why Ma's Diner has been so successful in predicting elections, we run the risk of supporting a fluke.

From *The Practice of Social Research with InfoTrac® College Edition*, 9th edition, by E. Babbie © 2001. Reprinted with permission of Wadsworth.

If we know why it has happened, we can anticipate whether or not it will work in the future.

Second, theories make sense of observed patterns in a way that can suggest other possibilities. If we understand the reasons why broken homes produce more juvenile delinquency than do intact homes—lack of supervision, for example—we can take effective action, such as after-school youth programs.

Finally, theories shape and direct research efforts, pointing toward likely discoveries through empirical observation. If you were looking for your lost keys on a dark street, you could whip your flashlight around randomly, hoping to chance upon the errant keys—or you could use your memory of where you had been to limit your search to more likely areas. Theories, by analogy, direct researchers' flashlights where they are most likely to observe interesting patterns of social life....

This reading explores some specific ways theory and research work hand in hand during the adventure of inquiry into social life. We'll begin by looking at some fundamental frames of reference, called *paradigms*, that underlie social theories and inquiry.

SOME SOCIAL SCIENCE PARADIGMS...

Macrotheory and Microtheory

Let's begin with a difference concerning focus that stretches across many of the paradigms we'll discuss. Some social theorists focus their attention on society at large, or at least on large portions of it. Topics of study for such **macrotheory** include the struggle between economic classes in a society, international relations, or the interrelations among major institutions in society, such as government, religion, and family. Macrotheory deals with large, aggregate entities of society or even whole societies.

Some scholars have taken a more intimate view of social life. **Microtheory** deals with issues of social life at the level of individuals and small groups. Dating behavior, jury deliberations, and student–faculty interactions are apt subjects for a microtheoretical perspective. Such studies often come

close to the realm of psychology, but whereas psychologists typically focus on what goes on inside humans, social scientists study what goes on between them....

Early Positivism

When the French philosopher Auguste Comte (1798–1857) coined the term *sociologie* in 1822, he launched an intellectual adventure that is still unfolding today. Most importantly, Comte identified society as a phenomenon that can be studied scientifically....

Prior to Comte's time, society simply was. To the extent that people recognized different kinds of societies or changes in society over time, religious paradigms generally predominated in explanations of such differences. The state of social affairs was often seen as a reflection of God's will. Alternatively, people were challenged to create a "City of God" on earth to replace sin and godlessness.

Comte separated his inquiry from religion. He felt that religious belief could be replaced with scientific study and objectivity. His "positive philosophy" postulated three stages of history. A "theological stage" predominated throughout the world until about 1300. During the next five hundred years, a "metaphysical stage" replaced God with philosophical ideas such as "nature" and "natural law."

Comte felt he was launching the third stage of history, in which science would replace religion and metaphysics by basing knowledge on observations through the five senses rather than on belief or logic alone. Comte felt that society could be observed and then explained logically and rationally and that sociology could be as scientific as biology or physics....

Social Darwinism...

In 1858, when Charles Darwin published his *The Origin of Species*, he set forth the idea of evolution through the process of natural selection. Simply put, the theory states that as a species coped with its environment, those individuals most suited to success would be the most likely to survive long enough to reproduce. Those less well suited would perish. Over time the traits of the survivor would come to dominate the species. As later Darwinians put it, species evolved into different forms through the "survival of the fittest."

As scholars began to study society analytically, it was perhaps inevitable that they would apply Darwin's ideas to changes in the structure of human affairs. The journey from simple hunting-and-gathering tribes to large, industrial civilizations was easily seen as the evolution of progressively "fitter" forms of society....

Conflict Paradigm

...Karl Marx (1818–1883) suggested that social behavior could best be seen as the process of conflict: the attempt to dominate others and to avoid being dominated. Marx focused primarily on the struggle among economic classes. Specifically, he examined the way capitalism produced the oppression of workers by the owners of industry....

The conflict paradigm proved to be fruitful outside the realm of purely economic analyses. Georg Simmel (1858–1918) was especially interested in small-scale conflict, in contrast to the class struggle that interested Marx. Simmel noted, for example, that conflicts among members of a tightly knit group tended to be more intense than those among people who did not share feelings of belonging and intimacy.

In a more recent application of the conflict paradigm, when Michael Chossudovsky's (1997) analysis of the International Monetary Fund and World Bank suggested that these two international organizations were increasing global poverty rather than eradicating it, he directed his attention to the competing interests involved in the process. In theory, the chief interest being served should be the poor people of the world or perhaps the impoverished, Third World nations. The researcher's inquiry, however, identified many other interested parties who benefited: the commercial lending institutions that made loans in conjunction with the IMF and World Bank and multinational corporations seeking cheap labor and markets for their goods, for example. Chossudovsky's analysis concluded that the interests of the banks and corporations tended to take precedence over those of the poor people, who were the intended beneficiaries. Moreover, he found many policies were weakening national economies in the Third World, as well as undermining democratic governments.

Whereas the conflict paradigm often focuses on class, gender, and ethnic struggles, it would be appropriate to apply it whenever different groups have competing interests. For example, it could be fruitfully applied to understanding relations among different departments in an organization, fraternity and sorority rush weeks, or student–faculty–administrative relations, to name just a few.

Symbolic Interactionism...

Simmel was one of the first European sociologists to influence the development of U.S. sociology. His focus on the nature of interactions particularly influenced George Herbert Mead (1863–1931), Charles Horton Cooley (1864–1929), and others who took up the cause and developed it into a powerful paradigm for research.

Cooley, for example, introduced the idea of the "primary group," those intimate associates with whom we share a sense of belonging, such as our family, friends, and so forth. Cooley also wrote of the "looking-glass self" we form by looking into the reactions of people around us. If everyone treats us as beautiful, for example, we conclude that we are. Notice how fundamentally the concepts and theoretical focus inspired by this paradigm differ from the society-level concerns of...Marx.

Mead emphasized the importance of our human ability to "take the role of the other," imagining how others feel and how they might behave in certain circumstances. As we gain an idea of how people in general see things, we develop a sense of what Mead called the "generalized other."

Mead also showed a special interest in the role of communications in human affairs. Most interactions, he felt, revolved around the process of individuals reaching common understanding through the use of language and other such systems, hence the term *symbolic interactionism*.

This paradigm can lend insights into the nature of interactions in ordinary social life, but it can also help us understand unusual forms of interaction....

Structural Functionalism

Structural functionalism, sometimes also known as "social systems theory," grows out of a notion introduced by Comte and Spencer: A social entity, such as an organization or a whole society, can be viewed as an organism. Like other organisms, a social system is made up of parts, each of which contributes to the functioning of the whole.

By analogy, consider the human body. Each component—such as the heart, lungs, kidneys, skin, and brain—has a particular job to do. The body as a whole cannot survive unless each of these parts does its job, and none of the parts can survive except as a part of the whole body. Or consider an automobile. It is composed of the tires, the steering wheel, the gas tank, the spark plugs, and so forth. Each of the parts serves a function for the whole; taken together, that system can get us across town. None of the individual parts would be very useful to us by itself, however.

The view of society as a social system, then, looks for the "functions" served by its various components. Social scientists using the structural functional paradigm might note that the function of the police, for example, is to exercise social control—encouraging people to abide by the norms of society and bringing to justice those who do not. Notice, though, that they could just as reasonably ask what functions criminals serve in society. Within the functionalist paradigm, we might say that criminals serve as job security for the police. In a related observation, Emile Durkheim (1858–1917) suggested that crimes and their punishment provide an opportunity to reaffirm society's values. By catching and punishing thieves, we reaffirm our collective respect for private property.

To get a sense of the structural-functional paradigm, suppose you were interested in explaining how your college or university works. You might thumb through the institution's catalog and begin assembling a list of the administrators and support staff (such as president, deans, registrar, campus security, maintenance personnel). Then you might figure out what each of them does and relate their roles and activities to the chief functions of your college or university, such as teaching or research. This way of looking at an institution of higher learning would clearly suggest a different line of inquiry than, say, a conflict paradigm, which might emphasize the clash of interests between people who have power in the institution and those who don't.

People often discuss "functions" in everyday conversations. Typically, however, the alleged functions are seldom tested empirically. Some people argue, for example, that welfare, intended to help the poor, actually harms them in a variety of ways. It is sometimes alleged that welfare creates a deviant, violent subculture in society, at odds with the mainstream. From this viewpoint, welfare programs actually result in increased crime rates.

Lance Hannon and James Defronzo (1998) decided to test this last assertion. Working with data drawn from 406 urban counties in the United States, they examined the relationship between levels of welfare payments and crime rates. Contrary to the beliefs of some, their data indicated that higher welfare payments were associated with lower crime rates. In other words, welfare programs have the function of decreasing rather than increasing lawlessness.

Feminist Paradigms...

When feminists first began questioning the use of masculine pronouns and nouns whenever gender was ambiguous, their concerns were often viewed as petty, even silly. At most, many felt the issue was one of women having their feelings hurt, their egos bruised...

In a similar way, researchers looking at the social world from a feminist paradigm have called attention to aspects of social life that are not revealed by other paradigms. In part, feminist theory and research have focused on gender differences and how they relate to the rest of social organization. These lines of inquiry have drawn attention to the oppression of women in many societies, which in turn has shed light on oppression generally.

Feminist paradigms have also challenged the prevailing notions concerning consensus in society. Most descriptions of the predominant beliefs, values, and norms of a society are written by people representing only portions of society. In the United States, for example, such analyses have typically been written by middle-class white men—not surprisingly, they have written about the beliefs, values, and norms they themselves share. Though George Herbert Mead spoke of the "generalized other" that each of us becomes aware of and can "take the role of," feminist paradigms question whether such a generalized other even exists.

Further, whereas Mead used the example of learning to play baseball to illustrate how we learn about the generalized other, Janet Lever's research suggests that understanding the experience of boys may tell us little about girls.

> Girls' play and games are very different. They are mostly spontaneous, imaginative, and free of structure or rules. Turn-taking activities like jumprope may be played without setting explicit goals. Girls have far less experience with interpersonal competition. The style of their competition is indirect, rather than face to face, individual rather than team affiliated. Leadership roles are either missing or randomly filled.

Social researchers' growing recognition of the general intellectual differences between men and women led the psychologist Mary Field Belenky and her colleagues to speak of *Women's Ways of Knowing*. In-depth interviews with 45 women led the researchers to distinguish five perspectives on knowing that should challenge the view of inquiry as obvious and straightforward:

Silence: Some women, especially early in life, feel themselves isolated from the world of knowledge, their lives largely determined by external authorities.

Received knowledge: From this perspective, women feel themselves capable of taking in and holding knowledge originating with external authorities.

Subjective knowledge: This perspective opens up the possibility of personal, subjective knowledge, including intuition.

Procedural knowledge: Some women feel they have mastered the ways of gaining knowledge through objective procedures.

Constructed knowledge: The authors describe this perspective as "a position in which women view all knowledge as contextual, experience themselves as creators of knowledge, and value both subjective and objective strategies for knowing."

"Constructed knowledge" is particularly interesting in the context of paradigms. The positivistic paradigm of Comte would have a place neither for "subjective knowledge" nor for the idea that truth might vary according to its context. The ethnomethodological paradigm, on the other hand, would accommodate these ideas.

Rational Objectivity Reconsidered

We began this discussion of paradigms with Comte's assertion that society can be studied rationally and objectively. Since his time, the growth of science and technology, together with the relative decline of superstition, have put rationality more and more in the center of social life. As fundamental as rationality is to most of us, however, some contemporary scholars have raised questions about it.

For example, positivistic social scientists have sometimes erred in assuming that social reality can be explained in rational terms because humans always act rationally. I'm sure your own experience offers ample evidence to the contrary. Yet many modern economic models fundamentally assume that people will make rational choices in the economic sector: They will choose the highest-paying job, pay the lowest price, and so forth. This assumption ignores the power of tradition, loyalty, image, and other factors that compete with reason and calculation in determining human behavior.

A more sophisticated positivism would assert that we can rationally understand and predict even nonrational behavior. An example is the famous "Asch

Experiment." In this experiment, a group of subjects is presented with a set of lines on a screen and asked to identify the two lines that are equal in length.

Imagine yourself a subject in such an experiment. You are sitting in the front row of a classroom in a group of six subjects. A set of lines is projected on the wall in front of you. The experiment asks each of you, one at a time, to identify the line to the right (A, B, or C) that matches the length of line X. The correct answer (B) is pretty obvious to you. To your surprise, however, you find that all the other subjects agree on a different answer!

The experimenter announces that all but one of the group has gotten the correct answer. Since you are the only one who chose B, this amounts to saying that you've gotten it wrong. Then a new set of lines is presented, and you have the same experience. What seems to be the obviously correct answer is said by everyone else to be wrong.

As it turns out, of course, you are the only real subject in this experiment—all the others are working with the experimenter. The purpose of the experiment is to see whether you will be swayed by public pressure to go along with the incorrect answer. In his initial experiments, all of which involved young men, Asch found that a little over one-third of his subjects did just that.

Choosing an obviously wrong answer in a simple experiment is an example of nonrational behavior. But as Asch went on to show, experimenters can examine the circumstances that lead more or fewer subjects to go along with the incorrect answer. For example, in subsequent studies, Asch varied the size of one group and the number of "dissenters" who chose the "wrong" (that is, the correct) answer. Thus, it is possible to study nonrational behavior rationally and scientifically.

More radically, we can question whether social life abides by rational principles at all....

The contemporary challenge to positivism, however, goes beyond the question of whether people behave rationally. In part, the criticism of positivism challenges the idea that scientists can be as objective as the positivistic ideal assumes. Most scientists would agree that personal feelings can and do influence the problems scientists choose to study, what they choose to observe, and the conclusions they draw from their observations.

There is an even more radical critique of the ideal of objectivity. As we glimpsed in the discussions of feminism and ethnomethodology, some contemporary researchers suggest that subjectivity might actually be preferable in some situations....

To begin, all our experiences are inescapably subjective. There is no way out. We can see only through our own eyes, and anything peculiar to our eyes will shape what we see. We can hear things only the way our particular ears and brain transmit and interpret sound waves. You and I, to some extent, hear and see different realities. And both of us experience quite different physical "realities" than, say ... scientists on the planet Xandu who might develop theories of the physical world based on a sensory apparatus that we humans can't even imagine. Maybe they see X rays or hear colors.

Despite the inescapable subjectivity of our experience, we humans seem to be wired to seek an agreement on what is really real, what is objectively so. Objectivity is a conceptual attempt to get beyond our individual views. It is ultimately a matter of communication, as you and I attempt to find a common ground in our subjective experiences. Whenever we succeed in our search, we say we are dealing with objective reality....

Whereas our subjectivity is individual, our search for objectivity is social. This is true in all aspects of life, not just in science. While you and I prefer different foods, we must agree to some extent on what is fit to eat and what is not, or else there could be no restaurants or grocery stores. The same argument could be made regarding every other form of consumption. Without agreement reality, there could be no movies or television, no sports.

Social scientists as well have found benefits in the concept of a socially agreed-upon objective reality. As people seek to impose order on their experience of life, they find it useful to pursue this goal as a collective venture. What are the causes and cures of prejudice? Working together, social researchers have uncovered some answers that hold up to intersubjective scrutiny. Whatever your subjective experience of things, for example, you can discover for yourself that as education increases, prejudice generally tends to decrease. Because each of us can discover this independently, we say that it is objectively true....

Some say that the ideal of objectivity conceals as much as it reveals. As we saw earlier in years past much of what was regarded as objectivity in Western social science was actually an agreement primarily among white, middle-class European men. Equally real experiences common to women, to ethnic minorities, to non-Western cultures, or to the poor were not necessarily represented in that reality....

Ultimately, we will never be able to distinguish completely between an objective reality and our subjective experience. We cannot know whether our concepts correspond to an objective reality or are simply useful in allowing us to predict and control our environment. So desperate is our need to know what is really real, however, that both positivists and postmodernists are sometimes drawn into the belief that their own view is real and true. There is a dual irony in this. On the one hand, the positivist's belief that science precisely mirrors the objective world must ultimately be based on faith; it cannot be proven by "objective" science since that's precisely what's at issue. And the postmodernists, who say nothing is objectively so and everything is ultimately subjective, do at least feel that that is really the way things are.

Fortunately, as social researchers we are not forced to align ourselves entirely with either of these approaches. Instead, we can treat them as two distinct arrows in our quiver. Each approach compensates for the weaknesses of the other by suggesting complementary perspectives that can produce useful lines of inquiry.

In summary, a rich variety of theoretical paradigms can be brought to bear on the study of social life. With each of these fundamental frames of reference, useful theories can be constructed....

3

What Should Students Understand After Taking Introduction to Sociology?

CAROLINE HODGES PERSELL, KATHRYN M. PFEIFFER, AND ALI SYED

Most students have heard the term "sociology" but do not really understand what this field in the social sciences entails or why it is such an important endeavor. For example, few people connect such familiar terms as interviews and public opinion polling to sociological methodologies. The authors of this article did a survey—another sociological method—and uncovered what you should have learned after taking the Introductory Sociology course. You will probably be surprised.

As you read this selection, consider the following questions as guides:

1. *Why is it important to know what is meant by the "social construction of ideas"?*
2. *Why is it necessary to understand that the central idea of society is inequality?*
3. *The authors note that one of the basic topics of sociology is to improve the world. Why is this an important task?*

Long before it was called SoTL (the Scholarship of Teaching and Learning), there was a tradition in sociology of discussing what should be taught in sociology. In the early years of the *American Journal of Sociology*, scholarly leaders in sociology published their syllabi and discussions of what they covered in various courses. While the venue changed over time, the creation of *Teaching Sociology*, the founding of the Section on Undergraduate Education in the American Sociological Association (now called the Section on Teaching and Learning in Sociology), the institutionalization of the Teaching Resources Group in ASA (now the Departmental Resources Group), the leadership of Hans Mauksch, Carla Howery, and many others all contributed to

important dialogue through time about what is or should be taught in sociology. While these and related developments helped institutionalize the scholarship of teaching and learning, they may also have contributed to an increased perception of specialization in the fields....

METHODS AND DATA

To obtain data, we defined a sample of leaders in the field, interviewed them, and compared their responses to three recent major works published by leading scholars of teaching and learning....

Teaching Sociology, Vol. 35, 2007 (October 300–314).

RESULTS

Here we discuss the results of our interviews with scholarly leaders, then compare those results with the views of the sub-sample who won the ASA award for Distinguished Contributions to Teaching. Finally we compare the interview results with the themes in the three major SoTL publications....

What Do Leaders Think Students Should Understand After Taking Introduction to Sociology?

When coding the interviews, we reached consensus on nine major themes. While we ranked these understandings according to the frequency with which they were mentioned, we did not consider close differences in the rankings to be significant due to the small sample size. We consider each of them in turn, including representative quotes that show what was coded in the category and illustrate how the leaders thought about the theme.

(1) The "Social" Part of Sociology, or Learning to Think Sociologically.   This was by far the most frequently mentioned principle that leaders wanted students to understand after taking an introductory course, and they articulated four dimensions. First was that students understand the importance of getting beyond the individual when trying to understand and explain the social world. One of the leaders wanted students to understand "the existence of social factors [because] most students come in assuming psychological explanations" (18)....

Second, various responses elaborated on the meaning of the "social." "It isn't just individuals. There are groups and institutions.... [I] expect a new and broader perspective on the world...that goes beyond the individual" (3)....

Third, the most frequently stressed idea within this theme was that macro-level factors and individuals are interconnected. One interviewee said, "I want my students to be able to see beyond themselves and think about the groups they belong to and how these groups have an effect on individual

characters. The individual is always in constant interaction with the [social] environment. There is agency but there are [also] constraints" (25). Similarly, another stressed that it was important for students to understand "that they live in a social world in which they are constrained by a variety of social forces but in which they [also] have individual ability to carve out directions for themselves. While pushed by forces, they can adapt.... Relatedly, we live in a very fast-changing world. It is important to understand what those changes are and chart one's course and direction in relation to those changes" (39)....

Finally, some hoped students would push their understanding of the social even deeper, to "be able to see behind the surface appearance [of social phenomena] in the way that we as sociologists typically try to do in our work" (35). Another put it this way: "Understanding the sociological lens, or the sociological imagination [involves realizing that] what often appears natural isn't. There are often paradoxes in social life. For example, inequality comes from abundance, not scarcity. Deviance serves as a social bond" (30). Thus, these leaders stressed the importance of looking beyond the obvious to uncover seemingly incongruous relationships.

They also thought that it was critically important for undergraduates to understand that their own lives, not just other people's lives, are affected by various social factors. This is a threshold principle of sociology, but one that is difficult for highly individualized, often middle class, college students to understand and accept.

(2) The Scientific Nature of Sociology.   The scientific aspect of sociology was the second most frequently mentioned element that leaders wanted students to get from an introductory sociology course. As one said, "Thinking about the social requires self-conscious attention to methods. How do we know what we know? If they read something in the paper, I want them to ask, 'How do they know that?' When we want to draw a conclusion, on what basis do we do that?"(17).... Similarly, another said it was important for students to learn that "sociology is a science, and how to distinguish sciences from other ways of thinking about the world, and why certain things that they clearly understand as science, such as chemistry and biology, are science. And why sociology, ideally, is also a science, and how it fits into the other sciences" (29). Reflected another, "We can actually explain things people take for granted… and the reason we are able to do that is because sociology is scientific and systematic and other forms of social commentary are not, such as journalism" (20)….

A few leaders explicitly stressed causality: "I want them to understand causal relationships and the logic of social science. How to argue and present a case, not just by doing it more loudly and often but by presenting evidence" (26). Another noted, "It is important for students to understand the puzzle-solving aspects of sociology, to understand how we approach the scientific study of social life. I want them to learn that sociology has a methodological way of analyzing the world that is useful in all kinds of situations. They tend to reason from a single example, 'my grandmother…' rather than considering all possible causal mechanisms that could be operating in a situation. I want them to be able to think through a problem and how they might be able to answer it. What would you need to know to answer a particular question? I want them to be able to identify that" (16).

In sum, leaders wanted students to appreciate the scientific or systematic nature of data collection and analysis used in sociological research, rather than seeing sociology as simply a bundle of different opinions. They wanted students to grasp the importance of marshalling evidence to support an argument. Some refined this further to emphasize an understanding of causal relationships. One respondent hoped students would gain some awareness of "the potential, possibilities, and limitations of research as a form of inquiry" (1).

(3) Complex and Critical Thinking. Nearly a third (12) of the respondents identified complex and critical thinking as important. As one elaborated, "By critical thinking I mean the ability to not necessarily accept beliefs or ideas just because they were raised to think a certain way, or even because the professor says it in class. Instead, I hope they will ask, what are the important questions to ask? Should the question be a different one from the one being raised here? How do we use evidence to think about this question? How do we get new evidence?" (36)…. One noted, "I think analytical thinking is really hard to teach… but if you never teach it, you never give the people who want to think that way the opportunity to do so" (20).

For many respondents, the understanding that there are multiple perspectives on any given question or issue was key to their conception of critical thinking. Complex and critical thinking involves approaching social issues and problems with a nuanced view that takes multiple perspectives into account and raises new questions….

(4) The Centrality of Inequality. Eight respondents stressed the importance of having students understand more about social stratification and inequality. In the words of one respondent, it is "really important to integrate issues of stratification in a way that clearly ties into the theme of social structures, that ties to issues of equal opportunity and also a person's location in a social [structure]" (34)…. Others mentioned the importance of students understanding that "inequality is all about power" (19) or what someone else called "constructions of power" (24).

(5) A Sense of Sociology as a Field. Eight respondents also indicated that they hoped the introductory course would help students understand something about sociology as a field, as well as to prepare them to major in it… One leader emphasized the importance of "understanding that there are theoretical underpinnings" to sociology (8), and several discussed the importance of teaching the main theoretical traditions in sociology, including symbolic interactionism, structural functionalism, and conflict theory…. Another aimed for "an understanding of the way that the discipline is organized in subdivisions and that there is an overlap in those sub-disciplines with other disciplines" (8).

(6) The Social Construction of Ideas. At least two respondents directly, and three others implicitly, stressed the importance of the social construction of ideas, in gender and race, for example. As one noted, "The things we take for granted as natural are really socially constructed, e.g., human nature. But ideas about human nature differ widely in different societies. [Take] the example of love and marriage. To us it seems natural that they go together. But, just feeling it is natural doesn't make it so. The feeling is real, but we can explain where that feeling comes from and why it might be different in other societies. Our sense of time is another example of this, as are categories or ideas about race, gender. They differ across societies....

(7) The Difference Between Sociology and Other Social Sciences. Four respondents wanted students to recognize the difference between sociology and other social sciences—for example, to "understand 'the sociological perspective' and how it differs from the approaches taken in other fields" (11), or to "understand the ways that sociology interfaces with other disciplines" (10). As one said, "I cover the institutions of society because that makes clear the links to other social sciences and gives a whole view [of the] breadth of sociology and the things we study" (34). Another indicated the desire "to engage multidisciplinary arguments, to bring in economics, anthropology, as well as sociology...[to help students] see the distinctiveness of sociology but also...be able to read the media from a broad social science perspective" (4).

(8) The Importance of Trying to Improve the World. Four respondents indicated that it was important that students use their understanding of sociology to relate to the world and even improve it. One of the interviewees said, "I want [students] to be able to use some of what they've learned in sociology...in the way they approach problems, read the newspaper, and apply it to their everyday lives" (3).... Still others emphasized activism, as in the case of one who hoped students would be able to "link the basic sociological concepts and theories with social activism or 'public sociology' and bring

in the principles of humanity, equality, the humanitarian spirit" (27)....

(9) The Importance of Social Institutions in Society. Three respondents specifically discussed the salience of key social institutions in society. One respondent summarized this sentiment when he expressed the desire for students to gain "a general understanding of the important institutions in society; that would include everything from the family to the economy to the polity to...health care, the important institutional sectors that sociologists—of course, the bulk of the sociology work force—actually devote all their time to studying" (40)....

DISCUSSION

Overall, we were somewhat surprised by the amount of general agreement among leaders and publications on teaching and learning in sociology.... These findings underscore the importance of Kain's question about how to build structures and cultures that affirm "the interrelationships between the different parts of being a teacher/scholar" (2006:338)....

CONCLUSIONS AND SUGGESTIONS FOR FURTHER RESEARCH

...This study raises a number of other questions for further research. One is, how do leaders teach the understandings they want students to obtain? We are currently examining this question. An anonymous reviewer of this article suggested that since the word "understand" may connote cognitive learning more than the learning of values, norms, or social roles, future research might ask specifically about various kinds of understandings—cognitive as well as values and roles. Future research might also explore how instructors assess whether students obtain the understandings they seek. Another

A Comparison of Learning Goals in Introductory Sociology

Scholarly Leaders' Views of "One or Two Most Important Principles You Would Like Students to Understand" After an Introductory Course	Wagenaar Survey (2004) (In Top 5 for Introductory Sociology From a List of 72 "Core Concepts, Topics and Skills")	*Liberal Learning and the Sociology Major* (2004), A Report of the ASA Task Force on the Undergraduate Sociology Major	Grauerholz and Gibson's 2006 Analysis of 402 Syllabi for Most Commonly Taught Courses in Sociology
(1) The "social" part of sociology, or learning to think sociologically	■ "Sociological imagination" (9.8%, p. 9) ■ "Think like a sociologist" (3.7%, p. 9) ■ "Applications to students' lives" (4%, p. 9)	■ Understand "the importance of social structure and culture—the sociological perspective" (p. 1)	■ "Appreciate concept of structure" (61%) ■ "Think sociologically" (54%) ■ "Connect personal and social" (23%) ■ "Theoretical sophistication" (11%, p. 14)
(2) The scientific nature of sociology	■ "How to use and assess research" (3.5%, p. 9)	■ "Infuse the empirical base of sociology throughout the curriculum" (p. 8)	■ "Data analysis or methodological skills" (12%, p. 14)
(3) Complex and critical thinking	■ "Sociological critical thinking" (6.8%, p. 9)	■ "Offer community and classroom-based learning experiences that develop students' critical thinking skills and prepare them for lives of civic engagement" (p. 22)	■ "Critical thinking" (40%, p. 14)
(4) The centrality of inequality	■ "Stratification-general" (8.4%) ■ "Intersections of race/class/gender" (2.6%, p. 9)	■ "Underscore the centrality of race, class, and gender in society" (p. 5)	■ "Race/class/gender" (29%, p. 14)
(5) A sense of sociology as a field	■ "Sociology as a discipline" (2.9%, p. 9)		
(6) The social construction of ideas	■ "Culture" (5%, p. 9) [although culture includes more than the idea of social construction]		

(7) The difference between sociology and other social sciences	■ "Recognize explicitly the intellectual connections between sociology and other fields" (p. 19)	
(8) The importance of trying to improve the world	■ "Offer community and classroom-based learning experiences that develop students' critical thinking skills and prepare them for lives of civic engagement" (p. 22)	■ "Other (e.g., social change)" (15%, p. 14) ■ "Service learning or community building" (2%, p. 14)
(9) The important social institutions in society		■ "Social structure" (6%, p. 9) [although the concept of social structure includes more than social institutions] ■ "Socialization" (4%, p. 9)
	■ "Increase students' exposure to multicultural, cross-cultural, and cross-national content" (p. 19)	■ "Socio-historical awareness" (35%) ■ "Cross-cultural/cross-national awareness" (34%) ■ "Multi-cultural awareness" (9%, p. 14)
		■ "Written communication skills" (11%) ■ "Oral communication skills" (9%) ■ "Technological literacy" (2%, p. 14)

reviewer suggested further dialogue on these issues among scholarly leaders and authors of SoTL publications. Finally, it would be interesting to see how the results for sociology compare with what exists in other fields with respect to convergence on larger understandings. Are all the social sciences similar? Are humanities and science fields similar within themselves? Do they differ from each other or from the social sciences? What about fields such as law and medicine? Such research would illuminate whether sociology is similar to, or different from, other fields of knowledge with respect to the degree of agreement between scholarly and pedagogical leaders.

4

The Promise

C. WRIGHT MILLS

Here Mills completes the task of defining what the sociologist does, how he or she does it, and how all subjects of human behavior are in the sociological purview. Mills does this by indicating what the sociological mind helps individuals accomplish—what is going on in the world and happening to themselves.

As you read, ask yourself the following questions:

1. *What does the sociological imagination allow people to accomplish?*
2. *Which changes in your personal milieu might be better understood by looking at changes in the society?*
3. *How are the prior articles in this section related to the promise of sociology?*

GLOSSARY **Sociological imagination** The capacity to understand the most impersonal and remote changes in terms of their effect on the human self and to see the relationship between the two. **Personal trouble** A private matter that occurs within the character of an individual and within the range of that individual's immediate relations with others. **Public issue** A matter that transcends the local environment of an individual and the range of that individual's inner life.

Nowadays men* often feel their private lives are a series of traps. They sense that within their everyday worlds, they cannot overcome their troubles, and in this feeling, they are often quite correct: What ordinary men are directly aware of and what they try to do are bounded by the private orbits in which they live; their visions and their powers are limited to the close-up scenes of job, family, neighborhood; in other milieux, they move vicariously and remain spectators. And the more aware they become, however vaguely, of ambitions and of threats which transcend their immediate locales, the more trapped they seem to feel.

Underlying this sense of being trapped are seemingly impersonal changes in the very structure of continent-wide societies. The facts of contemporary history are also facts about the success and the failure of individual men and women. When a society is industrialized, a peasant becomes a worker; a feudal lord is liquidated or becomes a businessman. When

Source: Abridged from "The Promise," *The Sociological Imagination* by C. Wright Mills. Copyright © 2000 by Oxford University Press, Inc. Reprinted by permission of the publisher.

classes rise or fall, a man is employed or unemployed; when the rate of investment goes up or down, a man takes new heart or goes broke. When wars happen, an insurance salesman becomes a rocket launcher; a store clerk, a radar man; a wife lives alone; a child grows up without a father. Neither the life of an individual nor the history of a society can be understood without understanding both.

Yet men do not usually define the troubles they endure in terms of historical change and institutional contradiction. The well-being they enjoy, they do not usually impute to the big ups and downs of the societies in which they live. Seldom aware of the intricate connection between the patterns of their own lives and the course of world history, ordinary men do not usually know what this connection means for the kinds of men they are becoming and for the kinds of history-making in which they might take part. They do not possess the quality of mind essential to grasp the interplay of man and society, of biography and history, of self and world. They cannot cope with their personal troubles in such ways as to control the structural transformations that usually lie behind them.

Surely it is no wonder. In what period have so many men been so totally exposed at so fast a pace to such earthquakes of change? That Americans have not known such catastrophic changes as have the men and women of other societies is due to historical facts that are now quickly becoming "merely history." The history that now affects every man is world history. Within this scene and this period, in the course of a single generation, one-sixth of mankind is transformed from all that is feudal and backward into all that is modern, advanced, and fearful. Political colonies are freed; new and less visible forms of imperialism installed. Revolutions occur; men feel the intimate grip of new kinds of authority. Totalitarian societies rise, and are smashed to bits—or succeed fabulously. After two centuries of ascendancy, capitalism is shown up as only one way to make society into an industrial apparatus. After two centuries of hope, even formal democracy is restricted to a quite small portion of mankind. Everywhere in the underdeveloped world, ancient ways of life are broken up, and

vague expectations become urgent demands. Everywhere in the overdeveloped world, the means of authority and of violence become total in scope and bureaucratic in form. Humanity itself now lies before us, the super-nation at either pole concentrating its most coordinated and massive efforts upon the preparation of World War III.

The very shaping of history now outpaces the ability of men to orient themselves in accordance with cherished values. And which values? Even when they do not panic, men often sense that older ways of feeling and thinking have collapsed and that newer beginnings are ambiguous to the point of moral stasis. Is it any wonder that ordinary men feel they cannot cope with the larger worlds with which they are so suddenly confronted? That they cannot understand the meaning of their epoch for their own lives? That—in defense of selfhood—they become morally insensible, trying to remain altogether private men? Is it any wonder that they come to be possessed by a sense of the trap?

It is not only information that they need—in this Age of Fact, information often dominates their attention and overwhelms their capacities to assimilate it. It is not only the skills of reason that they need—although their struggles to acquire these often exhaust their limited moral energy.

What they need, and what they feel they need, is a quality of mind that will help them to use information and to develop reason in order to achieve lucid summations of what is going on in the world and of what may be happening within themselves. It is this quality, I am going to contend, that journalists and scholars, artists and publics, scientists and editors are coming to expect of what may be called the sociological imagination.

I

The sociological imagination enables its possessor to understand the larger historical scene in terms of its meaning for the inner life and the external career of a variety of individuals. It enables him to take into account how individuals, in the welter of their daily experience, often become falsely conscious of their social positions. Within that welter, the framework of modern society is sought, and within that framework the psychologies of a variety of men and women are formulated. By such means the personal uneasiness of individuals is focused upon explicit troubles and the indifference of publics is transformed into involvement with public issues.

The first fruit of this imagination—and the first lesson of the social science that embodies it—is the idea that the individual can understand his own experience and gauge his own fate by locating himself within his period; that he can know his own chances in life only by becoming aware of those of all individuals in his circumstances. In many ways it is a terrible lesson; in many ways a magnificent one. We do not know the limits of man's capacities for supreme effort or willing degradation, for agony or glee, for pleasurable brutality or the sweetness of reason. But in our time we have come to know the limits of "human nature" are frighteningly broad. We have come to know that every individual lives, from one generation to the next, in some society; that he lives out a biography, and that he lives it out within some historical sequence. By the fact of his living he contributes, however minutely, to the shaping of this society and to the course of its history, even as he is made by society and by its historical push and shove.

The sociological imagination enables us to grasp history and biography and the relations between the two within society. That is its task and its promise. To recognize this task and this promise is the mark of the classic social analyst.... And it is the signal of what is best in contemporary studies of man and society.

No social study that does not come back to the problems of biography, of history, and of their intersections within a society has completed its intellectual journey. Whatever the specific problems of the classic social analysts, however limited or however broad the features of social reality they have examined, those who have been imaginatively aware of the promise of their work have consistently asked three sorts of questions:

1. What is the structure of this particular society as a whole? What are its essential components,

and how are they related to one another? How does it differ from other varieties of social order? Within it, what is the meaning of any particular feature for its continuance and for its change?

2. Where does this society stand in human history? What are the mechanics by which it is changing? What is its place within and its meaning for the development of humanity as a whole? How does any particular feature we are examining affect, and how is it affected by, the historical period in which it moves? And this period—what are its essential features? How does it differ from other periods? What are its characteristic ways of history-making?

3. What varieties of men and women now prevail in this society and in this period? And what varieties are coming to prevail? In what ways are they selected and formed, liberated and repressed, made sensitive and blunted? What kinds of "human nature" are revealed in the conduct and character we observe in this society in this period? And what is the meaning of "human nature" of each and every feature of the society we are examining?

Whether the point of interest is a great power state or a minor literary mood, a family, a prison, a creed—these are the kinds of questions the best social analysts have asked. They are the intellectual pivots of classic studies of man in society—and they are the questions inevitably raised by any mind possessing the sociological imagination. For that imagination is the capacity to shift from one perspective to another—from the political to the psychological; from examination of a single family to comparative assessment of the national budgets of the world; from the theological school to the military establishment; from considerations of an oil industry to studies of contemporary poetry. It is the capacity to range from the most impersonal and remote transformations to the most intimate features of the human self—and to see the relations between the two. Back of its use there is always the urge to know the social and historical meaning of the individual in the society and in the period in which he has his quality and his being.

That, in brief, is why it is by means of the sociological imagination that men now hope to grasp what is going on in the world, and to understand what is happening in themselves as minute points of the intersections of biography and history within society. In large part, contemporary man's self-conscious view of himself as at least an outsider, if not a permanent stranger, rests upon an absorbed realization of social relativity and of the transformative power of history. The sociological imagination is the most fruitful form of this self-consciousness. By its use men whose mentalities have swept only a series of limited orbits often come to feel as if suddenly awakened in a house with which they had only supposed themselves to be familiar. Correctly or incorrectly, they often come to feel that they can now provide themselves with adequate summations, cohesive assessments, comprehensive orientations. Older decisions that once appeared sound now seem to them products of a mind unaccountably dense. Their capacity for astonishment is made lively again. They acquire a new way of thinking, they experience a transvaluation of values; in a word, by their reflection and by their sensibility, they realize the cultural meanings of the social sciences.

II

Perhaps the most fruitful distinction with which the sociological imagination works is between "the personal troubles of milieu" and "the public issues of social structure." This distinction is an essential tool of the sociological imagination and a feature of all classic work in social science.

Troubles occur within the character of the individual and within the range of his immediate relations with others; they have to do with his self and with those limited areas of social life of which he is directly and personally aware. Accordingly, the statement and the resolution of troubles properly lie within the individual as a biographical entity and within the scope of his immediate milieu—the social setting that is directly open to his personal experience and to some extent his willful activity. A trouble is a private matter: values cherished by an individual are felt by him to be threatened.

Issues have to do with matters that transcend these local environments of the individual and the range of his inner life. They have to do with the organization of many such milieux into the institutions of an historical society as a whole, with the ways in which various milieux overlap and interpenetrate to form the larger structure of social and historical life. An issue is a public matter: some value cherished by publics is felt to be threatened. Often there is a debate about what that value really is and about what it is that really threatens it. This debate is often without focus if only because it is the very nature of an issue, unlike even widespread trouble, that it cannot very well be defined in terms of the immediate and everyday environment of ordinary men. An issue, in fact, often involves a crisis in institutional arrangements, and often too it involves what Marxists call "contradictions" or "antagonisms."

In these terms, consider unemployment. When, in a city of 100,000, only one man is unemployed, that is his personal trouble, and for its relief we properly look to the character of the man, his skills, and his immediate opportunities. But when in a nation of 50 million employees, 15 million men are unemployed, that is an issue, and we may not hope to find its solution within the range of opportunities open to any one individual. The very structure of opportunities has collapsed. Both the correct statement of the problem and the range of possible solutions require us to consider the economic and political institutions of the society, and not merely the personal situation and character of a scatter of individuals.

Consider war. The personal problem of war, when it occurs, may be how to survive it or how to die in it with honor; how to make money out of it; how to climb into the higher safety of the military apparatus; or how to contribute to the war's termination. In short, according to one's values, to find a set of milieux and within it to survive the war or make one's death in it meaningful. But the structural issues of war have to do with its causes; with what types of men it throws up into command;

with its effects upon economic and political, family and religious institutions, with the unorganized irresponsibility of a world of nation-states.

Consider marriage. Inside a marriage a man and a woman may experience personal troubles, but when the divorce rate during the first four years of marriage is 250 out of every 1000 attempts, this is an indication of a structural issue having to do with the institutions of marriage and the family and other institutions that bear upon them.

Or consider the metropolis—the horrible, beautiful, ugly, magnificent sprawl of the great city. For many upper-class people, the personal solution to "the problem of the city" is to have an apartment with a private garage under it in the heart of the city, and forty miles out, a house by Henry Hill, garden by Garrett Eckbo, on a hundred acres of private land. In these two controlled environments—with a small staff at each end and a private helicopter connection—most people could solve many of the problems of personal milieux caused by the facts of the city. But all this, however splendid, does not solve the public issues that the structural fact of the city poses. What should be done with this wonderful monstrosity? Break it all up into scattered units, combining residence and work? Refurbish it as it stands? Or, after evacuation, dynamite it and build new cities according to new plans in new places? What should those plans be? And who is to decide and to accomplish whatever choice is made? These are structural issues; to confront them and to solve them require us to consider political and economic issues that affect innumerable milieux.

Insofar as an economy is so arranged that slumps occur, the problem of unemployment becomes incapable of personal solution. Insofar as war is inherent in the nation-state system and in the uneven industrialization of the world, the ordinary individual in his restricted milieu will be powerless—with or without psychiatric aid—to solve the troubles this system or lack of system imposes upon him. Insofar as the family as an institution turns women into darling little slaves and men into their chief providers and unweaned dependents, the problem of a satisfactory marriage remains incapable of purely private solution. Insofar as the overdeveloped megalopolis

and the overdeveloped automobile are built-in features of the overdeveloped society, the issue of urban living will not be solved by personal ingenuity and private wealth.

What we experience in various and specific milieux, I have noted, is often caused by structural changes. Accordingly, to understand the changes of many personal milieux we are required to look beyond them. And the number and variety of such structural changes increase as the institutions within which we live become more embracing and more intricately connected with one another. To be aware of the idea of social structure and to use it with sensibility is to be capable of tracing such linkages among a great variety of milieux. To be able to do that is to possess the sociological imagination.

III

What are the major issues for publics and the key troubles of private individuals in our time? To formulate issues and troubles, we must ask what values are cherished yet threatened, and what values are cherished and supported, by the characterizing trends of our period. In the case both of threat and of support we must ask what salient contradictions of structure may be involved.

When people cherish some set of values and do not feel any threat to them, they experience *well-being*. When they cherish values but *do* feel them to be threatened, they experience a crisis—either as a personal trouble or as a public issue. And if all their values seem involved, they feel the total threat of panic.

But suppose people are neither aware of any cherished values nor experience any threat? That is the experience of *indifference*, which, if it seems to involve all their values, becomes apathy. Suppose, finally, they are unaware of any cherished values, but still are very much aware of a threat? That is the experience of *uneasiness*, of anxiety, which, if it is total enough, becomes a deadly unspecified malaise.

Ours is a time of uneasiness and indifference—not yet formulated in such ways as to permit the work of reason and the play of sensibility. Instead of troubles—defined in terms of values and threats—there is often the misery of vague uneasiness; instead of explicit issues there is often merely the beat feeling that all is somehow not right. Neither the values threatened nor whatever threatens them has been stated; in short, they have not been carried to the point of decision. Much less have they been formulated as problems of social science.

In the thirties there was little doubt—except among certain deluded business circles—that there was an economic issue which was also a pack of personal troubles. In these arguments about "the crisis of capitalism," the formulations of Marx and the many unacknowledged reformulations of his work probably set the leading terms of the issue, and some men came to understand their personal troubles in these terms. The values threatened were plain to see and cherished by all; the structural contradictions that threatened them also seemed plain. Both were widely and deeply experienced. It was a political age.

But the values threatened in the era after World War II are often neither widely acknowledged as values nor widely felt to be threatened. Much private uneasiness goes unformulated; much public malaise and many decisions of enormous structural relevance never become public issues. For those who accept such inherited values as reason and freedom, it is the uneasiness itself that is the trouble; it is the indifference that is the issue. And it is this condition, of uneasiness and indifference, that is the signal feature of our period.

All this is so striking that it is often interpreted by observers as a shift in the very kinds of problems that need now to be formulated. We are frequently told that the problems of our decade, or even the crisis of our period, have shifted from the external realm of economics and now have to do with the quality of individual life—in fact with the question of whether there is soon going to be anything that can properly be called individual life. Not child labor but comic books, not poverty but mass leisure, are at the center of concern. Many great public issues as well as many private troubles are described in terms of "the psychiatric"—often, it seems, in a pathetic attempt to avoid the large issues and problems of modern society. Often this

statement seems to rest upon a provincial narrowing of interest to the Western societies, or even to the United States—thus ignoring two-thirds of mankind; often, too, it arbitrarily divorces the individual life from the larger institutions within which that life is enacted, and which on occasion bear upon it more grievously than do the intimate environments of childhood.

Problems of leisure, for example, cannot even be stated without considering problems of work. Family troubles over comic books cannot be formulated as problems without considering the plight of the contemporary family in its new relations with the newer institutions of the social structure. Neither leisure nor its debilitating uses can be understood as problems without recognition of the extent to which malaise and indifference now form the social and personal climate of contemporary American society. In this climate, no problems of "the private life" can be stated and solved without recognition of the crisis of ambition that is part of the very career of men at work in the incorporated economy.

It is true, as psychoanalysts continually point out, that people do often have "the increasing sense of being moved by obscure forces within themselves which they are unable to define." But it is *not* true, as Ernest Jones asserted, that "man's chief enemy and danger is his own unruly nature and the dark forces pent up within him." On the contrary: "Man's chief danger" today lies in the unruly forces of contemporary society itself, with its alienating methods of production, its enveloping techniques of political domination, its international anarchy—in a word, its pervasive transformations of the very "nature" of man and the conditions and aims of his life.

It is now the social scientist's foremost political and intellectual task—for here the two coincide—to make clear the elements of contemporary uneasiness and indifference. It is the central demand made upon him by other cultural workmen—by physical scientists and artists, by the intellectual community in general. It is because of this task and these demands, I believe, that the social sciences are becoming the common denominator of our cultural period, and the sociological imagination our most needed quality of mind.

PART II

Becoming a Member of Society

To become a member of society requires a number of interactions. The first of these is socialization (Chapter 2). The infant must learn the accepted roles that go along with its sexual being—its gender (Chapter 3) and to learn the right and wrong ways to meet its needs and fulfill their societal responsibilities—its culture (Chapter 4). To become a social being, we are taught from infancy through childhood and adult years through the process of socialization. Interaction with significant individuals and our participation in groups are key to socialization.

Perhaps the most important of the socialization agents is the family. It is the first agent in the process, and among its roles is the necessity to care for and protect the infant, to provide this human being with the behavior that accompanies this species in a particular culture, and to train it so that it can survive and become an active member of its society. The socialization process is lifelong, as the human being goes through numerous stages throughout its life cycle—changes in the family circumstances, more education, advancement in occupational roles, and the different circumstances one encounters as one interacts with different groups, learns the facets of one's culture, and participates in society. Each stage requires new and more learning.

It may come as a surprise to many that the gender behavior that accompanies one's sex is not biologically determined. In fact, it is one of the many behaviors that one learns through the socialization process—first from parents, with different accepted behaviors for each sex. The education institution will continue this differentiation by channeling the different sexes into different learning endeavors and sports. The religious institution seemingly confirms the inferiority of women in their socialization process by having men residing in superior positions from God on down. The economic institution continues this gender differentiation with its "glass ceilings" on women in regard to their position and

31

salary. The political situation for women is best exemplified by the fact that it took approximately 225 years for a female to be considered presidential material both by herself and the public.

These lifelong socialization experiences and processes take place in our culture—the place where we learn to cherish certain behaviors and objects and to follow certain rules and behaviors. Instead of cherishing George Washington and the Stars and Stripes we could have been cherishing the hero and flag of some other country if you had been raised in that country's culture. Of course, during our life we encounter various groups and various cultures, and so our cultural beliefs can change even if we do not change countries. It is this fact that confirms the importance of the groups one interacts with during the life cycle.

As you read this part of the text, consider the process that trains the infant, creates the expected behavior to go along with sex, and teaches the desired norms and values of our cultures and groups. Think of how different we would be if we grew up as Aborigines in Australia, preferring to eat beetles rather than hot dogs.

Chapter 2

Socialization
A Lifelong Learning Process

Dr. Alan W. McEvoy

Did you ever wonder why your views on what is right and wrong, good and bad, are very different from many other people's views? You probably believe strongly in your ideas about religion and politics, but why? Where did those views come from? Had you been born into a different family, community, ethnic group, religion, or country, would your views be "wrong"? Are the billions who also believe their views are the right ones simply misinformed? Of course, those "others" might question your beliefs and values just as you may question theirs. The study of socialization is the next step in understanding how we learn and accept our culture's language, values, beliefs, behaviors, and ideologies.

How do we become who we are, with our particular values, attitudes, and behaviors? We do so through the process of socialization that takes place within a particular society such as the United States, France, Kenya, or Japan. Each society has a culture or way of life that has developed over time and that dictates appropriate, acceptable behavior. It is within our families that our initial socialization takes place. Here we develop a self, learn to be social beings, and develop into members of our culture. From the day we are born, socialization shapes us into social beings, teaching us the behaviors and beliefs that make social existence possible. Through interaction with others, we develop our self-concepts. This process begins at an early age, when young children interact with others in a process called *symbolic interaction*. Very simply, the individual (whether a baby or adult) initiates contact—a cry or words—and receives a response. By interpreting and reacting to that response, the individual learns what brings positive reactions; those actions that receive positive responses and rewards are likely to be continued. Actions receiving negative responses are likely to be dropped.

The process of socialization takes place through interaction with others—interaction that is vital to our social development. The initial *agent,* or transmitter of socialization, is the family. Here we begin to learn our roles for participation in the wider world, an important aspect of which is sex-role socialization. The process continues in educational and religious institutions. When children enter school, they face new challenges and expectations. No longer do they receive unconditional love as they do in most families; now they are judged in a competitive environment, their first introduction to the world outside the protection of home. Socialization typically takes place in a series of developmental stages from birth through old age. Some sociologists focus on childhood stages, others on male or female socialization, and others on middle to old age.

Informal agents of socialization—those whose primary purpose is something other than socialization, such as entertainment—can have a major impact on the process. For instance, the mass media, books, and advertising all send out powerful messages about desirable and appropriate behaviors by presenting role models and lifestyles. Peer groups also affirm or disapprove of behaviors of children. The boy who does not engage in "masculine" activities may be ridiculed, for instance.

You might ask if social development is a natural outgrowth of physical maturity. To answer that question, some social scientists have focused their studies on cases of social isolation. In rare instances, children do not experience early socialization in the family. For example, some cases of physical and mental abuse and neglect and some orphanages provide only minimal care and human contact; children growing up in such environments have been found to show a higher percentage of physical and mental retardation. Only a handful of cases of almost total isolation have been available for study. In the first article, Kingsley Davis describes the case histories of Anna and Isabelle and considers the effects of social isolation on their mental and physical development. He writes that severe retardation is likely to occur when consistent contact with other human beings is absent, and he concludes that even though socialization can take place after prolonged periods of isolation, some effects of isolation may be permanent.

Although family, educational institutions, and religious organizations have as their stated purposes to socialize their members, we learn from many sources. Informal agents of socialization, mentioned earlier, come in many forms, one of which is television. Television is one of the most powerful informal agents of socialization, especially for children, who may watch several hours each day. They learn both "good" and "bad," and different types of roles from gender roles to adult job roles. Concern over the impact of TV violence has prompted several studies, including those reported in this chapter. The Kaiser Family Foundation review of studies points out both major findings and methodologies used to study television violence.

The socialization process can be very rigid and difficult to change when it involves ascribed roles. Socialization into gender roles provides examples of both the socialization process and the acquisition and tenacious nature of stereotypes. In the third reading in this chapter, Clyde Franklin addresses the socialization experience of men and boys and how they learn masculine gender roles. Each agent of socialization adds to the process of learning attitudes and behaviors: The family provides role models and reinforcement for "proper" behaviors; teachers reward boys and girls for gender-typed behaviors; religious institutions support traditional role behaviors; peer groups pressure boys into acceptable male behaviors. Informal or *nonpurposive* agents of male socialization, including mass media, barbershops, bars, and business meetings, also contribute to the process.

Results of rigid and stereotypic socialization can be destructive for men and women. Young men in the United States may take steroids to "bulk up," and young women go on extreme diets or binge and purge to achieve the perfect body portrayed in the media. Even aspects of life that we would think of as part of the learning process are affected by socialization, including the foods we eat and the way we eat them, the clothes we choose to wear, the college majors we select even sex! The next article by Tracey Steele explains how sex can be seen as a *social construct,* something that has proper behaviors and is governed by norms. We learn these social constructs through the process of socialization.

As you read the selections in this chapter, consider aspects of socialization we have mentioned. What are the effects of isolation from normal human contact? What are the stages of socialization? How does socialization take place through formal and informal agents of socialization? What are the effects of the socialization process that we take for granted, such as gender stereotyping? And what are aspects of our lives that we take for granted that are, in fact, shaped by socialization (like sex)?

5

Final Note on a Case of Extreme Isolation

KINGSLEY DAVIS

Socialization takes place in stages throughout our lives. We learn through agents of social-
ization such as family and education, and we are socialized into proper sex roles and other
specialized roles as in athletics. Socialization requires contact with other human beings.
What would we be like if we were raised in isolation with limited or no contact with other
humans? Could socialization take place? Rare cases exist in which humans have grown up
in partial or total isolation. This reading discusses two such cases and the results of isolation
for these two girls.

> *As you read, ask yourself the following:*
>
> 1. *Why do humans need other humans to develop "normally"?*
> 2. *What is missing in the experience of isolated children, and how does this affect the*
>   *children?*
> 3. *What happens if a child experiences neglect or abuse during socialization?*
> 4. *What message does this article provide about what we can do to help children grow*
>   *up in healthy environments?*

GLOSSARY **Socialization** The process of learning cooperative group living.
Learning stage The knowledge and ability individuals are expected to have
attained at a particular age.

Early in 1940 there appeared ... an account of a girl called Anna.[1] She had been deprived of normal contact and had received a minimum of human care for almost the whole of her first six years of life. At this time observations were not complete and the report had a tentative character. Now, however, the girl is dead, and with more information available,[2] it is possible to give a fuller and more definitive description of the case from a sociological point of view.

Anna's death, caused by hemorrhagic jaundice, occurred on August 6, 1942. Having been born on March 1 or 6,[3] 1932, she was approximately ten and a half years of age when she died. The previous report covered her development up to the age of almost eight years; the present one recapitulates the

Source: Reprinted from *American Journal of Sociology*, Vol. III, No. 5, March 1947, pp. 432–437 by permission of the author.
© 1947 by the University of Chicago Press.

earlier period on the basis of new evidence and then covers the last two and a half years of her life.

EARLY HISTORY

The first few days and weeks of Anna's life were complicated by frequent changes of domicile.... She was an illegitimate child, the second such child born to her mother, and her grandfather, a widowed farmer in whose house the mother lived, strongly disapproved of this new evidence of the mother's indiscretion. This fact led to the baby's being shifted about.

Two weeks after being born in a nurse's private home, Anna was brought to the family farm, but the grandfather's antagonism was so great that she was shortly taken to the house of one of her mother's friends. At this time a local minister became interested in her and took her to his house with an idea of possible adoption. He decided against adoption, however, when he discovered that she had vaginitis. The infant was then taken to a children's home in the nearest large city. This agency found that at the age of only three weeks she was already in a miserable condition, being "terribly galled and otherwise in very bad shape." It did not regard her as a likely subject for adoption but took her in for a while anyway, hoping to benefit her. After Anna had spent nearly eight weeks in this place, the agency notified her mother to come to get her. The mother responded by sending a man and his wife to the children's home with a view to their adopting Anna, but they made such a poor impression on the agency that permission was refused. Later the mother came herself and took the child out of the home and then gave her to this couple. It was in the home of this pair that a social worker found the girl a short time thereafter. The social worker went to the mother's home and pleaded with Anna's grandfather to allow the mother to bring the child home. In spite of threats, he refused. The child, by then more than four months old, was next taken to another children's home in a nearby town. A medical examination at this time revealed

that she had impetigo, vaginitis, umbilical hernia, and a skin rash.

Anna remained in this second children's home for nearly three weeks, at the end of which time she was transferred to a private foster-home. Since, however, the grandfather would not, and the mother could not, pay for the child's care, she was finally taken back as a last resort to the grandfather's house (at the age of five and a half months). There she remained, kept on the second floor in an attic-like room because her mother hesitated to incur the grandfather's wrath by bringing her downstairs.

The mother, a sturdy woman weighing about 180 pounds, did a man's work on the farm. She engaged in heavy work such as milking cows and tending hogs and had little time for her children. Sometimes she went out at night, in which case Anna was left entirely without attention. Ordinarily, it seems, Anna received only enough care to keep her barely alive. She appears to have been seldom moved from one position to another. Her clothing and bedding were filthy. She apparently had no instruction, no friendly attention.

It is little wonder that, when finally found and removed from the room in the grandfather's house at the age of nearly six years, the child could not talk, walk, or do anything that showed intelligence. She was in an extremely emaciated and undernourished condition, with skeletonlike legs and a bloated abdomen. She had been fed on virtually nothing except cow's milk during the years under her mother's care.

Anna's condition when found, and her subsequent improvement, have been described in the previous report. It now remains to say what happened to her after that.

LATER HISTORY

In 1939, nearly two years after being discovered, Anna had progressed, as previously reported, to the point where she could walk, understand simple commands, feed herself, achieve some neatness, remember people, etc. But she still did not speak, and, though she was much more like a normal infant of something over one year of age in mentality, she was far from normal for her age.

On August 30, 1939, she was taken to a private home for retarded children, leaving the country home where she had been for more than a year and a half. In her new setting she made some further progress, but not a great deal. In a report of an examination made November 6 of the same year, the head of the institution pictured the child as follows:

> Anna walks about aimlessly, makes periodic rhythmic motions of her hands, and, at intervals, makes guttural and sucking noises. She regards her hands as if she had seen them for the first time. It was impossible to hold her attention for more than a few seconds at a time—not because of distraction due to external stimuli but because of her inability to concentrate. She ignored the task in hand to gaze vacantly about the room. Speech is entirely lacking. Numerous unsuccessful attempts have been made with her in the hope of developing initial sounds. I do not believe that this failure is due to negativism or deafness but that she is not sufficiently developed to accept speech at this time.... The prognosis is not favorable.

More than five months later, on April 25, 1940, a clinical psychologist, the late Professor Francis N. Maxfield, examined Anna and reported the following: large for her age; hearing "entirely normal"; vision apparently normal; able to climb stairs; speech in the "babbling stage" and "promise for developing intelligible speech later seems to be good." He said further that "on the Merrill-Palmer scale she made a mental score of 19 months. On the Vineland social maturity scale she made a score of 23 months."[4]

Professor Maxfield very sensibly pointed out that prognosis is difficult in such cases of isolation. "It is very difficult to take scores on tests standardized under average conditions of environment and experience," he wrote, "and interpret them in a case where environment and experience have been so unusual." With this warning he gave it as his opinion at that time that Anna would eventually "attain an adult mental level of six or seven years."[5]

The school for retarded children, on July 1, 1941, reported that Anna had reached 46 inches in height and weighed 60 pounds. She could bounce and catch a ball and was said to conform to group socialization, though as a follower rather than a leader. Toilet habits were firmly established. Food habits were normal, except that she still used a spoon as her sole implement. She could dress herself except for fastening her clothes. Most remarkable of all, she had finally begun to develop speech. She was characterized as being at about the two-year level in this regard. She could call attendants by name and bring in one when she was asked to. She had a few complete sentences to express her wants. The report concluded that there was nothing peculiar about her, except that she was "feeble-minded—probably congenital in type."[6]

A final report from the school made on June 22, 1942, and evidently the last report before the girl's death, pictured only a slight advance over that given above. It said that Anna could follow directions, string beads, identify a few colors, build with blocks, and differentiate between attractive and unattractive pictures. She had a good sense of rhythm and loved a doll. She talked mainly in phrases but would repeat words and try to carry on a conversation. She was clean about clothing. She habitually washed her hands and brushed her teeth. She would try to help other children. She walked well and could run fairly well, though clumsily. Although easily excited, she had a pleasant disposition.

INTERPRETATION

Such was Anna's condition just before her death. It may seem as if she had not made much progress, but one must remember the condition in which she had been found. One must recall that she had no glimmering of speech, absolutely no ability to walk, no sense of gesture, not the least capacity to feed herself even when the food was put in front of her, and no comprehension of cleanliness. She was so apathetic that it was hard to tell whether or not she could hear. And all this at the age of nearly six years. Compared with this condition, her capacities at the time of her death seem striking indeed, though they do not amount to much more than a two-and-a-half year mental level. One conclusion therefore seems safe, namely, that her isolation prevented a considerable amount of mental development that was undoubtedly part of her capacity. Just what her original capacity was, of course, is hard to say; but her development after her period of confinement (including the ability to walk and run, to play, to dress, fit into a social situation, and, above all, to speak) shows that she had at least this capacity—capacity that never could have been realized in her original condition of isolation.

A further question is this: What would she have been like if she had received a normal upbringing from the moment of birth? A definitive answer would have been impossible in any case, but even an approximate answer is made difficult by her early death. If one assumes, as was tentatively surmised in the previous report, that it is "almost impossible for any child to learn to speak, think, and act like a normal person after a long period of early isolation," it seems likely that Anna might have had a normal or near-normal capacity, genetically speaking. On the other hand, it was pointed out that Anna represented "a marginal case [because] she was discovered before she had reached six years of age," an age "young enough to allow for some plasticity."[7] While admitting, then, that Anna's isolation *may* have been the major cause (and was certainly a minor cause) of her lack of rapid mental progress during the four-and-a-half years following her rescue from neglect, it is necessary to entertain the hypothesis that she was congenitally deficient.

In connection with this hypothesis, one suggestive though by no means conclusive circumstance needs consideration, namely, the mentality of Anna's forebears. Information on this subject is easier to obtain, as one might guess, on the mother's than on the father's side. Anna's maternal grandmother, for example, is said to have been college educated and wished to have her children receive a good education, but her husband, Anna's stern grandfather, apparently a shrewd, hard-driving, calculating farm-owner, was so penurious that her ambitions in this direction were thwarted. Under the circumstances her daughter (Anna's mother) managed, despite having to do hard work on the farm, to complete the eighth grade in a country school. Even so, however, the daughter was evidently not very smart. "A schoolmate of [Anna's mother] stated that she was retarded in school work; was very gullible at this age; and that her morals even at this time were discussed by other students." Two tests administered to her on March 4, 1938, when she was thirty-two years of age, showed that she was mentally deficient. On the Stanford Revision of the Binet-Simon Scale her performance was equivalent to that of a child of eight years, giving her an I.Q. of 50 and indicating mental deficiency of "middle-grade moron type."[8]

As to the identity of Anna's father, the most persistent theory holds that he was an old man about seventy-four years of age at the time of the girl's birth. If he was the one, there is no indication of mental or other biological deficiency, whatever one may think of his morals. However, someone else may actually have been the father.

To sum up: Anna's heredity is the kind that *might* have given rise to innate mental deficiency, though not necessarily.

COMPARISON WITH ANOTHER CASE

Perhaps more to the point than speculation about Anna's ancestry would be a case for comparison. If a child could be discovered who had been isolated about the same length of time as Anna but had achieved a much quicker recovery and a greater mental development, it would be a stronger indication that Anna was deficient to start with.

Such a case does exist. It is the case of a girl found at about the same time as Anna and under strikingly similar circumstances. A full description of the details of this case has not been published, but in addition to newspaper reports, an excellent preliminary account by a speech specialist, Dr. Marie K. Mason, who played an important role in the handling of the child, has appeared.[9] Also the late Dr. Francis N. Maxfield, clinical psychologist at Ohio State University, as was Dr. Mason, has written an as yet unpublished but penetrating analysis of the case.[10] Some of his observations have been included in Professor Zingg's book on feral man.[11] The following discussion is drawn mainly from these enlightening materials. The writer, through the kindness of Professors Mason and Maxfield, did have a chance to observe the girl in April 1940, and to discuss the features of her case with them.

Born apparently one month later than Anna, the girl in question, who has been given the pseudonym Isabelle, was discovered in November 1938, nine months after the discovery of Anna. At the time she was found she was approximately six-and-a-half years of age. Like Anna, she was an illegitimate child and had been kept in seclusion for that reason. Her mother was a deaf-mute, having become so at the age of two, and it appears that she and Isabelle had spent most of their time together in a dark room shut off from the rest of the mother's family. As a result Isabelle had no chance to develop speech; when she communicated with her mother, it was by means of gestures. Lack of sunshine and inadequacy of diet had caused Isabelle to become rachitic. Her legs in particular were affected; they "were so bowed that as she stood erect the soles of her shoes came nearly flat together, and she got about with a skittering gait."[12] Her behavior toward strangers, especially men, was almost that of a wild animal, manifesting much fear and hostility. In lieu of speech she made only a strange croaking sound. In many ways she acted like an infant. "She was apparently utterly unaware of relationships of

any kind. When presented with a ball for the first time, she held it in the palm of her hand, then reached out and stroked my face with it. Such behavior is comparable to that of a child of six months."[13] At first it was even hard to tell whether or not she could hear, so unused were her senses. Many of her actions resembled those of deaf children.

It is small wonder that, once it was established that she could hear, specialists working with her believed her to be feeble-minded. Even on nonverbal tests her performance was so low as to promise little for the future. Her first score on the Stanford-Binet was 19 months, practically at the zero point of the scale. On the Vineland social maturity scale her first score was 39, representing an age level of two-and-a-half years.[14] "The general impression was that she was wholly uneducable and that any attempt to teach her to speak, after so long a period of silence, would meet with failure."[15]

In spite of this interpretation, the individuals in charge of Isabelle launched a systematic and skillful program of training. It seemed hopeless at first. The approach had to be through pantomime and dramatization, suitable to an infant. It required one week of intensive effort before she even made her first attempt at vocalization. Gradually she began to respond, however, and, after the first hurdles had at least been overcome, a curious thing happened. She went through the usual stages of learning characteristic of the years from one to six not only in proper succession but far more rapidly than normal. In a little over two months after her first vocalization she was putting sentences together. Nine months after that she could identify words and sentences on the printed page, could write well, could add to ten, and could retell a story after hearing it. Seven months beyond this point she had a vocabulary of 1500–2000 words and was asking complicated questions. Starting from an educational level of between one and three years (depending on what aspect one considers), she had reached a normal level by the time she was eight-and-a-half years old. In short, she covered in two years the stages of learning that ordinarily require six.[16] Or, to put it another way, her I.Q. trebled in a year and a half.[17] The speed with which she reached the normal level of mental development seems analogous to the recovery of body weight in a growing child after an illness, the recovery being achieved by an extra fast rate of growth for a period after the illness until normal weight for the given age is again attained.

When the writer saw Isabelle a year-and-a-half after her discovery, she gave him the impression of being a very bright, cheerful, energetic little girl. She spoke well, walked and ran without trouble, and sang with gusto and accuracy. Today she is over fourteen years old and has passed the sixth grade in a public school. Her teachers say that she participates in all school activities as normally as other children. Though older than her classmates, she has fortunately not physically matured too far beyond their level.[18]

Clearly the history of Isabelle's development is different from that of Anna's. In both cases there was an exceedingly low, or rather blank, intellectual level to begin with. In both cases it seemed that the girl might be congenitally feeble-minded. In both a considerably higher level was reached later on. But the Ohio girl achieved a normal mentality within two years, whereas Anna was still markedly inadequate at the end of four-and-a-half years. This difference in achievement may suggest that Anna had less initial capacity. But an alternative hypothesis is possible.

One should remember that Anna never received the prolonged and expert attention that Isabelle received. The result of such attention, in the case of the Ohio girl, was to give her speech at an early stage, and her subsequent rapid development seems to have been a consequence of that. "Until Isabelle's speech and language development, she had all the characteristics of a feeble-minded child." Had Anna, who, from the standpoint of psychometric tests and early history, closely resembled this girl at the start, been given a mastery of speech at an earlier point by intensive training, her subsequent development might have been much more rapid.[19]

The hypothesis that Anna began with a sharply inferior mental capacity is therefore not established. Even if she were deficient to start with, we have no way of knowing how much so. Under ordinary conditions she might have been a dull normal or, like her mother, a moron. Even after the blight of her isolation, if she had lived to maturity, she might have finally reached virtually the full level of her capacity, whatever it may have been. That her isolation did have a profound effect upon her mentality, there can be no doubt. This is proved by the substantial degree of change during the four-and-a-half years following her rescue.

Consideration of Isabelle's case serves to show, as Anna's case does not clearly show, that isolation up to the age of six, with failure to acquire any form of speech and hence failure to grasp nearly the whole world of cultural meaning, does not preclude the subsequent acquisition of these. Indeed, there seems to be a process of accelerated recovery in which the child goes through the mental stages at a more rapid rate than would be the case in normal development. Just what would be the maximum age at which a person could remain isolated and still retain the capacity for full cultural acquisition is hard to say. Almost certainly it would not be as high as age fifteen; it might possibly be as low as age ten. Undoubtedly various individuals would differ considerably as to the exact age.

Anna's is not an ideal case for showing the effects of extreme isolation, partly because she was possibly deficient to begin with, partly because she did not receive the best training available, and partly because she did not live long enough. Nevertheless, her case is instructive when placed in the record with numerous other cases of extreme isolation. This and the previous article about her are meant to place her in the record. It is to be hoped that other cases will be described in the scientific literature as they are discovered (as unfortunately they will be), for only in these rare cases of extreme isolation is it possible "to observe *concretely separated* two factors in the development of human personality which are always otherwise only analytically separated, the biogenic and the sociogenic factors."[20]

NOTES

1. Kingsley Davis, "Extreme Social Isolation of a Child," *American Journal of Sociology*, XLV (January 1940), 554–65.

2. Sincere appreciation is due to the officials in the Department of Welfare, commonwealth of Pennsylvania, for their kind cooperation in making available the records concerning Anna and discussing the case frankly with the writer. Helen C. Hubbell, Florentine Hackbusch, and Eleanor Mecklenburg were particularly helpful, as was Fanny L. Matchette. Without their aid neither of the reports on Anna could have been written.

3. The records are not clear as to which day.

4. Letter to one of the state officials in charge of the case.

5. *Ibid.*

6. Progress report of the school.

7. Davis, *op. cit.*, p. 564.

8. The facts set forth here as to Anna's ancestry are taken chiefly from a report of mental tests administered to Anna's mother by psychologists at a state hospital where she was taken for this purpose after the discovery of Anna's seclusion. This excellent report was not available to the writer when the previous paper on Anna was published.

9. Marie K. Mason, "Learning to Speak after Six and One-Half Years of Silence," *Journal of Speech Disorders*, VII (1942), 295–304.

10. Francis N. Maxfield, "What Happens When the Social Environment of a Child Approaches Zero." The writer is greatly indebted to Mrs. Maxfield and to Professor Horace B. English, a colleague of Professor Maxfield, for the privilege of seeing this manuscript and other materials collected on isolated and feral individuals.

11. J. A. L. Singh and Robert M. Zingg, *Wolf-Children and Feral Man* (New York: Harper & Bros., 1941), pp. 248–251.

12. Maxfield, unpublished manuscript cited above.

13. Mason, *op. cit.*, p. 299.

14. Maxfield, unpublished manuscript.

15. Mason, *op. cit.*, p. 299.

16. *Ibid.*, pp. 300–304.

17. Maxfield, unpublished manuscript.

18. Based on a personal letter from Dr. Mason to the writer, May 13, 1946.

19. This point is suggested in a personal letter from Dr. Mason to the writer, October 22, 1946.

20. Singh and Zingg, *op. cit.*, pp. xxi–xxii, in a foreword by the writer.

6

Key Facts on TV Violence

The Henry J. Kaiser Family Foundation

Socialization takes place through formal agents (family, school, religion) and informal agents whose primary purpose is not socialization but who nonetheless affect the socialization process. One powerful informal agent of socialization is television, which has a major impact on who we are. The following reading presents findings from studies on the effect of violence depicted on television, especially on children. It presents summaries of studies using different methods of data collection: laboratory experiments, field experiments, cross-sectional studies, and longitudinal studies.

Consider the following questions as you read this article:

1. *How does television violence influence the socialization of girls and boys?*
2. *Which methods have been used to study violence on TV? How would you conduct such a study?*
3. *What are the implications of this report for children's TV shows and TV viewing?*
4. *What do you think should be done about television violence?*

GLOSSARY **TV violence** Any overt depiction of usc of physical force or credible threat of such force intended to harm people physically.

Since the advent of television, the effect of TV violence on society has been widely studied and vigorously debated. Based on the cumulative evidence of studies conducted over several decades the scientific and public health communities overwhelmingly conclude that viewing violence poses a harmful risk to children. Critics of the research challenge this conclusion and dispute claims that exposure to TV violence leads to real-life aggression. As we move into the digital era with enhanced images and sound, media violence will undoubtedly continue to be a focus of public concern and scientific research.

PREVALENCE OF VIOLENCE ON TV

The National Television Violence Study is the largest content analysis undertaken to date. It analyzed

Source: The Henry J. Kaiser Foundation, *Key Facts on TV Violence*, Spring 2003. www.kff.org/entmedia/

programming over three consecutive TV seasons from 1994 to 1997.[1] Among the findings:

- Nearly 2 out of 3 TV programs contained some violence,[2] averaging about 6 violent acts per hour.[3]

- Fewer than 5% of these programs featured an anti-violence theme or prosocial message emphasizing alternatives to or consequences of violence.[4]

- Violence was found to be more prevalent in children's programming (69%) than in other types of programming (57%). In a typical hour of programming children's shows featured more than twice as many violent incidents (14) than other types of programming (6).[5]

- The average child who watches 2 hours of cartoons a day may see nearly 10,000 violent incidents each year, of which the researchers estimate that at least 500 pose a high risk for learning and imitating aggression and becoming desensitized to violence.[6]

- The number of prime-time programs with violence increased over the three years of the study, from 53% to 67% on broadcast television and from 54% to 64% on basic cable. Premium cable networks have the highest percentage of shows with violence, averaging 92% since 1994.[7]

The UCLA Television Violence Monitoring Report also analyzed three years of programming from 1994 to 1997. This study relied on the qualitative judgments of a team of student monitors and staff researchers rather than a systematic content analysis, to determine whether individual violent depictions "raised concern" for viewers.[8] Among the findings:

- Children's Saturday morning TV shows that feature "sinister combat violence" raised the most serious concerns for these researchers. These are fantasy live-action shows and animated cartoons in which violence is central to the storyline, the villains and superheroes use violence as an acceptable and effective way to get what they want, and the perpetrators are valued for their combatability. Among the most popular shows for children, the number of troubling shows in this genre decreased from seven to four over the three years of the study.[9]

- The number of prime-time series that raised frequent concerns about violence steadily declined over the three years, from nine such series in 1994–95 to just two in 1996–97.[10]

- TV specials was the only category that raised new concerns at the end of the three years. In the second year five live-action reality shows featured real or re-created graphic images of animals attacking and sometimes killing people. By the third year the number of such shows had increased again.[11]

SCIENTIFIC STUDIES OF TV VIOLENCE EFFECTS

Researchers hypothesize that viewing TV violence can lead to three potentially harmful effects: increased antisocial or aggressive behavior, desensitization to violence (becoming more accepting of violence in real life and less caring about other people's feelings), or increased fear of becoming a victim of violence.[12] Many researchers believe that children age 7 and younger are particularly vulnerable to the effects of viewing violence because they tend to perceive fantasy and cartoon violence as realistic.[13]

Since the 1960s, a body of research literature has been accumulating on the effects of TV violence. Taken together, the studies conclude that TV violence is one of many factors that contribute to aggressive behavior. Following are examples of the various types of research studies that have been conducted:[14]

Laboratory experiments are conducted in a controlled setting in order to manipulate media exposure and assess the short-term effects. Participants are randomly assigned to view either a violent or nonviolent film clip and their subsequent behavior is observed.

- A series of classic experiments conducted in the 1960s provided the earliest evidence of a link between TV violence and aggression. In these studies, children who were exposed to a TV clip of an actor hitting an inflatable doll were more likely than children who did not see the clip to imitate the action in their play, especially if the aggressive actions in the film clip were rewarded.[15]

- Other laboratory experiments have indicated that exposure to media violence increases children's tolerance for real-life aggression. For example, when third- and fourth-graders were left in charge of two younger children they could see on a TV monitor, the ones who viewed an aggressive film were much more reluctant than those who had not seen the film to ask an adult for help when the younger children began to fight, even though the fight was becoming progressively aggressive.[16]

Field experiments are conducted in a more naturalistic setting. As with the laboratory studies, children are shown video clips and their short-term post-viewing behavior is monitored by researchers. Over the past 30 years, numerous field studies have indicated that some children behave more aggressively after viewing violence.

- In one study, researchers showed children episodes of either *Batman* and *Spiderman* or *Mister Rogers' Neighborhood* over several weeks and then observed their behavior for two weeks afterwards. The children who viewed violent cartoons were more likely to interact aggressively with their peers, while those children who watched *Mister Rogers' Neighborhood* became more cooperative and willing to share toys.[17]

- In another study, researchers exposed children to an episode of *Mighty Morphin Power Rangers* and then observed their verbal and physical aggression in the classroom. Compared to children who had not seen the episode, viewers committed seven times as many aggressive acts such as hitting, kicking, shoving, and insulting a peer.[18]

Cross-sectional studies survey a large and representative sample of viewers at one point in time. Since the 1970s, a large number of these studies have concluded that viewing TV violence is related to aggressive behavior and attitudes.[19] These studies are correlational and do not prove causality; that is, it is difficult to know whether watching violence on TV is causing the increase in aggression or whether viewers who are already aggressive prefer watching violent content.

- In one study, 2,300 junior and senior high school students were asked to list their four favorite programs, which were analyzed for violent content, and to provide a self-reported checklist of activities that ranged from fighting at school to serious delinquency. Researchers found that teens whose favorite programs were violent tended to report a higher incidence in overall aggressive and delinquent behavior.[20]

- A recent study demonstrated a relationship between children's bullying and their exposure to media violence. Third-, fourth-, and fifth-graders who were identified by their peers as being the ones who spread rumors, exclude and insult peers, and behave in ways that hurt others, were more likely to view violence than nonaggressive children.[21]

Longitudinal studies offer the best way to study long-term effects of exposure to violent TV content. These studies survey the same group of individuals at several different times over many years to determine whether viewing violence is related to subsequent aggressive behavior. This method is designed to detect causal relationships and statistically control for environmental, family, and personal characteristics that might otherwise account for aggression.

- One study demonstrated that TV habits of children in the 1960s were a significant predictor of adult aggression, even criminal behavior, regardless of children's initial aggressiveness, IQ, social status, or parenting style. In this study, which spans more than 20 years, boys who preferred and viewed more violent programming at age 8 were more likely to be aggressive as teenagers and have arrests and convictions as adults for interpersonal crimes such as spousal and child abuse, murder, and aggravated assault.[22]

■ Television exposure during adolescence has also been linked to subsequent aggression in young adulthood. A 17-year longitudinal study concluded that teens who watched more than one hour of TV a day were almost four times as likely as other teens to commit aggressive acts in adulthood (22% versus 6%), taking into account prior aggressiveness, psychiatric disorders, family income, parental education, childhood neglect, and neighborhood violence.[23]

Meta-analyses use a statistical procedure to combine the results from many different studies.

■ The largest meta-analysis on TV violence analyzed 217 studies conducted between 1957 and 1990 and found that viewing violence was significantly linked to aggressive and antisocial behavior, especially among the youngest viewers. The overall effect size was 31, meaning that exposure to TV violence was estimated to account for 10% of the variance in antisocial behavior.[24]

OPPOSING VIEWPOINT

A small number of critics of the scientific evidence have concluded that TV violence does not contribute to real-life aggression. For the most part, they do not base their conclusions on studies with contrary findings, but argue that the studies that have been conducted are flawed.[25]

NOTES

1. Center for Communication and Social Policy, University of California, Santa Barbara (UCSB), *National Television Violence Study*, Executive Summary, Volume 3, 1998. Commissioned by the National Cable Television Association, the study analyzed almost 10,000 hours of broadcast and cable programming randomly selected from 23 channels over the course of three TV seasons from 1994 to 1997.

2. Ibid., 30. Researchers defined three main types of violent depictions: credible threats, behavioral acts, and harmful consequences.

3. Ibid., 33.

4. Ibid.

5. Barbara Wilson et al., "Violence in Children's Television Programming: Assessing the Risks," *Journal of Communication* 52 (2002): 5–35.

6. Center for Communication and Social Policy, UCSB, 33–34.

7. Ibid., 32.

8. UCLA Center for Communication Policy, *UCLA Television Violence Monitoring Report*, 1998, <http://ccp.ucla.edu/Webreport96/tableof.htm>

(28 September 2002). Commissioned by the National Association of Broadcasters, the researchers analyzed more than 3,000 hours of TV over three consecutive TV seasons from 1994 to 1997. TV shows with violence were divided into four different categories based on the level of concern, ranging from high levels of violence and serious concern to no serious concern because the context is appropriate.

9. Ibid., <http://ccp.ucla.com/Webreport96/network.htm#kids> (28 September 2002).

10. Ibid., <http://ccp.ucla.com/Webreport96/network.htm#Series> (28 September 2002).

11. Ibid., <http://ccp.ucla.com/Webreport96/network.htm#specials> (28 September 2002).

12. American Psychological Association, *Report of the American Psychological Association Commission on Violence and Youth, Volume 1*, 1993, 33; Stacy L. Smith and Edward Donnerstein, "Harmful Effects of Exposure to Media Violence: Learning of Aggression, Emotional Desensitization, and Fear," *Human Aggression: Theories, Research, and Implications for Social Policy*, eds. R. Geen and E. Donnerstein (New York: Academic Press, 1998), 167–202.

13. Brad Bushman and L. Rowell Huesmann, "Effects of Televised Violence on Aggression," *Handbook of Children and the Media*, eds. D. Singer and J. Singer (Thousand Oaks, CA: Sage Publications, 2001), 223–268.

14. For summaries of these studies, see Bushman and Huesmann, 2001; Victor Strasburger and Barbara Wilson, "Media Violence," *Children, Adolescents & the Media* (Thousand Oaks, CA: Sage Publications, 2002), 73–116; W. James Potter, *On Media Violence* (Thousand Oaks, CA: Sage, 1999).

15. Albert Bandura, Dorothea Ross, and Sheila Ross, "Transmission of Aggression through Imitation of Aggressive Models," *Journal of Abnormal and Social Psychology* 63 (1961): 575–582; Albert Bandura, Dorothea Ross, and Sheila Ross, "Imitation of Film-Mediated Aggressive Models," *Journal of Abnormal and Social Psychology* 66 (1963): 3–11; Albert Bandura, Dorothea Ross, and Sheila Ross, "Vicarious Reinforcement and Imitative Learning," *Journal of Abnormal and Social Psychology* 67 (1963): 601–607

16. Ronald S. Drabman and Margaret Hanratty Thomas, "Does Media Violence Increase Children's Toleration of Real-Life Aggression?" *Developmental Psychology* 10 (1974): 418–421; Ronald S. Drabman and Margaret Hanratty Thomas, "Does Watching Violence on Television Cause Apathy?" *Pediatrics* 57 (1976): 329–331.

17. Aletha Huston-Stein and L.K. Friedrich, "Television Content and Young Children's Social Behavior," *Television and Social Behavior, Volume II, Television and Social Learning*, eds. J. Murray, E. Rubinstein, and G. Comstock (Washington, D.C.: U.S. Government Printing Office, 1972), 207–317.

18. Chris J. Boyatzis, Gina M. Matillo, and Kristen M. Nesbitt, "Effects of the Mighty Morphin Power Rangers' on Children's Aggression with Peers," *Child Study Journal* 25: 1 (1995): 45–55.

19. See, for example, Bushman and Huesmann, 2001; Strasburger and Wilson, 2002; Potter, 1999.

20. Jennie McIntyre and James Teevan Jr.,"Television Violence and Deviant Behavior," *Television and Social Behavior, Volume III, Television and Adolescent Aggressiveness*, eds. G. Comstock and E. Rubinstein (Washington, D.C.: U.S. Government Printing Office, 1972), 383–435.

21. Audrey Buchanan et al., "What Goes in Must Come Out: Children's Media Violence Consumption at Home and Aggressiveness at School," <http://www.mediafamily.org/research/reports/issbd.shtml> (28 September 2002).

22. L. Rowell Huesmann et al., "Stability of Aggression Over Time and Generations," *Developmental Psychology* 20 (1984): 1120–1134. This study was begun in 1960 with a sample of 875 youths in New York state; ten years later 427 were re-interviewed, and twelve years later 409 were interviewed and criminal justice data was collected on 632 of the original subjects.

23. Jeffrey Johnson et al., "Television Viewing and Aggressive Behavior During Adolescence and Adulthood," *Science* 295 (March 29, 2002): 2468–2471. The longitudinal study was conducted over a 17-year time span with a sample of 707 families. Criminal arrest and charge data (assault or physical fights resulting in injury, robbery, threats to injure someone, or weapon used to commit a crime) were obtained.

24. Haejung Paik and George Comstock, "The Effects of Television Violence on Antisocial Behavior: A Meta-Analysis," *Communication Research* 21: 4 (August 1994). 516–546.

25. See, for example, Jonathan Freedman, *Media Violence and Its Effect on Aggression* (Toronto: University of Toronto Press, 2002); this review was commissioned by the Motion Picture Association of America. See also Jib Fowles, *The Case for Television Violence* (Newbury Park, CA: Sage Publications, 1999).

7

Becoming "Boys," "Men," "Guys," and "Dudes"

CLYDE W. FRANKLIN II

Being female or male is not a simple matter of being born with the distinct anatomy. One must learn the behavior that a particular culture assigns to that sex. Learning one's gender roles, then, is a key component of the socialization process. This process takes place throughout the stages of socialization. Franklin discusses this process for males, and indirectly for females, by describing several agents of socialization—family, education, religion, and peer groups.

As you read, consider the following questions:

1. *How do babies become boys become men?*
2. *Who and what determines the socialization process boys go through?*

GLOSSARY **Socialization** The process of learning to become a social being.
Agents Groups and organizations that help in the socialization process.
Nonpurposive agents Those activities and organizations that do not have the explicit purpose of socializing but that influence socialization.

AGENTS OF MALE SOCIALIZATION

The human male undergoes a long socialization process whereby he becomes aware of himself as a male and develops sex role skills necessary for full functioning as a social male in society. [As important as self-development] is the sex role which the biological male must learn if he is to fulfill self and others' expectations of himself as a boy, a man, a guy, or dude. What is the nature of this sex role which must be performed if the biological male is to function fully as a social male?...

A glance at some of the agents responsible for the development and maintenance of male

socialization should contribute to an understanding of this process. The agents of socialization to be discussed are divided into those formally charged with the responsibility for male sex role socialization and those that have informal (and often latent) responsibility.

PURPOSEFUL AGENTS OF MALE SOCIALIZATION

Each newborn male in society is expected to undergo a lengthy learning process to acquire appropriate male behaviors. Responsibility for this process historically lay with such societal agents as the family, religious institutions, educational institutions, the mass media, and adolescent and young

Source: From *Men and Society,* Chicago: Nelson-Hall, 1988.

adult male peer groups. A discussion of the roles of these agents in male socialization is presented in this section. Let us begin with the family.

The Family   The family is a vital agent involved in teaching males appropriate attitudes and behaviors. From the moment the newborn is identified as male, a set of cultural expectations unfolds dictating what behaviors may and may not be displayed. The agent charged with initial responsibility for insuring sex role conformity with societal expectations is the family. Studies have suggested (e.g., Schau et al. 1980 and Fu and Leach 1980) that if the newborn is male, rather rigid cultural expectations exist for him to learn to give "male performances" in social interactions. This means that the socialization process for males is likely to be especially constraining, allowing little deviation.

Even more critical for many young males learning male sex role requirements is the presence of older males within the family who serve as role models. Often such males are the fathers of these young males, although all that seems necessary for partial male socialization is that older males are seen by the younger males performing certain roles within the family setting. Seeing older males' role performances within the family provides younger males with the opportunity to learn vicariously cultural expectations for their own behaviors.

Some studies have found that parents treat children differently depending on the sex of the child (Schau et al. 1980). Differential treatment of children according to gender has been observed in fathers who are much more likely to "rough it up" with boys than with girls (Parke and Suomi 1980; Power and Parke 1983). These differences in fathers' behaviors toward their children depending on gender often follows stereotypical directions. Interestingly, fathers' stereotypical behaviors in interactions with their children follow parents' stereotypical descriptions of their newborns. Despite the lack of significant differences in birth length, weight, and APGAR scores, parents of daughters are more likely than parents of sons to give descriptions of their babies as "dainty," "pretty," "beautiful," and "cute" (Rubin et al.

1974). Certainly such differences in descriptions of children by gender may foretell parental behavioral differences by gender in parent-child interaction. We already know, for example, that parents of boys are much *less* directing of their offsprings' play than parents of girls. Such interferences by parents in the behaviors of girls may well affect girls' creativity and interests in ways inimical to their later independence and assertiveness. In the same vein, when parents of boys are less directing of their play, this begins to prepare boys for the independent and active male sex role many expect them to assume when they become adult males.

Other studies of parents' behaviors during the socialization of their children have produced mixed findings with respect to differential treatment of children by gender. Snow et al. (1983) found that parents responded differentially to some types of sex-typed behaviors in toddlers, but not others. In another study, fathers punished boys' cross-sex play behavior while mothers were found to punish and reward boys' cross-sex behaviors (Langlois and Downs 1980). These findings regarding fathers' lack of tolerance for boys' cross-sex behaviors are consistent with our contention that male sex role socialization tends to be more restrictive than female sex socialization, especially early in life. A final study which is instructive on this point is one by Eisenberg et al. (1985). In this study of mothers' and fathers' socialization of one- and two-year-olds' sex-typed play behaviors, several findings are notable. On the variables "parental choice of toys" and "parental reinforcement," parents of boys tended to choose neutral and masculine toys more than feminine toys, while parents of girls chose neutral toys more than the other two types. However, once parents had chosen toys for their children, they did not differentially reinforce them or neutrally respond to them for sex-typed or other-sex play. Eisenberg et al. concluded that "apparently, in the home, parents exert influence over their young children's play primarily via their selection of available toys" (p. 1512). Thus, parental opportunity to select and influence behavior may be a preferred method of socializing children's sex-typed behavior. Another finding from Eisenberg et al.'s study of interest is that parents reduced positive feedback for children's toy play with age. "Parents provided less positive feedback (and thus more neutral feedback) at age 26 to 33 months than at 19 to 26 months" (p. 1512). The reduction in parental reinforcement of play with age of the child occurred only for other-sex play activities, not neutral or sex-typed behaviors. This means that boys in all likelihood are aided in the development of gender constancy by continued parental reinforcement of sex-type play throughout childhood.

Findings regarding differential parental treatment by sex of child seem to be mixed at this point. Definite conclusions about differential parental behaviors by sex of child await further research. However, differentiated parental reinforcement may not be necessary for the development of sex-differentiated behavior in children. Simply attending to behavior differentially may be enough. Consider a study by Fagot and Hagan et al. (1985). This study of thirty-four children in infant play groups revealed no sex differences in assertive acts and attempts to communicate verbally with adults at ages 13 to 14 months. However, the authors observed learning center teachers attending more to boys' assertive behaviors and more to girls' *less intense* communication attempts. The result was that eleven months later twenty-nine of the same children exhibited sex-differentiated behavior: boys were more assertive and girls talked more with adults. Thus, caregivers seemed to be responsible, in part, for the development of boys' and girls' sex-differentiated behavior by guiding infant behaviors in stereotypical directions.

Educational Institutions It is well documented that there is a significant difference in what adults observe depending on whether the persons being observed are described as males or females (Condry and Ross 1985). Purported reasons for adult differential perceptions of children's behavior by sex vary. Some feel that adults may be differentially responsive to certain types of behaviors by girls and boys. For example, because girls are expected to be more verbal than boys, are teachers more attentive to girls' verbal behaviors than boys'?

By the same token, because boys are expected to have more assertive interchanges in peer activities than girls, do adults attend more to boys' assertive behaviors than girls' assertive behaviors? If the answers to both questions are yes, then differential attention to certain behaviors of boys and girls result in adults' differential perception of boys' and girls' behaviors.

Another common assumption stemming from social learning theory is that adults directly socialize children to behave in sex-typed ways through differential reinforcement and punishment. This assumption is supported by Beverly Fagot's (1981) findings that teachers differentially reinforce boys and girls for high activity levels. Even the large school context seems to be more supportive of males than females. Males continue to hold the more prestigious positions in the school system, schoolyards remain sex segregated, and in general gender differences remain in confidence, self-concept, and problem solving behaviors. Certainly such differences are related to the educational system's reinforcement of gender differences and traditional sex role behaviors. For example, findings from Phillips' playground study (1982) were quite consistent with those of Janet Lever (1976), who had found in her analysis of boys' and girls' spontaneous games on play-grounds that the games were sex differentiated. Lever concluded that boys' games were less structured than girls' games, with less emphasis on "turn-raking" and invariable procedural rules. Moreover, girls played with fewer participants while boys' games emphasized more initiative, improvisation, and extemporaneity, encouraging within-group cooperation and between-group competition. Phillips also found in her study of school playground activities that school spaces provided for boys and girls encouraged sex-differentiated play activities. Boys had large play spaces supportive of large competitive groups for competitive games. Girls' play spaces were small and generally supportive of cooperative, dyadic, and/or triadic activities. The major play space for girls in Phillips' study was on the playground apparatus, which could be easily invaded by boys and on occasion *was* invaded by boys, with the girls submissively leaving the equipment until the boys no longer used it. Phillips concluded: "Boys' play was preparing them for future work roles that would consist of the networks of competitively based groups necessary for success and achievement in the work place" (Franklin 1984, p. 43).

Jeanne Block's (1981) summary of the effects of sex-differentiated socialization in educational institutions suggests that male socialization in the education institution (which encourages curiosity, independence, initiative, etc.) extends male experiences, while female socialization in the educational institution (which discourages exploration, emphasizes class supervision, stresses proprieties, etc.) restricts the experiences of females. While some changes in the educational institution have occurred in recent years, males and females still have sex-differentiated experiences throughout their tenures in educational institutions.

Religious Institutions Almost as influential in teaching males to assume the male sex role is another agent, the religious institution. The only reason the religious institution does not assume a more critical role in male socialization is that the typical child does not spend an inordinate amount of time in religious settings. The time that is spent, however, generally is time when gender distinctions are emphasized. Such distinctions, within Christianity for example, are seen as divinely inspired in that they support the ideal relationship between husband and wife. On this point, Patricia M. Lengermann and Ruth A. Wallace (1985, p. 239) state that calling for sex role equality, questioning patriarchy, and critiquing traditional male dominance and female submissiveness in marriage and family life are antithetical to the divine plan as visualized by many Christians. Such a posture on the part of religious agents supports traditional sex roles against gender equality. In the 1980s with the rise of evangelical Christian movements and retrenchment in Roman Catholicism, we can only conclude that the religious institutions in the United States remain staunchly supportive of traditional female and male sex roles.

Support for the above position is seen in "God Goes Back to College," an article appearing in Newsweek's "On Campus" edition, November 1986, noting the fervor with which college students on campuses across the nation (those mentioned included Brown University, Arizona State University, University of Illinois—Champaign-Urbana, University of Texas, Duke University, Washington University, and Northwestern University) are embracing fundamentalist religious beliefs. Two striking implications for sex role changes are discussed. These implications center around a great deal of sentiment among religious groups in these settings to deny gays equal rights and to thwart women's attempts to pursue careers. Increased religious proselytizing on college campuses in the 1980s frequently has resulted in support for homophobia and traditional sex roles for females and males.

Peer Groups A consistent finding in the literature on children is that American children show a preference for same-sex peers by the beginning of their sixth year. This tendency toward peer-group sex segregation increases during middle childhood and reaches its peak right before adolescence (Hartup 1983). In addition, as Thompson (1985) found in his study, males in preadolescence are more peer oriented than females. Part of the reason for greater peer orientation among males undoubtedly is linked to greater encouragement of independence in males at an earlier age. What this means for male socialization is that boys at a relatively early age are more subject to peer-group influence than girls. Just as important is that such influence may be perceived positively and supported by parents as indicative of boys' independence.

If boys are susceptible to early peer-group influence, this also means that males' early-age peer groups may be responsible for a great deal of those sex role performances by boys. This is to be expected if Fagot's (1981) findings that boys who exhibit feminine behaviors receive negative feedback from peers are generalizable. Some support for peer-group influence on boys' sex role performances derive also from Eisenberg et al.'s (1985) study of a stronger match for males than females between same-sex peer interactions and neutral or sex-typed toy play for fifty-one four-year-olds. When boys play with boys they prefer sex-typed toy play. Eisenberg et al. (1985) felt that this is consistent with the notion that there is more pressure for males than females to avoid sex-inappropriate activities:

> Although initiation of and/or continuation of interaction per se may not be used consciously as a positive reinforcer by children, it could function as one. Thus, unintentionally as well as intentionally (Lamb et al., 1980), children, especially boys, may socialize peers into sex-stereotypic play behaviors. They may do so not only by initiating play with others in possession of sex appropriate toys, but also by inducing other children to engage in same sex play. (p. 1049)

There seems to be a logical relationship between children's play behavior and their everyday role performances. Indeed, when boys' play behaviors are channeled in a decidedly stereotypical masculine direction, certainly they learn that these same behaviors are expected of them by significant others in everyday situations. After all, parental brokering, approval, support, and reinforcement by early age peer groups function to inform the boy of the importance of this early socialization agent.

An early study by Fling and Manosevitz (1972) on male socialization found that young males are encouraged to participate in activities that teach and reinforce male stereotyped roles. There is little reason to think that such participation has declined in the 1980s. Interestingly, peer-group influence over males tends to decline as the young male approaches late adolescence. While in late childhood and early adolescence, male peer groups are quite influential in boys learning competitiveness, aggression, violence, and antifemininity, young males also learn that they must become independent, self-reliant, and detached from the peer group. This latter socialization, in a sense, prepares young males for the role which must be assumed in adulthood, a role which minimizes male-male relationships. Yet, adolescent peer groups, for most males, are kinds of reference groups providing information which the sixteen or seventeen-year-old male actively filters,

alters, and modifies to fit his own perspective. Typically, peer group information, standards, and values are some variant of those from other socialization agents, including nonpurposive ones discussed in the next section. Most young males experience a kind of socialization which teaches them societally approved sex roles, dysfunctional ones as well as functional ones.

The Mass Media   Mass media influences on sex role socialization are thought by many to be critical in the development and support of sex role stereotypes specifically and sex role inequality generally. The link between sex role stereotypes and sex role inequality is a direct one. Sex role stereotypes (expectations about and attitudes toward the sexes) lead to sex role inequality (in-equitable actions toward a person based on the sex of that person). This linkage is consistent with findings in social psychological literature suggesting that stereotypes are better predictors of behavior than of attitudes.

Yet, how do sex role stereotypes relate to male socialization? Recalling that male socialization is a dual process, involving male self development and the learning of societal "shoulds" and "should nots" for males, one can see that much male socialization actually involves learning conceptions of males' "makeup" and "places" and females' "makeup" and "places." Undoubtedly, the mass media play critical roles in this process. When females in television commercials usually perform household duties and pamper men while males typically perform active roles outside the home and do not perform domestic duties, a message is given to viewers that housework is women's work and work outside the home is men's work (Mamay and Simpson 1981).

That television may play a powerful role in gender socialization is suggested by Drabman et al. (1979) since their findings indicate that young children (first-graders), when shown videotaped presentations of males and females in counterstereotyped occupations (such as male nurse and female doctor), tended to reverse sex role information in the stereotyped direction. For Drabman et al. this finding meant that television should be used in a specific way to modify sex role socialization in a more equitable manner for boys and girls. They state: "Television programming which directly informs the child nearly all life roles are available to both sexes might prove more fruitful in attempts to alter traditional gender stereotypes" (p. 388).

Not only is television a potentially powerful agent in the sex role socialization of males and females, but newspapers, comics, movies, and popular songs may also influence conceptions of gender roles. Lengermann and Wallace (1985) feel that such mass media are "a forum where critical views on gender equality are heard and aired, can affect the thinking and beliefs of men as well as women, and can be for both a resource for new meanings" (p. 222). To the extent that men and women are affected by such mass media changes and also participate in the teaching of males, the effects on male socialization are obvious.

To be sure, there have been some changes in the mass media toward presenting male and female images in a manner more consistent with sex role equality. Lengermann and Wallace point to the inclusion of women columnists like Ellen Goodman and Mary McGrory in daily newspapers as evidence of changes in newspapers which can modify a man's thinking about a woman's place. They note also the emergence of Alan Alda, a popular television entertainer and self-described feminist, as a role model in the media for nontraditional men in contrast to more conventional male images like Bob Hope and John Wayne. Men's magazines, too, are thought to be sources of sex role changes in the mass media. Magazines such as *Esquire* and *Sports Illustrated* are thought by Lengermann and Wallace to reflect "new meanings." They cite the November 1982 issue of *Esquire* with a feature article entitled "Father Love" by Anthony Brandt and *Sports Illustrated*'s (Feb. 28, 1983) coverage of Louisiana Tech women's basketball team (a sign that women's sports are making strides toward parity) as evidence of further change in sex role meanings....

In summary, some mass media changes in the last decade or so have been in the direction of sex role equality which would eventually lessen male dominance, male violence, male destruction, competition, and so on. Simultaneously,

however, forces have arisen in the mass media which either support traditional sex role distinctions or at least suggest that change in men's behavior cannot occur or is trivial and of dubious value for society when it does occur. As we move into the 1990s, it is hoped that the mass media will come to portray and reflect gender, especially the valued male sex role, in a realistic manner, that is, as sets of cultural expectations that are socially constructed.

NONPURPOSIVE AGENTS OF MALE SOCIALIZATION

Families, boys' groups, educational institutions, churches, newspapers, magazines, television, and radio are not the only socializing agents teaching males to be dominant, aggressive, violent, competitive, nonintimate, and non-nurturant. There are other agents in American society which are not charged with a learning function but which, nevertheless, teach males conceptions of themselves, other males, and what males should and should not do. Some of these agents are male-centered barbershops, sports events, taverns, and business meetings, where primarily males engage in social interaction. These are the same agents forming the core of men's culture. Two latent socialization consequences of the above agents are emphasized: (1) indirect socialization of young males and (2) reinforcement and support of traditional conceptions of the male sex role.

With some exceptions (e.g., unisex hair salons), male barbershops, topless taverns, male-dominated business meetings, sports events, and the like function as social settings/negotiation contexts where men negotiate masculinity. While the negotiation of masculinity is a complex process involving numerous contextual and social psychological variables, the emphasis here is on the process used by men in certain settings to arrive at conceptions of who men are and how men should and should not behave. At the same time, they also form conceptions about persons who are not men and masculine—women and others perceived as feminine.

The "particulars" of masculinity negotiations in various social settings will not be discussed here; however, a broad description of such processes includes verbal and nonverbal behavior by male participants in social settings which define appropriate male attitudes and behaviors. In such settings as barbershops and male-centered sports events, frequently young, impressionable males are present during the negotiation process. The young males learn not only what behaviors are expected from the primary male participants, but also the attitudes that they should hold about the negotiation process and what outcomes from the negotiation process are most desirable. For example, a young male attending a professional football game learns not only that the more "manly" team wins—the team that is more competitive, more aggressive subdues—but also how he is to respond to such characteristics. The young boy leaves the stadium *knowing* that dominance is a desirable trait for men to have.

After all, an entire group of men have just been rewarded by a host of other men for displaying the dominance trait. Just as important, too, for the young boy is the low esteem many others hold for the losing team—the one that has been subdued.

Young boys generally do not go to topless taverns where women are seen in various stages of undress. Nevertheless, masculinity negotiations and male socialization are features of such settings. Men receive support and reinforcement from other men for certain behaviors they display. The swaggers, the yells, the obscenities, the sexual references all become permanently etched in their little minds as appropriate behavior for men. Those men who do not engage in the behavior nevertheless learn that if they want others to think of them as manly, all they have to do is display similar behavior. The "new" male in the topless tavern setting will know that he is engaging in appropriate male behavior when other males slap him on the back or shake his hand as he, too, screams and yells, "Take it off, baby."

REFERENCES

Block, J. H., 1983. Differential premises arising from differential socialization of the sexes: Some conjectures. *Child Development* 54:1335–54.

Drabman, R. S., S. J. Robertson, J. N. Patterson, G. J. Javie, D. Hammer, and G. Gordua, 1979. Children's perception of media portrayal of sex roles. *Sex Roles* 7:379–89.

Eisenberg, N., S. A. Wolchik, R. Hernandez, and J. F. Pasternack, 1985. Parental socialization of young children's play: A short-term longitudinal study. *Child Development* 56:1506–13.

Fagot, B. I., 1981. Male and female teachers: Do they treat boys and girls differently? *Sex Roles* 7:263–71.

Fagot, B. I., R. Hagan, M. D. Leinbach, and S. Kronsberg, 1985. Differential reaction to assertive and communicative acts of toddler boys and girls. *Child Development* 56:1499–1505.

Fling, S., and M. Manosevitz, 1972. Sex typing in nursery school children's play interests. *Developmental Psychology* 7:146–52.

Franklin, C. W., 1984. *The changing definition of masculinity*. New York: Plenum Press.

Fu, V. R., and D. J. Leach, 1980. Sex role preferences among elementary school children in rural America. *Psychological Reports* 46:555–60.

Hartup, W. W., 1983. The peer system. In *Handbook of child psychology*. Vol. 4: *Socialization, personality, and social development*, edited by E. M. Hetherington and P. H. Mussen (series ed.). New York: Wiley.

Langlois, J. H., and A. C. Downs, 1980. Mothers, fathers, and peers as socialization agents of sex typed play behavior in young children. *Child Development* 51:1217–47.

Lengermann, P. M., and R. A. Wallace, 1980. *Gender in America: Social control and social change*. Engle-wood Cliffs, N.J.: Prentice-Hall.

Lever, J., 1978. Sex differences in the complexity of children's play and games. *American Sociological Review* 43:471–83.

Mamay, P. D., and R. I. Simpson, 1981. Three female roles in television commercials. *Sex Roles* 7:1223–32.

Parke, R. D., and S. J. Suomi, 1980. Adult male-infant relationships: Human and non-primate evidence. In *Behavioral development: The Bielefeld interdisciplinary project*, edited by K. Immelmann, G. Barlow, M. Main, and L. Petrinovitch. New York: Cambridge University Press.

Phillips, B. D., 1982. Sex role socialization and play behavior on a rural playground. Unpublished master's thesis, Department of Sociology, Ohio State University, Columbus, Ohio.

Power, T. G., and R. D. Parke, 1982. Play as a context for early learning: Lab and home analysis. In *The family as a learning environment*, edited by I. Sigel and M. Laosa. New York: Plenum Press.

Rubin, J. Z., E. J. Provenzano, and Z. Luria, 1976. The eye of the beholder: Parents' views on sex of newborns. *American Journal of Orthopsychiatry* 44:512–19.

Schau, C. G., I. Kahn, J. H. Diepold, and F. Cherry, 1980. The relationship of parental expectations and pre-school children's verbal sex typing to their sex-typed toy play behavior. *Child Development* 51:266–70.

Snow, M. E., C. N. Jacklin, and E. E. Maccoby, 1981. Sex-of-child differences in father-child interaction at one year of age. *Child Development* 54:227–32.

Thompson, D. N., 1985. Parent-peer compliance in a group of preadolescent youths. *Adolescence* 20(79): 501–7.

8

"Doing It": The Social Construction of S-E-X

TRACEY STEELE

Sex is a natural, normal, enjoyable activity, right? Not so simple, says Tracey Steele as she shows how humans construct sex. This involves a number of processes and differs across cultures and with motivation of those involved. Steele discusses some of the ways humans view and construct sex to meet various goals: love, money, pleasure, exercise, companionship—the list goes on as do the methods for defining sex.

As you read, consider the following questions:

1. *Why is "sex" complicated to define?*
2. *Which factors should one consider when defining sex?*
3. *Is homosexual sex really sex?*
4. *Why is sex a social construction?*

GLOSSARY **Social construction** Behaviors that are learned and produced within specific cultural contexts. **Deconstructing** Examining the hidden assumptions built into how a situation is defined and understood.

What is sex? And by this question I do not mean the dichotomous morphological division of the species into male and female. Rather, I mean what is "sex," **S–E–X?** Is it the sacred joining of two souls on a spiritual and physical plane? A commodity bought, sold, and traded in the cultural marketplace? A way to sustain the species? A delightful muscle spasm? A conjugal obligation? A weapon used to objectify, humiliate, and subjugate?

Sex is all of these things and many more. Its meanings and functions vary from epoch to epoch, and from culture to culture. It can simultaneously serve a variety of private motives and social aims, which are themselves shaped by the vicissitudes of personal desire and the sexual possibilities articulated through existing social conventions. In short, the meaning of sex is a product of both individual and social factors. What often goes unnoticed, however, is that while the purpose and ultimate significance of our sexual interactions may differ, the underlying sense that we know what sex *is,* generally does not. I may be in it for love, and you may be in it for money, pleasure, revenge, pity, exercise, or countless other reasons, but we both "*know*" what "*it*" is. After all, it's sex, it's natural, and *everybody* knows what sex is.

Source: Steele, Tracey. 2008. Revised version of previously published paper in Steele, Tracey. 2005. *Sex, Self and Society: The Social Context of Sexuality.* Belmont, CA: Wadsworth Publishing.

But do we? Curiously, we seem to forget, or wholly disregard, the fact that the road to adulthood is routinely paved with intense insecurity, doubt, and angst concerning that great mystery of mysteries—sex. Playing "doctor" with the next-door-neighbor, furtive scrutiny of purloined pornography, tentative adolescent fumblings, as well as those "mature" jokes told by friends that induced our hesitant laughter, silent confusion, and nervously feigned sophistication all testify to a decided uncertainty and *lack* of knowledge regarding sex. The truth is sex is *not* something innate, something we instinctively know; it is something we learn about as we grow to adulthood.

In fact, like most of our social behaviors, we gain our knowledge about what sex is from others, including key agents of socialization such as parents, peers, and the media. These groups convey cultural constructions—they pass on ideas about what "normal" sex is supposed to be, including who is supposed to be involved, for how long, in what locations, and toward what ends. And because sexual socialization begins quite early, it typically becomes so internalized that by the time we reach adulthood we fail to realize it was something we ever learned at all.

In the following pages I will utilize what is known as the social-constructionist perspective to explore the meaning of sex in contemporary American society. This perspective posits that sex, rather than being natural and instinctual is, in fact, principally a *learned* behavior that is produced within specific cultural contexts. I begin by **deconstructing** the meaning of sex in contemporary American society. Deconstructing sex involves examining the hidden assumptions built into how sex is defined and understood. In addition, the deconstruction of sex will reveal both the possibilities and the limits of existing definitions (Rubin, 1984). It will also bring to light how existing definitions work to privilege some social groups at the expense of others. Together these discussions will show that sex is largely shaped and defined by factors external to the individual and is, fundamentally, a *social* enterprise.

THE SOCIAL CONSTRUCTIONIST PARADIGM

Modern Western societies are not the first to explore sexuality as a topic of intellectual inquiry. Philosophers, scholars, and clergy of countless generations have pondered its significance and come to radically different conclusions about how best to define, express, circumscribe, and manage libidinal energies and behaviors.

Lynne Segal (1994) has identified three historical intellectual traditions that have dominated Western thinking about sex. These frameworks are the spiritual, the biological, and the social. In pre-industrial Europe, societal views about sexuality were largely shaped by spiritual and religious beliefs. With the emergence of the Industrial Revolution, however, the dominance of these views were supplanted by models grounded in "science" and scientific thinking. In these early stages, modern science was heavily influenced by Darwinian logic; thus, scientific explorations of human behavior tended to be flavored with a decidedly biological bent. It is within this historical moment that the idea that sex is a biological phenomenon became a centrally accepted truism in Western thought (Segal, 1994).

Generally, such *essentialist* approaches to sexuality hold that sex is a matter of biological essence, that it is a product of biological force "seeking expression in ways that are preordained" (Epstein, 1987:15). For essentialists, sexuality is fundamentally a product of organic and biochemical processes, a function of our "nature." Essentialists consider sexual identities to be "cognitive realizations of genuine, underlying differences" (Epstein, 1992:241). For example, essentialists maintain that individuals are *by nature* heterosexual or homosexual. We are *born* a top or a bottom. Men are *naturally* more sexually aggressive and promiscuous than women [ostensibly because they have been biologically programmed to spread their seed and maximize their procreative potential (Buss, 1998)]. All of these claims are grounded in the fundamental notion that sex is a category of human existence dictated by nature that is eternally "unchanging, asocial and transhistorical" (Rubin, 1984:275).

The association of sex with nature in these models is not unintentional. There is great rhetorical power in terminology such as *sexual nature* because it "sounds like something solid and valid, not human-made" (Tiefer, 1995:33). In other words, by invoking nature, the *claims* of scientists regarding the etiology of sex and sexual behavior are consequently elevated to seeming "*facts*" and larger "*truths*" conveniently eluding the role of social forces in their creation.

In recent years, a formidable challenge to essentialist models of sex and sexual behavior has emerged. Though biological models still hold considerable sway in both popular and scientific thinking, new thinking about sex, sexuality, and the erotic assert that each is, in fact, socially constructed and not simply a matter of biological mandates.

Social constructionism is a theoretical perspective which argues that our perception of what is real is defined only by the meaning that we attribute to a given situation (Berger and Luckman, 1967; Blumer, 1969). Things do not have their own intrinsic meaning; we impose meaning upon them. These meanings are created through social interaction that occurs in particular social environments. Simply put, it is through our interaction with others that we learn how to interpret and evaluate the world around us. Constructionists also reveal that social hierarchies play an important role in ascription of meaning; individuals and groups at the top of social hierarchies are those most likely to have their definitions and views imposed and enforced as "reality."

Applied to sexuality this perspective holds that sexuality has no "inherent essence" (Harding, 1998:9). Ideas about sexuality are not hard-wired or "natural"; they arise in particular social–historical contexts (Foucault, 1980). For constructionists, sexuality is an arrangement of cultural norms, values, and expectations, which themselves are fundamentally shaped by hierarchies and matrices of social power relations (Foucault, 1980; Gagnon and Parker, 1995; Harding, 1998; Vance, 1984).

Constructionists argue that if sex and the realm of the sexual *were* natural, we would not have to be taught what sex was and it would take consistent forms across societies (Tiefer, 1995). But, evidence shows that sex is learned, and it does vary. Robert Padgug observes: "The forms, content, and context of sexuality always differ. There is no abstract and universal category of 'the erotic' or 'the sexual' applicable without change to all societies" (1992:54). Sex and the erotic are highly malleable constructs that take innumerable forms across the globe—in fact, we can (and do) attach erotic desires to almost anything. For example, in some cultures, small feet may be considered sexually appealing in women, in another culture it is large earlobes that are particularly "sexy," and in yet another culture it is women's breasts that are supposed to make (heterosexual) men "hot." In one culture the term "sex" might be reserved exclusively for heterosexual intercourse, while in another "sex" might refer to any sexual encounter that produced orgasm in at least one of the participants. The bottom line is— sex varies.

The definition of sex, who can engage in it, at what age, for how long, and from what position are just a few examples of sites of sexual variation. Our culture provides the template upon which our erotic desires are channeled and shaped. Because we share the same cultural sexual indoctrination as other members of society, we generally share the same sexual beliefs, values, and desires as those around us. So, for all intents and purposes, sex *appears* to be invariant and natural when it is not. In fact, because sex *is* culturally constituted, what we regard as "natural" will often be viewed as "unnatural" in other societies.

The implications of this perspective challenge many contemporary notions about sexuality. For example, a constructionist perspective would assert that sexual aggression in men is learned rather than inborn, that women may not "naturally" wish to bear children, and that public nudity, promiscuity, and premarital sex are not "inherently" wrong or sinful. Rather, constructionists contend, ideas and values surrounding sexual "reality" are learned, and they can, and do, change. In the following section, I adopt a constructionist framework to critically examine, identify, explore, and "deconstruct" cultural assumptions about S-E-X that are embedded within contemporary American society. This construct is essential to examine because it serves as a primary focal point for most scientific inquiry and public debate—for most of us, sexuality is primarily about S-E-X. A critical examination of the social construction of S-E-X can therefore provide considerable conceptual entrée into a wider analysis of the sexual realm.

ISSUES OF DEFINITION

I begin my course on sexuality by having students write their definition of sex down on a small index card. "Let's say someone from Mars came to Earth and asked you to explain what this thing called sex is that s/he keeps hearing about in movies, songs, and barroom conversations," I begin. "What would you say? What is it? What is it for? How do you know when you've had it? How would you explain to this ultimate outsider what this activity 'sex' is all about?"

Quickly they set to the task, most of them confident and eager. A few finish quickly. "The man puts his penis in the woman's vagina." "Intercourse." "It is something adults do when they are in love." "It's the way we reproduce the species." A few take longer to complete the task. Slowly, steadily, frustration grows. Brows wrinkle, heads tilt, lips purse. Many struggle trying to put into words something they believe they "know, " something "natural," something that, as they put pen to paper, suddenly does not seem so simple, or so automatic. Typically, clarification is requested: "Do we need to discuss gay sex?" "Does the Martian know what genitalia are?" "Should I limit it to 'vanilla' sex?" "Sex like in the movies or 'real' sex?" "Should I explain about how men want it all the time?" "The Martian knows NOTHING about sex," I reply. "Start from ground zero." Invariably, someone remarks, "This is hard!"

And indeed it is. When we speak of sex and matters sexual, we may be referring to a single specific act, a group of acts, or something that can encompass an enormously wide variety of thoughts, emotions, and behaviors. Does sex inhere in certain affective states (e.g., love), physiological responses (e.g., excitement, orgasm) specific actors (e.g., males with females), or specific acts (e.g., vaginal penetration by an erect penis)?

The ambiguity inherent in these questions was perhaps best illustrated in 1997 with the eruption of a sexual political scandal involving former President Bill Clinton and White House intern Monica Lewinsky. Amid intense political pressure and media scrutiny, the president made a public address in which he vehemently denied "having sex" with Ms. Lewinsky. However, later, the public learned that the President and the intern *had* been involved in *oral* sex. While many members of the general public may have felt understandably deceived, the President was by no means alone in making this kind of categorical distinction.

Several recent studies indicate that the American public does differentiate between oral sex and what it considers "real" sex (Bogart, Cecil, Wagstaff, Pinkerton, and Abramson, 2000; Bogart, Pinkerton, Myaskovsky, Wagstaff, and Abramson, 1999; Sanders and Reinish, 1999). The work of Bogart et al. (2000), for example, indicates that while a vast majority of the college students they sampled (97%) considered heterosexual vaginal intercourse to be "sex," slightly fewer (93%) were willing to label anal heterosexual sex as "sex," and less than half (44%) considered that oral sex constituted real "sex."

Similar findings have been echoed in national opinion surveys. Of particular interest are reports that youth may be engaging in significantly higher rates of oral sex and other sexual behaviors, believing that by doing so they have not "had sex." Further, many indicate that by engaging in this particular form of erotic expression they have not "technically" lost their virginity (Sanders and Reinisch, 1999; Indigo, 2000) In short, despite the existence of a wide variety of socially identified sexual behaviors, in modern American society, the designation of having "had sex" is typically limited to a quite narrow range of erotic expression.

A critical examination of S-E-X reveals that the term represents a particularized constellation of socially determined sexual norms and expectations. In other words, sex, *real* sex, means something very specific in this culture, but the content of that meaning lies cloaked beneath layers of taken-for-granted assumptions that are so embedded in our cultural framework that they are rendered conceptually invisible. So, what *is* the content of S-E-X? What exactly *does* S-E-X mean? I raise this and subsequent questions not in search of definitive answers or solutions, but rather, in order to stimulate awareness and provoke discovery—to peel back the ornate and intricate layers of the social fabric to expose both the limits and possibilities of human sexual interaction.

SO, S-E-X IS...

To discern the basic assumptions that go into defining sex in this culture, it is useful to strip away rhetoric and focus on the basics. Though there are several ways to do this, let's try pantomimes. Take a moment and, using only your hands, make a gesture for sex. Go ahead; it won't work nearly as well if you don't play along. C'mon—it won't hurt, really. Great! Now, repeat this gesture and pay close attention as you do it. This simple exercise goes a long way toward capturing many of the essential elements that constitute the modern conceptualization of S-E-X. What did you notice about your gesture? Think about the shapes you formed, the movements you utilized and how these are meant to represent S-E-X. For example, did your gesture involve one hand or finger actively breaching the boundaries of the other in a mock penetrative motion? This, and other gestures for sex, reveals many of the unspoken requisite components of S-E-X that will be described in the paragraphs below.

Penetration and Male Agency

As symbolized in the exercise above, active penile penetration is one of the most essential components of the social construction of S-E-X in our culture. Its centrality is well illustrated in the research

conducted by Bogart and her colleagues described above (Bogart et al., 2000). In both of the cases involving penile penetration (anal and vaginal sex), a majority of the respondents concluded that the activities involved constituted "sex." This was not the case for the non-penetrative activity described in the scenarios (i.e., oral sex). Similar findings were echoed in Sanders and Reinisch's 1999 study, which indicated that 99.5% of their Midwestern college student sample defined penile–vaginal intercourse as sex, while somewhat fewer (81%) reported that penile–anal intercourse would qualify. However, only about 40% of the respondents indicated that oral sex constituted "having sex" (1999).

These findings point to the cultural privileging of sex as penile penetration, revealing a conceptualization of sexuality that is tacitly connected to male agency. Sex in America is a male purview; women *can* have sex but to be a "real man," men *must* have "real," aggressive, penetrative sex (Stoltenberg, 1990). In our culture's hegemonic articulation of sex, it is men who are defined as the sexual actors, as those who seek sex, as those around whom notions of sexuality are constructed, and to which active female agency is the notable (though increasing) exception. It is also significant that, in this construction, all qualitative aspects of penetration are rendered irrelevant. The question is not, for example, the depth, frequency, or duration of penetration, but whether or not penile penetration has occurred at all.

Heterosexuality

Another important assumption built into our cultural system is that S-E-X is a *heterosexual* activity—it is penetrative, male agentic, and it takes place between a man and a woman. But, *must* sex involve a man and a woman? If so, are we asserting that gay men and lesbians do not have sex? Why, then, all the concern about homosexuality if we don't really count same-sex sexual activities as "real" sex? What *would* we call these activities? Near-sex? Pseudo-sex? Sex-like? And what of bisexuals? Does it only count as sex when a bisexual participates in penetrative, male agentic sexual activity with some-one of the "opposite" sex?

Do we consider mutual digital genital stimula-tion between two lesbians "sex"? If so, then why would we not consider this same activity performed between a man and a woman to qualify as "sex"? In grade-school vernacular this is merely "third base," clearly short of making it "home" or "all the way." How can we say Veronica had sex with Betty but Archie "only got to third" if they engaged in iden-tical sexual practices?

Continuing in this same vein, we might ask if there is a need to call gay male sexual activities something entirely different than lesbian sexual ac-tivities. After all, gay men have penises: they can engage in penile penetrative activity. But herein lies the problem of constructing a definition of sex as contingent on the relative genitalia of the actors involved; identical acts may be evaluated differently depending upon the biological sex of the partici-pants. For example, it is typically considered to be sex when a man penetrates a woman anally with his penis [93% believed as much in Bogart et al.'s (2000) research and 81% in Sanders and Reinisch's (1999)]. Would as many people consider that this same act counted as sex if it were conducted by a man upon a man? However, even if we consider expanding Americans' tacitly heterosexist defini-tional criterion to include same-sex sexual expres-sion, we are still excluding valid and important alternative conceptualizations. For example, must the erotic activities involve two people? What about masturbation? Is this sex? Multiple partners? Non-human partners?

It is also important to note that our culture gives us important cues as to the culturally legiti-mated forms of sex. The presence of a qualifier or hyphen differentiates subordinated forms from the hegemonic: hence the denotations of *lesbian sex* and *gay-male sex* rather than simply "*sex*." Linguistic devices such as these signal that the qualified forms are different from, and inferior to, the presumed norm. These linguistic devices operate in a similar manner as when they are used to qualify occupa-tions in gendered ways such as "male nurse," "female firefighter," and the "**W**"NBA.

Orifice Specification

Another essential component of the hegemonic con struction of sex in contemporary American society follows logically from the three preceding presump-tive elements—if sex is constituted as a penetrative heterosexual male activity, what, we may ask, is the site of this activity? Where is all this effort supposed to be directed? To answer this, we can again gain considerable insight by looking at the linguistic cues; take, for example, anal sex and oral sex. Both are sex acts that specify the orifice or bodily site of sexual expression. Both modify the presumptive case. What orifice is involved in plain ole "vanilla" S-E-X? What is the presumed site of erotic expression when we say he, she, or they had "sex"? The answer is the vagina. The vagina disappears from view linguistically because its presence is presupposed in much the same way that the heterosexuality of the participants in S-E-X remains unacknowledged.

Why such a focused concern surrounding the penetration of a vagina by a penis? Modern con-structions of S-E-X are derivate of historical con-structions which held that sex was an activity directed toward procreation (Seidman, 1996) In other words, our understanding of what sex is has evolved over time from definitions of sex which were tied to human reproduction. *Real* sex was what could get you pregnant—historically this has necessitated vaginal penile penetration. However, the majority of Americans today do not have sex for procreative purposes. Most typically, sex is a recreational activity pursued with an eye toward pleasure rather than procreation (Seidman, 1996; Vance, 1984). In fact, great pains are typically taken to *avoid* pregnancy. This shift away from reproduc-tively oriented constructions of sex renders the het-erosexual imperative embedded in contemporary constructions particularly anachronistic.

Further, medical technology has now made it possible for the human species to reproduce without any direct contact between male and female genera-tive parts. In fact, technology has so far removed procreation from sex that a woman can give birth to a child harvested from another woman's eggs that have been fertilized with a complete stranger's sperm. There is no *sex* (as we define it) in this reproductive equation. The issue of human cloning moves the reproduction of the species even further into the laboratory and away from the embodied and imperfect process of human sexual intercourse.

Orgasm, Pleasure, and Love

Another issue that is closely tied to the issue of reproduction is orgasm. What is the role of orgasm in S-E-X? Typically, orgasm is viewed as the culmination of a sexual encounter. However, not all orgasms are created equal. Male orgasm in heterosexual coital relations is the event that is most closely associated with human reproduction (Laumann, Gagnon, Michael, and Michaels, 1994; Masters and Johnson, 1966) and *both* males and females indicate that it is chiefly *male* orgasms which signal whether or not a particular sexual interaction qualifies as sex (Segal, 1994; Bogart et al., 2000). And, although many females are capable of multiple orgasms, it is typically the number of male climaxes that heterosexual partners count when describing how many "times" they have had sex. So accustomed are we to conceiving of sex in this manner that it is difficult to imagine alternate conceptualizations. Consider Marilyn Frye's discussion of the "number of times" question for lesbians:

> Some might have counted a two- or three-cycle evening as one "time" they "had sex"; some might have counted it as two or three "times." Some may have counted as "times" only the times both partners had orgasms; some may have counted as "times" occasions on which at least one had an orgasm; those who do not have orgasms or have them far more rarely than they "have sex" may not have figured orgasms into the calculations; perhaps some counted as a "time" every episode in which both touched the other's vulva more than fleetingly and not for something like a health examination. For some, to count every reciprocal touch of the vulva would have made them count as "having sex" more than most people with a job or

work would dream of having time for; how do we suppose those individuals counted "times"? Is there any good reason why they should *not* count all those as "times"? Does it depend on how fulfilling it was? Was anybody else counting by occasions of fulfillment? (1990:308)

What of fulfillment? What about love? Unlike the factors mentioned previously, neither of these concepts is intrinsically bound up in prevailing constructions of sex. Though often desired, neither is generally considered to be a conditional prerequisite of the claim of having had "sex." It may be the cultural ideal, and the most culturally *legitimated* form, but sex that occurs between two people who are in love is by no means the *hegemonic* norm. S-E-X is about physical pleasure, and is measured (male) orgasm by (male) orgasm.

A few considerations remain. For example, what do we call cases where there is orgasm between two people who have no actual physical contact? Is that sex? Is, for example, phone sex "sex"? Is cyber sex "sex"? If the informal student polls I have taken in my sexuality courses are any indication, then it appears the answer is "not really"—S-E-X is, apparently, a contact sport.

Ironically, however, when these same students are asked if these activities should be considered "cheating" if engaged in by someone in a committed relationship with someone outside of that relationship, most say yes. While these acts will not typically qualify as S-E-X, for many, they still feel like sexual betrayal. Structurally, S-E-X and romantic love may be defined as separate and distinct categories of human activity, but at the interpersonal level such normative distinctions appear to dissolve quite readily.

Consent

Finally, we must address the issue of consent. Even in cases that fit all of the criteria tacitly embedded in our cultural construction of S-E-X, one would hope that most Americans would be loathe to consider that a forced sexual assault could be construed as sex. It's likely most survivors of sexual assault don't. However, even though S-E-X is generally constructed as a pleasurable pursuit, sexual pleasure is more strongly associated with its male participants. Because sexual agency is ascribed to males in contemporary conceptualizations, S-E-X does not necessarily require enjoyment or consent on the part of the sexual "recipient" (be they male or female). Further, not only are issues of power, domination, and force not precluded, but some authorities contend that contemporary definitions of sexuality proceed from a foundation of male domination and aggression. From this perspective; S-E-X *is* violence within patriarchal regimes. (Dworkin, 1981; MacKinnon, 1987; Stoltenberg, 1990).

The dangers inherent in contemporary constructions of S-E-X begin to emerge quite clearly here. S-E-X is something one "gets," something (or some*one*) one "has"; the subjective needs and desires of the "object of affection" are largely irrelevant. Nor is this the only example of how the social construction of S-E-X manifests and propagates social inequality. The examples discussed above demonstrate quite clearly that our culture privileges some forms of sexuality (e.g., penetrative, male agentic, heterosexual) while simultaneously marginalizing others (e.g., non-penetrative, female-agentic, gay, lesbian, transgendered). From this we can begin to discern not only the benefits that accrue to those members of society whose beliefs and activities fall within prescribed expectations, but also the costs of our sexual constructions for subordinated groups— those whose sexual values and practices fall outside normative boundaries. Vance notes:

> Our ability to think about sexual difference is limited ... by a cultural system that organizes sexual differences in a hierarchy in which some acts and partners are privileged

and others are punished. Privileged forms of sexuality ... are protected and rewarded by the state and subsidized through social and economic incentives. Those engaging in privileged acts, or pretending to do so, enjoy good name and good fortune. Less privileged forms of sexuality are regulated and interdicted by the state, religion, medicine, and public opinion. Those practicing less privileged forms of sexuality ... suffer from stigma and invisibility although they also resist. (1984:19)

The benefits of conformity are as enriching as the price of nonconformity is costly. Yet, as Vance notes, many do resist the confines of social strictures and normative boundaries. And, it is in resistance, struggle, and defiance that we find the avenues to social change. Social constructions are far from static: As ideological silhouettes they shift and change, ebb and flow, remaining ever-powerful, yet as historically ephemeral as the social structures they serve.

The examination I have presented here is by no means meant to be perceived as exhaustive; there are many more aspects of S-E-X that remain for analysis. We could, for example, discuss the construction of S-E-X as shameful, S-E-X as sin, S-E-X as signifier of adulthood, and so on, as well as the social inequalities that each of these categorizations engenders. Nonetheless, it is my hope that this brief analysis has demonstrated the importance of critically examining the hidden assumptions and varying functions of sex that are deeply rooted within our cultural framework, and has fostered an appreciation of how an examination of alternate understandings of sex and sexual behaviors across a diversity of cultural contexts may help us better understand our own.

REFERENCES

Berger, Peter, and Thomas Luckman. 1967. *The Social Construction of Reality: A Treatise in the Sociology of Knowledge*. Garden City, NY: Anchor Books.

Blumer, Herbert. 1969. *Studies in Symbolic Interaction*. Englewood Cliffs, NJ: Prentice Hall.

Bogart, Laura M., Heather Cecil, David A. Wagstaff, Steven D. Pinkerton, and Paul R. Abramson. 2000. "Is It 'Sex'?: College Students' Interpretations of Sexual Behavior Terminology," *Journal of Sex Research* 37(2).

Bogart, L. M., S. D. Pinkerton, H. Cecil, L. Myaskovsky, D. A. Wagstaff, and P. R. Abramson. 1999. "Attitudes Toward and Definitions of Having Sex" [Letter], *Journal of the American Medical Association* 282:1917–1918.

Buss, D. M. 1998. "Sexual Strategies Theory: Historical Origins and Current Status," *Journal of Sex Research* 35(1).

Dworkin, Andrea. 1981. *Pornography: Men Possessing Women*. London: Women's Press.

Epstein, Steven. 1987. "Gay Politics, Ethnic Identity: The Limits of Social Constructionism," *Socialist Review* 93/94.

Epstein, Steven. 1992. "Gay Politics, Ethnic Identity: The Limits of Social Constructionism," in Edward Stein (ed.), *Forms of Desire: Sexual Orientation and the Social Constructionist Controversy*. New York: Routledge.

Foucault, Michel. 1980. *The History of Sexuality. Volume I: An introduction*. New York: Vintage Books.

Frye, Marilyn. 1990. "Lesbian 'Sex,'," in Jeffner Allen (ed.), *Lesbian Philosophies*. Albany NY: Statue University of New York Press.

Gagnon, John H., and Richard G. Parker, 1995. "Conceiving Sexuality," in Richard G. Parker and John H. Gagnon (eds.), *Conceiving Sexuality. Approaches to Sex Research in a Postmodern World*. New York: Routledge.

Harding, Jennifer. 1998. *Sex Acts: Practices of Femininity and Masculinity*. London: Sage.

Indigo, Susannah. 2000. "Blow Jobs and Other Boring Stuff," xxxxx. Published December 14, 2000. Accessed January 6, 2001.

Laumann, E. J. Gagnon, R. Michael, and S. Michaels. 1994. *The Social Organization of Sexuality*. Chicago: University of Chicago Press.

MacKinnon, Catherine. 1987. *Feminism Unmodified*: Discourses on Life and Law. Cambridge, MA: Harvard University Press.

Masters, W. H., and V. E. Johnson. 1966. *Human Sexual Response*. Boston: Little, Brown.

Padgug, Robert. 1992. "Sexual Matters: On Conceptualizing Sexuality in History," in Edward Stein (ed.), *Forms of Desire: Sexual Orientation and the Social Constructionist Controversy*. New York: Routledge.

Rubin, Gayle S. 1984. "Thinking Sex: Notes for a Radical Theory of the Politics of Sexuality," in Carole Vance (ed.), *Pleasure and Danger: Exploring Female Sexuality*. Boston: Routledge & Kegan Paul.

Sanders, S. A., and J. M. Reinisch. 1999. "Would You Say You 'Had Sex' if…?", *Journal of the American Medical Association* 281(2).

Segal, Lynne. 1994. *Straight Sex: The Politics of Pleasure*. London: Virago.

Seidman, Steven. 1996. "The Sexualization of Love," in Steven Seidman (ed.), *Queer Theory/Sociology*. Cambridge, MA: Blackwell.

Stoltenberg, John. 1990. "How Men Have (a) Sex," in *Refusing to Be a Man*. Meridian Books.

Tiefer, Lenore. 1995. *Sex Is Not a Natural Act and Other Essays*. Boulder, CO: Westview Press.

Vance, Carole. 1984. *Pleasure and Danger: Exploring Female Sexuality*. Boston MA and London: Routledge and Kegan Paul.

Chapter 3

Gender

The Social Meaning of the Sexual Division

Dr. Alan W. McEvoy

The sociological imagination that Mills proposes in the first chapter will also help you to understand many factors about gender that you thought you knew. Most of what the public knows about gender is usually based on stereotypes that have little or no backing. Like most stereotypes, gender stereotypes are exaggerated and oversimplified beliefs about femininity and masculinity that have little correlation with being men and women. For example, many stereotypes are based on supposed innate differences between the sexes due to biology. People who hold a strong biological position will argue that whatever happens in regard to behavior of the sexes is inevitable because your sexual related behavior is determined by our biological makeup. Thus, men are considered to be natural leaders because of their greater aggressiveness and women are considered to be more nurturing because of their need to take care of children. In reality, none of these supposed innate behaviors of men and women has been supported by research (Tavris, 1992). What the research evidence actually reveals is that gender is "humanmade" (Schwalbe, 2001). Understanding the fact that gender is a social process by which gender is defined, the two first articles in this chapter (by Spade and Valentine; Taub and McLorg) deal with this prism in which gender is seen and a major outcome of one of its beliefs.

Chapter 1 discussed the role of social theory in predicting outcomes. In regard to gender, several theories have been proposed as to how this comes about: socialization—boys and girls are socialized into their respective gender identities and practices; interaction—this socialization into gender roles takes place in interaction with the various groups encountered; and institutional—in Part III of this text we examine the roles played by the five major institutions of marriage and family, education, religion, the economy, and politics. Each of these institutions has a major part in shaping gender behavior and expectations. These theories indicate that gender is a social process that involves incorporating others into our own perspectives via socialization (West and Zimmerman, 1987). As noted in Chapter 2, socialization is the process of teaching members of a society the norms and values of the existing culture. This is not to imply that the individual is born as a blank slate upon which we are socialized with the existing culture through our group interactions. Indeed, as individuals we also create and respond to social stimuli.

The next chapter in this text deals with culture—the beliefs, material objects, and practices that are created and shared within a group of people and thereby constitute their way of life. As we will see in Chapter 4, culture and gender are inexplicably intertwined—in relation to a particular culture, specific gender beliefs and practices are taught. For example, there are cultures in which men are taught to be modest and women are taught to be strong so as to perform "masculine" roles. This distinction tells us that the behavior that we consider normal for men and women is really

our ethnocentric belief in our culture and its values. The examining of other cultures soon awakens us to the fact that there is no universal definition or expression of gender because gender is socially constructed and is constantly evolving.

The problem with gender classification and behavior is not that each sex has gender-related roles, but rather that these roles shape and confine people to specific behaviors that are believed to be biological to that sex. Thus, your behavior may have been referred to as "just like a boy" or "just like a girl." The opposite behavior by a girl would mean that she was a "tomboy" but that she will grow out of it. These beliefs lead, for example, to the channeling of future occupations—servers of cosmetics in a department store rather than hawkers of items in a sports stadium; a flight attendant rather than a flight engineer (pilot). A glance at these occupational examples awakens us to the fact that gender channeling leads to inequality. United States society, like most societies, is organized so that men usually have greater access to and control over valued resources—authority, prestige, and wealth. For example, in the United States women were considered so inferior in logical thinking that they were denied the right to vote when the country was founded. Another example is seen in the phrase "glass ceiling" in income and jobs, which meant that women were seen as less capable or important even in comparable tasks. Granted, gender inequality in employment is lessening due to the actions of such government agencies as the Equal Employment Opportunity Commission, but such changes have come slowly despite the agitating of the women's liberation movement. Turning to another culture, this discrimination is seen in the preference for sons over daughters in China. It is this inequality in gender roles and expectations that lead to the questions raised in the final two articles of this chapter: How can we solve the problem of gender inequality by unraveling its knot (Johnson) and is it really possible to have a degendered society? (Kimmel)

Perhaps the best way to understand this gender label is to ask yourself if you can recall the first time you knew that you were a boy or a girl—the first time you were instructed to do a boy thing or a girl thing.

9

The Prism of Gender

JOAN Z. SPADE AND CATHERINE G. VALENTINE

One of the more confusing terminologies occurs when people speak of sex and gender. They believe that sex is related to the genitalia of birth as a means of labeling you a female or a male. Gender is what most people believe is the biological behavior—feminine and masculine—that goes with that sex distinction. As this article notes, neither of these beliefs is totally accurate.

As you read this selection, consider the following questions as guides:

1. *We usually do not doubt what we learn about sex and gender. Why?*
2. *On what basis do the authors claim that we are not stereotypes?*
3. *The authors claim that gender is not totally dependent on biology. What are its components?*

Two questions are addressed in this chapter to help us think more clearly about the complexity of gender: (1) how does Western culture condition us to think about gender, especially in relation to sex? and (2) how does social scientific research challenge Western beliefs about gender and sex?

WESTERN BELIEFS ABOUT GENDER AND SEX

Most people in Western cultures grow up learning that there are two and only two sexes, male and female, and two and only two genders, feminine and masculine (Bem, 1993; Wharton, 2005). We are taught that a real woman is feminine, a real man is masculine, and that any deviation or variation is strange or unnatural. Most people also learn that femininity and masculinity flow from biological sex characteristics (e.g., hormones, secondary sex characteristics, external and internal genitalia). We are taught that testosterone, a beard, big muscles, and a penis make a man, while estrogen, breasts, hairless legs, and a vagina make a woman. Many of us never question what we have learned about sex and gender, so we go through life assuming that gender is a relatively simple matter: a person who wears lipstick, high heel shoes, and a skirt is a feminine female, while a person who plays rugby, belches in public, and walks with a swagger is a masculine male (Lorber, 1994; Ridgeway & Correll, 2004)....

Americans habitually overemphasize biology and underestimate the power of social facts to explain sex and gender (O'Brien, 1999). For instance, Americans tend to equate aggression with biological maleness and vulnerability with femaleness;

Source: Pine Forge publications, 2008.

natural facility in physics with masculinity and natural facility in childcare with femininity; lace and ribbons with girlness and rough and tumble play with boyness (Glick & Fiske, 1999; Ridgeway & Correll, 2004). The notions of natural sex and gender difference, duality, and even opposition and inequality permeate our thinking, color our labeling of people and things in our environment, and affect our practical actions (Bem, 1993; Wharton, 2005).

We refer to the American tendency to assume that biological sex differences cause gender differences as "the pink and blue syndrome." This syndrome is deeply lodged in our minds and feelings, and reinforced through everyday talk, performance, and experience. It's everywhere. Any place, object, discourse, or practice can be gendered. Children's birthday cards come in pink and blue. Authors of popular books assert that men and women are from different planets. People love PMS and alpha male jokes. In "The Pink Dragon Is Female"… Adie Nelson's research reveals that even children's fantasy costumes tend to be gendered as masculine and feminine. The "pink and blue syndrome" is so embedded within our culture, and consequently within individual patterns of thinking and feeling, that most of us cannot remember when we learned gender stereotypes and expectations or came to think about sex and gender as natural, immutable, and fixed. It all seems so simple and natural. Or is it?

What is gender? What is sex? How are gender and sex related? Why do most people in our society believe in the "pink and blue syndrome"? Why do so many of us attribute one set of talents, temperaments, skills, and behaviors to women and another set to men? These are the kinds of questions social scientists in sociology, anthropology, psychology, and other disciplines have been asking and researching for almost fifty years. Thanks to decades of good work by an array of scientists, we now understand that gender and sex are not so simple. Social scientists have discovered that the gender landscape is complicated, shifting, and contradictory. Among the beliefs that have been called into question by research are

- The notion that there are two and only two sexes and, consequently, two and only two genders

- The assumption that men and women are the same everywhere and all the time
- The belief that biological factors cause the "pink and blue syndrome"

USING OUR SOCIOLOGICAL RADAR

Before we look at how social scientists answer questions such as "What is gender," let's do a little research of our own. Try the following: relax, turn on your sociological radar, and examine yourself and the people you know carefully. Do all the men you know fit the ideal of masculinity all the time, in all relationships, and in all situations? Do all the women in your life consistently behave in stereotypical feminine fashion? Do you always fit into one as opposed to the other culturally approved gender box? Or are most of the people you know capable of "doing" both masculinity and femininity, depending on the interactional context? Our guess is that none of the people we know are aggressive all the time, nurturing all the time, sweet and submissive all the time, or strong and silent all the time. Thankfully, we are complex and creative. We stretch and grow and develop as we meet the challenges, constraints, and opportunities of different and new situations and life circumstances. Men can do mothering; women can "take care of business." Real people are not stereotypes.

Yet even in the face of real gender fluidity and complexity, the belief in gender dichotomy and opposition continues to dominate almost every aspect of the social worlds we inhabit. For example, recent research shows that even though men's and women's roles have changed and blended, the tendency of Americans to categorize and stereotype people based on the simple male/female dichotomy persists (Glick & Fiske, 1999). As Glick and Fiske (1999) put it, "we typically categorize people by sex effortlessly, even nonconsciously, with diverse and profound effects on social interactions" (p. 368). To reiterate, many Americans perceive humankind as divided into mutually exclusive, nonoverlapping groups: males/masculine/men and females/feminine/women

(Bem, 1993; Wharton, 2005). The culturally created image of gender, then, is nonkaleidoscopic: no spontaneity, no ambiguity, no complexity, no diversity, no surprises, no elasticity, and no unfolding growth.

SOCIAL SCIENTIFIC UNDERSTANDINGS OF SEX AND GENDER

Modern social science offers us a very different image of gender. It opens the door to the richness and diversity of human experience, and it resists the tendency to reduce human behavior to single factors. Research shows that the behavior of real women and men depends on time and place, and context and situation, not on fixed gender differences (Lorber, 1994; Tavris, 1992). For example, just a few decades ago in the United States, cheerleading was a men's sport because it was considered too rigorous for women (Dowling, 2000), women were thought to lack the cognitive and emotional "stuff" to pilot flights into space, and medicine and law were viewed as too intellectually demanding for women. As Carol Tavris (1992) says, research demonstrates that perceived gender differences turn out to be a matter of "now you see them, now you don't" (p. 288).

If we expand our sociological examination of gender to include cultures outside North America, the real-life fluidity of gender comes fully alive.... In some cultures (for example, the Aka hunter-gatherers) fathers as well as mothers suckle infants (Hewlett, 2001). In other cultures, such as the Agta Negritos hunter-gatherers, women as well as men are the hunters (Estioko-Griffin & Griffin, 2001)....

Context, which includes everything in the environment of a person's life such as work, family, social class, race—and more—is the real source of gender definitions and practices. Gender is flexible, and "in its elasticity it stretches and unfolds in manifold ways so that depending upon its contexts, including the life progress of individuals, we see it and experience it differently" (Sorenson, 2000, p. 203). Most of us "do" both masculinity and femininity,

and what we do is situationally dependent and institutionally constrained.

Let's use sociological radar again and call upon the work of social scientists to help us think more precisely and "objectively" about what gender and sex are. It has become somewhat commonplace to distinguish between gender and sex by viewing sex as a biological fact, meaning that it is noncultural, static, scientifically measurable and unproblematic, while we see gender as a cultural attribute, a means by which people are taught who they are, how to behave, and what their roles will be (Sorenson, 2000). However, this mode of distinguishing between sex and gender has come under criticism, largely because new studies have begun to reveal the cultural dimensions of sex itself. That is, the physical characteristics of sex cannot be separated from the cultural milieu in which they are labeled and given meaning.... In other words, the relationship between biology and behavior is reciprocal, both inseparable and intertwined (Yoder, 2003).

Sex, as it turns out, is not a clear-cut matter of DNA, chromosomes, external genitalia, and the like, factors which produce two and only two sexes—females and males. First, there is considerable biological variation. Sex is not fixed in two categories. There is overlap. For example, all humans have estrogen, prolactin, and testosterone, but in varying and changing levels (Abrams, 2002). Think about this. In our society, people tend to associate breasts and related phenomena, such as breast cancer and lactation, with women. However, men have breasts. Indeed, some men have bigger breasts than some women, some men lactate, and some men get breast cancer. Also, in our society, people associate facial hair with men. What's the real story? All women have facial hair and some have more of it than some men. Indeed, recent hormonal and genetic studies (e.g., Abrams, 2002; Beale, 2001) are revealing that, biologically, women and men are far more similar than distinct.

In fact, variations in and complexities of sex development produce *intersexed* people whose bodies do not fit the two traditionally understood sex categories (Fausto-Sterling, 2000; Fujimora, 2006). Fujimora (2006) examined recent research on sex genes and concluded that "there is no single pathway through which sex is genetically determined" and that we might consider sex variations, such as intersex, as resulting from "multiple developmental pathways that involve genetic, protein, hormonal, environmental, and other agents, actions, and interactions" (p. 71). Lorber and Moore (2007) argue that intersexed people are akin to multiracial people. They point out that just as scientists have demonstrated through DNA testing that almost all of us are genetically interracial, similarly, "if many people were genetically sex-typed, we'd also find a variety of chromosomal, hormonal, and anatomical patterns unrecognized" in our rigid, two sex system (p. 138)....

Biology is complicated business, and that should come as no surprise. The more we learn about biology, the more elusive and complex sex becomes. What seemed so obvious—two, opposite sexes—turns out to be an oversimplification. Humans are not unambiguous, clearly demarcated, biologically distinct, nonoverlapping, invariant groups of males and females.

So, again, what is gender? First, gender is not sex. Biological sex characteristics do not cause specific gender behaviors or activities. As discussed earlier, biological sex is virtually meaningless outside the social context in which it develops and is expressed (Yoder, 2003). Second, gender is not an essential identity. It "does not have a locus nor does it take a particular form" (Sorenson, 2000, p. 202). In other words, individuals do not possess a clearly defined gender that is the same everywhere and all the time. At this point, you may be thinking, what in the world are these authors saying? We are saying that gender is a human invention, a means by which people are sorted (in our society, into two genders), a basic aspect of how our society organizes itself and allocates resources (e.g., certain tasks assigned to people called women and other tasks to those termed men), and a fundamental ingredient in how individuals understand themselves and others ("I feel feminine." "He's manly." "You're androgynous.").

One of the fascinating aspects of gender is the extent to which it is negotiable and dynamic. In effect, masculinity and femininity exist because people believe that women and men are distinct groups and, most important, because people "do gender," day in and day out.... Some social scientists call gender a performance, while others term it a masquerade. The terms "performance" and "masquerade" emphasize that it is through the ways in which we present ourselves in our daily encounters with others that gender is created and recreated.

We even do gender by ourselves and sometimes quite self-consciously. Have you ever tried to make yourself look and act more masculine or feminine? What is involved in "putting on" femininity or masculinity?... Although most people have deeply learned gender and view the gender box they inhabit as natural or normal, intersex and transgender activists attack the boundaries of "normal" by refusing to choose a traditional sex, gender, or sexual identity (Lorber & Moore, 2007). In so doing, cultural definitions of sex and gender are destabilized and expanded.

You may be wondering why we have not used the term *role*, as in *gender role*, to describe "doing gender." The problem with the concept of roles is that typical social roles, such as those of teacher, student, doctor, or nurse, involve situated positions and identities. However, gender, like race, is a status and identity that cuts across many situations and institutional arenas. In other words, gender does not "appear and disappear from one situation to another" (Van Ausdale & Feagin, 2001, p. 32). People are always doing gender. They rarely let their guard down. In part, this is a consequence of the pressures that other people exert on us to "do gender" no matter the situation in which we find ourselves. Even if an individual would like to "give up gender," others will define and interact with that individual in gendered terms. If you were a physician, you could "leave your professional role behind you" when you left the hospital or office and went shopping or vacationing. Gender is a different story. Could you leave gender at the office and go shopping or vacationing? What would that look like, and what would it take to make it happen?

So far, we have explored gender as a product of our interactions with others. It is something we do, not something we inherit. Gender is also built into the larger world we inhabit, including its institutions, images and symbols, organizations, and material objects. For example, jobs, wages, and hierarchies of dominance and subordination in workplaces are gendered. Even after decades of substantial increase in women's workforce participation, occupations continue to be allocated by gender (e.g., secretaries are overwhelmingly women; men dominate construction work) and a wage gap between men and women persists (Bose & Whaley, 2001; Steinberg, 2001)... In addition, men are still more likely to be bosses and women to be bossed. The symbols and images with which we are surrounded and by which we communicate are another part of our society's gender story. Our language speaks of difference and opposition in phrases such as "the opposite sex" and the absence of any words, except awkward medical terms (e.g., transsexual) or epithets (e.g., pervert), to refer to sex and gender variants. In addition, the swirl of gendered images in the media is almost overwhelming. Blatant gender stereotypes still dominate TV, film, magazines, and billboards (Lont, 2001). Gender is also articulated, reinforced, and transformed through material objects and locales (Sorenson, 2000). Shoes are gendered, body adornments are gendered, public restrooms are gendered, weapons are gendered, ships are gendered, wrapping paper is gendered, and deodorants are gendered. The list is endless. The point is that these locales and objects are transformed into a medium for gender to operate within (Sorenson, 2000). They make gender seem "real," and they give it material consequences (Sorenson, 2000, p. 82).

In short, social scientific research underscores the complexity of the prism of gender and demonstrates how gender is constructed at multiple, interacting levels of society.... We are literally and figuratively immersed in a gendered world—a world in which difference, opposition, and inequality are the culturally defined themes. And yet, that world is kaleidoscopic in nature. The lesson of the kaleidoscope is that "nothing in life is immune to change" (Baker, 1999, p. 29). Reality is in flux; you never know what's coming next. The metaphor of the kaleidoscope reminds us to keep seeking the shifting meanings as well as the recurring patterns of gender (Baker, 1999).

We live in an interesting time of kaleidoscopic change. Old patterns of gender difference and inequality keep reappearing, often in new guises, while new patterns of convergence, equality, and self-realization have emerged. Social science research is vital in helping us to stay focused on understanding the prism of gender as changeable, and responding to its context—as a social dialogue about societal membership and conventions—and "as the outcome of how individuals are made to understand their differences and similarities" (Sorenson, 2000, p. 203–204). With that focus in mind, we can more clearly and critically explore our gendered society.

REFERENCES

Abrams, D. C. (2002). Father nature: The making of a modern dad. *Psychology Today*, March/April, 38–42.

Baker, C. (1999). *Kaleidoscopes: Wonders of wonder*. Lafayette, CA: C&T Publishing.

Beale, B. (2001). The sexes: New insights into the X and Y chromosomes. *The Scientist*, 15(15), 18. Retrieved July 23, 2001, from http://www.the-scientist.com/yr2001/jul/research1_010723.html

Bem, S. L. (1993). *The lenses of gender*. New Haven, CT: Yale University Press.

Bose, C. E., & Whaley, R. B. (2001). Sex segregation in the U.S. labor force. In D. Vannoy (Ed.), *Gender mosaics* (pp. 228–239). Los Angeles: Roxbury Publishing.

Dowling, C. (2000). *The frailty myth*. New York: Random House.

Epstein, C. F. (1988). *Deceptive distinctions*. New Haven, CT: Yale University Press.

Estioko-Griffin, A., & Griffin, P. B. (2001). Woman the hunter: The Agta. In C. Brettell & C. Sargent (Eds.),

Gender in cross-cultural perspective (3rd ed., pp. 238–239). Upper Saddle River, NJ: Prentice Hall.

Fausto-Sterling, A. (2000). *Sexing the body: Gender politics and the construction of sexuality*. New York: Basic Books.

Fujimora, J. H. (2006). Sex genes: A critical sociomaterial approach to the politics and molecular genetics of sex determination. *Signs*, 32(1), 49–81.

Glick, P., & Fiske, S. T. (1999). Gender, power dynamics, and social interaction. In M. Ferree, J. Lorber, & B. Hess (Eds.), *Revisioning gender* (pp. 365–398). Thousand Oaks, CA: Sage.

Lont, C. M. (2001). The influence of the media on gender images. In D. Vannoy (Ed.), *Gender mosaics*. Los Angeles, CA: Roxbury Publishing.

Lorber, J. (1994). *Paradoxes of gender*. New Haven, CT: Yale University Press.

Lorber, J., & Moore, L. J. (2007). *Gendered bodies*. Los Angeles: Roxbury Publishing.

O'Brien, J. (1999). *Social prisms: Reflections on the everyday myths and paradoxes*. Thousand Oaks, CA: Pine Forge Press.

Ridgeway, C. L., & Correll, S. J. (2004). Unpacking the gender system: A theoretical perspective on gender beliefs and social relations. *Gender & Society*, *18*(4), 510–531.

Sorenson, M. L. S. (2000). *Gender archaeology*. Cambridge, England: Polity Press.

Tavris, C. (1992). *The mismeasure of woman*. New York: Simon & Schuster.

Van Ausdale, D., & Feagin, J. R. (2001). *The first r: How children learn race and racism*. Lanham, MD: Rowman & Littlefield Publishers.

Wharton, A. S. (2005). *The sociology of gender*. Malden, MA: Blackwell.

Yoder, J. D. (2003). *Women and gender: Transforming psychology*. Upper Saddle River, NJ: Prentice Hall.

10

Influences of Gender Socialization and Athletic Involvement on the Occurrence of Eating Disorders

DIANE E. TAUB AND PENELOPE A. MCLORG

Not everything we learn through the process of socialization has positive results. Consider eating disorders that result from the preoccupation many people, especially young women, have with body shape and weight. As Taub and McLorg note, to reduce this prevalent problem, we must consider the source of the negative body images and how to change the emphasis in the socialization process on ideal images. The authors also discuss the role of athletics in eating disorders.

As you read, think about the following:

1. *What messages do women and men receive about ideal body types, and from where?*
2. *What role do athletics play in contributing to or discouraging eating disorders?*
3. *What could be done to change the negative results of socialization, such as abuse, neglect, and unrealistic body images?*

GLOSSARY **Anorexia and bulimia nervosa** Eating disorders that involve (1) self-starving and (2) binging and purging behaviors. **Internalization** The process of making ideas and behavior patterns an integral part of one's repertoire of behaviors. **Gender socialization** Learning the gender roles expected in society. **Agents of gender socialization** Ways gender expectations are passed on.

The eating disorders of anorexia nervosa (self-starvation) and bulimia nervosa (binge-purge syndrome) are a major health and social problem, with the reported occurrence increasing steadily over the past 30 years (Gordon, 1988; Harrison and Cantor, 1997; Wiseman and others, 1992). As a risk group, females are much more likely than males to be affected, comprising approximately 90 percent of reported cases (American Psychiatric Association, 1994; Haller, 1992). This gender difference can be

Source: Original article. Reprinted by permission of the authors, 2008.

clarified by examining factors of gender socialization that relate to physical appearance.

Traditionally, more emphasis has been placed on the appearance of females than of males (Lovejoy, 2001; Thornton and Maurice, 1997). Women show awareness of this focus by being more concerned with their appearance than are men (Pliner, Chaiken, and Flett, 1990; Thompson and Heinberg, 1999). Physical appearance is also more crucial to self-concept among females than among males (Lovejoy, 2001; Rodin, Silberstein, and Striegel-Moore, 1985). Whereas in men self-image is associated with skill and achievement, among women it is linked to physical characteristics (Hesse-Biber, Clayton-Matthews, and Downey, 1987). In terms of bodily appeal, the physical attractiveness of males is related to physical abilities, with their bodies valued for being active and functional. In contrast, the bodies of females are judged on the basis of beauty (Rodin and others, 1985; Thompson and Heinberg, 1999).

One important consideration in appearance is body shape. Concerns about body shape expressed by women range from mild weight consciousness at one extreme to eating disorders at the other (Gordon, 1988; Rodin and others, 1985; Twamley and Davis, 1999). As part of the socialization in the "cult of thinness" (Hesse-Biber and others, 1987, p. 512), females accept an ideal body shape and a corresponding need for weight control. Dissatisfaction from a failure to meet slim appearance standards and subsequent dieting behavior have been identified as risk factors for the development of eating disorders (Drewnowski and Yee, 1987; Polivy and Herman, 1986; Striegel-Moore, Silberstein, and Rodin, 1986; Twamley and Davis, 1999).

This reading examines the relationship between eating disorders and gender socialization. Our purpose is to demonstrate the important contribution female and male socialization makes to the social context of anorexia nervosa and bulimia nervosa, including the gender difference in occurrence. To illustrate the connections between gender socialization and eating disorders, we use the following framework: (1) ideal body shape, as a representation of gender norms, (2) role models and mass media messages, as

agents of gender socialization, and (3) dieting, as an expression of gender socialization. Ideal body shape affects agents of socialization, which in turn reinforce ideal body shape. Acceptance of ideal body shape norms and exposure to agents of gender socialization are expressed through dieting behavior; eating disorders are an extreme response. Also discussed in this reading is the influence of athletic involvement on eating disorders. Participation in certain sport activities converges with female body norms and heightens females' vulnerability to eating disorders.

IDEAL BODY SHAPE

Current appearance expectations specify thinness for women (Garner and others, 1980; Lovejoy, 2001; Thompson and Heinberg, 1999). Slim bodies are regarded as the most beautiful and worthy ones; overweight is seen as not only unhealthy but also offensive and disgusting (Harrison and Cantor, 1997; Schwartz, Thompson, and Johnson, 1982). Although both males and females are socialized to devalue fatness, women are more exposed to the need to be thin (Raudenbush and Zellner, 1997; Rodin and others, 1985).

In contrast, males are socialized to be muscular and not skinny or weak (Harrison, 2000; Leon and Finn, 1984). In ratings by preadolescent, adolescent, and college-aged males, the mesomorphic or muscular male body type is associated with socially favorable behaviors and personality traits (Raudenbush and Zellner, 1997). Compared with endomorphic (plump) and ectomorphic (slender) individuals, mesomorphs are judged more likely to assume leadership and be assertive, smart, and most wanted as a friend. The devaluing of a thin body for males is also reflected in the frequent desire of preadolescent and teenage boys as well as college males to gain weight and/or size (Collins, 1991; Harrison, 2000; Raudenbush and Zellner, 1997; Striegel-Moore and others, 1986).

Among females, the orientation toward slimness is so established that even when they are not overweight, they frequently perceive themselves as such (Connor-Greene, 1988; Raudenbush and Zellner, 1997; Wiseman, Harris, and Halmi, 1998). For example, although over 60 percent of college females believe that they are overweight, only 2 percent actually are (Connor-Greene, 1988). In addition, college females underestimate the occurrence of being underweight. While 31 percent of college women are measured as underweight, only 13 percent think they weigh below weight norms (Connor-Greene, 1988). Further, over four out of five college women report that they want to lose weight (Hesse-Biber and others, 1987). Even among college women whose weight is within the normal range, three-fourths want to be thinner (Raudenbush and Zellner, 1997).

Other results indicate that the ideal body shape of college women is significantly thinner than both their actual body type (Fallon and Rozin, 1985; Raudenbush and Zellner, 1997) and the body they perceive as most attractive to males (Fallon and Rozin, 1985). In addition, Raudenbush and Zellner (1997) report that women who engage in disordered eating behavior desire a body shape that is thinner than what they believe is attractive to males. In a study of families, Rozin and Fallon (1988) show that mothers and daughters both want slimmer bodies than they currently have. Furthermore, the shape these females believe is most attractive to males is thinner than what males actually prefer.

Collins (1991) demonstrates that gender-based ideas of attractive bodies develop in children as young as six or seven years. First- through third-grade girls select illustrations of their ideal figures that are significantly thinner than their current figures; this pattern is found across all levels of actual weight. Moreover, girls choose significantly slimmer figures than boys do for the ideal girl, ideal female adult, and ideal male adult (Collins, 1991).

Among females, learning to desire thinness begins at an early age. In general, females want to be slim and are critical of their weight, regardless of their actual body size and weight. A similar concern for thinness in male bodies is not common among males (Collins, 1991; Connor-Greene, 1988; Fallon and Rozin, 1985; Harrison, 2000; Hesse-Biber and others, 1987; Rozin and Fallon, 1988). The inaccuracy with which females of all ages perceive their body shapes (Rodin and others, 1985) parallels the

distorted body images held by individuals with eating disorders.

AGENTS OF GENDER SOCIALIZATION

Images of ideal body shape affect agents of gender socialization, such as role models and mass media. In turn, these influences support expectations of body size. Reflecting gender socialization, traditional female role models and mass media messages express thinness norms for females.

Role Models

Examining patterns of ideal body shape, Garner and colleagues (1980) study the measurements of Miss America contestants over the 20-year span from 1959 to 1978. Mazur (1986) and Wiseman and colleagues (1992) conduct similar analyses of contestants' dimensions covering the period of 1979 to the mid-1980s. Both Garner and coworkers (1980) and Mazur (1986) report decreases in bust and hip measurements of Miss America contestants over the study periods. However, waist dimensions demonstrate periods of increase, suggesting a less hourglass standard. Further, weight for height of Miss America contestants progressively declines (Garner and others, 1980; Mazur, 1986), with a trend from 1970 to 1978 for pageant winners to be thinner than the average contestant (Garner and others, 1980). Analyzing the weight of contestants in relation to expected weight for their height and age, Wiseman and colleagues (1992) additionally find a significant decrease in the women's percentage of expected weight from 1979 to 1985.

Garner and colleagues (1980), Mazur (1986), Wiseman and coworkers (1992), and Katzmarzyk and Davis (2001) also examine the beauty ideal represented by *Playboy* centerfolds. As with Miss America contestants, bust and hip dimensions decline and waist measurements rise from 1959 to 1978 (Garner and others, 1980). During the early 1980s, bust, waist, and hip dimensions of centerfolds decrease (Mazur, 1986). Centerfolds also show declines in weight for height between 1959 and the early 1980s (Garner and others, 1980; Mazur, 1986). Further, *Playboy* centerfolds continue the diminished body size through the 1990s, with 70 percent of centerfolds from 1978 to 1998 being classified as underweight (Katzmarzyk and Davis, 2001).

The trends of slenderization exhibited by both *Playboy* centerfolds and Miss America contestants illustrate the slimness norm. In fact, from 1979 through the mid-1980s, approximately two-thirds of these ideals of female beauty weighed 15 percent or more below their expected weight (Wiseman and others, 1992). The pattern of at least two-thirds of *Playboy* centerfolds weighing 15 percent below expected weight norms persists through the 1990s (Katzmarzyk and Davis, 2001). Maintaining such a weight level is one of the criteria for anorexia nervosa (Wiseman and others, 1992). Thus, the declining size of female figures considered admirable represents a body size reflective of an eating disorder.

Other female role models, such as movie stars and magazine models, have become less curvaceous over the latter half of the 20th century (Silverstein and others, 1986). Hence, portrayals of females in media geared toward women as well as men demonstrate the thinness norm. In a study that examines cover models appearing on the four most popular American fashion magazines for the years 1959–1999, overall body size decreases significantly during the 1980s and 1990s (Sypeck, Gray, and Ahrens, 2004). From the 1960s to the 1990s, there is a dramatic increase in the frequency with which the covers depict the models' upper bodies and hips. Further, in comparison with the earliest period investigated, the latest covers reveal more of the models' bodies (Sypeck and others, 2004). Thus, over recent decades, figures of magazine cover models have become thinner and more emphasized and revealed.

The "anorectic body type" of models in major women's fashion magazines illustrates "an idealized standard of beauty and high fashion" (Gordon, 1988, p. 157). As preferred bodies, fashion models set an example of slimness that is unrealistic for most women (Thompson and Heinberg, 1999; Wiseman and others, 1998). College women viewing pictures of catalog models who typify the idealized thin female

figure exhibit lowered self-esteem and increased self-consciousness, anxiety about their body, and body dissatisfaction (Thornton and Maurice, 1997). In general, through exposure to female role models, the majority of women are continually reminded of their inadequacy (Pliner and others, 1990) and encouraged to obsess about thinness and beauty (Lovejoy, 2001).

Further, observing thin female celebrities and models pictured in magazines is related to anorexic symptoms in girls between the ages of 11 and 18, and to bulimic tendencies among girls between 15 and 18 years old (Harrison, 2000). For college women, reading fashion and fitness magazines is associated with eating disorder attitudes and behavior (Harrison, 1997; Harrison and Cantor, 1997). The relationship between exposure to these media sources and disordered eating patterns is maintained whether or not the women have an initial interest in fashion and fitness. Fitness magazines are particularly associated with anorexic rather than bulimic tendencies because they provide examples of diet and exercise behavior characteristic of anorexics, but do not offer illustrations of the binging and purging behavior of bulimics (Harrison, 1997; Harrison and Cantor, 1997).

Beyond simple exposure to magazines, the attraction of college women to role models in magazines relates to eating disorder symptoms. Factors such as feeling similar to, wanting to be like, or liking thin figures increases both anorexic and bulimic behavior and attitudes (Harrison, 1997). Images of thinness in magazines are more likely to be associated with body dissatisfaction and eating disturbance among college women who accept the thin body ideal (Twamley and Davis, 1999). In general, internalization of the thin body an as ideal has been shown to be a causal risk factor for body-image and eating disturbances. This risk increases when the internalization coexists with other risk factors, such as body dissatisfaction, dieting, and negative emotions (Thompson and Stice, 2001).

Research on newspaper and magazine advertisements indicates that male figures are generally portrayed as bigger than female figures (Goffman, 1979). Such representation of the size of men symbolizes their "social weight," in power, authority, or rank (Goffman, 1979, p. 28), as well as the positive valuing of a larger body in males (Raudenbush and Zellner, 1997). In contrast, females in print advertisements provide additional exposure to the thin female standard.

Women's socialization to be slim is also demonstrated by role models on television over the past 25 years. For example, in prime-time, top-10 Nielson-rated television shows and their commercials during the 1970s, females are more likely to be thin than heavy and to be thinner than males (Kurman, 1978). A related study (Silverstein and others, 1986) of most-watched television programs demonstrates that 69 percent of the actresses and only 17.5 percent of the actors are slim. In addition, 5 percent of the women are evaluated as heavy, while over a quarter of the men are rated as such. These contrasts remain over a range of ages of the performers (Silverstein and others, 1986). Most female television characters are thinner than the average woman, while fewer that 10 percent of the women on television are overweight (Thompson and Heinberg, 1999).

In a recent study of girls and boys between 11 and 18 years old, watching television shows with overweight characters is associated with bulimic-like behavior and attitudes among girls. For boys, only the youngest demonstrate a relationship between exposure to overweight characters on television and body dissatisfaction (Harrison, 2000). Research on college women reports that viewing television shows with overweight characters is related to body dissatisfaction, while exposure to television shows with thin characters is associated with drive for thinness (Harrison, 1997; Harrison and Cantor, 1997). Similar to findings with magazine figures, identification with thin characters on television is additionally associated with both anorexic and bulimic behavior and attitudes (Harrison, 1997). Further, college women who accept the thin body ideal represented by television characters are more likely to demonstrate body dissatisfaction and eating disturbance (Twamley and Davis, 1999).

Mass Media Messages

In addition to displaying slim female role models for imitation, agents of gender socialization promote

consciousness of weight and diet. Mass media messages encourage virtually uniform standards of beauty (Mazur, 1986); messages directed toward females emphasize the thin ideal (Thompson and Heinberg, 1999). The presence of electronic as well as print media has increased the accessibility of cultural standards of appearance (Thompson and Heinberg, 1999).

Among children and adolescents, media messages about physical beauty and the thin body ideal influence self-perceptions, body image, weight concerns, and eating behaviors (Morris and Katzman, 2003). In an analysis of 25 children's videos, messages emphasize the importance of physical attractiveness as well as support body stereotypes. Video characters with thin or muscular body figures are found to possess desirable traits, whereas obese characters have negative traits. Compared with videos, however, 20 children's books for ages 4–8 do not exhibit as many body-related messages (Herbozo and others, 2004).

Surveying major women's magazines, Garner and colleagues (1980) find a significant increase in diet articles from 1959 to 1978. From 1979 to 1988, the same women's magazines show a leveling off in number of diet articles but an increase in articles on exercise as a strategy for weight loss (Wiseman and others, 1992). In addition, dieting and weight control listings in the *Reader's Guide to Periodical Literature* almost double from 1977 to 1986 (Hesse-Biber and others, 1987).

Media emphasis on a slim body standard for females is also illustrated in comparisons of female- and male-directed magazines. An analysis (Silverstein and others, 1986) of the most popular women's and men's magazines indicates significant differences in content of articles and advertisements. In women's magazines, ads for diet foods, and articles and advertisements dealing with body shape or size, appear 63 times and 12 times more often, respectively, than in men's magazines (Silverstein and others, 1986).

Similarly, Andersen and DiDomenico (1992) find that the most popular magazines among women aged 18 to 24 contain 10 times more articles and advertisements on dieting or losing weight than do the most popular magazines among men aged 18 to 24. The focus on weight control in young women's magazines may have particular importance, as adolescence is the primary period of onset for both anorexia nervosa and bulimia nervosa (American Psychiatric Association, 1994; Haller, 1992; Harrison, 2000).

In addition to advertisements and articles on weight control, women's magazines surpass men's magazines in material concerning food. Articles on food and ads for food (excluding those for diet foods) in women's magazines exceed those in men's magazines by 71 to 1 (Silverstein and others, 1986). Thus, through this printed medium, females are being presented with conflicting messages. While food advertisements and articles encourage the consumption and enjoyment of food, diet aids and body shape ads and articles reinforce control of eating and weight. Popular magazines effectively maintain women's weight control preoccupation through their dual messages of eat and stay slim (McLorg and Taub, 1987). Moreover, although exposure to magazine advertising may be similar for individuals with and without eating disorders, anorexics and bulimics are especially likely to believe that advertisements promote the desirability of slimness (Peterson, 1987).

As agents of socialization, mass media and role models present a consistent portrayal of thin females in material directed toward both female and male audiences. Media preference for slimness in women is additionally shown in messages encouraging the weight control efforts of females. Through their selective representation of thin women, these sources not only reflect the slim ideal body shape for females, but also strengthen this gender norm for appearance. The influence of mass media and role models is affected by the extent to which females identify with the role models (Harrison, 1997) and accept the messages about thinness (Thompson and Heinberg, 1999; Twamley and Davis, 1999). Role models and mass media effectively reinforce the weight consciousness of females. The impact of these influences is to encourage and perpetuate women's repeated attempts to conform to the thin standard. A similar promotion or expression of a slimness ideal is not an aspect of the socialization experience of males (Andersen and DiDomenico, 1992; Drewnowski and Yee, 1987; Harrison, 2000; Silverstein and others, 1986).

DIETING

Concerns about thin body size reflect gender socialization of females. With the role obligation of being visually attractive, women alter their bodies to conform to an appearance ideal. Dieting can be viewed as a response to the gender norm of slimness in females. Worrying about weight and weight-loss efforts are so common among females that they have become normative (Rodin and others, 1985; Striegel-Moore and others, 1986). Nasser (1988, p. 574) terms dieting a "cultural preoccupation" among females, with concerns about weight and weight control persisting even into women's elderly years (Pliner and others, 1990). Further, females' continual efforts toward the thinness ideal are usually unsuccessful (Silverstein and others, 1986; Wiseman and others, 1998).

Frequency of dieting is related to actual body size, as well as to ideal body size and the emphasis a woman places on the importance of attractiveness (Silverstein and Perdue, 1988). A history of dieting, beginning with the teen years, is common among anorexics and bulimics (McLorg and Taub, 1987; Wiseman and others, 1998). In fact, researchers consider dieting a "precondition" (Polivy and Herman, 1986, p. 328) or a "chief risk factor" (Drewnowski and Yee, 1987, p. 633) of eating disorders.

As shown in a Nielson survey over 20 years ago, 56 percent of all women aged 24 to 54 dieted during the course of the year; 76 percent did so for cosmetic, rather than health, reasons (Schwartz and others, 1982). Among high school females, over half had dieted by the time they entered high school; and nearly 40 percent were currently dieting (Johnson and others, 1983). More recently, over half of women report that they are dieting (Haller, 1992). Another study demonstrates an even greater occurrence of dieting in preadolescent girls. Half of 9-year-olds and nearly 80 percent of 10- and 11-year-olds indicate that they have dieted (Stein, 1986). With dieting so common among girls and women (Lovejoy, 2001), it is not surprising that "serious dieting" in females is seen as "normal" (Leon and Finn, 1984, p. 328).

Compared with young females, young males are much less likely to diet. In one sample, 10 percent of boys versus 80 percent of girls had been on a diet before the age of 13 (Hawkins and others, cited in Striegel-Moore and others, 1986). Similarly, 64 percent of first-year college women, but only 29 percent of first-year males, followed a reduced-calorie diet in the previous month (Drewnowski and Yee, 1987). Compared with their male classmates, high school senior females are much more likely to use prescription amphetamines without a physician's orders for the purpose of losing weight (Taub, 1986). Further, among a sample of high school females, 34 percent had fasted to control weight, 16 percent had used diet pills, and 10 percent had taken diuretics during the past year, with 7 percent indicating daily use of diet pills and diuretics (Taub and Blinde, 1994).

Overall, studies demonstrate that females of differing ages frequently diet or engage in other weight loss behavior. In addition, females are much more likely than males to pursue weight loss through various methods. The extent and persistence of the dieting efforts of females indicate acceptance of the thin ideal. As an expression of gender socialization, dieting also reveals exposure to both role model and mass media promotion of slimness. Anorexia nervosa and bulimia nervosa represent extreme responses to female socialization toward thinness, with dieting usually preceding an eating disorder.

THE FACTOR OF ATHLETIC INVOLVEMENT

As noted, eating disorders are health problems in which social and cultural factors are very important. Body norms for women are shown to play a significant role in the genesis and maintenance of eating disorders. These cultural influences include the thinness norm, role models, mass media messages, and dieting/weight reducing behavior. The risk of eating disorders is increased when gender role expectations and athletic involvement interact. In particular, athletes are more prone to eating disorders than non-athletes when they participate in organized physical activities that have an aesthetic focus or that emphasize body shape and control (Benson and Taub, 1993; Hausenblas and Carron, 2002; Smolak, Murnen, and Ruble, 2000). Athletes not only internalize gender socialization about the thin body ideal, but they also hear from coaches and fellow participants that excessive leanness is essential for superior athletic performance. A high percentage of body fat is assumed to slow movement and hinder performance (Thornton, 1990). The sports most mentioned for increasing vulnerability to eating disorders are swimming, diving, figure skating, cross country running, dance, and gymnastics (Benson and Taub, 1993; Combs, 1982; Hausenblas and Carron, 2002; Smolak and others, 2000; Thornton, 1990).

In general, particular features of sport have been found to be associated with susceptibility to eating disorders. For example, eating disorders tend to be more prevalent in sports that are individual rather than team in nature, in which individual performance is more discernible. In addition, the risk of eating disorders is heightened in sports in which appearance is highlighted in the overall performance evaluation. Thirdly, sports that involve airborne movements exacerbate the need for thinness and increase vulnerability to eating disorders. And lastly, eating disorders tend to be more common in sports that involve a competitive uniform that exposes body size and contours (Combs, 1982; Hausenblas and Carron, 2002; Smolak and others, 2000; Thornton, 1990). Disordered eating is promoted when athletes experience public weighing or measuring, assessments of weight or percent fat at team practice, or pressure from coaches to lose weight (Benson and Taub, 1993; Thornton, 1990). Overall, occurrence of eating disorders increases as the competitive level of the sport environment increases (Smolak and others, 2000).

CONCLUSION

The influence of gender socialization in anorexia nervosa and bulimia nervosa is suggested by the gender distribution of the syndromes, with occurrence at least 10 times higher among females than males. In their connections with appearance expectations, eating disorders illustrate normative elements of female socialization. For example, weight loss efforts of females can be attributed to the greater importance of appearance in evaluations of women than of men; women's figures are more emphasized and more critically assessed (Lovejoy, 2001; Rodin and others, 1985).

As agents of gender socialization, role models and mass media are affected by, and support, notions of ideal body shape. Beauty queens, *Playboy* centerfolds, fashion models, and female television and movie characters are predominantly slender (e.g., Silverstein and others, 1986; Sypeck and others, 2004; Wiseman and others, 1992). Such role models serve as ideals for the female body shape. In addition to promoting the slimness standard, nude layouts and beauty contests epitomize the viewing of women as objects, with women's bodies judged according to narrow beauty standards (Lovejoy, 2001).

Also supporting the expectation of thinness in females are the numerous articles and advertisements on diet aids and body size in women's magazines (Andersen and DiDomenico, 1992; Silverstein and others, 1986). These media messages especially promote dieting behavior when accompanied by ample food articles and advertisements that encourage individuals to eat. Although females may enjoy forbidden food, they are continually reminded of the need to be thin. This double-edged message of enjoy eating but control your weight reinforces dieting (McLorg and Taub, 1987).

A low rate of success for dieting (Silverstein and others, 1986; Wiseman and others, 1998), combined with consistent pressure to be slim, results in repeated weight-loss efforts by females. Such manipulations of eating and body shape illustrate the tendency of females to view their bodies as objects, subject to modification for an attractiveness standard (Lovejoy, 2001; Rodin and others, 1985). While dieting reflects acceptance of gender norms for body shape, anorexia nervosa and bulimia nervosa represent extreme examples of gender socialization toward slimness.

Women's preoccupation with body shape ranges from mild weight consciousness to fully developed eating disorders (Gordon, 1988; Rodin and others, 1985; Twamley and Davis, 1999).

Individuals with eating disorders exemplify extreme concern about one's weight (Twamley and Davis, 1999), and can be viewed as extensions of the slim body ideal for females (Nasser, 1988; Rodin and others, 1985; Thompson and Heinberg, 1999). Female athletes are especially at risk of eating disorders if their athletic participation includes an aesthetic focus or emphasis on body shape and control (Benson and Taub, 1993; Smolak and others, 2000). Understanding of the gender distribution of anorexia nervosa and bulimia nervosa can be expanded by examining linkages with elements of gender socialization. Analysis of these factors is crucial for explaining the social context in which eating disorders occur.

REFERENCES

American Psychiatric Association, 1994. *Diagnostic and Statistical Manual of Mental Disorders*, 4th ed. Washington, DC: Author.

Andersen, Arnold E., and Lisa DiDomenico, 1992. Diet vs. shape content of popular male and female magazines: A dose-response relationship to the incidence of eating disorders? *International Journal of Eating Disorders* 11: 283–297.

Benson, Rose Ann, and Diane E. Taub, 1993. Using the PRECEDE model for causal analysis of bulimic tendencies among elite women swimmers. *Journal of Health Education* 24: 360–368.

Collins, M. Elizabeth, 1991. Body figure perceptions and preferences among preadolescent children. *International Journal of Eating Disorders* 10: 199–208.

Combs, Margaret R., 1982. By food possessed. *Women's Sports* 4(2): 12–13, 16–17.

Connor-Greene, Patricia Anne, 1988. Gender differences in body weight perception and weight-loss strategies of college students. *Women and Health* 14(2): 27–42.

Drewnowski, Adam, and Doris K. Yee, 1987. Men and body image: Are males satisfied with their body weight? *Psychosomatic Medicine* 49: 626–634.

Fallon, April E., and Paul Rozin, 1985. Sex differences in perceptions of desirable body shape. *Journal of Abnormal Psychology* 94: 102–105.

Garner, David M., Paul E. Garfinkel, Donald Schwartz, and Michael Thompson, 1980. Cultural expectations of thinness in women. *Psychological Reports* 47: 483–491.

Goffman, Erving, 1979. *Gender Advertisements*. New York: Harper & Row.

Gordon, Richard A., 1988. A sociocultural interpretation of the current epidemic of eating disorders. In Barton J. Blinder, Barry F. Chaitin, and Renee S. Goldstein (eds.), *The Eating Disorders: Medical and Psychological Bases of Diagnosis and Treatment*. New York: PMA.

Haller, Ellen, 1992. Eating disorders: A review and update. *Western Journal of Medicine* 157: 658–662.

Harrison, Kristen, 1997. Does interpersonal attraction to thin media personalities promote eating disorders? *Journal of Broadcasting & Electronic Media* 41: 478–500.

Harrison, Kristen, 2000. The body electric: Thin-ideal media and eating disorders in adolescents. *Journal of Communication* 50: 119–143.

Harrison, Kristen, and Joanne Cantor, 1997. The relationship between media consumption and eating disorders. *Journal of Communication* 47: 40–67.

Hausenblas, Heather A., and Albert V. Carron, 2002. Assessing eating disorder symptoms in sport groups: A critique with recommendations for future research. *International Sports Journal* 6: 65–74.

Herbozo, Sylvia, Stacey Tantleff-Dunn, Jessica Gokee-Larose, and J. Kevin Thompson, 2004. Beauty and thinness messages in children's media: A content analysis. *Eating Disorders* 12: 21–34.

Hesse-Biber, Sharlene, Alan Clayton-Matthews, and John A. Downey, 1987. The differential importance of weight and body image among college men and women. *Genetic, Social, and General Psychology Monographs* 113: 511–528.

Johnson, Craig L., Chris Lewis, Susan Love, Marilyn Stuckey, and Linda Lewis, 1983. A descriptive survey of dieting and bulimic behavior in a female high school population. In *Understanding Anorexia Nervosa and Bulimia: Report of the Fourth Ross Conference on Medical Research*. Columbus, OH: Ross Laboratories.

Katzmarzyk, Peter T., and Caroline Davis, 2001. Thinness and body shape of *Playboy* centerfolds from 1978 to 1998. *International Journal of Obesity* 25: 590–592.

Kurman, Lois, 1978. An analysis of messages concerning food, eating behaviors, and ideal body image on prime-time American network television. *Dissertation Abstracts International* 39: 1907A–1908A.

Leon, Gloria R., and Stephen Finn, 1984. Sex-role stereotypes and the development of eating disorders. In Cathy Spatz Widom (ed.), *Sex Roles and Psychopathology*. New York: Plenum.

Lovejoy, Meg, 2001. Disturbances in the social body: Differences in body image and eating problems among African American and white women. *Gender & Society* 15: 239–261.

Mazur, Allan, 1986. U.S. trends in feminine beauty and overadaptation. *Journal of Sex Research* 22: 281–303.

McLorg, Penelope A., and Diane E. Taub, 1987. Anorexia nervosa and bulimia: The development of deviant identities. *Deviant Behavior* 8: 177–189.

Morris, Anne M., and Debra K. Katzman, 2003. The impact of the media on eating disorders in children and adolescents. *Paediatrics & Child Health* 8: 287–289.

Nasser, Mervat, 1988. Culture and weight consciousness. *Journal of Psychosomatic Research* 32: 573–577.

Peterson, Robin T., 1987. Bulimia and anorexia in an advertising context. *Journal of Business Ethics* 6: 495–504.

Pliner, Patricia, Shelly Chaiken, and Gordon L. Flett, 1990. Gender differences in concern with body weight and physical appearance over the life span. *Personality and Social Psychology Bulletin* 16: 263–273.

Polivy, Janet, and C. Peter Herman, 1986. Dieting and binging reexamined: A response to Lowe. *American Psychologist* 41: 327–328.

Raudenbush, Bryan, and Debra A. Zellner, 1997. Nobody's satisfied: Effects of abnormal eating behaviors and actual and perceived weight status on body image satisfaction in males and females. *Journal of Social and Clinical Psychology* 16: 95–110.

Rodin, Judith, Lisa Silberstein, and Ruth Striegel-Moore, 1985. Women and weight: A normative discontent. In Theo B. Sonderegger (ed.), *Nebraska Symposium on Motivation. Vol. 32: Psychology and Gender*. Lincoln: University of Nebraska.

Rozin, Paul, and April Fallon, 1988. Body image, attitudes to weight, and misperceptions of figure preferences of the opposite sex: A comparison of men and women in two generations. *Journal of Abnormal Psychology* 97: 342–345.

Schwartz, Donald M., Michael G. Thompson, and Craig L. Johnson, 1982. Anorexia nervosa and bulimia: The socio-cultural context. *International Journal of Eating Disorders* 1(3): 20–36.

Silverstein, Brett, and Lauren Perdue, 1988. The relationship between role concerns, preferences for slimness, and symptoms of eating problems among college women. *Sex Roles* 18: 101–106.

Silverstein, Brett, Lauren Perdue, Barbara Peterson, and Eileen Kelly, 1986. The role of the mass media in promoting a thin standard of bodily attractiveness for women. *Sex Roles* 14: 519–532.

Smolak, Linda, Sarah K. Murnen, and Anne E. Ruble, 2000. Female athletes and eating problems: A meta-analysis. *International Journal of Eating Disorders* 27: 371–380.

Stein, Jeannine, 1986. Why girls as young as 9 fear fat and go on diets to lose weight. *Los Angeles Times*, October 29, Part v: 1, 10.

Striegel-Moore, Ruth, Lisa R. Silberstein, and Judith Rodin, 1986. Toward an understanding of risk factors for bulimia. *American Psychologist* 41: 246–263.

Sypeck, Mia Foley, James J. Gray, and Anthony H. Ahrens, 2004. No longer just a pretty face: Fashion magazines' depictions of ideal female beauty from 1959 to 1999. *International Journal of Eating Disorders* 36: 342–347.

Taub, Diane E., 1986. *Amphetamine usage among high school senior women, 1976–1982: An evaluation of social bonding theory.* Unpublished doctoral dissertation, Lexington: University of Kentucky.

Taub, Diane E., and Elaine M. Blinde, 1994. Disordered eating and weight control among adolescent female athletes and performance squad members. *Journal of Adolescent Research* 9: 483–497.

Thompson, J. Kevin, and Leslie J. Heinberg, 1999. The media's influence on body image disturbance and eating disorders: We've reviled them, now can we rehabilitate them? *Journal of Social Issues* 55: 339–353.

Thompson, J. Kevin, and Eric Stice, 2001. Thin-ideal internalization: Mounting evidence for a new risk factor for body-image disturbance and eating pathology. *Current Directions in Psychological Science* 10: (5) 181–183.

Thornton, Bill, and Jason Maurice, 1997. Physique contrast effect: Adverse impact of idealized body images for women. *Sex Roles* 37: 433–439.

Thornton, James S., 1990. Feast or famine: Eating disorders in athletes. *Physician and Sportsmedicine* 18(4): 116, 118–122.

Twamley, Elizabeth W., and Mary C. Davis, 1999. The sociocultural model of eating disturbance in young women: The effects of personal attributes and family environment. *Journal of Social and Clinical Psychology* 18: 467–489.

Wiseman, Claire V., James J. Gray, James E. Mosimann, and Anthony H. Ahrens, 1992. Cultural expectations of thinness in women: An update. *International Journal of Eating Disorders* 11: 85–89.

Wiseman, Claire V., Wendy A. Harris, and Katherine A. Halmi, 1998. Eating disorders. *Medical Clinics of North America* 82: 145–159.

11

Unraveling the Gender Knot

ALLAN JOHNSON

Growing up in the age of "Women's Lib," we are likely to believe that there has been rapid change in gender behavior. There has, indeed, been change—but it has been less than rapid. The main reason for this slow change is the belief that gender behavior is mostly related to biology and we would be going against nature. The author calls this belief into question and makes suggestions as to what can be done to bring about greater equality in regard to gender behavior.

As you read this selection, consider the following questions as guides:

1. *In regard to gender behavior, is it true that it has always been this way?*
2. *Do you really want to change the gender behavior system? Why? Why not?*
3. *What kind of gender relations do you see now? In 10 years? In 20 years?*

What is the knot we want to unravel? In one sense, it is the complexity of patriarchy as a system—the tree, from its roots to the smallest outlying twig. It is misogyny and sexist ideology that keep women in their place. It is the organization of social life around core patriarchal principles of control and domination. It is the powerful dynamic of fear and control that keeps the patriarchal engine going. But the knot is also about our individual and collective paralysis around gender issues. It is everything that prevents us from seeing patriarchy and our participation in it clearly, from the denial that patriarchy even exists to false gender parallels, individualistic thinking, and cycles of blame and guilt. Stuck in this paralysis, we can't think or act to help undo the legacy of oppression.

To undo the patriarchal knot we have to undo the knot of our paralysis in the face of it. A good place to begin is with two powerful myths about how change happens and how we can contribute to it.

Myth #1: "It's Always Been This Way, and It Always Will Be"

Given thousands of years of patriarchal history, it's easy to slide into the belief that things have always been this way. Even thousands of years, however, are a far cry from what "always" implies unless we ignore the more than 90 percent of humanity's time on Earth that preceded it. Given all the archaeological evidence pointing to the existence of goddess-based civilizations and the lack of evidence for perpetual patriarchy, there are plenty of reasons to doubt that life has always been organized around male dominance or any other form of oppression.... So, when it comes to human social life,

Source: Johnson, A. *Unraveling the Gender Knot*, copyright © 1997. Reprinted with permission of Temple University Press.

the smart money should be on the idea that nothing has always been this way or any other.

This should suggest that nothing *will* be this way or any other, contrary to the notion that patriarchy is here to stay. If the only thing we can count on is change, then it's hard to see why we should believe for a minute that patriarchy or any other kind of social system is permanent....

Social systems are also fluid. A society isn't some hulking *thing* that sits there forever as it is. Because a system only happens as people participate in it, it can't help but *be* a dynamic process of creation and recreation from one moment to the next.... Since we can always choose paths of greater resistance or create new ones entirely, systems can only be as stable as the flow of human choice and creativity, which certainly isn't a recipe for permanence. In the short run, patriarchy may look stable and unchangeable. But the relentless process of social life never produces the exact same result twice in a row, because it's impossible for everyone to participate in any system in an unvarying and uniform way. Added to this are the dynamic interactions that go on among systems—between capitalism and the state, for example, or between families and the economy—that also produce powerful and unavoidable tensions, contradictions, and other currents of change. Ultimately, systems can't help but change, whether we see it or not....

As the illusion of control becomes more apparent, men start doubting their ability to measure up to patriarchal standards of manhood. We have been here before. At the turn of the twentieth century, there was widespread white male panic in the United States about the "feminization" of society and the need to preserve masculine toughness. From the creation of the Boy Scouts to Teddy Roosevelt's Rough Riders, a public campaign tried to revitalize masculinity as a cultural basis for revitalizing a male-identified society and, with it, male privilege. A century later, the masculine backlash is again in full bloom. The warrior image has re-emerged as a dominant masculine ideal, from *Rambo, Diehard*, and *Under Siege* to right-wing militia groups to corporate takeovers to regional militarism to New Age Jungian archetypes in the new men's movement.[1]

Neither patriarchy nor any other system will last forever. Patriarchy is riddled with internal contradiction and strain. It is based on the false and self-defeating assumption that control is the answer to everything and that the pursuit of more control is always better than contenting ourselves with less. The transformation of patriarchy has been unfolding ever since it emerged seven thousand years ago, and it is going on still. We can't know what will replace it, but we can be confident that patriarchy will go, that it *is* going at every moment. It's only a matter of how quickly, by what means, and toward what alternatives, and whether each of us will do our part to make it happen sooner rather than later and with less rather than more human suffering in the process.

Myth #2: The Myth of No Effect and Gandhi's Paradox

Whether we help change patriarchy depends on how we handle the belief that nothing we do can make a difference, that the system is too big and powerful for us to affect it. In one sense the complaint is valid: if we look at patriarchy as a whole, it's true that we aren't going to make it go away in our lifetime. But if changing the entire system through our own efforts is the standard against which we measure the ability to do something, then we've set ourselves up to feel powerless. It's not unreasonable to want to make a difference, but if we have to see the final result of what we do, then we can't be part of change that's too gradual and long term to allow that. We also can't be part of change that's so complex that we can't sort out our contribution from countless others that combine in ways we can never grasp. Problems like patriarchy are of just that sort, requiring complex and long-term change coupled with short-term work to soften some of its worst consequences. This means that if we're going to be part of the solution to such problems, we have to let go of the idea that change doesn't happen unless we're around to see it happen and that what we do matters only if we make it happen. In other words, if we free ourselves of the expectation of being in control of things, we free ourselves to act and participate in the kind of fundamental change that transforms social life....

[W]e need to get clear about how our choices matter and how they don't. Gandhi once said that nothing we do as individuals matters, but that it's vitally important that we do it anyway. This touches on a powerful paradox in the relationship between society and individuals. In terms of the patriarchyas-tree metaphor, no individual leaf on the tree matters; whether it lives or dies has no effect on much of anything. But collectively, the leaves are essential to the whole tree because they photosynthesize the sugar that feeds it. Without leaves, the tree dies. So, leaves both matter and they don't, just as we matter and we don't. What each of us does may not seem like much, because in important ways, it *isn't* much. But when many people do this work together, they can form a critical mass that is anything but insignificant, especially in the long run. If we're going to be part of a larger change process, we have to learn to live with this sometimes uncomfortable paradox rather than going back and forth between momentary illusions of potency and control and feelings of helpless despair and insignificance.

A related paradox is that we have to be willing to travel without knowing where we're going. We need faith to do what seems right without necessarily knowing the effect that will have. We have to think like pioneers who may know the *direction* they want to move in or what they would like to find, without knowing where they will wind up. Because they are going where they've never been before, they can't know whether they will ever arrive at anything they might consider a destination, much less what they had in mind when they first set out. If pioneers had to know their destination from the beginning, they would never go anywhere or discover anything. In similar ways, to seek out alternatives to patriarchy, it has to be enough to move *away* from social life organized around dominance and control and to move *toward* the certainty that alternatives are possible, even though we may not have a clear idea of what those are or ever experience them ourselves. It has to be enough to question how we think about and experience different forms of power, for example, how

we see ourselves as gendered people, how oppression works and how we participate in it, and then open ourselves to experience what happens next. When we dare ask core questions about who we are and how the world works, things happen that we can't foresee; but they don't happen unless we *move*, if only in our minds. As pioneers, we discover what's possible only by first putting ourselves in motion, because we have to move in order to change our position—and hence our perspective—on where we are, where we've been, and where we *might* go. This is how alternatives begin to appear: to imagine how things might be, we first have to get past the idea that things will always be the way they are.

In relation to Gandhi's paradox, the myth of no effect obscures the role we can play in the long-term transformation of patriarchy. But the myth also blinds us to our own power in relation to other people. We may cling to the belief that there is nothing we can do precisely because we know how much power we do have and are afraid to use it because people may not like it. If we deny our power to affect people, then we don't have to worry about taking responsibility for how we use it or, more significant, how we don't. This reluctance to acknowledge and use power comes up in the simplest everyday situations, as when a group of friends starts laughing at a sexist joke and we have to decide whether to go along. It's a moment in a sea of countless such moments that constitutes the fabric of all kinds of oppressive systems. It is a crucial moment, because the group's seamless response to the joke reaffirms the normalcy and unproblematic nature of it and the sexism behind it. It takes only one person to tear the fabric of collusion and apparent consensus....

Our power to affect other people isn't simply about making them feel uncomfortable. Systems shape the choices that people make primarily by providing paths of least resistance. We typically follow those paths because alternatives offer greater resistance or because we aren't even aware that alternatives exist. Whenever we openly choose a different path, however, we make it possible for people to see both the path of least resistance they're following and the possibility of choosing something else. This is both radical and simple. When most people get on an elevator, for example, they turn and face front without ever thinking why. We might think it's for purely practical reasons—the floor indicators and the door we'll exit through are at the front. But there's more going on than that, as we'd discover if we simply walked to the rear wall and stood facing it while everyone else faced front. The oddness of what we were doing would immediately be apparent to everyone, and would draw their attention and perhaps make them uncomfortable as they tried to figure out why we were doing that. Part of the discomfort is simply calling attention to the fact that we make choices when we enter social situations and that there are alternatives, something that paths of least resistance discourage us from considering. If the possibility of alternatives in situations as simple as where to stand in elevator cars can make people feel uncomfortable, imagine the potential for discomfort when the stakes are higher, as they certainly are when it comes to how people participate in oppressive systems like patriarchy.

If we choose different paths, we usually won't know if we affect other people, but it's safe to assume that we do. When people know that alternatives exist and witness other people choosing them, things become possible that weren't before. When we openly pass up a path of least resistance, we *increase* resistance for other people around that path because now they must reconcile their choice with what they've seen us do, something they didn't have to deal with before. There's no way to predict how this will play out in the long run, and certainly no good reason to think it won't make a difference.

The simple fact is that we affect one another all the time without knowing it…. This suggests that the simplest way to help others make different choices is to make them myself, and to do it openly so they can see what I'm doing. As I shift the patterns of my own participation in patriarchy, I make it easier for others to do so as well, *and harder for them not to*. Simply by setting an example—rather than trying to change them—I create the possibility of their participating in change in their own time and in their own way. In this way I can widen the circle of change without provoking the kind of defensiveness that perpetuates paths of least resistance and the oppressive systems they serve.

It's important to see that in doing this kind of work we don't have to go after people to change their minds. In fact, changing people's minds may play a relatively small part in changing systems like patriarchy. We won't succeed in turning diehard misogynists into practicing feminists. At most, we can shift the odds in favor of new paths that contradict core patriarchal values. We can introduce so many exceptions to patriarchal rules that the children or grandchildren of diehard misogynists will start to change their perception of which paths offer the least resistance. Research on men's changing attitudes toward the male provider role, for example, shows that most of the shift occurs *between* generations, not within them.[2] This suggests that rather than trying to change people, the most important thing we can do is contribute to the slow sea change of entire cultures so that patriarchal forms and values begin to lose their "obvious" legitimacy and normalcy and new forms emerge to challenge their privileged place in social life.

In science, this is how one paradigm replaces another.[3] For hundreds of years, for example, Europeans believed that the stars, planets, and sun revolved around Earth. But scientists such as Copernicus and Galileo found that too many of their astronomical observations were anomalies that didn't fit the prevailing paradigm: if the sun and planets revolved around Earth, then they wouldn't move as they did. As such observations accumulated, they made it increasingly difficult to hang on to an Earth-centered paradigm. Eventually the anomalies became so numerous that Copernicus offered a new paradigm, for which he, and later Galileo, were persecuted as heretics. Eventually, however, the evidence was so overwhelming that a new paradigm replaced the old one.

In similar ways, we can think of patriarchy as a system based on a paradigm that shapes how we think about gender and how we organize social life in relation to it. The patriarchal paradigm has been under attack for several centuries and the defense has been vigorous, with feminists widely regarded as heretics who practice the blasphemy of "male bashing." The patriarchal paradigm weakens in the face of mounting evidence that it doesn't work, and that it produces unacceptable consequences not only for women but, increasingly, for men as well. We help to weaken it by openly choosing alternative paths in our everyday lives and thereby providing living anomalies that don't fit the prevailing paradigm. By our example, we contradict patriarchal assumptions and their legitimacy over and over again. We add our choices and our lives to tip the scales toward new paradigms that don't revolve around control and oppression. We can't tip the scales overnight or by ourselves, and in that sense we don't amount to much. But on the other side of Gandhi's paradox, it is crucial where we "choose to place the stubborn ounces of [our] weight."[4] It is in such small and humble choices that patriarchy and the movement toward something better actually happen.

STUBBORN OUNCES: WHAT CAN WE DO?

What can we do about patriarchy that will make a difference? I don't have the answers, but I do have some suggestions.

Acknowledge That Patriarchy Exists

A key to the continued existence of every oppressive system is people being unaware of what's going on, because oppression contradicts so many basic human values that it invariably arouses opposition when people know about it. The Soviet Union and its East European satellites, for example, were riddled with contradictions that were so widely known among their people that the oppressive regimes fell apart with barely a whimper when given half a chance. An awareness of oppression compels people to speak out, breaking the silence on which continued oppression depends. This is why most oppressive cultures mask the reality of oppression by denying its existence, trivializing it, calling it something else, blaming it on those most victimized by it, or drawing attention away from it to other things....

It's one thing to become aware and quite another to stay that way. The greatest challenge when we first become aware of a critical perspective on the world is simply to hang on to it. Every system's

paths of least resistance invariably lead *away* from critical awareness of how the system works. Therefore, the easiest thing to do after reading a book like this is to forget about it. Maintaining a critical consciousness takes commitment and work; awareness is something we either maintain in the moment or we don't. And the only way to hang on to an awareness of patriarchy is to make paying attention to it an ongoing part of our lives.

Pay Attention

Understanding how patriarchy works and how we participate in it is essential for change. It's easy to have opinions; it takes work to know what we're talking about. The easiest place to begin is by reading, and making reading about patriarchy part of our lives. Unless we have the luxury of a personal teacher, we can't understand patriarchy without reading, just as we need to read about a foreign country before we travel there for the first time, or about a car before we try to work under the hood. Many people assume they already know what they need to know about gender since everyone has a gender, but they're usually wrong. Just as the last thing a fish would discover is water, the last thing we'll discover is society itself and something as pervasive as gender dynamics. We have to be open to the idea that what we think we know about gender is, if not wrong, so deeply shaped by patriarchy that it misses most of the truth. This is why feminists talk with one another and spend time reading one another's work—seeing things clearly is tricky business and hard work. This is also why people who are critical of the status quo are so often self-critical as well: they know how complex and elusive the truth really is and what a challenge it is to work toward it. People working for change are often accused of being orthodox and rigid, but in practice they are typically among the most self-*critical* people around....

Reading, though, is only a beginning. At some point we have to look at ourselves and the world to see if we can identify what we're reading about. Once the phrase "paths of least resistance" entered my active vocabulary, for example, I started seeing them all over the place. Among other things,

I started to see how easily I'm drawn to asserting control as a path of least resistance in all kinds of situations. Ask me a question, for example, and the easiest thing for me to do is offer an answer whether or not I know what I'm talking about. "Answering" is a more comfortable mode, an easier path, than admitting I don't know or have nothing to say.[5] The more aware I am of how powerful this path is, the more I can decide whether to go down it each time it presents itself. As a result, I listen more, think more, and talk less than I used to....

Little Risks: Do Something

The more we pay attention to what's going on, the more we will see opportunities to do something about it. We don't have to mount an expedition to find those opportunities; they're all over the place, beginning in ourselves. As I became aware of how I gravitated toward controlling conversations, for example, I also realized how easily men dominate group meetings by controlling the agenda and interrupting, without women objecting to it. This pattern is especially striking in groups that are mostly female but in which most of the talking nonetheless comes from a few men. I would find myself sitting in meetings and suddenly the preponderance of male voices would jump out at me, an unmistakable hallmark of male privilege in full bloom. As I've seen what's going on, I've had to decide what to do about this little path of least resistance and my relation to it that leads me to follow it so readily. With some effort, I've tried out new ways of listening more and talking less. At times it's felt contrived and artificial, like telling myself to shut up for a while or even counting slowly to ten (or more) to give others a chance to step into the space afforded by silence. With time and practice, new paths have become easier to follow and I spend less time monitoring myself. But awareness is never automatic or permanent, for patriarchal paths of least resistance will be there to choose or not as long as patriarchy exists.

As we see more of what's going on, questions come up about what goes on at work, in the media, in families, in communities, in religion, in government, on the street, and at school—in short, just about

everywhere. The questions don't come all at once (for which we can be grateful), although they sometimes come in a rush that can feel overwhelming. If we remind ourselves that it isn't up to us to do it all, however, we can see plenty of situations in which we can make a difference, sometimes in surprisingly simple ways. Consider the following possibilities:

- *Make noise, be seen.* Stand up, volunteer, speak out, write letters, sign petitions, show up. Like every oppressive system, patriarchy feeds on silence. Don't collude in silence....

- *Find little ways to withdraw support from paths of least resistance and people's choices to follow them, starting with ourselves.* It can be as simple as not laughing at a sexist joke or saying we don't think it's funny; or writing a letter to the editor objecting to sexism in the media....

- *Dare to make people feel uncomfortable, beginning with ourselves.* At the next local school board meeting, for example, we can ask why principals and other administrators are almost always men (unless your system is an exception that proves the rule), while the teachers they control are mostly women. Consider asking the same thing about church, workplaces, or local government....

It may seem that such actions don't amount to much until we stop for a moment and feel our resistance to doing them—our worry, for example, about how easily we could make people feel uncomfortable, including ourselves. If we take that resistance to action as a measure of power, then our potential to make a difference is plain to see. The potential for people to feel uncomfortable is a measure of the power for change inherent in such simple acts as of not going along with the status quo.

Some will say that it isn't "nice" to make people uncomfortable, but oppressive systems like patriarchy do a lot more than make people feel uncomfortable, and it certainly isn't "nice" to allow them to continue unchallenged. Besides, discomfort is an unavoidable part of any meaningful process of education. We can't grow without being willing to challenge our assumptions and take ourselves to the edge of our competencies, where we're bound to feel uncomfortable.

If we can't tolerate ambiguity, uncertainty, and discomfort, then we'll never go beneath the superficial appearance of things or learn or change anything of much value, including ourselves.

- *Openly choose and model alternative paths.* As we identify paths of least resistance—such as women being held responsible for child care and other domestic work—we can identify alternatives and then follow them openly so that other people can see what we're doing. Patriarchal paths become more visible when people choose alternatives, just as rules become more visible when someone breaks them. Modeling new paths creates tension in a system, which moves toward resolution....

- *Actively promote change in how systems are organized around patriarchal values and male privilege.* There are almost endless possibilities here because social life is complicated and patriarchy is everywhere. We can, for example,

  - Speak out for equality in the workplace.
  - Promote diversity awareness and training.
  - Support equal pay and promotion for women.
  - Oppose the devaluing of women and the work they do, from the dead-end jobs most women are stuck in to the glass ceilings that keep women out of top positions.
  - Support the well-being of mothers and children and defend women's right to control their bodies and their lives.
  - Object to the punitive dismantling of welfare and attempts to limit women's access to reproductive health services.
  - Speak out against violence and harassment against women wherever they occur, whether at home, at work, or on the street.
  - Support government and private support services for women who are victimized by male violence.
  - Volunteer at the local rape crisis center or battered women's shelter.
  - Call for and support clear and effective sexual harassment policies in workplaces, unions, schools, professional associations, churches,

and political parties, as well as public spaces such as parks, sidewalks, and malls.
  - Join and support groups that intervene with and counsel violent men.
  - Object to theaters and video stores that carry violent pornography....
  - Ask questions about how work, education, religion, family, and other areas of family life are shaped by core patriarchal values and principles....

- *Because the persecution of gays and lesbians is a linchpin of patriarchy, support the right of women and men to love whomever they choose.* Raise awareness of homophobia and heterosexism....

- *Because patriarchy is rooted in principles of domination and control, pay attention to racism and other forms of oppression that draw from those same roots.*

[P]atriarchy isn't problematic just because it emphasizes *male* dominance, but because it promotes dominance and control as ends in themselves. In that sense, all forms of oppression draw support from common roots, and whatever we do that draws attention to those roots undermines *all* forms of oppression. If working against patriarchy is seen simply as enabling some women to get a bigger piece of the pie, then some women probably will "succeed" at the expense of others who are disadvantaged by race, class, ethnicity, and other characteristics.... [I]f we identify the core problem as *any* society organized around principles of control and domination, then changing *that* requires us to pay attention to all of the forms of oppression those principles promote. Whether we begin with race or gender or ethnicity or class, if we name the problem correctly, we'll wind up going in the same general direction.

- *Work with other people.* This is one of the most important principles of participating in social change. From expanding consciousness to taking risks, it makes all the difference in the world to be in the company of people who support what we are trying to do. We can read and talk about books and issues and just plain hang out with other people who want to understand and do something about patriarchy. Remember that the modern women's movement's roots were in consciousness-raising groups in which women did little more than sit around and talk about themselves and their lives and try to figure out what that had to do with living in patriarchy. It may not have looked like much at the time, but it laid the foundation for huge social movements. One way down this path is to share a book like this one with someone and then talk about it. Or ask around about local groups and organizations that focus on gender issues, and go find out what they're about and meet other people. ...Make contact; connect to other people engaged in the same work; do whatever reminds us that we aren't alone in this.

- *Don't keep it to ourselves.* A corollary of looking for company is not to restrict our focus to the tight little circle of our own lives. It isn't enough to work out private solutions to social problems like patriarchy and other forms of oppression and keep them to ourselves. It isn't enough to clean up our own acts and then walk away, to find ways to avoid the worst consequences of patriarchy at home and inside ourselves and think that's taking responsibility. Patriarchy and oppression aren't personal problems and they can't be solved through personal solutions. At some point, taking responsibility means acting in a larger context, even if that means just letting one other person know what we're doing. It makes sense to start

with ourselves; but it's equally important not to *end* with ourselves.

If all of this sounds overwhelming, remember again that we don't have to deal with everything. We don't have to set ourselves the impossible task of letting go of everything or transforming patriarchy or even ourselves. All we can do is what *we* can *manage* to do, secure in the knowledge that we're making it easier for other people—now and in the future—to see and do what *they* can do. So, rather than defeat ourselves before we start:

- *Think small, humble, and doable rather than large, heroic, and impossible.* Don't paralyze yourself with impossible expectations. It takes very little to make a difference....

- *Don't let other people set the standard for us.*

- *Start where we are and work from there.... set reasonable goals* ("What small risk for change will I take *today*?"). As we get more experienced at taking risks, we can move up our lists....

In the end, taking responsibility doesn't have to be about guilt and blame, about letting someone off the hook or being on the hook ourselves. It is simply to acknowledge our obligation to make a contribution to finding a way out of patriarchy, and to find constructive ways to act on that obligation. We don't have to do anything dramatic or Earth-shaking to help change happen. As powerful as patriarchy is, like all oppressive systems, it cannot stand the strain of lots of people doing something about it, beginning with the simplest act of speaking its name out loud.

NOTES

1. See James William Gibson, *Warrior Dreams: Violence and Manhood in Post-Vietnam America* (New York: Hill and Wang, 1994).

2. J. R. Wilkie, "Changes in U.S. Men's Attitudes Towards the Family Provider Role, 1972–1989." *Gender & Society* 7, no. 2 (1993): 261–279.

3. The classic statement of how this happens is by Thomas S. Kuhn, *The Structure of Scientific*

Revolutions (Chicago: University of Chicago Press, 1970).

4. This is a line from a poem by Bonaro Overstreet that was given to me by a student many years ago. I have not been able to locate the source.

5. Or, as someone once said to me about a major corporation that valued creative thinking, "It's not OK to say you don't know the answer to a question here."

12

A Degendered Society?

MICHAEL KIMMEL

The author of this article claims that the future of gender differences is intimately tied to the future of gender inequality. What does this statement mean? If the statement is true, what is possible in regard to future gender differences?

As you read this selection, consider the following questions as guides:

1. *Why do you believe that the lessening of gender inequality is moving too slowly? Too rapidly?*
2. *What kinds of adjustments will men and women have to make to deal with gender equality?*
3. *Is it possible to have a totally degendered society? Why not?*

We sit perched at the beginning of a new millennium, looking over into an uncharted expanse of the future. What kind of society do we want to live in? What will be the gender arrangements of that society?

To see gender differences as intransigent leads also to a political resignation about the possibilities of social change and increased gender equality. Those who proclaim that men and women come from different planets would have you believe that the best we can hope for is a sort of interplanetary détente, an uneasy truce in which we exasperatingly accept the inherent and intractable foibles of the other sex, a truce mediated by ever-wealthier psychological interpreters who can try and decode their impenetrable language.

I think the evidence is clear that women and men are far more alike than we are different, and that we

need far fewer cosmic interpreters and far more gender equality to enable both women and men to live the lives we want to live. The future of gender differences is intimately tied to the future of gender inequality. As gender inequality is reduced, the differences between women and men will shrink.

And besides, the interplanetary model of gender differences entirely ignores the historical record. For the past century, we have steadily moved to lessen gender inequality—by removing barriers to women's entry into all arenas of the public sphere, protecting women who have been victimized by men's violent efforts to delay, retard, or resist that entry. And as we have done so, we have found that women can perform admirably in arenas once believed to be only suitable for men, and that men can perform admirably in arenas once held to be exclusively women's domain. Don't believe me; ask

Source: The Gendered Society, Oxford University Press, 2004.

those women surgeons, lawyers, and pilots. And ask those male nurses, teachers, and social workers, as well as all those single fathers, if they are capable of caring for their children.

In this book, I've made several arguments about our gendered society. I've argued that women and men are more alike than we are different, that we're not at all from different planets. I've argued that it is gender inequality that produces the differences we do observe, and that also produces the cultural impulse to search for such differences, even when there is little or no basis for them in reality. I've also argued that gender is not a property of individuals that is accomplished by socialization, but a set of relationships produced in our social interactions with one another and within gendered institutions, whose formal organizational dynamics reproduce gender inequality and produce gender differences.

I've also pointed to evidence of a significant gender convergence taking place over the past half-century. Whether we looked at sexual behavior, friendship dynamics, efforts to balance work and family life, or women's and men's experiences and aspirations in education or the workplace, we find the gender gap growing ever smaller. (The lone exception to this process.... is violence.)

To celebrate this gender convergence in behavior and attitudes is not to advocate degendering people. It's not a plea for androgyny. Some psychologists have proposed androgyny as a solution to gender inequality and gender differences. It implies a flattening of gender differences, so that women and men will think, act, and behave in some more "neutralized" gender nonspecific ways. "Masculinity" and "femininity" will be seen as archaic constructs as everyone becomes increasingly "human."

Such proposals take a leap beyond the ultimately defeatist claims of immutable difference offered by the intergalactic theorists. After all, proponents of androgyny at least recognize gender differences as socially constructed and that change is possible.

But androgyny remains unpopular as a political or psychological option because it would eliminate differences between people, mistaking equality for sameness. To many of us, the idea of sameness feels coercive, a dilution of difference into a bland, tasteless amalgam in which individuals would lose their distinctiveness. It's like Hollywood's vision of communism as a leveling of all class distinctions into a colorless, amorphous mass in which everyone would look, act, and dress the same—as in those advertisements that feature poorly but identically dressed Russians. Androgyny often feels like it would enforce life on a flat, and ultimately barren degendered landscape. Is the only way for women and men to be equal to become the same? Can we not imagine a vision of equality based on respect for and embracing of difference?

Fears about androgyny confuse gendered people with gendered traits. It's not that women and men need to be more like each other than we already are but that all the psychological traits and attitudes and behaviors that we, as a culture, label as "masculine" or "feminine" need to be redefined. These traits and attitudes, after all, also carry positive and negative values, and it is through this hierarchy, this unequal weighting, that gender inequality becomes so deeply entwined with gender difference. To degender people does not by itself eliminate gender inequality.

In fact, calls for androgyny paradoxically reify the very gender distinctions that they seek to eliminate. Advocates frequently urge men to express more of their "feminine" sides; women, to express more of their "masculine" sides. Such exhortations, frankly, leave me deeply insulted.

Let me give you an example: As I was sitting in my neighborhood park a couple of years ago with my newborn son in my arms, a passerby commented, "How wonderful it is to see men these days expressing their feminine sides." I growled, underneath my conspicuously false smile. While I tried to be pleasant, what I wanted to say was this: "I'm not expressing anything of the sort, ma'am. I'm being tender, and loving and nurturing toward my child. As far as I can tell, I'm expressing my *masculinity*!"

Why, after all, are love, nurturing, and tenderness defined as feminine? Why do I have to be expressing the affect of the other sex in order to have access to what I regard as human emotions? Being a man, everything I do expresses my masculinity. And

I'm sure my wife would be no less insulted if, after editing a particularly difficult article, or writing a long, involved essay, she were told how extraordinary and wonderful it is to see women expressing their masculine sides—as if competence, ambition, and assertiveness were not human properties to which women *and* men could equally have access.

Love, tenderness, nurturing: competence, ambition, assertion—these are *human* qualities, and all human beings—both women and men—should have equal access to them. And when we do express them, we are expressing, respectively, our gender identities, not the gender of the other. What a strange notion indeed, that such emotions should be labeled as masculine or feminine, when they are so deeply human, and when both women and men are so easily capable of so much fuller a range of feelings.

Strange, and also a little sad. "Perhaps nothing is so depressing an index of the inhumanity of the male supremacist mentality as the fact that the more genial human traits are assigned to the underclass: affection, response to sympathy, kindness, cheerfulness," was the way feminist writer Kate Millett put it in her landmark book, *Sexual Politics*, first published in 1969.[1]

So much has changed since then. The gendered world that I inhabit is totally unlike that of my parents. My father could have gone to an all-male college, served in an all-male military, and spent his entire working life in an all-male work environment. Today that world is but a memory. Women have entered every workplace, the military and its training academies (both federal and state supported), and all but three or four colleges today admit women. Despite persistent efforts from some political quarters to turn back the clock to the mid-nineteenth century, those changes are permanent; women will not go back to the home where some people think they belong.

These enormous changes will only accelerate in the next few decades. The society of the third millennium will increasingly degender traits and behaviors without degendering people. We will still be women and men, equal yet capable of appreciating our differences, different yet unwilling to use those differences as the basis for discrimination.

Imagine how quickly the pace of that change might accelerate if we continue to degender traits, not people. What if little boys and girls saw their mothers and their fathers go off to work in the morning, with no compromise to their masculinity or femininity. Those little boys and girls would grow up thinking that having a job—being competent, earning a living, striving to get ahead—was something that *grownups* did, regardless of whether they were male or female grownups. Not something that men did, and women did only with guilt, social approbation, and sporadically and irregularly depending on their fertility. "And when I grow up," those children would say, "I'm going to have a job also."

And when both mothers and fathers are equally loving and caring and nurturing toward their children, when nurture is something that *grownups* do—and not something that mothers do routinely and men do only during halftime on Saturday afternoon—then those same children will say to themselves, "And when I get to be a grownup, I'm going to be loving and caring toward my children."

Such a process may sound naively optimistic, but the signs of change are everywhere around us. In fact, the historical evidence points exactly in that direction. It was through the dogged insistence of that nineteenth-century ideology, the separation of spheres, that two distinct realms for men and women were imposed, with two separate sets of traits and behaviors that accompany each sphere. This was the historical aberration, the anomaly—its departure from what had preceded it and from the "natural" propensity of human beings goes a long way in explaining the vehemence with which it was imposed. Nothing so natural or biologically determined has to be so coercive.

The twentieth century witnessed the challenge to separate spheres, undertaken, in large part, by those who were demoted by its ideological ruthlessness—women. That century witnessed an unprecedented upheaval in the status of women, possibly the most significant transformation in gender relations in world history. From the rights to vote and work, asserted early in the century, to the rights to enter every conceivable workplace, educational institution, and the military, in the latter half, women have shaken the foundations of the gendered society. And they were left at the end of the century having accomplished half a revolution—a transformation of their opportunities to be workers and mothers.

This half-finished revolution has left many women frustrated and unhappy. For some reason, they remain unable to "have it all"—to be good mothers and also to be effective and ambitious workers. With astonishing illogic, some pundits explain women's frustrations as stemming not from the continued resistance of men, the intransigence of male-dominated institutions to accept them, or the indifference of politicians to enact policies that would enable these women to balance their work and family lives, but rather to the effort of women to expand their opportunities and to claim a full share of humanity. It is a constant source of amazement how many women have full-time jobs exhorting women not to take full-time jobs.

The second half of the transformation of gender is just beginning, and will be, I suspect, far more difficult to accomplish than the first. That's because there was an intuitively obvious ethical imperative attached to enlarging the opportunities for and eliminating discrimination against women. But the transformation of the twenty-first century involves the transformation of men's lives.

Men are just beginning to realize that the "traditional" definition of masculinity leaves them unfulfilled and dissatisfied. While women have left the home, from which they were "imprisoned" by the ideology of separate spheres, and now seek to balance work and family lives, men continue to search for a way back into the family, from which they were exiled by the same ideology. Some men express their frustration and confusion by hoping and praying for a return to the old gender regime, the very separation of spheres that made both women *and* men unhappy. Others join various men's movements, such as PromiseKeepers, the Million Man March, or troop off to a mythopoetic men's retreat in search of a more resonant, spiritually fulfilling definition of masculinity.

The nineteenth-century ideology of separate spheres justified gender inequality based on putative natural differences between the sexes. What was normative—enforced by sanction—was asserted to be normal, as part of the nature of things. Women have spent the better part of a century making clear that such an ideology did violence to their experiences, effacing the work outside the home that women actually performed, and enforcing a definition of femininity that allowed only partial expression of their humanity.

It did the same for men, of course—valorizing some emotions and experiences, discrediting others. As with women, it left men with only partially fulfilled lives. Only recently, though, have men begun to chafe at the restrictions that such an ideology placed on their humanity.

At the turn of the twenty-first century, it might be wise to recall the words of a writer at the turn of the twentieth century. In a remarkable essay written in 1917, the New York City writer Floyd Dell spelled out the consequences of separate spheres for both women and men.

> When you have got a woman in a box, and you pay rent on the box, her relationship to you insensibly changes character. It loses the fine excitement of democracy. It ceases to be a companionship, for companionship is only possible in a democracy. It is no longer a sharing of life together—it is a breaking of life apart. Half a life—cooking, clothes and children; half a life—business, politics and baseball. It doesn't make much difference which is the poorer half. Any half, when it comes to life, is very near to none at all.

Like feminist women, Dell understands that these separate spheres that impoverish the lives of both women and men are also built upon gender inequality. (Notice how he addresses his remarks to men who "have got a woman in a box.") Gender inequality produced the ideology of separate spheres, and the ideology of separate spheres, in turn, lent legitimacy to gender inequality. Thus, Dell argues in the opening sentence of his essay that "feminism will make it possible for the first time for men to be free."[2]

The direction of the gendered society in the new century and the new millennium is not for women and men to become increasingly *similar,* but for them to become more *equal*. For those traits and behaviors heretofore labeled as masculine and feminine—competence and compassion, ambition and affection—to be accepted as distinctly human qualities, accessible to both women and men who are grown-up enough to claim them. It suggests a form of gender proteanism—a temperamental and psychological flexibility, the ability to adapt to one's environment with a full range of emotions and abilities. The protean self, articulated by psychiatrist Robert Jay Lifton, is a self that can embrace difference, contradiction, and complexity, a self that is mutable and flexible in a rapidly changing world.[3] Such a transformation does not require that men and women become more like each other, but, rather, more deeply and fully themselves.

Personally, I'm optimistic. Not long ago, I was playing a game with my then three-year-old son Zachary, which we call "opposites." You know the game: I say a word, and he tells me the opposite. It's simple and fun, and we have a great time playing it. One evening, my mother was visiting, and the three of us were walking in our neighborhood park playing Opposites. Scratchy/smooth, tall/short, high/low, fast/slow—you get the idea. Then my mother asked, "Zachary, what's the opposite of boy?"

My whole body tensed. Here it comes, I thought, Mars and Venus, gender binary opposition, all the things I argue against in this book.

Zachary looked up at his grandmother and said, "Man."

Here, at last, on Planet Earth, there's one small voice that knows we're not from Mars and Venus, after all. And, as he ages, he'll also learn that the differences we see are created by the inequalities we have inherited.

NOTES

1. Kate Millett, *Sexual Politics* (New York: Random House, 1969).

2. Floyd Dell, "Feminism for Men," *The Masses*, February 1917; reprinted in *Against the Tide: Profeminist Men in the United States, 1776–1990* (*A Documentary History*), ed. M. S. Kimmel and T. Mosmiller (Boston: Beacon Press, 1992).

3. Robert Jay Lifton, *The Protean Self* (New York: Basic Books, 1994). See also Cynthia Fuchs Epstein, "The Multiple Realities of Sameness and Difference: Ideology and Practice," *Journal of Social Issues* 53, no. 2 (1997).

Chapter 4

Culture

Our Way of Life

Dr. Alan W. McEvoy

Culture—everyone has it! In fact, humans wouldn't be able to survive without it. Culture refers to our *total way of life*. It is both concrete and abstract. Material culture refers to the "things" we can see and touch—our clothes, food, cars, houses, books. Nonmaterial culture includes our language, beliefs, values, politics, and religious ideologies; we may not see them, but they guide everything we do and think.

Cultures around the world differ greatly because they evolved in different times and geographic places. Groups of humans developed patterns that seemed necessary to survive at the particular time and place. We catch a glimpse of this idea in the photo of cave paintings, depicting important elements of culture in the lives of these early humans. Although patterns may outlive their original purpose or usefulness, they often persist over centuries, holding a group together because of a shared way of life to which people adhere.

Thus human behavior is both patterned and orderly because within our society we are taught to follow similar rules of behavior (*norms*) and to cherish similar objects and behaviors (*values*). The importance of culture is indicated by the fact that most human behavior is learned within a cultural context. It is through *cultural relativism*—looking at other cultures in an objective manner—that we attempt to understand learned cultural patterns and behaviors by considering the functions they serve for society.

In the first reading in this chapter, Michael Jindra sets the stage with his discussion of culture—how and why it matters to humans. His focus is on diversity of cultures and what this means for human groups that come in contact with one another.

Next, Horace Minor gives us an anthropological look at what appears to be a "primitive" group by examining the societal needs served by the unusual attitudes and practices of the Nacirema toward the human body. If, as you read this selection, you feel glad to be an American while wondering about the "silly" actions of the Nacirema, then your ethnocentrism is showing. (Read the article carefully, especially the italicized words. You should find it amusing.) The reading by Miner is also a good example of the points made by Mills in Chapter 1. That is, our beliefs are represented by the social system of which we are a part, and they affect subsequent behavior.

Because we are taught the norms, values, language, and beliefs (folklore, legends, proverbs, religion) of our own culture, we frequently find it difficult to see our culture objectively. What we do routinely we accept as right without question, and possibly even without understanding. This *ethnocentrism*—the belief that one's own culture is superior to others—can make it difficult to accept

the different ways of others and to change our own ways. It is those two factors of learned behavior and ethnocentrism that lead to cultural constraints on our thoughts and behavior. Ethnocentrism is common despite the fact that many, if not most, of the material items used in any given culture were neither invented nor discovered in that culture but rather were adopted from other societies through *cultural diffusion*—the spread of cultural behavior and materials from one society to another.

Unfortunately, ethnocentrism can carry over into hostilities toward different groups of people. This hostility exists because groups feel strongly about the rightness of their own cultural beliefs, values, attitudes, and behaviors. A good example of cultural misunderstandings resulting from ethnocentrism is seen in the reading by Marvin Harris, "India's Sacred Cow." It makes the point that cultural practices have origins and can be explained, however strange they seem to outsiders.

As we see from the first article, "Culture Matters," societies develop cultural patterns over time, some of which last beyond their initial purposes. However, as the world changes due to advanced communication, transportation and travel, and new inventions (especially technology), traditional cultural values are challenged. We see evidence of this change in pressures on societies from globalization, with rich and powerful countries presenting alien values that threaten traditional societies and their value systems. Stephen Marglin discusses the threat to non-Western cultures from powerful ideologies in the world.

As you read this chapter, ask yourself these questions: What problems can cultural diversity raise? Why do the Nacirema have an apparently "pathological" concern with the body, and what are the implications of this concern? Why does our liking of the familiar sometimes lead to violent dislike of groups with dissimilar cultures? How can stereotypes shape our images, sometimes falsely, of cultural patterns? Why do cultural patterns vary so widely, sometimes to the disadvantage of members of society? And what happens when cultural values clash?

13

Culture Matters: Diversity in the United States and Its Implications

MICHAEL JINDRA

Culture is what determines much of our lives. Jindra both explains the role of culture in our lives and explores how the diversity in the world can result in misunderstandings and even conflicts.

As you read, ask yourself the following questions:

1. *Which parts of your life are influenced by culture?*
2. *How does culture influence your "emotional style"?*
3. *How do different "communication styles" illustrate the influence of culture?*
4. *How can cultural diversity help and hurt possibilities for diverse groups to get along?*

GLOSSARY **Focal institutions** Systems that embody long-standing customs and history and help outsiders to understand the important aspects of a culture.
Communication styles Differences in the ways people communicate within and across cultures.

WHAT IS CULTURE?

Culture is a complex topic, but simply put, culture is a social group's distinctive way of life, the beliefs and practices that members find "normal" and correct (Shweder 2003). Something as simple as a proper greeting turns out to reveal key features of a culture. Formal bowing in Japan tells us about the differing statuses of the individuals. Many cultures use different words when greeting someone, depending on whether one is meeting someone younger or older, a man or a woman, a friend or a stranger. Even languages such as Spanish, French, and German distinguish between formal and informal relationships, while Americans don't make such distinctions. This matches the American refusal to distinguish between "friends" and "acquaintances" in linguistic practice. When calling someone a "friend" in most other countries, it means we have a close, ongoing relationship with that person, while in the United States we use the term rather loosely, when we really mean "acquaintance" rather than "friend."

Source: Priest, Robert J., and Alvaro L. Nieves. 2007. *This Side of Heaven: Race, Ethnicity, and Christian Faith*, pp. 64–71. Oxford University Press.

This has caused confusion for international visitors who may assume something different from most Americans, and the topic is usually part of any "introduction to American culture" orientation for newcomers.

Likewise, many people from non-Western countries are shocked at the way many Americans pamper and even humanize pets, especially dogs and cats. Dogs in many cultures are not thought of as pets or companions, but instead may be used for hunting, protection, or as roving garbage collectors. In some cultures, dogs are what one eats. This is shocking to most Americans, who arbitrarily include dogs, cats, and horses in the category of "animals that are almost human," and thus should not be eaten (Sahlins 1976, 171). Most people in the world consider pet cemeteries, grooming salons, pet clothing, pet therapists, and even special food for pets as quite strange, and may critically note at the same time how senior citizens live in nursing homes instead of with family. Many are equally shocked to find that some senior citizens prefer this to living with family because they value their privacy and want to avoid being dependent on family. When we live abroad, we quickly learn that notions and practices of privacy and dependency can vary radically.

Cultures have developed in different locales, as geography and other factors facilitate the development of distinctive languages whose unique and key concepts and practices intimately tie back into and reinforce notions of cosmology and ethical practice. Differing key ideas and narratives (life is a circle, an eternal return; life is a linear path toward some better end; humans are intimately tied up with creation; humans are stewards or masters of creation) have crucial implications for practice and material culture (Smith 2003, 48ff.). Everything from art to the economy is affected. People who have close interactions will tend to learn and share similar patterns of discourse, behavior, and thought.

Different cultures and subcultures may value different emotional styles (Middleton 2003, 37–38). People in cultures differ over when to express emotions, and with what intensity. Many

Asian cultures, for instance, value social grace and cohesion over personal expressiveness and independence (Hsu 1963). Recent studies also show that East Asians are more aware of social context than most Americans, who tend to focus on individual action. When describing scenes they have observed, East Asians describe the surroundings, whereas Americans focus on the movement (Nisbett 2003). Again, this is not due to any genetic difference, but to processes of socialization into one's culture. In Africa, social interdependence is stressed, while Americans stress independence and self-reliance. This has enormous implications for behavior, and can lead to many misunderstandings when individuals from the two cultures interact, especially over money matters (Maranz 2001).

Cultural values also help us understand how societies are organized and develop. The dominant American values of individualism and self-reliance (Holmes and Holmes 2002; see Kusserow 2004 for a more complex view of individualism) have helped produce a vibrant economy, but also family instability, a bureaucratic state that needs to provide a safety net, and a focus on rights over responsibilities; orientations toward progress, science, and exploration have led to many accomplishments and improvements, but also to occasional aggression, as in the case of Native Americans; and a focus on youth, mobility, and productivity that serves the economy but can lead to the devaluation of the elderly and their segregation from the rest of society (Kottak and Kozaitis 2003, 11–12). In sum, when we understand the role of culture, we take seriously how people's values and beliefs are connected with their practices and ways of life.

Culture is not a simple or uniform thing, however, because one may have competing definitions and practices of what is correct. Even more generally agreed-upon values may be in tension with one another. In the United States, notions of "freedom" and "family" are strong, but they are in competition, since many Americans use their freedom to move away from their families and pursue careers. Also, most cultures today are not systems in the strict sense of the word, but composites or fragments of diverse origins. In the United States, different traditions (mostly from Europe) contributed to create the dominant culture (Huntington 2004). Most white Americans take these cultural values and practices for granted, and don't think of them as anything but "normal."

Cultures also change over time. While this has always happened, the process has been speeded up as technology has allowed more mobility, social groups come more in contact with other cultures, and the process of culture change becomes more intense. Borrowings of cultural elements occur, and the boundaries between cultures become more blurred (Bashkow 2004). Though intensified in recent times, this process has been occurring for a long time, especially in the Americas, with its fusion of European, Native American, and African histories (Walker 2001). Technology, social forces such as the media and advertising, and social movements also influence how cultures change. Over the last decades, white youth culture has continually appropriated portions of minority cultures, particularly inner-city black culture, in the belief that they are partaking of "black authenticity." Through media, we can consume other cultures or, more accurately, select portions of other cultures (Samuels 1991).

FOCAL INSTITUTIONS

One way to understand the culture of a distinct social group is to examine specific cultural activities or symbols that offer a "window" into the culture. A classic example from anthropology is the Balinese (Indonesia) cockfight, which reveals key notions about gender, status, and worldview (Geertz 1973, 412ff.). Observing how people from different cultures deal with death always reveals fundamental notions about relationships and worldview (Corr, Nabe, and Corr 2000).

One can't always identify a "cultural focus" (Herskovits 1964, 182) in every culture, but important events, such as the life cycle rituals of weddings or funerals, reveal central concerns of specific cultures. In the United-States, the court system

could be a focal institution (Vansina 1994, 297n; Hammond 1998, 108). It embodies the long-standing customs and history of English common law, and is rooted in a central document, the Constitution, to define the rules by which Americans operate and to which new Americans swear allegiance. The court's contemporary importance signifies a move away from local authorities to centralized, bureaucratic rule, and it is relied upon to settle key moral issues, definitions of life and death, and redress of injustice. It prioritizes the individual and focuses on rights, rather than responsibilities. It certainly doesn't tell us everything about American culture, but it does give us an insight into central themes. For someone seeking to understand the United States, the court system is certainly one of the places to examine.

Likewise, for Americans involved with an unfamiliar culture, it pays to look for focal institutions or central events. Among Latinos (especially Mexicans), an institution that reveals key themes is the *Quinceañera* (fifteenth birthday) celebration of Latina womanhood. The event serves to introduce the girl to society as a young woman. The *Quinceañera* is hosted by the family, incorporates the relatives and friends of the family, and usually involves a Catholic mass. Preparations for the mass and following reception begin well in advance of the event, and the expense can be tremendous. The event tells us about gender ideals and relations, family status and connections, religious values and, more recently, the influence of popular culture. Like most rituals, the *Quinceañera* is not a tradition-bound event that has never changed. Though it may have roots in a combination of indigenous Indian and Spanish custom (Deiter 2002, 36), it actually became popular only in the twentieth century. Thus, it also reveals how culture changes. American popular culture, for instance, has become increasingly important among Latino youth, and the *Quinceañera* shows the tension between popular culture and Latino Catholicism. The secular focus of the event as a "debut" and show of ostentatious consumption sometimes overwhelms the religious meaning of the event as a demonstration of fidelity, and some Catholic churches are reacting by attempting to put a stronger Christian focus on the event (Deiter 2002).

When looking at focal institutions, one must always be aware of the partial, perspectival view they give us. The *Quinceañera* focuses more on female ideals. One would need to look elsewhere to find out more about male life, such as the bullfight, which contains ideas of hierarchy, honor, valor, style, and machismo (Graña 1987). It is, however, popular only in certain Latino countries and among certain groups of people, so, like the *Quinceañera*, it only gives a partial view.

Possible focal points among other ethnic groups are certainly debatable. Among Chinese and Koreans, the Confucian influence has maintained key notions of filial piety that have kept families strong, but also promoted family tensions. For African Americans, the black church has certainly been a source and outlet for much expressive culture, and has had a strong influence on other cultures nationally and internationally through its musical forms. The spirituality found in the black church has played a major role in the lives of African Americans, and in the way they "interpret, inform and reshape their social conditions" (Frederick 2003, ix). The movement came largely out of the black church, most black political leaders have been nurtured in the church, and social and community topics and issues of personal transformation are also discussed chiefly through the church.

COMMUNICATION STYLES

When entering another culture, differences in communication styles are likely to be some of the first things to be encountered (Hall 2002). One may be surprised or even shocked by how vocal the locals are, or how reticent they are, depending on what one is used to. East Asians, for instance, tend to be less outspoken and talkative than most Americans, a practice tied into a different worldview and understanding of self and other (Kim and Markus 2002). White Americans tend to be more comfortable with impersonal work relationships and mere

acquaintances, while Latinos value developing closer friendships among people who work together. People in cultures that stress family loyalty will, in general, be less likely to sacrifice family relationships in order to further careers (Samovar, Porter, and Stefani 1998, 108f.). Along with this, many white Americans focus on efficiency and economy, sometimes to the neglect of social relationships.

Misunderstandings can easily occur unless one is aware of different communication styles. For instance, whites may regard blacks as less respectful when they are more vocal, while whites might seem to be insincere to blacks because they tend to hide their true feelings or ideas to maintain social peace (Kochman 1981; Hecht, Jackson, and Ribeau 2003). Another writer discussed how black and white women sometimes have "difficult" dialogues because of mutual stereotypes, but also how it is possible to develop good relationships (Houston 2000). Still another looked at other everyday interactions among people from different cultures that can cause hurt feelings or create misunderstandings (Williams 2000; cf. Martin 2000). Interactions between African Americans and Koreans, for example, can be tense, because African Americans want Korean grocers to be friendly and hire locals, while Koreans say it is not their culture to smile frequently with strangers. Koreans' strong family orientation means relatives are preferred over local people for jobs, which creates resentment (Kottak and Kozaitis 2003, 80). Competition for economic resources such as jobs and housing, and over political power, can cause tensions. Perceived discrimination and mutual stereotypes can create problems, for instance, between Latinos and African Americans (Niemann 1999).

Latino expressivity can be understood by comparing Latino notions of public space with Anglo-American ones, for example. Anglos generally construct invisible boundaries around their bodies, and apologies must be issued for even near bumping. Latinos would let this pass without comment, seeing it as normal. For Anglos, this space cannot be violated even verbally, since publicly flattering women is likely to be viewed as harassment, whereas it is a more ambiguous practice among Latinos.

Styles of public expressiveness vary among cultures. Misunderstandings and hurt feelings can easily result, especially in pluralistic work or social situations. Stereotypes about "rude" behavior can be reinforced, unless one has a deeper understanding of stylistic differences. As a first step, it is important to know the unique communication patterns and values of our own culture, and realize that not everyone shares them.

FAMILY

Differences in family dynamics are other things one may notice among cultures (Lynch and Hanson 2004). If you are a white American, you may be surprised at how much family members of many minority ethnic groups rely on one another. People from other cultures may be surprised at how individualistic some white American family members are, as each person busily attends to his or her own activities....

Family issues are central to all cultures. Compared with more individualistic white Americans, Latinos generally have a stronger focus on the extended family as a source of social and material support (Gangotena 2003; Marín and Marín 1991). Issues of *machismo* and *marianismo* (male dominance, and female submission and domesticity, respectively) can create tensions and distance among Latino males and females (Abalos 2002). Generational tensions among Latinos growing up in the United States and their foreign-born parents have also created problems in families. Latino children have found American popular culture attractive, but its glorification of sexuality and delinquency is at odds with their parents' more conservative family and work-oriented values (Suro 1999, 3–26). As Carlos Pozzi points out ... children of immigrants may take on a different identity than that of their parents.

Family issues among African Americans are framed by a different history. The tragic nature of the forced relocation of slaves meant that most African social institutions, such as marriage customs, were severely broken by slavery in the New World. Under slavery, family life suffered, as spouses and children were often kept apart, and men were essentially emasculated, denied normal roles of husband and provider. According to sociologist Orlando Patterson (1998), problems that beset the African American family—such as low marriage and high single-parenthood rates, and tension between the sexes—can largely be traced back to the time of slavery.

Asian Americans generally carry the "model minority" stereotype of high academic and economic achievement, but also pay a price with high levels of family pressure to succeed, which creates its own family tensions, and sometimes alienation and loneliness (Nam 2001; Ng 1998). Not all Asian groups fit this stereotype, however. Some of those from rural parts of Southeast Asia, such as the Hmong, struggle economically in the United States, partially because their cultures were small-scale, relatively egalitarian cultures that did not have a tradition of literacy.

In most Asian cultures, the individual is not an entirely independent entity, but is seen as being bound in relationships. There is strong social pressure to conform, while respecting and honoring parents and elders are held in higher regard. The focus on hierarchy contrasts and sometimes clashes with the American emphasis on egalitarianism (such as the common American practice of using first names with strangers rather than surnames). Understandings of ethics can differ. Americans tend to see relational issues (e.g., caring for relatives) as a personal choice, while an Indian's decisions are influenced more by culturally determined obligations (Miller 1991). Shame, dependent on social context, is a stronger motivator among Asians (and in other areas such as the Mediterranean), whereas guilt, more individualistic and rooted in Christian theology, is stronger among Americans.

Asian families also have their battles with the influence of American popular culture. American dating practices, for instance, are not acceptable to many Asian families, especially more recent immigrants. American teens are given much more independence than in Asian families, especially in meeting members of the opposite sex. Some Indian parents, for instance, see this influence as "cultural contamination of the worst sort" (Hegde 1998, 49) when their children begin to act more like autonomous Americans (Nam 2001).

Many Asian Americans, especially Chinese and Koreans, are now Christians. Many Chinese have found aspects of Christianity compatible with Confucian moral values, with its focus on attachment to family, and this helps reinforce some of the traditional values in the face of American popular culture pressures on their children. The result is that tensions, especially on family and gender issues, are played out among three identities: American, Christian, and Chinese (Yang 1999). Though Asians now face less prejudice than other minority groups, they still have to deal with unique family and identity questions....

DIVERSITY WITHIN AND BETWEEN CULTURES

Virtually all minorities in the United States display tremendous diversity of class, religion, ideology, and other factors. Not all members of an ethnic group share the central values of their own culture. Some Latinos ignore the *Quinceañera*. Some blacks are alienated from historic African American churches, are Muslim, or find liturgical or other traditions such as the Jehovah's Witnesses more attractive. Women and men may have different perspectives on practices or ideals.

Class is an important factor when considering cultural differences (Hall, Neitz, and Battani 2003, 43ff.). People in different ethnic groups but in the same social class may have commonalities. The upper classes of different ethnicities may share tastes in music, food, or religious or political beliefs. Lower-class whites may reject aspects of middle- or upper-class whites. But social classes within an ethnic group also share many ethnic traits (Landrine and Klonoff 1996, 88; Pattillo-McCoy 1999, 12; Abrahams 1970, 22ff.), so ethnicity is clearly important. And while the similar classes of different ethnic groups have commonalities, there

are significant differences among them, which can only be related to ethnicity.

The level of "acculturation" to dominant Euro-American values and practices varies widely among minority individuals. Acculturation can be understood through surveys and ethnographic research that examines the strength of beliefs and practices unique to ethnic group members (Landrine and Klonoff 1996, 62ff.). When compared with practices of the dominant culture, minorities can range from "traditional" to "acculturated" with "bicultural" in between (Landrine and Klonoff 1996; Padilla 1980). Most often, traditional cultural practices are maintained through family and religion, while acculturation to the dominant culture occurs through schools and popular culture. The speed of this process, however, has many variables, from family structure to societal reception (Portes and Rumbaut 2001, chap. 3).

Socially constructed notions of racial difference have made it harder for some groups, such as blacks, to enter the dominant culture (Yancey 2003, 156ff.). Other groups, such as some Asians, have found it easier. Some resist acculturation, and prefer oppositional identities, sometimes in response to discrimination or rejection from dominant cultures (Pattillo-McCoy 1999, 120ff.). If we try to treat all people just "as individuals" and ignore the different histories and cultures of peoples, and how they may be subtly stereotyped by others, we do not realize the obstacles that minorities often face (Plaut 2002). As mentioned above, cultures are not entirely unified or well defined, but are often sites of conflict. Dominant groups have used socially constructed notions about minority groups (such as "racial" tendencies) to control people and create divisions and social hierarchies among cultures in a bid to maintain positions of dominance (Landrine and Klonoff 1996, 8ff.). Minority groups of all kinds (not only ethnic but also religious) must deal with pressures to either assimilate to dominant group ideals, and thus give up part of their identity, or to reject some of those dominant ideals and thus risk continued stigmatization and marginalization (Young 2001; Shannon 2001).

Culture is something we all have whether we know it or not, while "ethnicity" is consciously adopted identity that can be changed (Waters 1990). It can also be imposed by others. African Americans, for instance, are sometimes considered "white" in Africa, and "black" in the United States, no matter what their skin color. Former tennis champion Yannick Noah, whose father and mother were from Cameroon and France, respectively, has mentioned that he is considered white in Africa, and black in France (Landrine and Klonoff 1996, 13). Out of the contact of different cultures may come an enormous amount of creativity, along with the destabilization of traditional identities as people combine cultural symbols/practices in unique ways to create new expressive forms, a process that postcolonial scholars refer to as "hybridity" (Bhabha 1990).

REFERENCES

Abalos, David T. 2002. The Latino male: A radical redefinition. Boulder, CO: Lynne Rienner.

Abrahams, Roger. 1970. *Positively black*. New York: Prentice Hall.

Bashkow, Ira. 2004. A Neo-Boasian conception of cultural boundaries. *American Anthropologist* 106(3): 443–58.

Bhabha, Homi. 1990. *Nation and narration*. London: Routledge.

Corr, Charles, Clyde Nabe, and Donna M. Corr. 2000. *Death and dying, life and living*. 3rd ed. Belmont, CA: Wadsworth.

Deiter, Kristen. 2002. From church blessing to Quinceañera Barbie: America as "spiritual benefactor" in La Quinceañera. *Christian Scholar's Review* 32 (1):31–48.

Frederick, Marla. 2003. *Between Sundays: Black women and everyday struggles of faith*. Berkeley and Los Angeles: University of California Press.

Gangotena, Margarita. 2003. The rhetoric of *La Familia* among Mexican Americans. In *Our voices: Essays in culture, ethnicity and communication*, ed. A. Gonzalez, M. Houston, and V. Chen. 72–73. Los Angeles: Roxbury.

Geertz, Clifford. 1973. *The interpretation of cultures*. New York: Basic.

Graña, César. 1987. The bullfight and Spanish national decadence. *Society* 24(5):33–37.

Hall, John R., Mary Jo Neitz, and Marshall Battani. 2003. *Sociology on culture*. New York: Routledge.

Hammond, Phillip. 1998. *With liberty for all*. Louisville: Westminster John Knox.

Hecht, Michael L., Ronald L. Jackson II, and Sidney A. Ribeau. 2003. *African American communication*. 2nd ed. Mahwah, NJ: Lawrence Erlbaum Associates.

Hegde, Radha S. 1997. Swinging the trapeze: The negotiation of identity among Asian Indian immigrant women in the United States. In *Communication and identity across cultures*, ed. A. Gonzalez and D. V. Tanno, 34–35. Thousand Oaks, CA: Sage.

Holmes, Lowell D., and Ellen Rhoads Holmes. 2002. The American cultural configuration. In *Distant mirrors: American as a foreign culture*. 3rd ed. P. DeVita and J. Armstrong. Belmont, CA: Wadsworth.

Houston, Marsha. 2000. When black women talk with white women. In *Our voices: Essays in culture, ethnicity and communication*, ed. A. Gonzalez, M. Houston, and V. Chen. Los Angeles: Roxbury.

Hsu, Francis L. K. 1963. *Clan, caste and club*. Princeton, NJ: D. Van Nostrand.

Huntington, Samuel. 2004. *Who are we? The challenges to America's national identity*. New York: Simon & Schuster.

Kim, Heejung S., and Hazel Rose Markus. 2002. Freedom of speech and freedom of silence: An analysis of talking as a cultural practice. In *Engaging cultural differences*, ed. R Shweder, M. Minow, and H. R. Markus. New York: Russell Sage Foundation.

Kochman, Thomas. 1981. *Black and white styles in conflict*. Chicago: University of Chicago.

Kottak, Conrad Phillip, and Kathryn A. Kozaitis. 2003. *On being different*. 2nd ed. Boston: McGraw Hill.

Kusserow, Adrie. 2004. *American individualisms: Child rearing and social class in three neighborhoods.* New York: Palgrave Macmillan.

Landrine, Hope, and Elizabeth A. Klonoff. 1996. *African American acculturation.* Thousand Oaks, CA: Sage.

Maranz, David. 2001. *African friends and money matters.* Dallas: SIL International.

Marín, G., and B. V. Marín. 1991. *Research with Hispanic populations.* Newbury Park, CA: Sage.

Martin, Judith. 2000. Everyone behaving badly. *New York Times Book Review.* October 15.

Middleton, Dwight R. 2003. *The challenge of human diversity.* 2nd ed. Prospect Heights, IL: Waveland.

Miller, Joan G. 1991. A cultural perspective on the morality of beneficence and interpersonal responsibility. In S. Ting-Toomey and F. Korzenny (eds.). *International and Intercultural Communication Annual* 15: 11–27.

Nam, Vickie, ed. 2001. *Yell-oh girls!: Emerging voices explore culture, identity and growing up Asian American.* New York: HarperCollins.

Ng. Franklin. 1998. *Asian American family life and community.* New York: Garland.

Niemann, Yolanda Flores. 1999. Social ecological contexts of prejudice between Hispanics and blacks. In *Race, ethnicity and nationality in the United States,* ed. P. Wong, 170–90. Boulder, CO: Westview.

Nisbett, Richard. 2003. *The geography of thought: How Asians and Westerners think differently…and why.* NewYork: Free Press.

Padilla, A. 1980. *Acculturation.* Boulder, CO: Westview.

Patterson, Orlando. 1998. *Rituals of blood: Consequences of slavery in two American centuries.* New York: Basic Civitas.

Pattillo-McCoy, Mary. 1999. *Black picket fences.* Chicago: University of Chicago Press.

Plaut, Victoria C. 2002. Cultural models of diversity in America: The psychology of difference and inclusion.

In *Engaging cultural differences,* ed. R. Shweder, M. Minow, and H. R. Markus, 365–95. New York: Russell Sage Foundation.

Portes, Alejandro, and Rubén G. Rumbaut. 2001. *Legacies: The story of the immigrant second generation.* Berkeley and Los Angeles: University of California Press.

Sahlins, Marshall. 1976. *Culture and practical reason.* Chicago: University of Chicago Press.

Samovar, Larry A., Richard Porter, and Lisa A. Stefani. 1998. *Communication between cultures.* 3rd ed. Belmont, CA: Wadsworth.

Samuels, David. 1991. The rap on rap. *The New Republic* 205(20):24–29.

Shannon, Christopher. 2001. *A world made safe for differences.* Lanham, MD: Rowman and Littlefield.

Smith, Christian. 2003. *Moral, believing animals: An essay on human personhood and culture.* Oxford: Oxford University Press.

Suro, Roberto. 1999. *Strangers among us: Latino lives in a changing America.* New York: Vintage.

Vansina, Jan. 1994. *Living with Africa.* Madison: University of Wisconsin Press.

Walker, Sheila S., ed. 2001. *African roots/American cultures.* Lanham, MD: Rowman and Littlefield.

Waters, Mary C. 1990. *Ethnic options: Choosing identities in America.* Berkeley and Los Angeles: University of California Press.

Williams, Lena. 2000. *It's the little things: The everyday interactions that get under the skin of blacks and whites.* New York: Harcourt.

Yancey, George. 2003. *Who is white?: Latinos, Asians, and the new black/nonblack divide.* Boulder, CO: Lynne Rienner.

Yang, Fenggang. 1999. *Chinese Christians in America.* Pittsburgh: Pennsylvania State University Press.

Young, Iris Marion. 2001. Justice and the politics of difference. In *The new social theory reader,* ed. S. Seidman. and J. Alexander. London: Routledge.

14

Body Ritual among the Nacirema

HORACE MINER

When sociologists and anthropologists study other cultures, they attempt to be objective in their observations, trying to understand the culture from that culture's point of view. Understanding other cultures can help us gain perspective on our own culture. Yet because individuals are socialized into their own culture's beliefs, values, and practices, other cultures may seem to have strange, even bizarre or immoral activities and beliefs.

As you read Miner's essay, keep in mind these questions:

1. *Which practices in the Nacirema culture appear strange to an outsider?*
2. *How does this reading illustrate ethnocentrism and cultural relativity?*
3. *Which problems might you encounter in accurately observing other cultures?*
4. *How might someone from a different culture view or interpret practices in your culture?*

GLOSSARY **Body ritual** Ceremonies focusing on the body or body parts.

The anthropologist has become so familiar with the diversity of ways in which different peoples behave in similar situations that he is not apt to be surprised by even the most exotic customs. In fact, if all of the logically possible combinations of behavior have not been found somewhere in the world, he is apt to suspect that they must be present in some yet undescribed tribe. This point has, in fact, been expressed with respect to clan organization by Murdock (1949:71). In this light, the magical beliefs and practices of the Nacirema present such unusual aspects that it seems desirable to describe them as an example of the extremes to which human behavior can go.

Professor Linton first brought the ritual of the Nacirema to the attention of anthropologists twenty years ago (1936:326), but the culture of this people is still very poorly understood. They are a North American group living in the territory between the Canadian Cree, the Yaqui and Tarahumare of Mexico, and the Carib and Ara-wak of the Antilles. Little is known of their origin, although tradition states that they came from the east. According to Nacirema mythology, their nation was originated by a culture hero, Notgnihsaw, who is otherwise known for two great feats of strength—the throwing of a piece of wampum across the river Pa-To-Mac and the chopping down of a cherry tree in which the Spirit of Truth resided.

Nacirema culture is characterized by a highly developed market economy which has evolved in

Source: From *American Anthropologist* 58(3), pp. 503–7, 1956. Not for further reproduction.

a rich natural habitat. While much of the people's time is devoted to economic pursuits, a large part of the fruits of these labors and a considerable portion of the day are spent in ritual activity. The focus of this activity is the human body, the appearance and health of which loom as a dominant concern in the ethos of the people. While such a concern is certainly not unusual, its ceremonial aspects and associated philosophy are unique.

The fundamental belief underlying the whole system appears to be that the human body is ugly and that its natural tendency is to debility and disease. Incarcerated in such a body, man's only hope is to avert these characteristics through the use of the powerful influences of ritual and ceremony. Every household has one or more shrines devoted to this purpose. The more powerful individuals in this society have several shrines in their houses and, in fact, the opulence of a house is often referred to in terms of the number of such ritual centers it possesses. Most houses are of wattle and daub construction, but the shrine rooms of the more wealthy are walled with stone. Poorer families imitate the rich by applying pottery plaques to their shrine walls.

While each family has at least one such shrine, the rituals associated with it are not family ceremonies but are private and secret. The rites are normally only discussed with children, and then only during the period when they are being initiated into these mysteries. I was able, however, to establish sufficient rapport with the natives to examine these shrines and to have the rituals described to me.

The focal point of the shrine is a box or chest which is built into the wall. In this chest are kept the many charms and magical potions without which no native believes he could live. These preparations are secured from a variety of specialized practitioners. The most powerful of these are the medicine men, whose assistance must be rewarded with substantial gifts. However, the medicine men do not provide the curative potions for their clients, but decide what the ingredients should be and then write them down in an ancient and secret language. This writing is understood only by the medicine men and by the herbalists who, for another gift, provide the required charm.

The charm is not disposed of after it has served its purpose, but is placed in the charm-box of the household shrine. As these magical materials are specific for certain ills, and the real or imagined maladies of the people are many, the charm-box is usually full to overflowing. The magical packets are so numerous that people forget what their purposes were and fear to use them again. While the natives are very vague on this point, we can only assume that the idea in retaining all the old magical materials is that their presence in the charm-box, before which the body rituals are conducted, will in some way protect the worshipper.

Beneath the charm-box is a small font. Each day every member of the family, in succession, enters the shrine room, bows his head before the charm-box, mingles different sorts of holy water in the font, and proceeds with a brief rite of ablution. The holy waters are secured from the Water Temple of the community, where the priests conduct elaborate ceremonies to make the liquid ritually pure.

In the hierarchy of magical practitioners, and below the medicine men in prestige, are specialists whose designation is best translated "holy-mouth-men." The Nacirema have an almost pathological horror of and fascination with the mouth, the condition of which is believed to have a supernatural influence on all social relationships. Were it not for the rituals of the mouth, they believe that their teeth would fall out, their gums bleed, their jaws shrink, their friends desert them, and their lovers reject them. They also believe that a strong relationship exists between oral and moral characteristics. For example, there is a ritual ablution of the mouth for children which is supposed to improve their moral fiber.

The daily body ritual performed by everyone includes a mouth-rite. Despite the fact that these people are so punctilious about care of the mouth, this rite involves a practice which strikes the uninitiated stranger as revolting. It was reported to me that the ritual consists of inserting a small bundle of hog hairs into the mouth, along with certain magical powders, and then moving the bundle in a highly formalized series of gestures.

In addition to the private mouth-rite, the people seek out a holy-mouth-man once or twice a year. These practitioners have an impressive set of paraphernalia, consisting of a variety of augers, awls, probes, and prods. The use of these objects in the exorcism of the evils of the mouth involves almost unbelievable ritual torture of the client. The holy-mouth-man opens the client's mouth and, using the above-mentioned tools, enlarges any holes which decay may have created in the teeth. Magical materials are put into these holes. If there are no naturally occurring holes in the teeth, large sections of one or more teeth are gouged out so that the supernatural substance can be applied. In the client's view, the purpose of these ministrations is to arrest decay and to draw friends. The extremely sacred and traditional character of the rite is evident in the fact that the natives return to the holy-mouth-man year after year, despite the fact that their teeth continue to decay.

It is to be hoped that, when a thorough study of the Nacirema is made, there will be careful inquiry into the personality structure of these people. One has but to watch the gleam in the eye of a holy-mouth-man, as he jabs an awl into an exposed nerve, to suspect that a certain amount of sadism is involved. If this can be established, a very interesting pattern emerges, for most of the population shows definite masochistic tendencies. It was to these that Professor Linton referred in discussing a distinctive part of the daily body ritual which is performed only by men. This part of the rite involves scraping and lacerating the surface of the face with a sharp instrument. Special women's rites are performed only four times during each lunar month, but what they lack in frequency is made up in barbarity. As part of this ceremony, women bake their heads in small ovens for about an hour. The theoretically interesting point is that what seems to be a preponderantly masochistic people have developed sadistic specialists.

The medicine men have an imposing temple, or *latipso*, in every community of any size. The more elaborate ceremonies required to treat very sick patients can only be performed at this temple. These ceremonies involve not only the thaumaturge but a permanent group of vestal maidens who move sedately about the temple chambers in distinctive costume and headdress.

The *latipso* ceremonies are so harsh that it is phenomenal that a fair proportion of the really sick natives who enter the temple ever recover. Small children whose indoctrination is still incomplete have been known to resist attempts to take them to the temple because "that is where you go to die." Despite this fact, sick adults are not only willing but eager to undergo the protracted ritual purification, if they can afford to do so. No matter how ill the supplicant or how grave the emergency, the guardians of many temples will not admit a client if he cannot give a rich gift to the custodian. Even after one has gained admission and survived the ceremonies, the guardians will not permit the neophyte to leave until he makes still another gift.

The supplicant entering the temple is first stripped of all his or her clothes. In everyday life the Nacirema avoids exposure of his body and its natural functions. Bathing and excretory acts are performed only in the secrecy of the household shrine, where they are ritualized as part of the body-rites. Psychological shock results from the fact that body secrecy is suddenly lost upon entry into the *latipso*. A man, whose own wife has never seen him in an excretory act, suddenly finds himself naked and assisted by a vestal maiden while he performs his natural functions into a sacred vessel. This sort of ceremonial treatment is necessitated by the fact that the excreta are used by a diviner to ascertain the course and nature of the client's sickness. Female clients, on the other hand, find their naked bodies are subjected to the scrutiny, manipulation, and prodding of the medicine men.

Few supplicants in the temple are well enough to do anything but lie on their hard beds. The daily ceremonies, like the rites of the holy-mouth-men, involve discomfort and torture. With ritual precision, the vestals awaken their miserable charges each dawn and roll them about on their beds of pain while performing ablutions, in the formal movements of which the maidens are highly trained. At other times they insert magic wands in the supplicant's mouth or force him to eat substances which are supposed to be healing. From time to time the medicine men come to their clients and jab magically treated needles into their flesh. The fact that these temple ceremonies may not cure, and may even kill the neophyte, in no way decreases the people's faith in the medicine men.

There remains one other kind of practitioner, known as a "listener." This witch-doctor has the power to exorcise the devils that lodge in the heads of people who have been bewitched. The Nacirema believe that parents bewitch their own children. Mothers are particularly suspected of putting a curse on children while teaching them the secret body rituals. The counter-magic of the witch-doctor is unusual in its lack of ritual. The patient simply tells the "listener" all his troubles and fears, beginning with the earliest difficulties he can remember. The memory displayed by the Nacirema in these exorcism sessions is truly remarkable. It is not uncommon for the patient to bemoan the rejection he felt upon being weaned as a babe, and a few individuals even see their troubles going back to the traumatic effects of their own birth.

In conclusion, mention must be made of certain practices which have their base in native esthetics but which depend upon the pervasive aversion to the natural body and its functions. There are ritual fasts to make fat people thin and ceremonial feasts to make thin people fat. Still other rites are used to make women's breasts larger if they are small, and smaller if they are large. General dissatisfaction with breast shape is symbolized in the fact that the ideal form is virtually outside the range of human variation. A few women afflicted with almost inhuman hypermammary development are so idolized that they make a handsome living by simply going from village to village and permitting the natives to stare at them for a fee.

Reference has already been made to the fact that excretory functions are ritualized, routinized, and relegated to secrecy. Natural reproductive functions are similarly distorted. Intercourse is taboo as a topic and scheduled as an act. Efforts are made to avoid pregnancy by the use of magical materials or by limiting intercourse to certain phases of the moon. Conception is actually very infrequent. When pregnant, women dress so as to hide their condition. Parturition takes place in secret, without friends or relatives to assist, and the majority of women do not nurse their infants.

Our review of the ritual life of the Nacirema has certainly shown them to be a magic-ridden people. It is hard to understand how they have managed to exist so long under the burdens which they have imposed upon themselves. But even such exotic customs as these take on real meaning when they are viewed with the insight provided by Malinowski when he wrote (1948:70):

> Looking from far and above, from our high places of safety in the developed civilization, it is easy to see all the crudity and irrelevance of magic. But without its power and guidance early man could not have mastered his practical difficulties as he has done, nor could man have advanced to the higher stages of civilization.

REFERENCES

Linton, Ralph, 1936. *The Study of Man*. New York: Appleton-Century.

Malinowski, Bronislaw, 1948. *Magic, Science, and Religion*. Glencoe, Ill. Free Press.

Murdock, George P., 1949. *Social Structure*. New York: Macmillan.

15

India's Sacred Cow

MARVIN HARRIS

Cultures vary dramatically in their beliefs and practices, yet each cultural practice has evolved with some reason behind it. One of these practices is cow worship. In India, the cultural practice among Hindus is to treat cows with great respect, even in the face of human hunger. Harris discusses this practice, which many find curious.

Consider the following as you read:

1. *Why are cows sacred? What is sacred in your country that might seem strange to others?*
2. *Which other practices in different cultures do you find strange or unusual, and what purpose might those practices serve for the culture?*
3. *How and why do cultural traditions evolve?*

GLOSSARY **Untouchables** Lowest group in the stratification (caste) system of India. **Hinduism** Dominant religious belief system in India.

News photographs that came out of India during the famine of the late 1960s showed starving people stretching out bony hands to beg for food while sacred cattle strolled behind undisturbed. The Hindu, it seems, would rather starve to death than eat his cow or even deprive it of food. The cattle appear to browse unhindered through urban markets eating an orange here, a mango there, competing with people for meager supplies of food.

By Western standards, spiritual values seem more important to Indians than life itself. Specialists in food habits around the world like Fred Simoons at the University of California at Davis consider Hinduism an irrational idealogy that compels people to overlook abundant, nutritious foods for scarcer, less healthful foods.

What seems to be an absurd devotion to the mother cow pervades Indian life. Indian wall calendars portray beautiful young women with bodies of fat white cows, often with milk jetting from their teats into sacred shrines.

Cow worship even carries over into politics. In 1966 a crowd of 120,000 people, led by holy men, demonstrated in front of the Indian House of Parliament in support of the All-Party Cow Protection Campaign Committee. In Nepal, the only contemporary Hindu kingdom, cow slaughter is severely punished. As one story goes, the car driven by an official of a United States agency struck and

Source: From *Human Nature Magazine* 1(2), pp. 28, 30–36, February 1978. Copyright © 1978 by Human Nature, Inc. Reprinted by permission of the publisher.

killed a cow. In order to avoid the international incident that would have occurred when the official was arrested for murder, the Nepalese magistrate concluded that the cow had committed suicide.

Many Indians agree with Western assessments of the Hindu reverence for their cattle, the zebu, or *Bos indicus*, a large-humped species prevalent in Asia and Africa. M. N. Srinivas, an Indian anthropologist, states: "Orthodox Hindu opinion regards the killing of cattle with abhorrence, even though the refusal to kill vast number of useless cattle which exist in India today is detrimental to the nation." Even the Indian Ministry of Information formerly maintained that "the large animal population is more a liability than an asset in view of our land resources." Accounts from many different sources point to the same conclusion: India, one of the world's great civilizations, is being strangled by its love for the cow.

The easy explanation for India's devotion to the cow, the one most Westerners and Indians would offer, is that cow worship is an integral part of Hinduism. Religion is somehow good for the soul, even if it sometimes fails the body. Religion orders the cosmos and explains our place in the universe. Religious beliefs, many would claim, have existed for thousands of years and have a life of their own. They are not understandable in scientific terms.

But all this ignores history. There is more to be said for cow worship than is immediately apparent. The earliest Vedas, the Hindu sacred texts from the second millennium B.C., do not prohibit the slaughter of cattle. Instead, they ordain it as part of sacrificial rites. The early Hindus did not avoid the flesh of cows and bulls; they ate it at ceremonial feasts presided over by Brahman priests. Cow worship is a relatively recent development in India; it evolved as the Hindu religion developed and changed.

This evolution is recorded in royal edicts and religious texts written during the last 3,000 years of Indian history. The Vedas from the first millennium B.C. contain contradictory passages, some referring to ritual slaughter and others to a strict taboo on beef consumption. A. N. Bose, in *Social and Rural Economy of Northern India, Cir. 600 B.C.–200 A.D.*, concludes that many of the sacred-cow passages were incorporated into the texts by priests of a later period.

By 200 A.D. the status of Indian cattle had undergone a spiritual transformation. The Brahman priesthood exhorted the population to venerate the cow and forbade them to abuse it or to feed on it. Religious feasts involving the ritual slaughter and the consumption of livestock were eliminated and meat eating was restricted to the nobility.

By 1000 A.D., all Hindus were forbidden to eat beef. Ahimsa, the Hindu belief in the unity of all life, was the spiritual justification for this restriction. But it is difficult to ascertain exactly when this change occurred. An important event that helped to shape the modern complex was the Islamic invasion, which took place in the eighth century A.D. Hindus may have found it politically expedient to set themselves off from the invaders, who were beefeaters, by emphasizing the need to prevent the slaughter of their sacred animals. Thereafter, the cow taboo assumed its modern form and began to function much as it does today.

The place of the cow in modern India is every place—on posters, in the movies, in brass figures, in stone and wood carvings, on the streets, in the fields. The cow is a symbol of health and abundance. It provides the milk that Indians consume in the form of yogurt and ghee (clarified butter), which contribute subtle flavors to much spicy Indian food.

This, perhaps, is the practical role of the cow, but cows provide less than half the milk produced in India. Most cows in India are not dairy breeds. In most regions, when an Indian farmer wants a steady, high-quality source of milk he usually invests in a female water buffalo. In India the water buffalo is the specialized dairy breed because its milk has a higher butterfat content than zebu milk. Although the farmer milks his zebu cows, the milk is merely a by-product.

More vital than zebu milk to South Asian farmers are zebu calves. Male calves are especially valued because from bulls come oxen, which are the mainstay of the Indian agricultural system.

Small, fast oxen drag wooden plows through late-spring fields when monsoons have dampened the dry, cracked earth. After harvest, the oxen break the grain from the stalk by stomping through mounds of cut wheat and rice. For rice cultivation in irrigated fields, the male water buffalo is preferred (it pulls better in deep mud), but for most other crops, including rainfall rice, wheat, sorghum, and millet, and for transporting goods and people to and from town, a team of oxen is preferred. The ox is the Indian peasant's tractor, thresher, and family car combined; the cow is the factory that produces the ox.

If draft animals instead of cows are counted, India appears to have too few domesticated ruminants, not too many. Since each of the 70 million farms in India requires a draft team, it follows that Indian peasants should use 140 million animals in the fields. But there are only 83 million oxen and male water buffalo on the subcontinent, a shortage of 30 million draft teams.

In other regions of the world, joint ownership of draft animals might overcome a shortage, but Indian agriculture is closely tied to the monsoon rains of late spring and summer. Field preparation and planting must coincide with the rain, and a farmer must have his animals ready to plow when the weather is right. When the farmer without a draft team needs bullocks most, his neighbors are all using theirs. Any delay in turning the soil drastically lowers production.

Because of this dependence on draft animals, loss of the family oxen is devastating. If a beast dies, the farmer must borrow money to buy or rent an ox at interest rates so high that he ultimately loses his land. Every year foreclosures force thousands of poverty-stricken peasants to abandon the countryside for the overcrowded cities.

If a family is fortunate enough to own a fertile cow, it will be able to rear replacements for a lost team and thus survive until life returns to normal. If, as sometimes happens, famine leads a family to sell its cow and ox team, all ties to agriculture are cut. Even if the family survives, it has no way to farm the land, no oxen to work the land, and no cows to produce oxen.

The prohibition against eating meat applies to the flesh of cows, bulls, and oxen, but the cow is the most sacred because it can produce the other two. The peasant whose cow dies is not only crying over a spiritual loss but over the loss of his farm as well.

Religious laws that forbid the slaughter of cattle promote the recovery of the agricultural system from the dry Indian winter and from periods of drought. The monsoon, on which all agriculture depends, is erratic. Sometimes, it arrives early, sometimes late, sometimes not at all. Drought has struck large portions of India time and again in this century, and Indian farmers and the zebus are accustomed to these natural disasters. Zebus can pass weeks on end with little or no food and water. Like camels, they store both in their humps and recuperate quickly with only a little nourishment.

During droughts the cows often stop lactating and become barren. In some cases the condition is permanent but often it is only temporary. If barren animals were summarily eliminated, as Western experts in animal husbandry have suggested, cows capable of recovery would be lost along with those entirely debilitated. By keeping alive the cows that can later produce oxen, religious laws against cow slaughter assure the recovery of the agricultural system from the greatest challenge it faces—the failure of the monsoon.

The local Indian governments aid the process of recovery by maintaining homes for barren cows. Farmers reclaim any animal that calves or begins to lactate. One police station in Madras collects strays and pastures them in a field adjacent to the station. After a small fine is paid, a cow is returned to its rightful owner when the owner thinks the cow shows signs of being able to reproduce.

During the hot, dry spring months most of India is like a desert. Indian farmers often complain they cannot feed their livestock during this period. They maintain the cattle by letting them scavenge on the sparse grass along the roads. In the cities the cattle are encouraged to scavenge near food stalls to supplement their scant diet. These are the wandering cattle tourists report seeing throughout India.

Westerners expect shopkeepers to respond to these intrusions with the deference due a sacred animal; instead, their response is a string of curses and the crack of a long bamboo pole across the beast's back or a poke at its genitals. Mahatma Gandhi was well aware of the treatment sacred cows (and bulls and oxen) received in India. "How we bleed her to take the last drop of milk from her. How we starve her to emaciation, how we ill-treat the calves, how we deprive them of their portion of milk, how cruelly we treat the oxen, how we castrate them, how we beat them, how we overload them" [Gandhi, 1954].

Oxen generally receive better treatment than cows. When food is in short supply, thrifty Indian peasants feed their working bullocks and ignore their cows, but rarely do they abandon the cows to die. When the cows are sick, farmers worry over them as they would over members of the family and nurse them as if they were children. When the rains return and when the fields are harvested, the farmers again feed their cows regularly and reclaim their abandoned animals. The prohibition against beef consumption is a form of disaster insurance for all India.

Western agronomists and economists are quick to protest that all the functions of the zebu cattle can be improved with organized breeding programs, cultivated pastures, and silage. Because stronger oxen would pull the plow faster, they could work multiple plots of land, allowing farmers to share their animals. Fewer healthy, well-fed cows could provide Indians with more milk. But pastures and silage require arable land, land needed to produce wheat and rice.

A look at Western cattle farming makes plain the cost of adopting advanced technology in Indian agriculture. In a study of livestock production in the United States, David Pimentel of the College of Agriculture and Life Sciences at Cornell University, found that 91 percent of the cereal, legume, and vegetable protein suitable for human consumption is consumed by livestock. Approximately three-quarters of the arable land in the United States is devoted to growing food for livestock. In the production of meat and milk, American ranchers use enough fossil fuel to equal more than 82 million barrels of oil annually.

Indian cattle do not drain the system in the same way. In a 1971 study of livestock in West Bengal, Stewart Odend'hal [1972] of the University of Missouri found that Bengalese cattle ate only the inedible remains of subsistence crops—rice straw, rice hulls, the tops of sugar cane, and mustard-oil cake. Cattle graze in the fields after harvest and eat the remains of crops left on the ground; they forage for grass and weeds on the roadsides. The food for zebu cattle costs the human population virtually nothing. "Basically," Odend'hal says, "the cattle convert items of little direct human value into products of immediate utility."

In addition to plowing the fields and producing milk, the zebus produce dung, which fires the hearths and fertilizes the fields of India. Much of the estimated 800 million tons of manure produced annually is collected by the farmers' children as they follow the cows and bullocks from place to place. And when the children see the droppings of another farmer's cattle along the road, they pick those up also. Odend'hal reports that the system operates with such high efficiency that the children of West Bengal recover nearly 100 percent of the dung produced by their livestock.

From 40 to 70 percent of all manure produced by Indian cattle is used as fuel for cooking; the rest is returned to the fields as fertilizer. Dried dung burns slowly, cleanly, and with low heat—characteristics that satisfy the household needs of Indian women. Staples like curry and rice can simmer for hours. While the meal slowly cooks over an unattended fire, the women of the household can do other chores. Cow chips, unlike firewood, do not scorch as they burn.

It is estimated that the dung used for cooking fuel provides the energy-equivalent of 43 million tons of coal. At current prices, it would cost India an extra 1.5 billion dollars in foreign exchange to replace the dung with coal. And if the 350 million tons of manure that are being used as fertilizer were replaced with commercial fertilizers, the expense would be even greater. Roger Revelle of the University of California at San Diego has calculated that 89 percent of the energy used in Indian agriculture (the equivalent of about 140 million tons of coal) is provided by local sources. Even if foreign loans were to provide the money, the capital outlay necessary to replace the Indian cow with tractors and fertilizers for the fields, coal for the fires, and transportation for the family would probably warp international financial institutions for years.

Instead of asking the Indians to learn from the American model of industrial agriculture, American farmers might learn energy conservation from the Indians. Every step in an energy cycle results in a loss of energy to the system. Like a pendulum that slows a bit with each swing, each transfer of energy from sun to plants, plants to animals, and animals to human beings involves energy losses. Some systems are more efficient than others; they provide a higher percentage of the energy inputs in a final, useful form. Seventeen percent of all energy zebus consume is returned in the form of milk, traction, and dung. American cattle raised on Western rangeland return only 4 percent of the energy they consume.

But the American system is improving. Based on techniques pioneered by Indian scientists, at least one commercial firm in the United States is reported to be building plants that will turn manure from cattle feedlots into combustible gas. When organic matter is broken down by anaerobic bacteria, methane gas and carbon dioxide are produced. After the methane is cleansed of the carbon dioxide, it is available for the same purposes as natural gas—cooking, heating, electric generation. The company constructing the biogasification plant plans to sell its product to a gas-supply company, to be piped through the existing distribution system. Schemes similar to this one could make cattle ranches almost independent of utility and gasoline companies, for methane can be used to run trucks, tractors, and cars as well as to supply heat and electricity. The relative energy self-sufficiency that the Indian peasant has achieved is a goal American farmers and industry are now striving for.

Studies like Odend'hal's understate the efficiency of the Indian cow, because dead cows are used for purposes that Hindus prefer not to acknowledge. When a cow dies, an Untouchable, a member of one of the lowest ranking castes in India, is summoned to haul away the carcass. Higher castes consider the body of the dead cow polluting; if they handle it, they must go through a rite of purification.

Untouchables first skin the dead animal and either tan the skin themselves or sell it to a leather factory. In the privacy of their homes, contrary to the teachings of Hinduism, untouchable castes cook the meat and eat it. Indians of all castes rarely acknowledge the existence of these practices to non-Hindus, but most are aware that beefeating takes place. The prohibition against beefeating restricts consumption by the higher castes and helps distribute animal protein to the poorest sectors of the population that otherwise would have no source of these vital nutrients.

Untouchables are not the only Indians who consume beef. Indian Muslims and Christians are under no restriction that forbids them beef, and its consumption is legal in many places. The Indian ban on cow slaughter is state, not national, law and not all states restrict it. In many cities, such as New Delhi, Calcutta, and Bombay, legal slaughterhouses sell beef to retail customers and to restaurants that serve steak.

If the caloric value of beef and the energy costs involved in the manufacture of synthetic leather were included in the estimate of energy, the calculated efficiency of Indian livestock would rise considerably. As well as the system works, experts often claim that its efficiency can be further improved. Alan Heston [et al., 1971], an economist at the University of Pennsylvania, believes that Indians suffer from an overabundance of cows simply because they refuse to slaughter the excess cattle. India could produce at least the same number of oxen and the same quantities of milk and manure with 30 million fewer cows. Heston calculates that only 40 cows are necessary to maintain a population of 100 bulls and oxen. Since India averages 70 cows for every 100 bullocks, the difference, 30 million cows, is expendable.

What Heston fails to note is that sex ratios among cattle in different regions of India vary tremendously, indicating that adjustments in the cow population do take place. Along the Ganges River, one of the holiest shrines of Hinduism, the ratio drops to 47 cows for every 100 male animals. This ratio reflects the preference for dairy buffalo in the irrigated sectors of the Gangetic Plains. In nearby Pakistan, in contrast, where cow slaughter is permitted, the sex ratio is 60 cows to 100 oxen.

Since the sex ratios among cattle differ greatly from region to region and do not even approximate the balance that would be expected if no females were killed, we can assume that some culling of herds does take place; Indians do adjust their religious restrictions to accommodate ecological realities.

They cannot kill a cow but they can tether an old or unhealthy animal until it has starved to death. They cannot slaughter a calf but they can yoke it with a large wooden triangle so that when it nurses it irritates the mother's udder and gets kicked to death. They cannot ship their animals to the slaughterhouse but they can sell them to Muslims, closing their eyes to the fact that the Muslims will take the cattle to the slaughterhouse.

These violations of the prohibition against cattle slaughter strengthen the premise that cow worship is a vital part of Indian culture. The practice arose to prevent the population from consuming the animal on which Indian agriculture depends. During the first millennium B.C., the Ganges Valley became one of the most densely populated regions of the world.

Where previously there had been only scattered villages, many towns and cities arose and peasants farmed every available acre of land. Kingsley Davis, a population expert at the University of California at Berkeley, estimates that by 300 B.C. between 50 million and 100 million people were living in India. The forested Ganges Valley became a windswept semidesert and signs of ecological collapse appeared; droughts and floods became commonplace, erosion took away the rich topsoil, farms shrank as population increased, and domesticated animals became harder and harder to maintain.

It is probable that the elimination of meat eating came about in a slow, practical manner. The farmers who decided not to eat their cows; who saved them for procreation to produce oxen, were the ones who survived the natural disasters. Those who ate beef lost the tools with which to farm. Over a period of centuries, more and more farmers probably avoided beef until an unwritten taboo came into existence.

Only later was the practice codified by the priesthood. While Indian peasants were probably aware of the role of cattle in their society, strong sanctions were necessary to protect zebus from a population faced with starvation. To remove temptation, the flesh of cattle became taboo and the cow became sacred.

The sacredness of the cow is not just an ignorant belief that stands in the way of progress. Like all concepts of the sacred and the profane, this one affects the physical world; it defines the relationships that are important for the maintenance of Indian society.

Indians have the sacred cow, we have the "sacred" car and the "sacred" dog. It would not occur to us to propose the elimination of automobiles and dogs from our society without carefully considering the consequences, and we should not propose the elimination of zebu cattle without first understanding their place in the social order of India.

Human society is neither random nor capricious. The regularities of thought and behavior called culture are the principal mechanisms by which we human beings adapt to the world around us. Practices and beliefs can be rational or irrational, but a society that fails to adapt to its environment is doomed to extinction. Only those societies that draw the necessities of life from their surroundings inherit the earth. The West has much to learn from the great antiquity of Indian civilization, and the sacred cow is an important part of that lesson.

REFERENCES

Gandhi, Mohandas K. 1954. *How to Serve the Cow.* Bombay: Navajivan Publishing House.

Heston, Alan et al., 1971. An Approach to the Sacred Cow of India. *Current Anthropology* 12, 191–209.

Odend'hal, Stewart. 1972. Gross Energetic Efficiency of Indian Cattle in Their Environment. *Journal of Human Ecology* 1, 1–27.

16

Development as Poison: Rethinking the Western Model of Modernity

STEPHEN A. MARGLIN

The West is best—or so many Westerners believe! Throughout recent centuries Western countries have exported their languages, religions, judicial and political systems, and other institutions, often by force, mostly to poorer countries. Marglin presents contrasts between Western and traditional community cultures. He argues that assuming Western culture is the best and good for everyone is problematic because it may result in breakdown of core traditional values such as community.

As you read, ask yourself the following questions:

1. *What is meant by "development as poison"?*
2. *What effect might the destruction of non-Western traditions have on the world?*
3. *What has led to the West's view that it is culturally superior?*
4. *Why do some leaders believe that Western values are best for a globalizing world?*

GLOSSARY **White man's burden** The portrayal of imperialism as an altruistic effort to bring benefits of the West to uncivilized peoples.
Externalities Lessening involvement of individuals in community efforts.

Very real gains ... have come with development. In the past three decades, infant and child mortality have fallen by 66 percent in Indonesia and Peru, by 75 percent in Iran and Turkey, and by 80 percent in Arab oil-producing states. In most parts of the world, children not only have a greater probability of surviving into adulthood, they also have more to eat than their parents did—not to mention better access to schools and doctors and a prospect of work lives of considerably less drudgery.

Nonetheless,...[m]alnutrition and hunger persist alongside the tremendous riches that have come with development and globalization. In South Asia almost a quarter of the population is undernourished and in sub-Saharan Africa, more than a third. The outrage of anti-globalization protestors in Seattle, Genoa, Washington, and Prague was directed against the meagerness of the portions, and rightly so.

But more disturbing than the meagerness of development's portions is its deadliness. Whereas other critics highlight the distributional issues that compromise development, my emphasis is rather on the terms of the project itself, which involve the destruction of indigenous cultures and communities. This result is more than a side effect of development; it is central to the underlying values and assumptions of the entire Western development enterprise.

THE WHITE MAN'S BURDEN

Along with the technologies of production, health-care, and education, development has spread the culture of the modern West all over the world, and thereby undermined other ways of seeing, understanding, and being. By culture I mean something more than artistic sensibility or intellectual refinement. "Culture" is used here the way anthropologists understand the term, to mean the totality of patterns of behavior and belief that characterize a specific society. Outside the modern West, culture is sustained through community, the set of connections that bind people to one another economically, socially, politically, and spiritually. Traditional communities are not simply about shared spaces, but about shared participation and experience in producing and exchanging goods and services, in governing, entertaining and mourning, and in the physical, moral, and spiritual life of the community. The culture of the modern West, which values the market as the primary organizing principle of life, undermines these traditional communities just as it has undermined community in the West itself over the last 400 years.

The West thinks it does the world a favor by exporting its culture along with the technologies that the non-Western world wants and needs. This is not a recent idea. A century ago, Rudyard Kipling, the poet laureate of British imperialism, captured this sentiment in the phrase "White Man's burden," which portrayed imperialism as an altruistic effort to bring the benefits of Western rule to uncivilized peoples. Political imperialism died in the wake of World War II, but cultural imperialism is still alive and well. Neither practitioners nor theorists speak today of the white man's burden—no

development expert of the 21st century hankers after clubs or golf courses that exclude local folk from membership. Expatriate development experts now work with local people, but their collaborators are themselves formed for the most part by Western culture and values and have more in common with the West than they do with their own people. Foreign advisers—along with their local collaborators—are still missionaries, missionaries for progress as the West defines the term. As our forbears saw imperialism, so we see development.

There are in fact two views of development and its relationship to culture, as seen from the vantage point of the modern West. In one, culture is only a thin veneer over a common, universal behavior based on rational calculation and maximization of individual self interest. On this view, which is probably the view of most economists, the Indian subsistence-oriented peasant is no less calculating, no less competitive, than the U.S. commercial farmer.

There is a second approach which, far from minimizing cultural differences, emphasizes them. Cultures, implicitly or explicitly, are ranked along with income and wealth on a linear scale. As the West is richer, Western culture is more progressive, more developed. Indeed, the process of development is seen as the transformation of backward, traditional, cultural practices into modern practice, the practice of the West, the better to facilitate the growth of production and income.

What these two views share is confidence in the cultural superiority of the modern West. The first, in the guise of denying culture, attributes to other cultures Western values and practices. The second, in the guise of affirming culture, posits an inclined plane of history (to use a favorite phrase of the Indian political psychologist Ashis Nandy) along which the rest of the world is, and ought to be, struggling to catch up with us. Both agree on the need for "development." In the first view, the Other is a miniature adult, and development means the tender nurturing by the market to form the miniature Indian or African into full-size Westerner. In the second, the Other is a child who needs structural transformation and cultural improvement to become an adult.

Both conceptions of development make sense in the context of individual people precisely because there is an agreed-upon standard of adult behavior against which progress can be measured. Or at least there was until two decades ago when the psychologist Carol Gilligan challenged the conventional wisdom of a single standard of individual development. Gilligan's book *In a Different Voice* argued that the prevailing standards of personal development were male standards. According to these standards, personal development was measured by progress from intuitive, inarticulate, cooperative, contextual, and personal modes of behavior toward rational, principled, competitive, universal, and impersonal modes of behavior—that is, from "weak" modes generally regarded as feminine and based on experience to "strong" modes regarded as masculine and based on algorithm.

Drawing from Gilligan's study, it becomes clear that on an international level, the development of nation-states is seen the same way. What appear to be universally agreed-upon guidelines to which developing societies must conform are actually impositions of Western standards through cultural imperialism. Gilligan did for the study of personal development what must be done for economic development: allowing for difference. Just as the development of individuals should be seen as the flowering of that which is special and unique within each of us—a process by which an acorn becomes an oak rather than being obliged to become a maple—so the development of peoples Should be conceived as the flowering of what is special and unique within each culture. This is not to argue for a cultural relativism in which all beliefs and practices sanctioned by some culture are equally valid on a moral, aesthetic, or practical plane. But it is to reject the universality claimed by Western beliefs and practices.

Of course, some might ask what the loss of a culture here or there matters if it is the price of material progress, but there are two flaws to this argument. First, cultural destruction is not necessarily a corollary of the technologies that extend life and improve its quality. Western technology can be decoupled from the entailments of Western culture. Second, if I am wrong about this, I would ask, as Jesus does in the account of Saint Mark, "[W]hat shall it profit a man, if he shall gain the whole world, and lose his own soul?" For all the material progress that the West has achieved, it has paid a high price through the weakening to the breaking point of communal ties. We in the West have much to learn, and the cultures that are being destroyed in the name of progress are perhaps the best resource we have for restoring balance to our own lives. The advantage of taking a critical stance, with respect to our own culture is that we become more ready to enter into a genuine dialogue with other ways of being and believing.

THE CULTURE OF THE MODERN WEST

Culture is in the last analysis a set of assumptions, often unconsciously held, about people and how they relate to one another. The assumptions of modern Western culture can be described under five headings: individualism, self-interest, the privileging of "rationality," unlimited wants, and the rise of the moral and legal claims of the nation-state on the individual.

Individualism is the notion that society can and should be understood as a collection of autonomous individuals, that groups—with the exception of the nation-state—have no normative significance as groups; that all behavior, policy, and even ethical judgment should be reduced to their effects on individuals. All individuals play the game of life on equal terms, even if they start with different amounts of physical strength, intellectual capacity, or capital assets. The playing field is level even if the players are not equal. These individuals are taken as given in many important ways rather than as works in progress. For example, preferences are accepted as given and cover everything from views about the relative merits of different flavors of ice cream to views about the relative merits of prostitution, casual sex, sex among friends, and sex within committed relationships. In an excess of democratic zeal, the children of the 20th century have extended the notion of radical subjectivism to the whole domain of preferences: one set of "preferences" is as good as another.

Self-interest is the idea that individuals make choices to further their own benefit. There is no room here for duty, right, or obligation, and that is a good thing, too. Adam Smith's best remembered contribution to economics, for better or worse, is the idea of a harmony that emerges from the pursuit of self-interest. It should be noted that while individualism is a prior condition for self-interest—there is no place for self-interest without the self—the converse does not hold. Individualism does not necessarily imply self-interest.

The third assumption is that one kind of knowledge is superior to others. The modern West privileges the algorithmic over the experiential, elevating knowledge that can be logically deduced from what are regarded as self-evident first principles over what is learned from intuition and authority, from touch and feel. In the stronger form of this ideology, the algorithmic is not only privileged but recognized as the sole legitimate form of knowledge. Other knowledge is mere belief, becoming legitimate only when verified by algorithmic methods.

Fourth is unlimited wants. It is human nature that we always want more than we have and that there is, consequently, never enough. The possibilities of abundance are always one step beyond our reach. Despite the enormous growth in production and consumption, we are as much in thrall to the economy as our parents, grandparents, and great-grandparents. Most U.S. families find one income inadequate for their needs, not only at the bottom of the distribution—where falling real wages have eroded the standard of living over the past 25 years—but also in the middle and upper ranges of the distribution. Economics, which encapsulates in stark form the assumptions of the modern West, is frequently defined as the study of the allocation of limited resources among unlimited wants.

Finally, the assumption of modern Western culture is that the nation-state is the preeminent social grouping and moral authority. Worn out by fratricidal wars of religion, early modern Europe moved firmly in the direction of making one's relationship to God a private matter—a taste or preference among many. Language, shared commitments, and a defined territory would, it was hoped, be a less divisive basis for social identity, than religion had proven to be.

AN ECONOMICAL SOCIETY

Each of these dimensions of modern Western culture is in tension with its opposite. Organic or holistic conceptions of society exist side by side with individualism. Altruism and fairness are opposed to self-interest. Experiential knowledge exists, whether we recognize it or not, alongside algorithmic knowledge. Measuring who we are by what we have has been continually resisted by the small voice within that calls us to be our better selves. The modern nation-state claims, but does not receive, unconditional loyalty.

So the sway of modern Western culture is partial and incomplete even within the geographical boundaries of the West. And a good thing too, since no society organized on the principles outlined above could last five minutes, much less the 400 years that modernity has been in the ascendant. But make no mistake—modernity is the dominant culture in the West and increasingly so throughout the world. One has only to examine the assumptions that underlie contemporary economic thought—both stated and unstated—to confirm this assessment. Economics is simply the formalization of the assumptions of modern Western culture. That both teachers and students of economics accept these assumptions uncritically speaks volumes about the extent to which they hold sway.

It is not surprising then that a culture characterized in this way is a culture in which the market is the organizing principle of social life. Note my choice of words, "the market" and "social life," not "markets" and "economic life." Markets have been with us since time out of mind, but the market, the idea of markets as a system for organizing production and exchange, is a distinctly modern invention, which grew in tandem with the cultural assumption of the self-interested, algorithmic individual who pursues wants without limit, an individual who owes allegiance only to the nation-state.

There is no sense in trying to resolve the chicken–egg problem of which came first. Suffice it to say that we can hardly have the market without the assumptions that justify a market system—and the market system can function acceptably only when the assumptions of the modern West are widely shared. Conversely, once these assumptions are prevalent, markets appear to be a "natural" way to organize life.

MARKETS AND COMMUNITIES

If people and society were as the culture of the modern West assumes, then market and community would occupy separate ideological spaces, and would coexist or not as people chose. However, contrary to the assumptions of individualism, the individual does not encounter society as a fully formed human being. We are constantly being shaped by our experiences, and in a society organized in terms of markets, we are formed by our experiences in the market. Markets organize not only the production and distribution of things; they also organize the production of people.

The rise of the market system is thus bound up with the loss of community. Economists do not deny this, but rather put a market friendly spin on the destruction of community: impersonal markets accomplish more efficiently what the connections of social solidarity, reciprocity, and other redistributive institutions do in the absence of markets....

If love is not scarce in the way that bread is, it is not sensible to design social institutions to economize on it. On the contrary, it makes sense to design social institutions to draw out and develop the community's stock of love. It is only when we focus on barns rather than on the people raising barns that insurance appears to be a more effective way of coping with disaster than is a community-wide

barn-raising. The Amish, who are descendants of 18th-century immigrants to the United States, are perhaps unique in the United States for their attention to fostering community; they forbid insurance precisely because they understand that the market relationship between an individual and the insurance company undermines the mutual dependence of the individuals that forms the basis of the community. For the Amish, barn-raisings are not exercises in nostalgia, but the cement which holds the community together.

Indeed, community cannot be viewed as just another good subject to the dynamics of market supply and demand that people can choose or not as they please, according to the same market test that applies to brands of soda or flavors of ice cream. Rather, the 'maintenance of community must be a collective responsibility for two reasons. The first is the so-called "free rider" problem. To return to the insurance example, my decision to purchase fire insurance rather than participate in the give and take of barn-raising with my neighbors has the side effect—the "externality" in economics jargon—of lessening my involvement with the community. If I am the only one to act this way, this effect may be small with no harm done. But when all of us opt for insurance and leave caring for the community to others, there will be no others to care, and the community will disintegrate. In the case of insurance, I buy insurance because it is more convenient, and—acting in isolation—I can reasonably say to myself that my action hardly undermines the community. But when we all do so, the cement of mutual obligation is weakened to the point that it no longer supports the community.

The free rider problem is well understood by economists, and the assumption that such problems are absent is part of the standard fine print in the warranty that economists provide for the market. A second, deeper, problem cannot so easily be translated into the language of economics. The market creates more subtle externalities that include effects on beliefs, values, and behaviors—a class of externalities which are ignored in the standard framework of economics in which individual "preferences" are assumed to be unchanging. An Amishman's decision to insure his barn undermines the mutual dependence of the Amish not only by making him less dependent on the community, but also by subverting the beliefs that sustain this dependence. For once interdependence is undermined, the community is no longer valued; the process of undermining interdependence is self-validating.

Thus, the existence of such externalities means that community survival cannot be left to the spontaneous initiatives of its members acting in accord with the individual maximizing model. Furthermore, this problem is magnified when the externalities involve feedback from actions to values, beliefs, and then to behavior. If a community is to survive, it must structure the interactions of its members to strengthen ways of being and knowing which support community. It will have to constrain the market when the market undermines community.

A DIFFERENT DEVELOPMENT

There are two lessons here. The first is that there should be mechanisms for local communities to decide, as the Amish routinely do, which innovations in organization and technology are compatible with the core values the community wishes to preserve. This does not mean the blind preservation of whatever has been sanctioned by time and the existing distribution of power. Nor does it mean an idyllic, conflict-free path to the future. But recognizing the value as well as the fragility of community would be a giant step forward in giving people a real opportunity to make their portions less meager and avoiding the poison.

The second lesson is for practitioners and theorists of development. What many Westerners see simply as liberating people from superstition, ignorance, and the oppression of tradition, is fostering values, behaviors, and beliefs that are highly problematic for our own culture. Only arrogance and a supreme failure of the imagination cause us to see them as universal rather than as the product of a particular history. Again, this is not to argue that "anything goes." It is instead a call for sensitivity for entering into a dialogue that involves listening instead of dictating—not so that we can better implement our own agenda, but so that we can genuinely learn that which modernity has made us forget.

Chapter 5

Social Interaction, Groups, and Bureaucracy

Life is with People

Dr. Alan W. McEvoy

Groups are the setting for socialization, the place in which the lifelong process of learning the ways of society is carried out. The first place is generally in the family, then in play groups, followed by more formal group associations in institutions of education and religion, and as adults in economic and political institutions. As individuals grow, so do their group contacts and activities within these groups. Familial relations expand to other "primary groups" such as play and peer groups, then to more formal groups such as the school. When individuals become adults, many of their interactions take place within secondary groups or organizations such as the workplace and characterized by more formal relations.

Individuals make choices about some of their group affiliations; other memberships are obligatory. Etzioni (1975) classifies these group and organizational affiliations into three primary, sometimes overlapping, categories based on how the organizations gain compliance or loyalty from their members: utilitarian, coercive, and normative. Utilitarian organizations provide benefits to members in the form of wages or profits; individuals usually need these benefits to live. Coercive organizations are involuntary; individuals are forced to be there, whether it be prison, the military, or school. Normative organizations are most often voluntary; people join because they believe in the organization's purposes or goals, such as religion, politics, or social causes.

Within these categories, we find different types of groups. Consider your own group affiliations. Some are *primary*, being composed of close and lasting relationships—your family and best friends. Others are *secondary*, consisting of formal, impersonal, businesslike relationships. In addition, some groups are *in-groups*, meaning that we feel a sense of loyalty and belonging to the group, as opposed to *out-groups*, to which we do not belong and may see as in competition or opposition to our group (Ballantine and Roberts 2009).

People interact within groups and organizations. We think of interaction as talking with others, but as you may have discovered if you traveled abroad, talking is not the only or even main part of interaction. A major part of our interaction with other individuals or within groups takes place through nonverbal communication. Edward and Mildred Hall describe the silent part of language—the nonverbal communication of unconscious body movements. Body language involves physical cues—glances, movement, facial expressions, and so on—as well as social distance and personal space. Because of its subtleties and the vast range of possible nonverbal rules, silent language is more difficult to master than verbal speech. Consider your interactions with a person you find attractive. The possibilities for misreading cues about their interest are enormous. To complicate matters, as Hall and Hall show, similar movements may have different meanings in different situations or cultures.

Most of our interactions take place in groups. A large body of sociological literature deals with group relations, from role relationships, leadership, and decision making to the tendency toward conformity in group situations because individuals strive to get along with others. In a classic discussion of in-groups and out-groups, William Graham Sumner points out that some groups bind members together because of

> relation to each other (kin, neighborhood, alliance, connubium and commercium) which draws them together and differentiates them from others. Thus a differentiation arises between ourselves, the we-group, or in-group, and everybody else, or the others-groups, out-groups. The insiders in a we-group are in a relation of peace, order, law, government, and industry to each other. Their relation to all outsiders, or others-groups, is one of war and plunder, except so far as agreements have modified it (Sumner 1904).

Within groups, we play "roles"—those behaviors expected of us because of our participation in groups and positions we hold: mother, brother, daughter, uncle, worker, student, friend. Each role demands certain expectations and behaviors. Sometimes these expectations conflict, causing "role conflict." An example is seen in the next article. The Adlers analyze the process of learning and carrying out the college athlete role. The intense pressure for performance, adulation from crowds and the press, and intimidation of classroom expectations combine to provide an example of group socialization into rigid role expectations.

Although the case of athletes presents a stressful kind of group for the members, Philip Zimbardo asks what happens when the situation facilitates or even encourages negative group behavior. He considers prison abuse in which bad behavior is allowed and even encouraged by analyzing the abuse scandals at Abu Ghraib prison in Iraq in the context of group norms and the context of setting.

One famous observer of groups and society was Max Weber. Writing at the turn of the century, he saw a modern form of groups emerging as societies industrialized. As businesses grew larger and more technical, owners no longer hired friends and relatives, but rather hired and fired people on the basis of their training, skills, and performance. Bureaucracies as Weber described them were organized as hierarchies with sets of rules and regulations, and individuals were formally contracted to carry out certain tasks in exchange for compensation. Today it would seem strange if organizations did not function in this manner. Characteristics of bureaucratic organizations include the formal contractual relationships, rules and regulations, and hiring and firing based on merit and competence, as discussed in the last reading.

Since Weber first wrote about bureaucracy, there have been changes in the way bureaucracies function. Technology, for instance, has created new relationships between organizations and individuals. At the same time, individual participation in groups has been changing. Today there is less personal involvement in groups and more technological involvement, which tends to isolate individuals in modern societies.

As you read the selections in this chapter, consider the definitions of primary and secondary groups and reflect on examples of these groups in modern society. What dynamics do we need to understand to interact and communicate effectively in groups? Are some problems in group dynamics outcomes of our need for belonging and acceptance? How are organizations changing? How is individual participation in groups changing?

REFERENCES

Ballantine, Jeanne H., and Keith A. Roberts. 2009. *Our Social World: Introduction to Sociology*. Thousand Oaks, CA: Pine Forge, p. 145.

Etzioni, Amitai. 1975. *A Comparative Analysis of Complex Organizations*. New York: The Free Press.

Sumner, William Graham. 1904. *Folkways*. New York: Ginn and Company, secs. 12, 13.

17

The Sounds of Silence

EDWARD T. HALL AND MILDRED REED HALL

A crucial element in human interaction and group behavior is language, both verbal and nonverbal. Nonverbal communication is quite as important in communicating as verbal language. When we study foreign language, we may learn the words, but the gestures and facial expressions are difficult to master unless we have been raised in the culture. The Halls discuss nonverbal communication patterns and differences across cultures.

Think about the following questions as you read this selection:

1. *What is meant by "the sounds of silence"?*
2. *How do individuals learn nonverbal language?*
3. *Can you think of times you have been uncomfortable because of misunderstandings in nonverbal communication?*
4. *What might be the consequences of misunderstood nonverbal language cues between men and women?*

GLOSSARY   **Silent language** Communication that takes place through gestures, facial expressions, and other body movements.   **Machismo** Assertive masculinity.

Bob leaves his apartment at 8:15 A.M. and stops at the corner drugstore for breakfast. Before he can speak, the counterman says, "The usual?" Bob nods yes. While he savors his Danish, a fat man pushes onto the adjoining stool and overflows into his space. Bob scowls and the man pulls himself in as much as he can. Bob has sent two messages without speaking a syllable.

Henry has an appointment to meet Arthur at 11 o'clock; he arrives at 11:30. Their conversation is friendly, but Arthur retains a lingering hostility. Henry has unconsciously communicated that he doesn't think the appointment is very important or that Arthur is a person who needs to be treated with respect.

George is talking to Charley's wife at a party. Their conversation is entirely trivial, yet Charley glares at them suspiciously. Their physical proximity and the movements of their eyes reveal that they are powerfully attracted to each other.

José Ybarra and Sir Edmund Jones are at the same party, and it is important for them to establish a cordial relationship for business reasons. Each is trying to be warm and friendly, yet they will part with mutual distrust and their business transaction will probably fall through. José, in Latin fashion, moved

Source: Excerpted from *The Sounds of Silence* by Edward T. Hall and Mildred Reed Hall. Originally appeared in *Playboy* Magazine. Copyright © 1981 by Edward T. Hall and Mildred Reed Hall. Reprinted with permission of the authors.

closer and closer to Sir Edmund as they spoke, and this movement was miscommunicated as pushiness to Sir Edmund, who kept backing away from this intimacy, and this was miscommunicated to José as coldness. The silent languages of Latin and English cultures are more difficult to learn than their spoken languages.

In each of these cases, we see the subtle power of nonverbal communication. The only language used throughout most of the history of humanity (in evolutionary terms, vocal communication is relatively recent), it is the first form of communication you learn. You use this preverbal language, consciously and unconsciously, every day to tell other people how you feel about yourself and them. This language includes your posture, gestures, facial expressions, costume, the way you walk, even your treatment of time and space and material things. All people communicate on several different levels at the same time but are usually aware of only the verbal dialog and don't realize that they respond to nonverbal messages. But when a person says one thing and really believes something else, the discrepancy between the two can usually be sensed. Nonverbal communication systems are much less subject to the conscious deception that often occurs in verbal systems. When we find ourselves thinking, "I don't know what it is about him, but he doesn't seem sincere," it's usually this lack of congruity between a person's words and his behavior that makes us anxious and uncomfortable.

Few of us realize how much we all depend on body movement in our conversation or are aware of the hidden rules that govern listening behavior. But we know instantly whether or not the person we're talking to is "tuned in" and we're very sensitive to any breach in listening etiquette. In white middle-class American culture, when someone wants to show he is listening to someone else, he looks either at the other person's face, or specifically, at his eyes, shifting his gaze from one eye to the other.

If you observe a person conversing, you'll notice that he indicates he's listening by nodding his head. He also makes little "Hmm" noises. If he agrees with what's being said, he may give a vigorous nod. To show pleasure or affirmation, he smiles; if he has some reservations, he looks skeptical by raising an eyebrow or pulling down the corners of his mouth. If a participant wants to terminate the conversation, he may start shifting his body position, stretching his legs, crossing or uncrossing them, bobbing his foot or diverting his gaze from the speaker. The more he fidgets, the more the speaker becomes aware that he has lost his audience. As a last measure, the listener may look at his watch to indicate the imminent end of the conversation.

Talking and listening are so intricately intertwined that a person cannot do one without the other. Even when one is alone and talking to oneself, there is part of the brain that speaks while another part listens. In all conversations, the listener is positively or negatively reinforcing the speaker all the time. He may even guide the conversation without knowing it, by laughing or frowning or dismissing the argument with a wave of his hand.

The language of the eyes—another age-old way of exchanging feelings—is both subtle and complex. Not only do men and women use their eyes differently but there are class, generation, regional, ethnic, and national cultural differences. Americans often complain about the way foreigners stare at people or hold a glance too long. Most Americans look away from someone who is using his eyes in an unfamiliar way because it makes them self-conscious. If a man looks at another man's wife in a certain way, he's asking for trouble, as indicated earlier. But he might not be ill-mannered or seeking to challenge the husband. He might be a European in this country who hasn't learned our visual mores. Many American women visiting France or Italy are acutely embarrassed because, for the first time in their lives, men really look at them—their eyes, hair, nose, lips, breasts, hips, legs, thighs, knees, ankles, feet, clothes, hairdo, even their walk....

Analyzing the mass of data on the eyes, it is possible to sort out at least three ways in which the eyes are used to communicate: dominance versus submission, involvement versus detachment, and positive versus negative attitude. In addition, there are three levels of consciousness and control, which can be categorized as follows: (1) conscious use of the eyes to communicate, such as the flirting blink and the intimate nose-wrinkling squint; (2) the very extensive category of unconscious but learned behavior governing where the eyes are directed and when (this unwritten set of rules dictates how and under what circumstances the sexes, as well as people of all status categories, look at each other); and (3) the response of the eye itself, which is completely outside both awareness and control—changes in the cast (the sparkle) of the eye and the pupillary reflex.

The eye is unlike any other organ of the body, for it is an extension of the brain. The unconscious pupillary reflex and the cast of the eye have been known by people of Middle Eastern origin for years—although most are unaware of their knowledge. Depending on the context, Arabs and others look either directly at the eyes or deeply *into* the eyes of their interlocutor. We became aware of this in the Middle East several years ago while looking at jewelry. The merchant suddenly started to push a particular bracelet at a customer and said, "You buy this one." What interested us was that the bracelet was not the one that had been consciously selected by the purchaser. But the merchant, watching the pupils of the eyes, knew what the purchaser really wanted to buy. Whether he specifically knew *how* he knew is debatable.

A psychologist at the University of Chicago, Eckhard Hess, was the first to conduct systematic studies of the pupillary reflex. His wife remarked one evening, while watching him reading in bed, that he must be very interested in the text because his pupils were dilated. Following up on this, Hess slipped some pictures of nudes into a stack of photographs that he gave to his male assistant. Not looking at the photographs but watching his assistant's pupils, Hess was able to tell precisely when the assistant came to the nudes. In further experiments, Hess retouched the eyes in a photograph of a woman. In one print, he made the pupils small, in another, large; nothing else was changed. Subjects who were given the photographs found the woman with the dilated pupils much more attractive. Any man who has had the experience of seeing a woman look at him as her pupils widen with reflex speed knows that she's flashing him a message.

The eye-sparkle phenomenon frequently turns up in our interviews of couples in love. It's apparently one of the first reliable clues in the other person that love is genuine. To date, there is no scientific data to explain eye sparkle; no investigation of the pupil, the cornea, or even the white sclera of the eye shows how the sparkle originates. Yet we all know it when we see it.

One common situation for most people involves the use of the eyes in the street and in public. Although eye behavior follows a definite set of rules, the rules vary according to the place, the needs and feelings of the people, and their ethnic background. For urban whites, once they're within definite recognition distance (16–32 feet for people with average eyesight), there is mutual avoidance of eye contact—unless they want something specific: a pickup, a handout, or information of some kind. In the West and in small towns generally, however, people are much more likely to look at and greet one another, even if they're strangers.

It's permissible to look at people if they're beyond recognition distance; but once inside this sacred zone, you can only steal a glance at strangers. You *must* greet friends, however; to fail to do so is insulting. Yet, to stare too fixedly even at them is considered rude and hostile. Of course, all of these rules are variable....

[A] very basic difference between people of different ethnic backgrounds is their sense of territoriality and how they handle space. This is the silent communication, or miscommunication, that caused friction between Mr. Ybarra and Sir Edmund Jones in our earlier example. We know from research that everyone has around himself an invisible bubble of space that contracts and expands depending on several factors: his emotional state, the activity he's performing at the time, and his cultural background. This bubble is a kind of mobile territory that he will defend against intrusion. If he is accustomed to close personal distance between himself and others, his bubble will be smaller than that of someone who's accustomed to greater personal distance. People of North European heritage—English, Scandinavian, Swiss, and German—tend to avoid contact. Those whose heritage is Italian, French, Spanish, Russian,

Latin American, or Middle Eastern like close personal contact.

People are very sensitive to any intrusion into their spatial bubble. If someone stands too close to you, your first instinct is to back up. If that's not possible, you lean away and pull yourself in, tensing your muscles. If the intruder doesn't respond to these body signals, you may then try to protect yourself, using a briefcase, umbrella, or raincoat.... As a last resort, you may move to another spot and position yourself behind a desk or a chair that provides screening. Everyone tries to adjust the space around himself in a way that's comfortable for him; most often, he does this unconsciously.

Emotions also have a direct effect on the size of a person's territory. When you're angry or under stress, your bubble expands and you require more space. New York psychiatrist Augustus Kinzel found a difference in what he calls Body-Buffer Zones between violent and nonviolent prison inmates. Dr. Kinzel conducted experiments in which each prisoner was placed in the center of a small room and then Dr. Kinzel slowly walked toward him. Nonviolent prisoners allowed him to come quite close, while prisoners with a history of violent behavior couldn't tolerate his proximity and reacted with some vehemence.

Apparently, people under stress experience other people as looming larger and closer than they actually are. Studies of schizophrenic patients have indicated that they sometimes have a distorted perception of space, and several psychiatrists have reported patients who experience their body boundaries as filling up an entire room. For these patients, anyone who comes into the room is actually inside their body, and such an intrusion may trigger a violent outburst.

Unfortunately, there is little detailed information about normal people who live in highly congested urban areas. We do know, of course, that the noise, pollution, dirt, crowding, and confusion of our cities induce feelings of stress in most of us, and stress leads to a need for greater space. [People who are] packed into a subway, jostled in the street, crowded into an elevator, and forced to work all day in a bull pen or in a small office without auditory or visual privacy [are] going to be very stressed at the end of [the] day. They need places that provide relief from constant overstimulation.... Stress from overcrowding is cumulative and people can tolerate more crowding early in the day than later; note the increased bad temper during the evening rush hour as compared with the morning melee. Certainly one factor in people's desire to commute by car is the need for privacy and relief from crowding (except, often, from other cars); it may be the only time of day when nobody can intrude.

In crowded public places, we tense our muscles and hold ourselves stiff, and thereby communicate to others our desire not to intrude on their space and, above all, not to touch them. We also avoid eye contact and the total effect is that of someone who has "tuned out." Walking along the street, our bubble expands slightly as we move in a stream of strangers, taking care not to bump into them. In the office, at meetings, in restaurants, our bubble keeps changing as it adjusts to the activity at hand.

Most white middle-class Americans use four main distances in their business and social relations: intimate, personal, social, and public. Each of these distances has a near and a far phase and is accompanied by changes in the volume of the voice. Intimate distance varies from direct physical contact with another person to a distance of six to eighteen inches and is used for our most private activities— caressing another person or making love. At this distance, you are overwhelmed by sensory inputs from the other person—heat from the body, tactile stimulation from the skin, the fragrance of perfume, even the sound of breathing—all of which literally envelop you. Even at the far phase, you're still within easy touching distance. In general, the use of intimate distance in public between adults is frowned on. It's also much too close for strangers, except under conditions of extreme crowding.

In the second zone—personal distance—the close phase is one and a half to two and a half feet; it's at this distance that wives usually stand from their husbands in public. If another woman moves into this zone, the wife will most likely be disturbed. The far phase—two and a half to four feet—is the distance used to "keep someone at arm's length" and is the most common spacing used by people in conversation.

The third zone—social distance—is employed during business transactions or exchanges with a clerk or repairman. People who work together tend to use close social distance—four to seven feet. This is also the distance for conversation at social gatherings. To stand at this distance from one who is seated has a dominating effect (for example, teacher to pupil, boss to secretary). The far phase of the third zone—seven to twelve feet—is where people stand when someone says, "Stand back so I can look at you." This distance lends a formal tone to business or social discourse. In an executive office, the desk serves to keep people at this distance.

The fourth zone—public distance—is used by teachers in classrooms or speakers at public gatherings. At its farthest phase—25 feet and beyond—it is used for important public figures. Violations of this distance can lead to serious complications. During his 1970 U.S. visit, the president of France, Georges Pompidou, was harassed by pickets in Chicago, who were permitted to get within touching distance. Since pickets in France are kept behind barricades a block or more away, the president was outraged by this insult to his person, and President Nixon was obliged to communicate his concern as well as offer his personal apologies.

It is interesting to note how American pitchmen and panhandlers exploit the unwritten, unspoken conventions of eye and distance. Both take advantage of the fact that once explicit eye contact is established, it is rude to look away, because to do so means to brusquely dismiss the other person and his needs. Once having caught the eye of his mark, the panhandler then locks on, not letting go until he moves through the public zone, the social zone, the personal zone, and finally, into the intimate sphere, where people are most vulnerable.

Touch also is an important part of the constant stream of communication that takes place between people. A light touch, a firm touch, a blow, a caress are all communications. In an effort to break down barriers among people, there's been a recent upsurge in group-encounter activities, in which strangers are encouraged to touch one another. In special situations such as these, the rules for not touching are broken with group approval and people gradually lose some of their inhibitions.

Although most people don't realize it, space is perceived and distances are set not by vision alone but with all the senses. Auditory space is perceived with the ears, thermal space with the skin, kinesthetic space with the muscles of the body, and olfactory space with the nose. And, once again, it's one's culture that determines how his senses are programmed—which sensory information ranks highest and lowest. The important thing to remember is that culture is very persistent. In this country, we've noted the existence of culture patterns that determine distance between people in the third and fourth generations of some families, despite their prolonged contact with people of very different cultural heritages.

Whenever there is great cultural distance between two people, there are bound to be problems arising from differences in behavior and expectations. An example is the American couple who consulted a psychiatrist about their marital problems. The husband was from New England and had been brought up by reserved parents who taught him to control his emotions and to respect the need for privacy. His wife was from an Italian family and had been brought up in close contact with all the members of her large family, who were extremely warm, volatile, and demonstrative.

When the husband came home after a hard day at the office, dragging his feet and longing for peace and quiet, his wife would rush to him and smother him. Clasping his hands, rubbing his brow, crooning over his weary head, she never left him alone. But when the wife was upset or anxious about her day, the husband's response was to withdraw completely and leave her alone. No comforting, no affectionate embrace, no attention—just solitude. The woman became convinced her husband didn't love her and, in desperation, she consulted a psychiatrist. Their problem wasn't basically psychological but cultural.

Why [have people] developed all these different ways of communicating messages without words? One reason is that people don't like to spell out certain kinds of messages. We prefer to find other ways of showing our feelings. This is especially true in relationships as sensitive as courtship.... We work out subtle ways of encouraging or discouraging each other that save face and avoid confrontations....

If a man sees a woman whom he wants to attract, he tries to present himself by his posture and stance as someone who is self-assured. He moves briskly and confidently. When he catches the eye of the woman, he may hold her glance a little longer than normal. If he gets an encouraging smile, he'll move in close and engage her in small talk. As they converse, his glance shifts over her face and body. He, too, may make preening gestures—straightening his tie, smoothing his hair, or shooting his cuffs.

How do people learn body language? The same way they learn spoken language—by observing and imitating people around them as they're growing up. Little girls imitate their mothers or an older female. Little boys imitate their fathers or a respected uncle or a character on television. In this way, they learn the gender signals appropriate for their sex. Regional, class, and ethnic patterns of body behavior are also learned in childhood and persist throughout life.

Such patterns of masculine and feminine body behavior vary widely from one culture to another. In America, for example, women stand with their thighs together. Many walk with their pelvis tipped sightly forward and their upper arms close to their body. When they sit, they cross their legs at the knee or cross their ankles. American men hold their arms away from their body, often swinging them as they walk. They stand with their legs apart (an extreme example is the cowboy, with legs apart and thumbs tucked into his belt). When they sit, they put their feet on the floor with legs apart and, in some parts of the country, they cross their legs by putting one ankle on the other knee.

Leg behavior indicates sex, status, and personality. It also indicates whether or not one is at ease or is showing respect or disrespect for the other person. Young Latin American males avoid crossing their legs. In their world of *machismo*, the preferred position for young males when with one another (if there is no old dominant male present to whom they must show respect) is to sit on the base of the spine with their leg muscles relaxed and their feet wide apart. Their respect position is like our military

equivalent: spine straight, heels and ankles together—almost identical to that displayed by properly brought up young women in New England in the early part of this century.

American women who sit with their legs spread apart in the presence of males are *not* normally signaling a come-on—they are simply (and often unconsciously) sitting like men. Middleclass women in the presence of other women to whom they are very close may on occasion throw themselves down on a soft chair or sofa and let themselves go. This is a signal that nothing serious will be taken up. Males, on the other hand, lean back and prop their legs up on the nearest object.

The way we walk, similarly, indicates status, respect, mood, and ethnic or cultural affiliation.... To white Americans, some French middle-class males walk in a way that is both humorous and suspect. There is a bounce and looseness to the French walk, as though the parts of the body were somehow unrelated. Jacques Tati, the French movie actor, walks this way; so does the great mime, Marcel Marceau....

All over the world, people walk not only in their own characteristic way but have walks that communicate the nature of their involvement with whatever it is they're doing. The purposeful walk of North Europeans is an important component of proper behavior on the job. Any male who has been in the military knows how essential it is to walk properly.... The quick shuffle of servants in the Far East in the old days was a show of respect. On the island of Truk, when we last visited, the inhabitants even had a name for the respectful walk that one used when in the presence of a chief or when walking past a chief's house. The term was *sufan*, which meant to be humble and respectful.

The notion that people communicate volumes by their gestures, facial expressions, posture and walk is not new; actors, dancers, writers, and psychiatrists have long been aware of it. Only in recent years, however, have scientists begun to make systematic observations of body motions. Ray L. Birdwhistell of the University of Pennsylvania is one of the pioneers in body-motion research and coined the term *kinesics* to describe this field. He developed an elaborate notation system to record both facial and body movement, using an approach similar to that of the linguist, who studies the basic elements of speech. Birdwhistell and other kinesicists such as Albert Shellen, Adam Kendon, and William Condon take movies of people interacting. They run the film over and over again, often at reduced speed for frame-by-frame analysis, so that they can observe even the slightest body movements not perceptible at normal interaction speeds. These movements are then recorded in notebooks for later analysis....

Several years ago in New York City, there was a program for sending children from predominantly black and Puerto Rican low-income neighborhoods to summer school in a white upper-class neighborhood on the East Side. One morning, a group of young black and Puerto Rican boys raced down the street, shouting and screaming and overturning garbage cans on their way to school. A doorman from an apartment building nearby chased them and cornered one of them inside a building. The boy drew a knife and attacked the doorman. This tragedy would not have occurred if the doorman had been familiar with the behavior of boys from low-income neighborhoods, where such antics are routine and socially acceptable and where pursuit would be expected to invite a violent response.

The language of behavior is extremely complex. Most of us are lucky to have under control one subcultural system—the one that reflects our sex, class, generation, and geographic region within the United States. Because of its complexity, efforts to isolate bits of nonverbal communication and generalize from them are in vain; you don't become an instant expert on people's behavior by watching them at cocktail parties. Body language isn't something that's independent of the person, something that can be donned and doffed like a suit of clothes.

Our research and that of our colleagues have shown that, far from being a superficial form of communication that can be consciously manipulated, nonverbal-communication systems are interwoven into the fabric of the personality and, as sociologist Erving Goffman has demonstrated, into society itself. They are the warp and woof of daily interactions with others, and they influence how one expresses oneself, how one experiences oneself as a man or a woman.

Nonverbal communications signal to members of your own group what kind of person you are, how you feel about others, how you'll fit into and work in a group, whether you're assured or anxious, the degree to which you feel comfortable with the standards of your own culture, as well as deeply significant feelings about the self, including the state of your own psyche. For most of us, it's difficult to accept the reality of another's behavioral system. And, of course, none of us will ever become fully knowledgeable of the importance of every nonverbal signal. But as long as each of us realizes the power of these signals, this society's diversity can be a source of great strength rather than a further—and subtly powerful—source of division.

18

Backboards & Blackboards
College Athletes and Role Engulfment

PATRICIA A. ADLER AND PETER ADLER

We all carry out roles in groups to which we belong. We are socialized into school, work, and other roles. Sometimes this socialization is intense and influences all other roles. Such is the case with the achieved role of college athletes. The Adlers provide an example from the world of NCAA basketball, describing the process of socialization, role engulfment, and abandonment of former roles, and formation of team cliques that reflect earlier socialization experiences. The athletes work hard to achieve their goals and in the process are shaped by pressures from adoring fans and intimidating professors. Their roles in the group are shaped by all these pressures.

Consider the following as you read:

1. *What are the pressures faced by high-profile athletes?*
2. *How do the roles of student athletes differ from roles of most other students?*
3. *What would you recommend doing to reduce role engulfment?*
4. *Which role responsibilities have you had that involve role engulfment and abandonment of former roles?*

GLOSSARY **Role engulfment** Demands and rewards of athletic role supersede other roles. **Role domination** Process by which athletes become engulfed in athletic roles to exclusion of other roles. **Role abandonment** Detachment from investment in other roles, letting go of alternative goals and priorities. **Statuses** Positions in organized groups related to other positions by set of normative expectations. **Roles** Activities of people of given status.

It was a world of dreams. They expected to find fame and glory, spotlights and television cameras. There was excitement and celebrity, but also hard work and discouragement, a daily grind characterized by aches, pains, and injuries, and an abundance of rules, regulations, and criticism. Their lives alternated between contacts with earnest reporters, adoring fans, and fawning women, and with intimidating professors, demanding boosters, and unrelenting coaches. There was secrecy and intrigue, drama and adulation, but also isolation and alienation, loss of freedom and personal autonomy, and overwhelming

Source: *Backboards & Blackboards: College Athletes and Role Engulfment* by Patricia A. Adler and Peter Adler. © 1991, Columbia University Press, New York. Reprinted with permission of the publisher.

demands. These conflicts and dualisms are the focus of this reading. This is a study of the socialization of college athletes.

For five years we lived in and studied the world of elite NCAA (National Collegiate Athletic Association) college basketball. Participant-observers, we fit ourselves into the setting by carving out evolving roles that integrated a combination of team members' expectations with our interests and abilities. Individually and together, we occupied a range of different positions including friend, professor, adviser, confidant, and coach. We observed and interacted with all members of the team, gaining an intimate understanding of the day-to-day and year-to-year character of this social world. From behind the scenes of this secretive and celebrated arena, we document the experiences of college athletes, focusing on changes to their selves and identities over the course of their college years....

THEORETICAL APPROACH

This is not only a study of college athletes, but also a study in the social psychology of the self. Our observations reveal a significant pattern of transformation experienced by all our subjects: *role engulfment.* Many of the individuals we followed entered college hoping to gain wealth and fame through their involvement with sport. They did not anticipate, however, the cost of dedicating themselves to this realm. While nearly all conceived of themselves as athletes first, they possessed other self-images that were important to them as well. Yet over the course of playing college basketball, these individuals found the demands and rewards of the athletic role overwhelming and became engulfed by it. However, in yielding to it, they had to sacrifice other interests, activities, and, ultimately, dimensions of their selves. They immersed themselves completely in the athletic role and neglected or abandoned their identities lodged in these other roles. They thus became extremely narrow in their focus. In this work we examine *role domination,* the process by which athletes became engulfed in their athletic role as it ascended to a position of prominence. We also examine

the concomitant process of *role abandonment*, where they progressively detached themselves from their investment in other areas and let go of alternative goals or priorities. We analyze the changes this dual process of selfengulfment had on their self-concepts and on the structure of their selves....

Role theory focuses on the systems, or institutions, into which interaction fits. According to its tenets, *statuses* are positions in organized groups or systems that are related to other positions by a set of normative expectations. Statuses are not defined by the people that occupy them, but rather are permanent parts of those systems. Each status carries with it a set of role expectations specifying how persons occupying that status should behave. *Roles* consist of the activities people of a given status are likely to pursue when following the normative expectations for their positions. *Identities* (or what McCall and Simmons 1978, call role-identities), are the self-conceptions people develop from occupying a particular status or enacting a role. The *self* is the more global, multi-role, core conception of the real person.

Because in modern society we are likely to be members of more than one group, we may have several statuses and sets of role-related behavioral expectations. Each individual's total constellation of roles forms what may be termed a *role-set,* characterized by a series of relationships with role-related others, or role-set members (Merton 1957). Certain roles or role-identities may be called to the fore, replacing others, as people interact with individuals through them. Individuals do not invest their core feelings of identity or self in all roles equally, however. While some roles are more likely to be called forth by the expectations of others, other roles are more salient to the individual's core, or "real self," than others. They are arranged along a *hierarchy of salience* from peripheral roles to those that "merge with the self" (Turner 1978), and their ranking may be determined by a variety of factors. Role theory enhances an understanding of both the internal structure of the self and the relation between self and society; it sees this relation as mediated by the concept of role and its culturally and structurally derived expectations.

The interpretive branch of symbolic interactionism focuses on examining agency, process, and change. One of the concepts most critical to symbolic interactionism is the self. Rather than merely looking at roles and their relation to society, symbolic interactionism looks at the individuals filling those roles and the way they engage not only in role-taking, but also in active, creative role-making (Turner 1962). The self is the thinking and feeling being connecting the various roles and identities individuals put forth in different situations (Cooley 1962; Mead 1934). Symbolic interactionism takes a dynamic view of individuals in society, believing that they go beyond merely reproducing existing roles and structures to collectively defining and interpreting the meaning of their surroundings. These subjective, symbolic assessments form the basis for the creation of new social meanings, that then lead to new, shared patterns of adaptation (Blumer 1969). In this way individuals negotiate their social order as they experience it (Strauss 1978). They are thus capable of changing both themselves and the social structures within which they exist. Symbolic interactionism enhances understandings of the dynamic processes characterizing human group life and the reciprocal relation between those processes and changes in the self.

In integrating these two perspectives we show how the experiences of college athletes are both bounded and creative, how athletes integrate structural, cultural, and interactional factors, and how they change and adapt through a dynamic process of action and reaction, forging collective adaptations that both affirm and modify existing structures.

THE SETTING

We conducted this research at a medium-size (6000 students), private university (hereafter referred to as "the University") in the south-western portion of the United States. Originally founded on the premise of a religious affiliation, the University had severed its association with the church several decades before, and was striving to make a name for itself as one of the finer private, secular universities in the region. For many years it had served the community chiefly as a commuter school, but had embarked on an aggressive program of national recruiting over the past five to ten years that considerably broadened the base of its enrollment. Most of the students were white, middle class, and drawn from the suburbs of the South, Midwest, and Southwest. Academically, the University was experimenting with several innovative educational programs designed to enhance its emerging national reputation. Sponsored by reforms funded by the National Endowment for the Humanities, it was changing the curriculum, introducing a more interdisciplinary focus, instituting a funded honors program, increasing the general education requirements, and, overall, raising academic standards to a fairly rigorous level.

Within the University, the athletic program overall had considerable success during the course of our research: the University's women's golf team was ranked in the top three nationally, the football team won their conference each season, and the basketball program was ranked in the top forty of Division I NCAA schools, and in the top twenty for most of two seasons. The basketball team played in post-season tournaments every year, and in the five complete seasons we studied them, they won approximately four times as many games as they lost. In general, the basketball program was fairly representative of what Coakley (1986) and Frey (1982a) have termed "big-time" college athletics. Although it could not compare to the upper echelon of established basketball dynasties or to the really large athletic programs that wielded enormous recruiting and operating budgets, its success during the period we studied it compensated for its size and lack of historical tradition. The University's basketball program could thus best be described as "up and coming." Because the basketball team (along with the athletic department more generally) was ranked nationally and sent graduating members into the professional leagues, the entire athletic milieu was imbued with a sense of seriousness and purpose.

The team's professionalism was also enhanced by the attention focused on it by members of the community. Located in a city of approximately 500,000 with no professional sports teams, the University's programs served as the primary source of athletic entertainment and identification for the local population. When the basketball program meteorically rose to prominence, members of the city embraced it with fanatical support. Concomitant with the team's rise in fortunes, the region—part of the booming oil and sun belts—was experiencing increased economic prosperity. This surging local pride and financial windfall cast its glow over the basketball team, as it was the most charismatic and victorious program in the city's history, and the symbol of the community's newfound identity. Interest and support were therefore lavished on the team members.

THE PEOPLE

Over the course of our research, we observed 39 players and seven coaches. Much like Becker et al. (1961), we followed the players through their recruitment and entry into the University, keeping track of them as they progressed through school. We also watched the coaches move up their career ladders and deal with the institutional structures and demands.

Players, like students, were recruited primarily from the surrounding region. Unlike the greater student population, though, they generally did not hail from suburban areas. Rather, they predominantly came from the farming and rural towns of the prairies or southlands, and from the ghetto and working-class areas of the mid-sized cities.

Demographics

Over the course of our involvement with the team, two-thirds of the players we studied were black and one-third white. White middle-class players accounted for approximately 23 percent of the team members. They came from intact families where fathers worked in such occupations as wholesale merchandising, education, and sales. Although several were from suburban areas, they were more likely to come from mid- to larger-sized cities or exurbs. The remaining white players (10 percent) were from working-class backgrounds. They came from small factory towns or cities and also from predominantly

(although not exclusively) intact families. Some of their parents worked in steel mills or retail jobs.

The black players were from middle-, working-, and lower-class backgrounds. Those from the middle class (15 percent) came from the cities and small towns of the Midwest and South. They had intact families with fathers who worked as police chiefs, ministers, high school principals, or in the telecommunications industry. A few came from families in which the mothers also worked. Several of these families placed a high premium on education; one player was the youngest of five siblings who had all graduated from college and gone on for professional degrees, while another's grandparents had received college educations and established professional careers. The largest group of players (33 percent) were blacks from working-class backgrounds. These individuals came from some small Southern towns, but more often from the mid- to larger-sized cities of the South, Midwest, and Southwest. Only about half came from intact families; the rest were raised by their mothers or extended families. Many of them lived in the ghetto areas of these larger cities, but their parents held fairly steady jobs in factories, civil service, or other blue-collar or less skilled occupations. The final group (18 percent) was composed of lower-class blacks. Nearly all of these players came from broken homes. While the majority lived with their mothers, one came from a foster family, another lived with his father and sister, and a third was basically reared by his older brothers. These individuals came from the larger cities in the Southwest, South, and Midwest. They grew up in ghetto areas and were street smart and tough. Many of their families subsisted on welfare; others lived off menial jobs such as domestic service. They were poor, and had the most desperate dreams of escaping.

Cliques

Moving beyond demographics, the players fell into four main coalitions that served as informal social groups. Not every single member of the team neatly fit into one of these categories or belonged to one of these groups, but nearly all players who stayed on the team for at least a year eventually drifted into a camp.

At the very least, individuals associated with the various cliques displayed many behavioral characteristics we will describe. Players often forged friendship networks within these divisions, because of common attitudes, values, and activities. Once in a clique, no one that we observed left it or shifted into another one. In presenting these cliques, we trace a continuum from those with the most "heart" (bravery, dedication, willingness to give everything they had to the team or their teammates), a quality highly valued by team members, to those perceived as having the least.

Drawing on the vernacular shared by players and coaches, the first group of players were the "*bad niggas*." All of these individuals were black, from the working or lower classes, and shared a ghetto upbringing. Members of what Edwards (1985) has called the underclass (contemporary urban gladiators), they possessed many of the characteristics cited by Miller (1958) in his study of delinquent gangs' lower-class culture: trouble, toughness, smartness, excitement, fate, and autonomy. They were street smart and displayed an attitude that bespoke their defiance of authority. In fact, their admiration for independence made it hard for them to adjust to domination by the coach (although he targeted those with reform potential as pet "projects"). Fighters, they would not hesitate to physically defend their honor or to jump into the fray when a teammate was in trouble. They worked hard to eke the most out of their athletic potential, for which they earned the respect of their teammates; they had little desire to do anything else. These were the players with the most heart. They may not have been "choir boys," but when the competition was fierce on the court, these were the kind of players the coach wanted out there, the kind he knew he could count on. Their on-court displays of physical prowess contributed to their assertions of masculinity, along with sexual conquests and drug use. They were sexually promiscuous and often boasted about their various "babes." With drugs, they primarily used marijuana, alcohol (beer), and cocaine. Their frequency of use varied from daily to occasional, although who got high and how often was a significant behavioral difference dividing the cliques. This type of social split, and the actual amounts of drugs team members used, is no

different from the general use characteristic of a typical college population (see Moffatt 1989).

Tyrone was one of the bad niggas. He came from the ghetto of a mid-sized city in the Southwest, from an environment of outdoor street life, illicit opportunity, and weak (or absent) parental guidance. He was basically self-raised: he saw little of his mother, who worked long hours as a maid, and he had never known his father. Exceptionally tall and thin, he walked with a swagger (to show his "badness"). His speech was rich with ghetto expressions and he felt more comfortable hanging around with "brothers" than with whites. He often promoted himself boastfully, especially in speaking about his playing ability, future professional chances, and success with women. His adjustment to life at the University was difficult, although after a year or so he figured out how to "get by"; he became acclimated to dorm life, classes, and the media and boosters. When it came to common sense street-smarts, he was one of the brightest people on the team. He neither liked nor was favored by many boosters, but he did develop a solid group of friends within the ranks of the other bad niggas.

Apollo was another bad nigga. His family upbringing was more stable than Tyrone's, as his parents were together and his father was a career government worker (first in the military, then in the postal service). They had never had much money though, and scraped by as best they could. He was the youngest of six children, the only boy, and was favored by his father because of this. Tall and handsome, he sported an earring (which the coach made him remove during the season) and a gold tooth. One of the most colorful players, Apollo had a charismatic personality and a way with words. He was a magnetic force on the team, a leader who related emotionally to his teammates to help charge and arouse them for big games. He spoke in the common street vernacular of a ghetto "brother," although he could converse in excellent "White English" when it was appropriate. He was appealing to women, and enjoyed their attention, even though he had a steady girlfriend on the women's basketball team. Like Tyrone, Apollo was intelligent and articulate; he was able to express his perspective on life in a way that was insightful, entertaining, and outrageous. He was eager to explore and experience the zest of life, traveling the world, partying heavily, and seizing immediate gratification. He disdained the boring life of the team's straighter members, and generally did not form close relations with them. Yet he managed to enjoy his partying and playing and still graduate in four years. He would never have thought about college except for basketball, and had to overcome several debilitating knee injuries, but he ended up playing professionally on four continents and learning a foreign language.

A second group of players were the "*candy-asses*." These individuals were also black, but from the middle and working classes. Where the bad niggas chafed under the authority of the coach, the candy-asses craved his attention and approval; they tended to form the closest personal ties with both him and his family. In fact, their strong ties to the coach made them the prime suspects as "snitches," those who would tell the coaches when others misbehaved. They "browned up" to the coaches and to the boosters and professors as well. The candy-asses were "good boys," the kind who projected the public image of conscientious, religious, polite, and quiet individuals. Several of them belonged to the Fellowship of Christian Athletes. They could be counted on to stay out of trouble. Yet although they projected a pristine image, they did not live like monks. They had girlfriends, enjoyed going to discos, and occasionally drank a few beers for recreation. They enjoyed parties, but, responsibly, moderated their behavior. The candy-asses enjoyed a respected position on the team because, like the bad niggas, they were good athletes and had heart. As much as they sought to be well rounded and attend to the student role, they did not let this interfere with their commitment to the team. They cared about the team first, and would sacrifice whatever was necessary—playing in pain, coming back too soon from an injury, relinquishing personal statistics to help the team win, diving to get the ball—for its benefit. Above all, they could be counted on for their loyalty to the coach, the team, and the game.

Rob was one such player from a large, extended working-class family in a sizable Southern city. He was friendly and easy-going, with a positive attitude that came out in most of his activities. Although he was black and most of his close friends on the team were as well, Rob interacted much more easily with white boosters and students than did Tyrone. Rob transferred to the University to play for the coach because he had competed against him and liked both his reputation and style of play. Once there, he devoted himself to the coach, and was adored by the coach's wife and children. He often did favors for the family; one summer he painted the house in his spare time, and ate many of his meals there during the off-season. Rob's family was very close-knit, and both his mother and brother moved to town to be near him while he was at the University. They grew close with the coach's family, and often did things together. They also became regulars at the games, and were courted by many boosters who wanted to feel as if they knew Rob. Rob kept his academic and social life on an even keel during his college years; he worked hard in class (and was often the favorite of the media when they wanted to hype the image of the good student-athlete), and had a steady girlfriend. She was also very visible, with her young child from a previous boyfriend, in the basketball stadium. Like one or two of the others on the team, Rob spent one summer traveling with a Christian group on an around-the-world basketball tour.

Another typical candy-ass was Darian. He lived next door to Rob for two of the years they overlapped at the University, and the two were very close. Darian came from a middle-class family in a nearby state and his family came to town for most of the home games. He was much more serious about life than Rob, and worked hard in everything he did. He was recruited by the team at the last minute (he had health problems that many thought would keep him from playing), yet he devoted himself to improvement. By his senior year he was one of the outstanding stars and had dreams of going pro. He wanted to make the most of his college education as well, and spent long hours in his room trying to study. Most people on the team looked up to him because he did what they all intended to do—work hard, sacrifice, and make the most of their college opportunity. His closest friends, then, were others with values like his, who were fairly serious, respectful, and who deferred their gratification in hopes of achieving a future career.

The "*whiners*" constituted a third category of players. Drawn from the middle and working classes and from both races, this group was not as socially cohesive as the previous two, yet it contained friendship cliques of mixed class and race. These individuals had neither the athletic prowess of the candy-asses nor the toughness of the bad niggas, yet they wanted respect. In fact the overriding trait they shared was their outspoken belief that they deserved more than they were getting: more playing time, more deference, more publicity. In many ways they envied and aspired to the characteristics of the two other groups. They admired the bad niggas' subculture, their independence, toughness, and disrespect for authority, but they were not as "bad." They wanted the attention (and perceived favoritism) the candy-asses received, but they could not keep themselves out of trouble. Like the bad niggas, they enjoyed getting high. While their athletic talent varied, they did not live up to their potential; they were not willing to devote themselves to basketball. They were generally not the kind of individuals who would get into fights, either on the court or on the street, and they lacked the heart of the previous two groups. Therefore, despite their potential and their intermittent complaining, they never enjoyed the same kind of respect among their teammates as the bad niggas, nor did they achieve the same position of responsibility on the team as the candy-asses.

Buck fell into this category. A young black from a rural, Southern town, Buck came from a broken home. He did not remember his father, and his mother, who worked in a factory, did not have the money to either visit him at the University or attend his games. Yet he maintained a close relationship with her over the phone. Buck had gone to a primarily white school back home and felt more comfortable around white people than most of the black players; in fact, several of his best friends on the team were white and he frequently dated several white girls. He had a solid academic background and performed well in his classes, although he was

not as devoted to the books as some of the candy-asses. He liked to party, and occasionally got in trouble with the coach for breaking team rules. He spent most of his time hanging around with other whiners and with some of the bad niggas, sharing the latter's critical attitude toward the coach's authoritarian behavior. He felt that the coach did not recognize his athletic potential, and that he did not get the playing time he deserved. He had been warned by the coach about associating too much with some of the bad niggas who liked to party and not study hard. Yet for all their partying, the bad niggas were fiercer on the court than Buck, and could sometimes get away with their lassitude through outstanding play. He could not, and always had the suspicion that he was on the coach's "shit list." Like the candy-asses, he wanted to defer his gratification, do well in school, and get a good job afterward, yet he was not as diligent as they were and ended up going out more. He redeemed himself in the eyes of his peers during his senior year by playing the whole season with a debilitating chronic back injury that gave him constant pain.

James was another player in this clique, who also fell somewhere in between the bad niggas and the candy-asses. Like Buck, he socialized primarily with whiners and with some of the bad niggas, although he was white. He came from a middle-class family in a small town near the school, and was recruited to play along with his brother (who was a year younger). James came to the University enthusiastic about college life and college basketball. He threw himself into the social whirl and quickly got into trouble with the coach for both his grades and comportment. He readily adapted to the predominantly black ambience of the players' peer subculture, befriending blacks and picking up their jargon. He occasionally dated black women (although this was the cause of one major fight between him and some of the football players, since black women were scarce on this campus). He had heart, and was willing to commit his body to a fight. The most famous incident erupted during a game where he wrestled with an opponent over territorial advantage on the court. Yet he had neither the speed nor the size of some other players, and only occasionally displayed flashes of the potential the coaches had seen in him.

The final group was known to their teammates as the "*L-7s*" (a "square," an epithet derived from holding the thumbs and forefingers together to form the "square" sign). Members of this mixed group were all middle class, more often white than black. They were the most socially isolated from other team members, as they were rejected for their squareness by all three other groups, and even among themselves seldom made friends across racial lines (the white and black L-7s constituted separate social groups). They came from rural, suburban, or exurban backgrounds and stable families. They were fairly moralistic, eschewing drinking, smoking, and partying. They attempted to project a studious image, taking their books with them on road trips and speaking respectfully to their professors. Compared with other players, they had a stronger orientation to the student population and booster community. They were, at heart, upwardly mobile, more likely to consider basketball a means to an end than an end in itself. Because of this orientation, several of the white players landed coveted jobs, often working at boosters' companies, at the end of their playing careers. They tended to be good technical players bred on the polished courts of their rural and exurban high schools rather than on the street courts of the cities; they knew how to play the game, but it did not occupy their full attention. They had varying (usually lesser) degrees of athletic ability, but they were even less likely than the whiners to live up to their potential. In contrast to the other groups, they had decidedly the least commitment, least loyalty, and least heart.

Mark was a dirty-blond-haired white boy from the West Coast with a strong upper body, built up from surfing. He was clean cut, respectful, and a favorite of the boosters. His sorority girlfriend was always on his arm, and helped create his desired image of a future businessman. He consciously worked to nurture this image, ostentatiously carrying books to places where he would never use them, and dressing in a jacket and tie whenever he went out in public. It was very important to him to make it financially, because he came from a working-class family without much money. He had arrived at the University highly touted, but his talent never materialized to the degree the coaches expected. He was somewhat bitter about this assessment, because he felt he could have "made it" if given more of a chance. Like Buck, part of his problem may have lain in the difference between his slow-down style and the fast-paced style favored by the coach. He roomed and associated with other L-7s on the road and at home, but he also spent a lot of his time with regular students. After graduation he got a job from a booster working for a life insurance company.

Constellation of Role-Set Members

Basketball players generally interacted within a circle that was largely determined by their athletic environment. Due to the obligations of their position, these other role-set members fell into three main categories: athletic, academic, and social. Within the athletic realm, in addition to their teammates, athletes related primarily to the coaching staff, secretaries, and athletic administrators. The coaching staff consisted of the head coach, his first assistant (recruiting, playing strategy), the second assistant (recruiting, academics, some scouting), a part-time assistant (scouting, tape exchange with other teams, statistics during games), a graduate assistant (running menial aspects of practices, monitoring study halls, tape analysis), the trainer (injuries, paramedical activities), and the team manager (laundry, equipment)....

Secondary members of the athletic role-set included boosters, fans, athletic administrators (the Athletic Director, Sports Information Director, and their staffs), and members of the media....

Within the academic arena, athletes' role-set members consisted of professors, tutors, and students in their classes, and, to a lesser extent, academic counselors and administrators. The players also tended to regard their families as falling primarily into this realm, although family members clearly cared about their social lives and athletic performance as well.

Socially, athletes related to girlfriends, local friends, and other students (non-athletes), but most especially to other college athletes: the teammates and dormmates (football players) who were members of their peer subculture.

REFERENCES

Becker, Howard, Blanche Geer, Everett Hughes, and Anselm Strauss, 1961. *Boys in White*. Chicago: University of Chicago Press.

Blumer, Herbert, 1969. *Symbolic Interactionism*. Englewood Cliffs, N.J.: Prentice-Hall.

Coakley, Jay J., 1986. *Sport in Society*. Third Edition. St. Louis: Mosby [Second edition, 1982.]

Cooley, Charles H., 1962. *Social Organization*. New York: Scribners.

Edwards, Harry, 1985. Beyond symptoms: Unethical behavior in American collegiate sport and the problem of the color line: *Journal of Sport and Social Issues* 9:3–11.

Frey, James H., 1982a. Boosterism, scarce resources and institutional control: The future of American intercollegiate athletics. *International Review of Sport Sociology* 17:53–70.

McCall, George J., and Jerry L. Simmons, 1978. *Identities and Interaction*. New York: Free Press.

Mead, George H., 1934. *Mind, Self and Society*. Chicago: University of Chicago Press.

Merton, Robert K., 1957. The role-set: Problems in sociological theory. *British Journal of Sociology* 8:106–20.

Miller, Walter B., 1958. Lower class culture as a generating milieu of gang delinquency. *Journal of Social Issues* 14:5–19.

Moffatt, Michael, 1989. *Coming of Age in New Jersey*. New Brunswick, N.J.: Rutgers University Press.

Strauss, Anselm, 1978. *Negotiations*. San Francisco: Jossey-Bass.

Turner, Ralph H., 1962. Role taking: Process versus conformity. In A. M. Rose, ed., *Human Behavior and Social Processes*, pp. 20–40. Boston: Houghton-Mifflin.

———, 1978. The role and the person. *American Journal of Sociology* 84:1–23.

19

You Can't Be a Sweet Cucumber in a Vinegar Barrel

PHILIP ZIMBARDO

Why do good people do bad things? This question has guided the work of social psychologist Zimbardo, founder of the National Center for the Psychology of Terrorism, for several decades. In this article he discusses questions underlying his research studies (especially his famous prison studies) and the applicability of his findings to the prison scandals at Abu Ghraib prison in Iraq and Guantanamo prison in Cuba. Zimbardo discusses why otherwise "nice kids" could do such awful things; social group and setting contribute greatly to behaviors, he says.

Consider the following as you read the selection:

1. *What were the key findings of Zimbardo's studies?*
2. *How can we apply Zimbardo's findings to current events?*
3. *Why do good kids do terrible things?*
4. *Have you ever been tempted to do something out of character for you? What were the circumstances?*

GLOSSARY **Dehumanization** Defining individuals as inferior, worthless.
Social modeling Someone takes the lead in an activity, providing "permission" for others to follow.   **Deindividualization** Sense of anonymity.

For years I've been interested in a fundamental question concerning what I call the psychology of evil: Why is it that good people do evil deeds? I've been interested in that question since I was a little kid. Growing up in the ghetto in the South Bronx, I had lots of friends who I thought were good kids, but for one reason or another they ended up in serious trouble. They went to jail, they took drugs, or they did terrible things to other people. My whole upbringing was focused on trying to understand what could have made them go wrong.

When you grow up in a privileged environment, you want to take credit for the success you see all around, so you become a dispositionalist. You look for character, genes, or family legacy to explain things, because you want to say your father did good things, you did good things, and your kid will do

Source: Edited from an interview with Philip Zimbardo at http://www.edge.org/3rd_culture/zimbardo05/zimbardo05_index.html. Reprinted by permission of Philip Zimbardo.

good things. Curiously, if you grow up poor, you tend to emphasize external situational factors when trying to understand unusual behavior. When you look around and you see that your father's not working, and you have friends who are selling drugs or their sisters are in prostitution, you don't want to say it's because there's something inside them that makes them do it, because then there's a sense in which it's in your line. Psychologists and social scientists who focus on situations more often than not come from relatively poor, immigrant backgrounds. That's where I came from.

Over the years I've asked that question in more and more refined ways. I began to investigate what specific kinds of situational variables or processes could make someone step across that line between good and evil. We all like to think that the line is impermeable—that people who do terrible things like commit murder, treason, or kidnapping are on the other side of the line—and we could never get over there. We want to believe that we're with the good people. My work began by saying, no, that line is permeable. The reason some people are on the good side of the line is that they've never really been tested. They've never really been put in unusual circumstances where they were tempted or seduced across that line. My research over the last 30 years has created situations in the laboratory or in field settings in which we take good, normal, average, healthy people—more often than not healthy college students—and expose them to these kinds of settings....

To investigate this I created an experiment. We took women students at New York University and made them anonymous. We put them in hoods, put them in the dark, took away their names, gave them numbers, and put them in small groups. And sure enough, within half an hour those sweet women were giving painful electric shocks to other women within an experimental setting. We also repeated that experiment on deindividuation with the Belgian military, and in a variety of formats, with the same outcomes. Any situation that makes you anonymous and gives permission for aggression will bring out the beast in most people. That was the start of my interest in showing how easy it is to

get good people to do things they say they would never do.

I also did research on vandalism. When I was a teacher at NYU, I noticed that there were hundreds and hundreds of vandalized cars on the streets throughout the city. I lived in Brooklyn and commuted to NYU in the Bronx, and I'd see a car in the street. I'd call the police and say, "You know, there's a car demolished on 167th and Sedgwick Avenue. Was it an accident?" When he told me it was vandals, I said, "Who were the vandals? I'd like to interview them." He told me that they were little black or Puerto Rican kids who come out of the sewers, smash everything, paint graffiti on the walls, break windows, and disappear.

So I created what ethologists would call "releaser cues." I bought used cars, took off license plates, and put the hood up, and we photographed what happened. It turns out that it wasn't little black or Puerto Rican kids, but white, middle-class Americans who happened to be driving by. We had a car near NYU in the Bronx. Within ten minutes the driver of the first car that passed by jacked it up and took a tire. Ten minutes later a little family would come. The father took the radiator, the mother emptied the trunk, and the kid took care of the glove compartment. In 48 hours we counted 23 destructive contacts with that car. In only one of those were kids involved. We did a comparison in which we set out a car a block from Palo Alto, where Stanford University is. The car was out for a week, and no one touched it until the last day when it rained and somebody put the hood down. God forbid that the motor should get wet.

This gives you a sense of what a community is. A sense of community means people are as concerned about any property or people on their turf because there's a sense of reciprocal concern. The assumption is that I am concerned because you will be concerned about me and my property. In an anonymous environment nobody knows who I am and nobody cares, and I don't care to know about anyone else. The environment can convey anonymity externally, or it can be put on like a Ku Klux Klan outfit.

And so I and other colleagues began to do research on dehumanization. What are the ways in which, instead of changing yourself and becoming the aggressor, it becomes easier to be hostile against other people by changing your psychological conception of them? You think of them as worthless animals. That's the killing power of stereotypes.

I put that all together with other research I did 30 years ago during the Stanford prison experiment. The question there was, what happens when you put good people in an evil place? We put good, ordinary college students in a very realistic, prison-like setting in the basement of the psychology department at Stanford. We dehumanized the prisoners, gave them numbers, and took away their identity. We also deindividuated the guards, calling them Mr. Correctional Officer, putting them in khaki uniforms, and giving them silver reflecting sunglasses like in the movie *Cool Hand Luke*. Essentially, we translated the anonymity of *Lord of the Flies* into a setting where we could observe exactly what happened from moment to moment.

What's interesting about that experiment is that it is really a study of the competition between institutional power versus the individual will to resist. The companion piece is the study by Stanley Milgram, who was my classmate at James Monroe High School in the Bronx. (Again, it is interesting that we are two situationists who came from the same neighborhood.) His study investigated the power of an individual authority: Some guy in a white lab coat tells you to continue to shock another person even though he's screaming and yelling. That's one way that evil is created as blind obedience to authority. But more often than not, somebody doesn't have to tell you to do something. You're just in a setting where you look around and everyone else is doing it. Say you're a guard and you don't want to harm the prisoners—because at some level you know they're just college students—but the two other guards on your shift are doing terrible things. They provide social models for you to follow if you are going to be a team player.

In this experiment we selected normal, healthy, good kids that we found through ads in the paper.... within a few days, if they were assigned to the guard role, they became abusive, red-necked prison guards.

Every day the level of hostility, abuse, and degradation of the prisoners became worse and worse and worse. Within 36 hours the first prisoner had an emotional breakdown, crying, screaming, and thinking irrationally. We had to release him, and each day after that we had to release another prisoner because of extreme stress reactions. The study was supposed to run for two weeks, but I ended it after six days because it was literally out of control. Kids we chose because they were normal and healthy were breaking down. Kids who were pacifists were acting sadistically, taking pleasure in inflicting cruel, evil punishment on prisoners....

There are stunning parallels between the Stanford prison experiment and what happened at Abu Ghraib, where some of the visual scenes that we have seen include guards stripping prisoners naked, putting bags over heads, putting them in chains, and having them engage in sexually degrading acts. And in both prisons the worst abuses came on the night shift. Our guards committed very little physical abuse. There was a prisoner riot on the second day, and the guards used physical abuse....

These are exact parallels between what happened in this basement at Stanford 30 years ago and at Abu Ghraib, where you see images of prisoners stripped naked, wearing hoods or masks as guards get them to simulate sodomy. The question is whether what we learned about the psychological mechanisms that transformed our good volunteers into these creatively evil guards can be used to understand the transformation of good American Army reservists into the people we see in these trophy photos in Abu Ghraib. And my answer is, yes, there are very direct parallels....

These terrible deeds form an interesting analog in America, because there are two things we are curious to understand about Abu Ghraib. First, how did the soldiers get so far out of hand? And secondly, why would the soldiers take pictures of themselves in positions that make them legally culpable? The ones that are on trial now are the ones in those pictures, although obviously there are many more people involved in various ways. We can understand why they did so not only by applying the basic social-psychological processes from the Stanford prison study, but also by analyzing what was unique in Abu Ghraib....

At Abu Ghraib you had the social modeling in which somebody takes the lead in doing something. You had the dehumanization, the use of labels of the other as inferior, as worthless. There was a diffusion of responsibility such that nobody was personally accountable.... The other thing, of course, is that you had low-level army reservists who had no "mission-specific" training in how to do this difficult new job. There was little or no supervision of them on the night and there was literally no accountability. This went on for months in which the abuses escalated over time.... And then there is the hidden factor of boredom. One of the main contributors to evil, violence, and hostility in all prisons that we underplay is the boredom factor. In fact, the worst things that happened in our prisons occurred during the night shifts....

Dehumanization also occurred because the prisoners often had no prison clothes available, or were forced to be naked as a humiliation tactic by the military police and higher-ups. There were too many of them; in a few months the number soared from 400 to over a thousand. They didn't have regular showers, did not speak English, and they stank. Under these conditions it's easy for guards to come to think of the prisoners as animals, and dehumanization processes set in.

When you put that set of horrendous work conditions and external factors together, it creates an evil barrel. You could put virtually anybody in it and you're going to get this kind of evil behavior. The Pentagon and the military say that the Abu Ghraib scandal is the result of a few bad apples in an otherwise good barrel. That's the dispositional analysis. The social psychologist in me, and the consensus among many of my colleagues in experimental social psychology, says that's the wrong analysis. It's not the bad apples, it's the bad barrels that corrupt good people. Understanding the abuses at this Iraqi prison starts with an analysis of both the situational and systematic forces operating on those soldiers working the night shift in that "little shop of horrors."

Coming from New York, I know that if you go by a delicatessen, and you put a sweet cucumber in the vinegar barrel, the cucumber might say, "No, I want to retain my sweetness." But it's hopeless. The barrel will turn the sweet cucumber into a pickle. You can't be a sweet cucumber in a vinegar barrel. My sense is that we have the evil barrel of war, into which we've put this evil barrel of this prison—it turns out actually all of the military prisons have had similar kinds of abuses—and what you get is the corruption of otherwise good people.

20

Characteristics of Bureaucracy

MAX WEBER

We've all stood in lines to transact some business, only to find after our wait that we were missing a necessary form or receipt. Bureaucracy can be frustrating, but it is also an efficient way to conduct business, relative to other systems of patronage or favoritism. Weber's classic work on bureaucracy analyzes the rise of this form of organization. In the following brief excerpt, Weber outlines the key components that define bureaucracy.

As you read, consider the following:

1. *How do the characteristics of bureaucracy discussed by Weber compare with the characteristics of an organization with which you are familiar?*
2. *Which parts of bureaucracy have you experienced?*
3. *Can you design a more efficient form of organization than that described by Weber?*

GLOSSARY **Bureaucracy** Formal organization in which rules and hierarchical rankings are used to achieve efficiency.

Modern officialdom functions in the following specific manner:

I. There is the principle of fixed and official jurisdictional areas, which are generally ordered by rules, that is, by laws or administrative regulations.

1. The regular activities required for the purposes of the bureaucratically governed structure are distributed in a fixed way as official duties.

2. The authority to give the commands required for the discharge of these duties is distributed in a stable way and is strictly delimited by rules concerning the coercive means, physical, sacerdotal, or otherwise, which may be placed at the disposal of officials.

3. Methodical provision is made for the regular and continuous fulfilment of these duties and for the execution of the corresponding rights; only persons who have the generally regulated qualifications to serve are employed.

In public and lawful government these three elements constitute "bureaucratic authority." In private economic domination, they constitute bureaucratic "management." Bureaucracy, thus understood, is fully developed in the private economy, only in the most advanced institutions of capitalism.

Source: From *Max Weber: Essays in Sociology*, pp. 196–204, edited by H. H. Gerth and C. Wright Mills, copyright 1946, by Oxford University Press. Reprinted by permission.

Permanent and public office authority, with fixed jurisdiction, is not the historical rule, but rather the exception. This is so even in large political structures such as those of the ancient Orient, the Germanic and Mongolian empires of conquest, or of many feudal structures of state. In all these cases, the ruler executes the most important measures through personal trustees, table-companions, or court-servants. Their commissions and authority are not precisely delimited and are temporarily called into being for each case.

II. The principles of office hierarchy and of levels of graded authority mean a firmly ordered system of super- and subordination in which there is a supervision of the lower offices by the higher ones. Such a system offers the governed the possibility of appealing the decision of a lower office to its higher authority in a definitely regulated manner. With the full development of the bureaucratic type, the office hierarchy is monocratically organized. The principle of hierarchical office authority is found in all bureaucratic structures: in state and ecclesiastical structures as well as in large party organizations and private enterprises. It does not matter for the character of bureaucracy whether its authority is called "private" or "public."

   When the principle of jurisdictional "competency" is fully carried through, hierarchical subordination—at least in public office—does not mean that the "higher" authority is simply authorized to take over the business of the "lower." Indeed, the opposite is the rule. Once established and having fulfilled its task, an office tends to continue in existence and be held by another incumbent.

III. The management of the modern office is based upon written documents ("the files"), which are preserved in their original or draught form.

There is, therefore, a staff of subaltern officials and scribes of all sorts. The body of officials actively engaged in a "public" office, along with the respective apparatus of material implements and the files, make up a "bureau." In private enterprise, "the bureau" is often called "the office."

In principle, the modern organization of the civil service separates the bureau from the private domicile of the official, and, in general, bureaucracy segregates official activity as something distinct from the sphere of private life. Public monies and equipment are divorced from the private property of the official....

IV. Office management, at least all specialized office management—and such management is distinctly modern—usually presupposes thorough and expert training. This increasingly holds for the modern executive and employee of private enterprises, in the same manner as it holds for the state official.

V. When the office is fully developed, official activity demands the full working capacity of the official, irrespective of the fact that his obligatory time in the bureau may be firmly delimited. In the normal case, this is only the product of a long development, in the public as well as in the private office. Formerly, in all cases the normal state of affairs was reversed: official business was discharged as a secondary activity.

VI. The management of the office follows general rules, which are more or less stable, more or less exhaustive, and which can be learned. Knowledge of these rules represents a special technical learning which the officials possess. It involves jurisprudence, or administrative or business management.

PART III

Major Institutions in Society

An *institution* is a formal relationship organized around common values to meet basic needs within the society. To survive, social systems require two opposite tendencies: stability and change. Stability is required so that behaviors and interactions will be relatively predictable. Thus, each social institution channels behavior in prescribed ways. Change is necessary to deal with new technology and the education required for the new technology. Although long-standing normative patterns are difficult to change, new behavior patterns are occurring constantly because of innovations in the material culture, such as the invention of the automobile, or because of challenges to expected behavior due to that invention—for example, the removal of dating supervision from parent control by removing dating from the home. In short, social institutions are social activities that form patterns which recur over and over again to maintain the social system. Institutions are usually justified in terms of the values of the social system they are designed to serve and the social problems they are designed to resolve.

To fulfill these prerequisites for survival, societies have developed five major institutions:

- The *family* sets the rules for procreation and socialization.
- *Education* provides for the transmission of attitudes, norms, values, knowledge, and skills from one generation to the next.
- *Religion* regulates our relations to the supernatural and legitimizes certain behaviors.
- The *economy* is in charge of the production, distribution, and consumption of goods and services to ensure the survival of the social system.
- The *political system* governs the legitimate use of power in society and enforces the norms.

Although their forms may differ greatly, each of these institutions can be found in every society.

In accomplishing the previously mentioned tasks, social institutions must meet certain prerequisites. According to Talcott Parsons, there are two major prerequisites: (1) The social system must be relatively compatible with both the individual members of the society and the cultural system as a whole; and (2) the social system requires the support of other systems around it.

The first of Parson's ideas indicates that institutions are essential in helping us develop and deal with our wants and needs. Institutions can also be dysfunctional—the means and ends can cause distortion and negative interrelations with negative results. This occurs because the effect of socialization means that transforming our institutions will be difficult and that the interrelationships among institutions will make it harder for them to accomplish their original ends. It is the second of these prerequisites that leads us to expect social institutions to be interdependent and interrelated. That is, what happens in one institution will affect others. For example, a recession occurring in the economic institution will affect the family institution through the loss of jobs by family members, the education institution by forcing people to leave school early, the religious institution by receiving fewer and smaller donations, and the political institution by politicians being defeated for reelection.

As you read through the various selections in the chapters of Part III, think about the changes that are ongoing in our major institutions—the large increase in single people within the marriage institution, the increase in educational attainment, the growing awareness of extremism in various religious groups, an economic institution undergoing a recession, and a political institution during an election. These are the major themes of each chapter in this part.

Chapter 6

Marriage and Family
Change and Diversity

Clay Ballantine

A glance at the early history of marriage in America would reveal that from the beginning it was a central feature of adult life. Most adults not only married but usually married relatively early in life. So important was marriage that almost all people stayed married until the death of a spouse and then re-married. Divorce was a rarity, and remarriage usually occurred after this event. This situation prevailed because there was a great dependency on the marital partner in dealing with the needs of everyday life. In a sense, the changes being noted now in the marriage institution are related to the fact that there is less dependency on a mate.

This system of family life began to change by the latter part of the twentieth century. The following changes occurred in the relatively short period of time between the end of the 1950s and the turn of the century:

- There was a postponement of marriage by several years.

- The number of people choosing cohabitation before marriage increased rapidly.

- Both divorce rates and remarriage rates increased, which in turn led to a large increase in married couple households with stepchildren.

- The percentage of wives in the labor force increased substantially, and spouses followed less traditional patterns in terms of the division of labor by gender in marriage.

With a lesser dependency on a mate, the choice of a mate switched from a formal, long-term commitment to one based on individualistic desires and so to a lesser commitment. Now marriage would be based on affection, friendship, and personal growth. Given that marriage also provides such benefits as better physical health, psychological well-being, and a higher standard of living (Cargan, 2007), most young adults will eventually marry. In short, the institution of marriage is changing—but this factor does not necessarily mean that it is in a state of decline.

The other half of the title of this chapter is "family." Is it a dying institution? The increasing rate of family disruption plus other institutional changes makes it appear that family life is in peril. Actually, what has changed is the basic family form, from the nuclear family consisting of a husband, a wife, and their children to a number of alternate forms. One form, cohabitation, has become a popular alternative to marriage. Unfortunately, the expectation of cohabiters that it will enhance the chances for a successful marriage does not necessarily occur; the growth in single-parent families has been another important trend. Since the 1970s, there has been an almost 30% increase in the number of families with one parent. The growth in dual-career families has been an even more rapid change. The rapidity of this change is seen in the fact that in the 1950s

about 27% of married women were employed outside the home, whereas currently more than 60% are thus employed. Although the trend in divorce has been increasing rather rapidly over the past 50 years, it is not as high as many people believe—24%. The major reason for this increasing rate was the initiation of no-fault divorce as the legal criterion for divorce. Perhaps a surprising trend is the number of births occurring among single women—more than one-third of all U.S. births now occur among single mothers.

Despite these changes in family forms, the nuclear family is not in dire straits: Most Americans will marry at some time in their lives, approximately three-fourths of those divorcing will remarry, the majority of Americans rate a good marriage and family life as extremely important to them (Whitehead and Popenoe, 2001), and an overwhelming majority are satisfied with their marriage and family (*Polling Report*, 2004). This positive response to marriage is seemingly shown in the fact that both married men and women report being happier and less stressed than their unmarried counterparts (Cargan, 2007). The importance of the family in society is emphasized by the fact that it is found in one form or another in every society due to the important functions that it performs.

The first reading in this chapter, by Arlene and Jerome Skolnick, examines in more detail our beliefs concerning the family. These authors conclude that the changes occurring in society are not changing the family as much as they are exposing the fact that the beliefs formerly held about the family were not based on reality in the first place. The Skolnicks do not claim that change does not affect the family, only that the many beliefs held about the family are not always true.

The second article by Cargan examines one of the major changes occurring in this institution—the rapid growth of the adult single population. Living single in a world dominated by the married has led to a number of beliefs regarding what it is like for singles. Not surprising, some of the beliefs turn out to be myths, but some of them are realities. Which of these do you think are myths: Singles are immature, are more affluent, are workaholics, have more freedom, are happier but also lonelier?

A belief held by many is that social change exacerbates family violence. Thus, it is suggested that family violence is a recent and growing outcome of working wives or that it is caused by its opposite, the economic strain produced by nonworking wives. Emery and Laumann-Billings examine the nature of abusive relations in the third article of this chapter and reveal both its causes and consequences.

Perhaps the most misunderstood changes have been those related to divorce. These changes involve the replacement of death with divorce as the early dissolver of marriage due to longer life spans, the growth in the rate of dissolution of unhappy marriages due to the passage of no-fault divorce laws in all states, and a misconception about divorce rates, which are actually much lower than 50%. Amato and Irving examine these historical divorce trends in the final article of this chapter.

As you read this chapter, consider the significance of the family for each of us and what the outcomes are when this institution is unable to fulfill its societal functions for us individually and for society as a whole. Also consider the means for dealing with this situation.

21

Family in Transition

ARLENE S. SKOLNICK AND JEROME H. SKOLNICK

The authors note that the seemingly staid institution of family is actually undergoing a constant and sometimes dramatic transformation.

As you read this article, consider the following questions as guides:

1. *What are some of the surprising changes that have taken place since the "golden age" of family life in the 1950s?*
2. *Why will the institution of family never disappear despite greater options as well as delay and decline in marriage?*
3. *Does the surge in the singles population among the over-30 age group mean that this age group has rejected marriage?*
4. *Does the increase in the numbers of singles in the over 30-age group pose a threat to the institution of marriage?*

GLOSSARY **Cloning** Asexual reproduction of a group of organisms by a single individual. **Flapper** A woman free from social and moral restraints in the post–World War I decade of the 1920s. **Genetic engineering** Practical application of pure science to reproduction. **Introspective** Inclined to examine one's own mental and emotional state.

Even before the attacks on September 11, 2001, it was clear that we had entered a period of profound and unsettling transformations. "Every once in a great while," states the introduction to a *Business Week* issue on the twenty-first century, "the established order is overthrown. Within a span of decades, technological advances, organizational innovations, and new ways of thinking transform economies" (Farrell, 1994, p. 16).

Whether we call the collection of changes *globalization, the information age,* or *post-industrial society,* they are affecting every area of the globe and every aspect of life, including the family. Although the state of the U.S. family has been the subject of great public attention in recent years, the discussion of family has been strangely disconnected from talk of the other transformations we are living through. Instead, family talk is suffused with nostalgia, confusion, and anxiety—what happened, we wonder, to the fifties family of Ozzie and Harriet?...

Of course, family life has changed massively in recent decades, and too many children and families are beset by serious stresses and troubles. But we can't understand these changes and problems without understanding the impact of large-scale changes in society on the small worlds of everyday family life. All the industrialized nations, and many of the emerging ones, have experienced the same changes the United States has—a transformation in women's roles, rising divorce rates, lower marriage and birth rates, and an increase in single-parent families. In no other country, however, has family change been so traumatic and divisive as in the United States.

For example, the two-earner family has replaced the breadwinner/housewife family as the norm in the United States, even when there are young children in the home. In the mid-nineties, more than 60 percent of married women with children under six were in the paid labor force (Han and Moen, 1999). Yet the question of whether mothers "should" work or stay home is still a hotly debated issue (except if the mother is on welfare).

A range of other, once-"deviant" lifestyles—from living with a partner while unmarried, to remaining single or childless, or even having a child while single (à la Murphy Brown)—has become both acceptable and controversial. It was once unthinkable that gay and lesbian families could be recognized or even tolerated, but despite persisting stigma and the threat of violence, they have been. Local governments and some leading corporations have granted these families increasing recognition as domestic partnerships entitled to spousal benefits.

The changes of recent decades have affected more than the forms of family life; they have been psychological as well. A major study of U.S. attitudes over two decades revealed a profound shift in how people think about family life, work, and themselves (Veroff, Douvan, and Kulka, 1981). In 1957 four-fifths of respondents thought that a man or woman who did not want to marry was sick, immoral, and selfish. By 1976 only one-fourth of respondents thought that choice was bad. Summing up many complex findings, the authors conclude that the United States underwent a "psychological revolution" in the two decades between surveys.

Ever since the 1970s, the mass media have been serving up stories and statistics that seemingly show the family is disintegrating, falling apart, or on the verge of disappearing....

A sudden blizzard of newspaper columns, magazine articles, and talk show "experts" warned that divorce and single parenthood are inflicting serious damage on children and on society in general. This family structure, they argued, is the single biggest problem facing the country, because it is the root cause of all the rest—poverty, crime, drugs, school failure, youth violence, and other social ills.

The proposed solution? Restore the "traditional" family. Make divorce and single parenthood socially unacceptable once again. Do away with welfare and make divorces more difficult to obtain. These arguments flooded the media as welfare reform was being debated in Congress: both sides of the debate cited social science evidence to bolster their case....

It's not surprising then that public debate about the family often sinks to the level of a "food fight" as it lurches from one hot topic to another—single mothers, divorce, gay marriage, nannies, and working mothers. Each issue has only two sides: Are you for or against the two-parent family? Is divorce bad or good for children? Should mothers of young children work or not? Is the family "in decline" or not?...

When one extreme position debates the opposite extreme position, it becomes difficult to realistically discuss the issues and problems facing the country. It doesn't describe the range of views among family scholars, and it doesn't fit the research evidence. For example, if someone takes the position that "divorce is damaging to children," the argument culture leads us to assume that there are people on the other "side" who will argue just the opposite—in other words, that divorce is "good," or at least not harmful. But as researcher Paul Amato suggests, the right question to ask is "Under what circumstances is divorce harmful or beneficial to children?" (1994). In most public debates about divorce, however, that question is never asked, and the public never hears the useful information they should.

Still another problem with popular discourse about the family is that it exaggerates the extent of change. For example, we sometimes hear that the traditional nuclear family no longer exists, or has shrunk to a tiny percentage of the population. But that statement depends on a very narrow definition of family—two biological parents, in their first marriage, with a full-time breadwinner husband and a full-time homemaker wife, and two or three children under the age of 18. Of course, that kind of family has declined—for the simple reason that most wives and mothers now work outside the home. It has also declined because there are more married couples with grown children than there used to be.

Similarly, we hear that divorce rates have shot up since the 1950s, but we are not told that the trend toward higher divorce rates started in the nineteenth century, with more marital breakups in each succeeding generation. Nor do we hear that despite the current high divorce rates (actually down from 1979), the United States has the highest marriage rates in the industrial world. About 90 percent of Americans marry at some point in their lives, and virtually all who do either have, or want to have, children. Further, surveys repeatedly show that family is central to the lives of most Americans. They find family ties their deepest source of satisfaction and meaning, as well as the source of their greatest worries (Mellman, Lazarus, and Rivlin, 1990). In sum, family life in the United States is a complex mixture of both continuity and change.

While the transformations of the past three decades do not mean the end of family life, they have brought a number of new difficulties. For example, most families now depend on the earnings of wives and mothers, but the rest of society has not caught up to the new realities. There is still an earnings gap between men and women. Employed wives and mothers still bear most of the workload in the home. For both men and women, the demands of the job are often at odds with family needs.

UNDERSTANDING THE CHANGING FAMILY

During the same years in which the family was becoming the object of public anxiety and political debate, a torrent of new research on the family was pouring forth…. Ironically, much of the new scholarship is at odds with the widespread assumption that the family had a long, stable history until

hit by the social "earthquake" of the 1960s and 1970s. We have learned from historians that the "lost" golden age of family happiness and stability we yearn for never actually existed….

Part of the confusion surrounding the current status of the family arises from the fact that the family is a surprisingly problematic area of study; there are few if any self-evident facts, even statistical ones. Researchers have found, for example, that when the statistics of family life are plotted for the entire twentieth century, or back into the nineteenth century, a surprising finding emerges: Today's young people—with their low marriage, high divorce, and low fertility rates—appear to be behaving in ways consistent with long-term historical trends (Cherlin, 1981; Masnick and Bane, 1980). The recent changes in family life only appear deviant when compared to what people were doing in the 1940s and 1950s. But it was the postwar generation that married young, moved to the suburbs, and had three, four, or more children that departed from twentieth-century trends. As one study put it, "Had the 1940s and 1950s not happened, today's young adults would appear to be behaving normally" (Masnick and Bane, 1980, p. 2).

Thus, the meaning of change as a particular indicator of family life depends on the time frame in which it is placed. If we look at trends over too short a period of time—say ten or twenty years—we may think we are seeing a marked change, when, in fact, an older pattern may be reemerging. For some issues, even discerning what the trends are can be a problem.

For example, whether we conclude that there is an "epidemic" of teenage pregnancy depends on how we define adolescence and what measure of illegitimacy we use. Contrary to the popular notion of skyrocketing teenage pregnancy, teenaged childbearing has actually been on the decline during the past two decades. It is possible for the *ratio* of illegitimate births to all births to go up at the same time as there are declines in the *absolute number* of births and in the likelihood that an individual will bear an illegitimate child. This is not to say that concern about teenage pregnancy is unwarranted; but the reality is much more complex than the simple and scary notion an "epidemic" implies. Given the complexities of interpreting data on the family, it is little wonder that, as Joseph Featherstone observes (1979), the family is a "great intellectual Rorschach blot" (p. 37).

1. The Myth of Universality

To say that the family is the same everywhere is in some sense true. Yet families vary in organization, membership, life cycles, emotional environments, ideologies, social and kinship networks, and economic and other functions. Although anthropologists have tried to come up with a single definition of family that would hold across time and place, they generally have concluded that doing so is not useful (Geertz, 1965; Stephens, 1963).

Biologically, of course, a woman and a man must unite sexually to produce a child—even if only sperm and egg meet in a test tube. But no social kinship ties or living arrangements flow inevitably from biological union. Indeed, the definition of marriage is not the same across cultures. Although some cultures have weddings and notions of monogamy and permanence, many cultures lack one or more of these attributes. In some cultures, the majority of people mate and have children without legal marriage and often without living together. In other societies, husbands, wives, and children do not live together under the same roof.

In our own society, the assumption of universality has usually defined what is normal and natural both for research and therapy and has subtly influenced our thinking to regard deviations from the nuclear family as sick or perverse or immoral. As Suzanne Keller (1971) once observed:

> The fallacy of universality has done students of behavior a great disservice. By leading us to seek and hence to find a single pattern, it has blinded us to historical precedents for multiple legitimate family arrangements.

2. The Myth of Family Harmony

To question the idea of the happy family is not to say that love and joy are not found in family life or that many people do not find their deepest satisfactions in their families. Rather, the happy-family assumption omits important, if unpleasant, aspects of family life. Intimate relations inevitably involve antagonism as well as love. This mixture of strong positive and negative feelings sets close relationships apart from less intimate ones....

In recent years, family scholars have been studying family violence such as child abuse and wife beating to better understand the normal strains of family life. Long-known facts about family violence have recently been incorporated into a general analysis of the family. More police officers are killed and injured dealing with family fights than in dealing with any other kind of situation; of all the relationships between murderers and their victims, the family relationship is most common. Studies of family violence reveal that it is much more widespread than had been assumed, cannot easily be attributed to mental illness, and is not confined to the lower classes. Family violence seems to be a product of psychological tensions and external stresses that can affect all families at all social levels.

The study of family interaction has also undermined the traditional image of the happy, harmonious family....

3. The Myth of Parental Determinism

The kind of family a child grows up in leaves a profound, lifelong impact. But a growing body of studies shows that early family experience is not the all-powerful, irreversible influence it has sometimes been thought to be. An unfortunate childhood does not doom a person to an unhappy adulthood. Nor does a happy childhood guarantee a similarly blessed future (Emde and Harmon, 1984; Macfarlane, 1964; Rubin, 1996).

First, children come into this world with their own temperamental and other individual characteristics. As parents have long known, child rearing is not like molding clay or writing on a blank slate. Rather, it's a two-way process in which both parent and child shape each other. Further, children are active perceivers and interpreters of the world. Finally, parents and children do not live in a social vacuum; children are also influenced by the world around them and the people in it—the kin group, the neighborhood, other children, the school, and the media....

4. The Myth of a Stable, Harmonious Past

Laments about the current state of decay of the family imply some earlier era when the family was more stable and harmonious. But unless we can agree what earlier time should be chosen as a baseline and what characteristics of the family should be specified, it makes little sense to speak of family decline. Historians have not, in fact, located a golden age of the family.

Indeed, they have found that premarital sexuality, illegitimacy, generational conflict, and even infanticide can best be studied as a part of family life itself rather than as separate categories of deviation....

The most shocking finding of recent years is the prevalence of infanticide throughout European history. Infanticide has long been attributed to primitive peoples or assumed to be the desperate act of an unwed mother. It now appears that infanticide provided a major means of population control in all societies lacking reliable contraception, Europe included, and that it was practiced by families on legitimate children (Hrdy, 1999).

Rather than being a simple instinctive trait, having tender feelings toward infants—regarding a baby as a precious individual—seems to emerge only when infants have a decent chance of surviving and adults experience enough security to avoid feeling that children are competing with them in a struggle for survival. Throughout many centuries of European history, both of these conditions were lacking.

Another myth about the family is that of changelessness—the belief that the family has been essentially the same over the centuries, until recently, when it began to come apart. Family life has always been in flux; when the world around them changes, families change in response. At periods when a whole society undergoes some major transformation, family change may be especially rapid and dislocating.

In many ways, the era we are living through today resembles two earlier periods of family crisis and transformation in U.S. history (see Skolnick, 1991). The first occurred in the early nineteenth century, when the growth of industry and commerce moved work out of the home. Briefly, the separation of home and work disrupted existing patterns of daily family life, opening a gap between the way people actually lived and the cultural blueprints for proper gender and generational roles (Ryan, 1981). In the older pattern, when most people worked on farms, a father was not just the head of the household, but also boss of the family enterprise. Mother and children and hired hands worked under his supervision. But when work moved out, father—along with older sons and daughters—went with it, leaving behind mother and the younger children. These dislocations in the functions and meaning of family life unleashed an era of personal stress and cultural confusion.

Eventually, a new model of family emerged that not only reflected the new separation of work and family, but also glorified it. No longer a workplace, the household now became idealized as "home sweet home," an emotional and spiritual shelter from the heartless world outside. Although father remained the head of the family, mother was now the central figure in the home. The new model celebrated the "true woman's" purity, virtue, and selflessness. Many of our culture's most basic ideas about the family in U.S. culture, such as "a woman's place is in the home," were formed at this time. In short, the family pattern we now think of as traditional was in fact the first version of the modern family.

Historians label this model of the family "Victorian" because it became influential in England and Western Europe as well as in the United States during the reign of Queen Victoria. It reflected, in idealized form, the nineteenth-century middle-class family. However, the Victorian model became the prevailing cultural definition of family. Few families could live up to the ideal in all its particulars; working-class, black, and ethnic families, for example, could not get by without the economic contributions of wives, mothers, and daughters. And even for middle-class families, the Victorian ideal prescribed a standard of perfection that was virtually impossible to fulfill (Demos, 1986).

Eventually, however, social change overtook the Victorian model. Beginning around the 1880s, another period of rapid economic, social, and cultural change unsettled Victorian family patterns, especially their gender arrangements. Several generations of so-called new women challenged Victorian notions of femininity. They became educated, pursued careers, became involved in political causes—including their own—and created the first wave of feminism. This ferment culminated in the victory of the women's suffrage movement. It was followed by the 1920s' jazz-age era of flappers and flaming youth—the first, and probably the major, sexual revolution of the twentieth century.

To many observers at the time, it appeared that the family and morality had broken down. Another cultural crisis ensued, until a new cultural blueprint emerged—the companionate model of marriage and the family. The new model was a revised, more relaxed version of the Victorian family; companionship and sexual intimacy were now defined as central to marriage.

This highly abbreviated history of family and cultural change forms the necessary backdrop for understanding the family upheavals of the late twentieth and early twenty-first centuries. As in earlier times, major changes in the economy and society have destabilized an existing model of family life and the everyday patterns and practices that have sustained it.

We have experienced a triple revolution: first, the move toward a post-industrial service and information economy; second, a life course revolution brought about by the reductions in mortality and fertility; and third, a psychological transformation rooted mainly in rising educational levels.

Although these shifts have profound implications for everyone in contemporary society, women have been the pacesetters of change. Most women's lives and expectations over the past three decades, inside and outside the family, have departed drastically from those of their own mothers. Men's lives today also are different from their fathers' generation, but to a much lesser extent.

THE TRIPLE REVOLUTION

The Post-industrial Family

The most obvious way the new economy affects the family is in its drawing women, especially married women, into the workplace. A service and information economy produces large numbers of jobs that, unlike factory work, seem suitable for women. Yet as Jessie Bernard (1982) once observed, the transformation of a housewife into a paid worker outside the home sends tremors through every family relationship. It creates a more "symmetrical" family, undoing the sharp contrast between men's and women's roles that marks the breadwinner/housewife pattern. It also reduces women's economic dependence on men, thereby making it easier for women to leave unhappy marriages.

Beyond drawing women into the workplace, shifts in the nature of work and a rapidly changing globalized economy have unsettled the lives of individuals and families at all class levels. The well-paying industrial jobs that once enabled a blue-collar worker to own a home and support a family are no longer available. The once secure jobs that sustained the "organization men" and their families in the 1950s and 1960s have been made shaky by downsizing, an unstable economy, corporate takeovers, and a rapid pace of technological change.

The new economic climate has also made the transition to adulthood increasingly problematic. The uncertainties of work are in part responsible for young adults' lower fertility rates and for women flooding the workplace. Further, the family formation patterns of the 1950s are out of step with the increased educational demands of today's post-industrial society. In the postwar years, particularly in the United States, young people entered adulthood in one giant step—going to work, marrying young and moving to a separate household from their parents, and having children quickly. Today, few young adults can afford to marry and have children in their late teens or early twenties. In an economy where a college degree is necessary to earn a living wage, early marriage impedes education for both men and women.

Those who do not go on to college have little access to jobs that can sustain a family. Particularly in the inner cities of the United States, growing numbers of young people have come to see no future for themselves at all in the ordinary world of work. In middle-class families, a narrowing opportunity structure has increased anxieties about downward mobility for offspring, and parents as well. The "incompletely launched young adult syndrome" has become common: Many young adults deviate from their parents' expectations by failing to launch careers and become successfully independent adults, and many even come home to crowd their parents' empty nest (Schnaiberg and Goldenberg, 1989).

The Life Course Revolution

The demographic transformations of the twentieth century were no less significant than the economic ones. We cannot hope to understand current predicaments of family life without understanding how radically the demographic and social circumstances of U.S. culture have changed. In earlier times, mortality rates were highest among infants, and the possibility of death from tuberculosis, pneumonia, or other infectious diseases was an ever-present threat to young and middle-aged adults. Before the turn of the twentieth century, only 40 percent of women lived through all the stages of a normal life course— growing up, marrying, having children, and surviving with a spouse to the age of 50 (Uhlenberg, 1980).

Demographic and economic change has had a profound effect on women's lives. Women today are living longer and having fewer children. When infant and child mortality rates fall, women no longer have to have five or seven or nine children to make sure that two or three will survive to adulthood. After rearing children, the average woman can look forward to three or four decades without maternal responsibilities. Because traditional assumptions about

women are based on the notion that they are constantly involved with pregnancy, child rearing, and related domestic concerns, the current ferment about women's roles may be seen as a way of bringing cultural attitudes in line with existing social realities.

As people live longer, they can stay married longer. Actually, the biggest change in contemporary marriage is not the proportion of marriages disrupted through divorce, but the potential length of marriage and the number of years spent without children in the home. By the 1970s the statistically average couple spent only 18 percent of their married lives raising young children, compared with 54 percent a century ago (Bane, 1976). As a result, marriage is becoming defined less as a union between parents raising a brood of children and more as a personal relationship between two individuals.

A Psychological Revolution

The third major transformation is a set of psychocultural changes that might be described as "psychological gentrification" (Skolnick, 1991). That is, cultural advantages once enjoyed only by the upper classes—in particular, education—have been extended to those lower down on the socioeconomic scale. Psychological gentrification also involves greater leisure time, travel, and exposure to information, as well as a general rise in the standard of living. Despite the persistence of poverty, unemployment, and economic insecurity in the industrialized world, far less of the population than in the historical past is living at the level of sheer subsistence.

Throughout Western society, rising levels of education and related changes have been linked to a complex set of shifts in personal and political attitudes. One of these is a more psychological approach to life—greater introspectiveness and a yearning for warmth and intimacy in family and other relationships (Veroff, Douvan, and Kulka, 1981). There is also evidence of an increasing preference on the part of both men and women for a more companionate ideal of marriage and a more democratic family. More broadly, these changes in attitude have been described as a shift to "post-materialist values," emphasizing self-expression, tolerance, equality, and a concern for the quality of life (Inglehart, 1990).

The multiple social transformations of our era have brought both costs and benefits: Family relations have become both more fragile and more emotionally rich; mass longevity has brought us a host of problems as well as the gift of extended life. Although change has brought greater opportunities for women, persisting gender inequality means women have borne a large share of the costs of these gains. But we cannot turn the clock back to the family models of the past.

Paradoxically, after all the upheavals of recent decades, the emotional and cultural significance of the family persists. Family remains the center of most people's lives and, as numerous surveys show, a cherished value. Although marriage has become more fragile, the parent–child relationship—especially the mother–child relationship—remains a core attachment across the life course (Rossi and Rossi, 1990). The family, however, can be both "here to stay" and beset with difficulties. There is widespread recognition that the massive social and economic changes we have lived through call for public and private-sector policies in support of families. Most European countries have recognized for some time that governments must play a role in supplying an array of supports to families—health care, children's allowances, housing subsidies, support for working parents and children (such as child care, parental leave, and shorter work days for parents), as well as an array of services for the elderly.

Each country's response to these changes, as we've noted earlier, has been shaped by its own political and cultural traditions. The United States remains embroiled in a cultural war over the family; many social commentators and political leaders have promised to reverse the recent trends and restore the "traditional" family. In contrast, other Western nations, including Canada and the other English-speaking countries, have responded to family change by establishing policies aimed at mitigating the problems brought about by economic and social transformations. As a result of these policies,

these countries have been spared much of the poverty and other social ills that have plagued the United States in recent decades.

Looking Ahead

The world at the beginning of the twenty-first century is vastly different from what it was at the beginning, or even the middle, of the twentieth century. Families are struggling to adapt to new realities. The countries that have been at the leading edge of family change still find themselves caught between yesterday's norms, today's new realities, and an uncertain future. As we have seen, changes in women's lives have been a pivotal factor in recent family trends. In many countries there is a considerable difference between men's and women's attitudes and expectations of one another. Even where both partners accept a more equal division of labor in the home, there is often a gap between beliefs and behavior. In no country have employers, the government, or men fully caught up to the changes in women's lives.

But a knowledge of family history reveals that the solution to contemporary problems will not be found in some lost golden age. Families have always struggled with outside circumstances and inner conflict. Our current troubles inside and outside the family are genuine, but we should never forget that many of the most vexing issues confronting us derive from benefits of modernization few of us would be willing to give up—for example, longer, healthier lives, and the ability to choose how many children to have and when to have them. There was no problem of the aged in the past, because most people never aged; they died before they got old. Nor was adolescence a difficult stage of the life cycle when children worked, education was a privilege of the rich, and a person's place in society was determined by heredity rather than choice. And when most people were hungry illiterates, only aristocrats could worry about sexual satisfaction and self-fulfillment.

In short, there is no point in giving in to the lure of nostalgia. There is no golden age of the family to long for, nor even some past pattern of behavior and belief that would guarantee us harmony and stability if only we had the will to return to it. Family life is bound up with the social, economic, and ideological circumstances of particular times and places. We are no longer peasants, Puritans, pioneers, or even suburbanites circa 1955. We face conditions unknown to our ancestors, and we must find new ways to cope with them....

REFERENCES

Amato, P. R. 1994. Life span adjustment of children to their parents' divorce. *The Future of Children* 4, no 1. (Spring).

Bane, M. J. 1976. *Here to Stay*. New York: Basic Books.

Bernard, J. 1982. *The Future of Marriage*. New York: Bantam.

Blake, J. 1978. Structural differentiation and the family: A quiet revolution. Presented at American Sociology Association, San Francisco.

Cherlin, A. J. 1981. *Marriage, Divorce, Remarriage*. Cambridge, Mass.: Harvard University Press.

Demos, John. 1986. *Past, Present, and Personal*. New York: Oxford University Press.

Emde, R. N., and R. J. Harmon, eds. 1984. *Continuities and Discontinuities in Development*. New York: Plenum Press.

Farrell, Christopher. 1994. Twenty-first century capitalism: The triple revolution. *Business Week* (November 18): 16–25.

Featherstone, J. 1979. Family matters. *Harvard Educational Review* 49, no. 1: 20–52.

Geertz, G. 1965. The impact of the concept of culture on the concept of man. In *New Views of the Nature of Man*, edited by J. R. Platt. Chicago: University of Chicago Press.

Han, S.-K., and P. Moen. 1999. Work and family over time: A life course approach. *The Annals of the American Academy of Political and Social Sciences* 562: 98–110.

Hrdy, Sarah, B. 1999. *Mother Nature*. New York: Pantheon Books.

Inglehart, Ronald. 1990. *Culture Shift*. New Jersey: Princeton University Press.

Keller, S. 1971. Does the family have a future? *Journal of Comparative Studies*, Spring.

Macfarlane, J. W. 1964. Perspectives on personality consistency and change from the guidance study. *Vita Humana* 7: 115–126.

Masnick, G., and M. J. Bane. 1980. *The Nation's Families: 1960–1990*. Boston: Auburn House.

Mellman, A., E. Lazarus, and A. Rivlin. 1990. Family time, family values. In *Rebuilding the Nest*, edited by D. Blankenhorn, S. Bayme, and J. Elshtain. Milwaukee: Family Service America.

Rossi, A. S., and P. H. Rossi. 1990. *Of Human Bonding: Parent–Child Relations Across the Life Course*. Hawthorne, New York: Aldine de Gruyter.

Rubin, L. 1996. *The Transcendent Child*. New York: Basic Books.

Ryan, M. 1981. *The Cradle of the Middle Class*. New York: Cambridge University Press.

Schnaiberg, A., and S. Goldenberg. 1989. From empty nest to crowded nest: The dynamics of incompletely launched young adults: *Social Problems* 36, no. 3 (June): 251–269.

Skolnick, A. 1991. *Embattled Paradise: The American Family in an Age of Uncertainty*. New York: Basic Books.

Stephens, W. N. 1963. *The Family in Cross-Cultural Perspective*. New York: World.

Tannen, D. 1998. *The Argument Culture*.

Uhlenberg, P. 1980. Death and the family. *Journal of Family History* 5, no. 3: 313–320.

Veroff, J., E. Douvan, and R. A. Kulka. 1981. *The Inner American: A Self-Portrait from 1957 to 1976*. New York: Basic Books.

22

Being Single on Noah's Ark

LEONARD CARGAN

The author describes what it is like to be an unmarried person in a world consisting of a majority of married people. By not marrying by a certain age, these unmarried people are violating the expected pattern of adult life. This deviation from the expected pattern leads to imagined reasons for their not following the pattern. Since little is really known about these deviants, their believed behavior is described in stereotypes such as "swingers" and "loneliness." Despite these negative applications, the singles population has been growing since the "golden age" of marriage in the 1950s due to an age delay in getting married and a rising divorce rate.

As you read this article, consider the following questions as guides:

1. *Do you believe that the situation described is temporary, will stay about the same, or will continue to grow in numbers? Why?*
2. *The increase in numbers of those who delay marriage or do not marry at all is also seen among men. Is it for the same reasons as women? If not, what reasons do men have for postponing marriage or staying single permanently?*
3. *Do you consider that the delayed marriage and nonmarriage phenomenon is a problem? Why or why not?*

All social systems must have norms and values. Norms are agreed-to rules of behavior that allow the group, the organization, or the community to attain its goals—those values that are considered important to achieve. However, if everyone pursued these valued items without regard to norms, there would soon be chaos. In America, a major value is that of individualism—being responsible for our own actions and ideas. An example of such values is seen in this country's great emphasis on competition. This emphasis on competition usually means that government proposed or run programs are usually considered inferior to those run by private enterprise. Another prized value is that of individuality and so we are encouraged to think for ourselves. In seeking these values it is necessary to follow norms—certain rules of behavior. For example, there are numerous norms that must be followed in order to drive safely and prevent accidents. Similarly, there are other norms to protect the public from those who do not follow the norms in seeking valued possessions. The importance of norms and values does not mean that they do not undergo change. For example, the value of a kingdom was exchanged for the value of a republic in the revolutionary war, the value of slavery was exchanged for freedom for all in the civil war, and the value of equal political rights was given to women when given the right to vote.

As America has grown larger in population and urbanized in the twentieth century, it has also become more of an organizational society. These changes have transformed all aspects of social life and required the country to put a greater emphasis on norms, on conforming to the rules whether written or simply understood as the way things are done. In regard to the prized value of individuality it means that it cannot be carried beyond certain limits without being restrained by the norms of conformity. To do so would bring about penalties either directly by, perhaps, going to jail or indirectly via the application of negative stereotypes regarding that behavior and a resulting shame.

This emphasis on conformity can lead to contradictions and even problems since values and norms change and may change rapidly due to rapidly changing technology. It appears that the penalties of such contradictions in the norms are more likely to be felt by minority groups since by definition they are different from the majority in the way things should be done. Thus, ethnic and racial groups may be discriminated against because they are different in their language, dress, food, and values. In a system of rapidly changing values and norms, it is possible for other groups to become recognized as minorities and subsequently discriminated against. For example, it really has been only recently that such groups as women, the poor, and the handicapped were recognized as minorities that are being discriminated against. Although the discrimination was obvious, it apparently was accepted since it was believed that little could be done about it by a minority group. Thus, it has been only relatively recently that efforts have been undertaken to lessen and perhaps eliminate discrimination against women and the handicapped. Obviously, such discrimination changes are slow. The result is that the unmarried are considered a minority in a society of mostly married and are discriminated against by that majority. This discrimination is seen in the development of beliefs designed to perpetuate the dominant system—couples. Thus, singleness is seen as a negative choice as compared to being married— a positive choice. As a means of encouraging this positive choice, singles have been discriminated against via acts—lack of promotions and beliefs— stereotypes concerning their behavior—a bunch of "old maids."

CHANGING FUNCTIONS

An important question to ask at this point is why? Why bother to discriminate against what seems to be a harmless group who mostly will eventually join the majority? Approximately 90 percent of all adult singles will eventually marry. To answer this question, it is necessary to examine the functions that marriage provides and that society believes are valuable and, therefore, should be perpetuated. Despite the loss of many functions in regard to education, protection, recreation, and work, marriage is still seen as the means for aiding and maintaining life: providing for such personal needs as affection and security, socializing the infant with human qualities, and providing the "matrix for the development of personality" (Ackerman, 1972:16). In addition, the values sought in marriage are among the best needed and desired and for which there are no reliable substitutes: love, loyalty, and stability (O'Brien, 1974:51). It is for these reasons that marriage is considered a valued part of the natural order of life's progression and that those who do not conform to this order are believed to threaten it. This being the belief, it becomes necessary to encourage conformity to this order via discrimination and stereotypical beliefs. Thus, marriage is presented as the normal and as the only healthy solution to life's dilemmas. This is the value presented despite the facts that show marriage to be an institution experiencing a number of problems. As a result, singleness has been relegated to the margins of society and being single is seen as a temporary position. There is also the added pressure of the family and society to get married. All of these factors contribute to the single feeling deviant and viewing life through this lens. Despite these factors, there has been a growing population of singles.

People are single for a number of reasons. Due to changing social conditions, these reasons are changing and as a result the numbers of people who are single are spending more time in this category. As noted, a major reason for this time spent as a single is the delay in age for a first marriage. In 1960, the number of unmarried men at age 24 was 53 percent but by 1990, it had risen by 25 percent to 78 percent. Similarly, the rates for unmarried women had spurted from 28 percent to 61 percent—a 33 percent increase. The average age for both men and women at first marriage is now at an all-time high. The main reason for this jump in age at first marriage appears to be a value change requiring increasing education to obtain work, especially "good" work.

Another factor adding to the population growth of singles is the rapid growth in those divorced. Apparently, divorce is no longer a rare and stigmatizing event. Currently, there are 1.2 million divorces a year. This figure has led to the mistaken belief that the divorce rate is a staggering 50 percent. This belief is based on the fact that there are about 2.4 million marriages per year. Hence, 50 percent of this figure is 1.2 million—the number of divorces occurring each year. This belief would be true if all these divorces came only from those 2.4 million marriages. However, they do not. These divorces come from all existing marriages and so the correct divorce rate is the number of divorces per thousands of existing marriages or 20 percent. Surprisingly, this is actually a lower percentage than that which occurred in the 1920s but much higher than what was believed to be the golden age for marriage— the 1950s. Another good question to ask at this point is why there is a growing rate of those being divorced? The factors believed to be involved include a declining influence in the role of religion, the liberalization of divorce laws, and the increased participation of women in the labor market. This latter item is believed to have made women less dependent on men. Thus, 60.7 percent of divorces filed were by wives versus about half that amount (32.5 percent) by husbands. Some think the divorce rate is not as bad as it seems since 75 percent of those getting a divorce will remarry. This may be wishful thinking since half of the remarried will re-divorce.

The changes noted that have made being single more common have also made it somewhat more acceptable but only up to a certain age bracket. Beyond 30, singleness remains uncomfortable for both men and women. The truth is that we still live in a world of couples and families. Singleness is still defined only in its relation to marriage—it is the absence of marriage. The status of being single is still seen as a transitional stage as one moves toward marriage. It is marriage that indicates intimacy and the sanctioned way of expressing that intimacy. Marriage is the norm for the community and also indicates the approved means for having children and becoming a part of the family-oriented community. As noted, this means that most people want to marry and do marry, usually by the age of 25 for men and 23 for women, and even remarry after a divorce.

Turning to those who remained single after the seemingly approved age of 30, an opposite picture is revealed. The single person is seen as one who is moving against the approved value that defines marriage as the most desirable state adults can attain. They are often perceived as one who, lacking a partner, are alone and not complete. They are, therefore, "unfit and deviant" (Bell, 1972:89).

THE STEREOTYPES

To not marry requires stereotypical beliefs regarding this failure to follow this approved path: you are socially inadequate and so failed in the dating game or are an alcoholic; you are immature and unwilling to assume responsibility; you are a homosexual and hostile to the opposite sex; you are over-focused on economics in that you believe you are too poor to marry or that marriage is a threat to the pursuance of your career; even if that career is a theological one, your single status is due to your being a religious fanatic or a recluse; or you are single due to geographic or occupational limitations. If none of these excuses for not marrying applies, then the failure not to marry must have been an oversight. The listing of why people do not marry is then a reflection of the belief that every one should marry and that everyone can if they really want to. For the stereotypes, singles are one of the fastest growing segments of American households—a record 40 million compose this category.

Freedom, Workaholic, Affluent

These three stereotypes are so interrelated that it is easier to deal with them as a whole rather than separately. Since the single is not burdened with a wife and family, they are perceived as being freer to have more time to do as they wish. This freedom also implies that the single is less responsible. They can, if they wish, be only concerned with having fun and looking after their own interests. Their lack of responsibilities and community ties also means that they are irresponsible and immature. Freedom is actually a feature that both the single and non-single population share—the freedom to divide their time between office work, housework, social responsibilities, and fun objectives. As noted, with their lesser obligations to deal with a wife and family, the single is freer to spend more time at work in an effort to gain promotions and affluence or more time having fun. Unfortunately, the stereotypes concerning singles provide a hindrance to his work ambition—it is the married who usually get the promotions since they are seen as more mature and responsible, they are the ones restrained by the obligations of the family. Thus, it is the married who may become workaholics since they have greater obligations.

Happiness and Loneliness

This is a mixed set of stereotypes. On the one hand, the singles are pictured as happier because they have more freedom to make choices, to get out more, and subsequently have more fun. This also means that they have fewer worries. All these beliefs indicate that the single should be happier than the non single. On the other hand, there is the likelihood of usually living alone and having no permanent person with whom to share. Therefore, the single person is lonelier and unhappy. In reply to

these absolutes regarding single life, singles note that life is not usually a one-way street for singles or the married. There are times when they are enjoying themselves and so are happy and there are times when they wished that they had company but also glad to have, at times, solitude. Thus, loneliness may be a problem at times but it is a problem willingly accepted for the perceived advantages of solitude and being single.

Deviance

Perhaps the single most approbation leveled at singles is that they are deviant. Since marriage is seen as the normal transition process for adults between certain ages, they should be married. Therefore, this charge is true—they are deviating from the norm of marriage. At the same time, singles are and always have been a substantial proportion of the population and it seems strange to label such a large minority as deviant. Population trends also indicate that this minority is rapidly growing due to a later age for a first marriage and a slowly increasing divorce rate. It is possible that before long the minority will be the majority. Still it is doubtful that the beliefs associated with being a minority will shift to the married.

Given these stereotypes, we are socialized with the belief that to be truly happy and fulfilled, it is necessary to marry and take on the responsibilities of a family. This norm is seen in the efforts by others to have you meet the right person and the joy displayed by parents when an engagement is announced and a wedding takes place. In these ways, marriage is proclaimed as society's norm. Despite all these beliefs and married, it appears beyond belief that remaining single might have been a rational choice.

The result of these beliefs are descriptions designed to degrade the single and their lifestyle: the single person is a "poseur, a squanderer, a narcissist, a wastrel." His lifestyle consists of dancing "the hustle in the apartment's house party room" and "loafing on his plastic horse in the … swimming pool." He "lives for lotions, balms, and sprays." He is also a "non-stop lover, drinker, laugher, and more (or less)" (Rosenblatt, 1977:14). Even when so-called positive images of singles are presented, they must be degraded. Thus, before marrying, Joe Namath is presented as a positive image of a single—the happy, swinging single yet, at the same time, it implies overtones of being immature, selfish, lechery, and social irresponsibility (Libby, 1978:165; Stein, 1978:2).

The fact that the preceding description of a single described a male does not mean that such stereotypes do not apply to females. According to the beliefs regarding single females, it is she who is "saved" by marriage since this is the only way that she can find love and sexual fulfillment. Thus, it is important for them to marry lest they face emotional and physical deterioration. Single women are such because they are unattractive, handicapped, or incompetent (Deegan, 1969:3) and so they are people to be pitied, ridiculed, disliked, and ascribed a low status except when needed such as in times of war. The single young woman may be intriguing and even challenging but prolonged singleness brings forth suspicions since it seems hard for married people to believe that these women are single because they do not want to be wives.

As a means of better understanding these stereotypical beliefs, it is necessary to sort them out into more specific groupings since their large number results in stereotypes that are entangled, overlapping, and contradictory. In order to be fair, a reply to these stereotypes will be attempted by those being condemned—the singles.

Immaturity

Since marriage is seen as a normal stage of adult development, the "failure" to marry must reflect some kind of immaturity. A man is seen as being tied to his mother's apron strings whereas the woman, being a spinster has failed to experience life's adventures. The sign of immaturity may be a reflection of their lack of altruism—they are selfish and unable to share. To contradict such beliefs, one merely need to point to the many singles like Isaac Newton and Ralph Nader who were experienced, selfless, and obviously mature.

Sexual Deviant

One of the most prevalent stereotypes of the single is seen in the approbation that one is a "swinger"—a lecher hopping from bed to bed. Or their sexual needs may be satisfied by an act of selfishness such as masturbation, an act of sexual inversion which is considered as unnatural as abstinence. What is being ignored in these allegations is that such behaviors have also been found to be quite extensive in the married population as well (Kinsey, 1948).

REFERENCES

Ackerman, Nathan W., in Harrold Hart (ed), *Marriage For and Against.* New York: Hart, 1972, pp. 10–26.

Bell, Robert, in Harrold Hart (ed), *Marriage For and Against.* New York: Hart, 1972, pp. 79–99.

Deegan, Dorothy, Y. *The Stereotypes of the Single Woman in American Novels: A Social Study with Implications for the Education of Women.* New York: Octagon, 1969.

Kinsey, A. C., W. B. Pomeroy, & C. E. Martin. *Sexual Behavior in the Human Male.* Philadelphia: W. B. Saunders, 1948.

Libby, Roger W. "Creative Singlehood as a Sexual Lifestyle: Beyond Marriage as a Rite of Passage," in Bernard I. Murster (ed), *Exploring Intimate Lifestyles.* New York: Springer, 1978.

O'Brien, Patricia. *The Woman Alone.* New York: Quadrangle, 1974.

Rosenblatt, Roger. "The Self as a Sybarite." *Harpers* (March 1977), pp. 12–14.

Stein, Peter. "The Lifestyle and Life Chances of the Never-Married." *Marriage and Family Review* (July/August, 1978), pp. 1–11.

23

An Overview of the Nature, Causes, and Consequences of Abusive Family Relationships

ROBERT E. EMERY AND LISA LAUMANN-BILLINGS

The authors note that the development of abusive relationships has a long and mostly overlooked history.

As you read, ask yourself the following questions:

1. *Which facts about family abuse surprised you the most?*
2. *What would you add to the means for dealing with family abuse?*

GLOSSARY   **Battered children** Physical, mental, or sexual abuse or neglect of a child constitutes child abuse.

From the vantage point of the present, the public outcry against family violence is shockingly recent. The Society for the Prevention of Cruelty to Children was formed only in 1875 (nine years after the formation of the Society for the Prevention of Cruelty to Animals), and the arousal of professional concern about battered children dates only to 1962 with the publication of Kempe's seminal article (Kempe, Silverman, Steele, Broegemueller, & Silver, 1962; more pervasive professional concern is far more recent). Spousal assault was made a crime in all states only in recent decades, and there still are no laws against spousal rape in some states. Professional concern about the children of battered women is less than two decades old, and sibling and elder abuse are just now beginning to receive clinical and research attention....

THE DEVELOPMENT OF ABUSIVE RELATIONSHIPS

Many factors contribute to the development of family violence, including individual personality factors, family interaction patterns, poverty and social disorganization, acute stressors, and the cultural context in which the family lives (Bronfenbrenner, 1979). Consistent with their comorbidity, a number of the same factors increase the risk for child and spouse abuse, which are the types of abuse that are the focus of our overview....

...Understanding Family Violence...

Individual Characteristics. Personality factors such as low self-esteem, poor impulse control,

Source: From Emery, R. E., & Laumann-Billings, L. (1998). An overview of the nature, causes, and consequences of abusive family relationships toward differentiating maltreatment and violence. *American Psychologist*, 53(2), 121–135. Copyright © 1998 by The American Psychological Association. Adapted with permission.

external locus of control, negative affectivity, and heightened response to stress all increase the likelihood that an individual will perpetrate family violence (Pianta, Egeland, & Erickson, 1989). Alcohol or drug dependence also plays a role both as a background risk factor and as an immediate precipitant of family violence (Kantor & Straus, 1990). Although the evidence is less consistent, some research has also suggested that the victims of family violence share some common characteristics, including poor physical or mental health, behavioral deviancies, and difficult temperament or personality features (Belsky, 1993). For child victims, age also seems to play a role, as younger children are more prone to be seriously injured as a result of family violence (Lung & Daro, 1996).

Immediate Context. Characteristics of the immediate social context, especially of the family system, have important implications for the etiology or perpetuation of violence in that family (Emery, 1989). Studies have examined structure and size, acute stressors such as the loss of a job or a death in the family, and characteristic styles of resolving conflicts or parenting.

Broader Ecological Context. Violence in the family also is related to qualities of the community in which the family is embedded, such as poverty, absence of family services, social isolation, and the lack of social cohesion in the community. As Garbarino and Kostelny (1992) noted, family violence is a social as well as a psychological indicator (p. 463).

Of course, most poor parents do not maltreat their children (Bronfenbrenner, 1979); thus, the relation between poverty and abuse is not simple or direct. Garbarino and colleagues have consistently shown that the principal difference between poor families who do and those who do not maltreat their children lies in the degree of social cohesion and mutual caring found in their communities (Garbarino & Kostelny, 1992).

Societal or Cultural Context. A number of commentators contend that family violence is perpetuated by broad cultural beliefs and values, such as the use of physical punishment, extremes in family privacy, and violence in the popular media (Garbarino, 1977). As noted earlier, our societal concern with violence in the family is historically recent, and our continued reluctance to intervene forcefully is evident in the reluctance to terminate parental rights even in egregious cases of child abuse (Besharov, 1996b). Societal policies may not cause family violence, but many of our practices appear to condone it.

CONSEQUENCES FOR VICTIMS...

Physical Injuries. Physical injury is a clear consequence of many instances of family violence, but information on the physical consequences of family violence is surprisingly incomplete. As noted earlier, an estimated 1,200 to 1,500 children, most of whom are under the age of five years, die each year as a result of either physical abuse (48%), neglect (37%), or both (15%) by a parent or parent figure (Lung & Daro, 1996). In addition, almost one-third of the 4,967 women murdered in 1995 were killed by a boyfriend or husband (Federal Bureau of Investigation, 1996). Official data may somewhat underestimate actual deaths due to family violence, however, because of misclassification of causes of death (McClain, Sacks, & Frohlke, 1993).

Information on nonlethal injuries is much more sketchy. According to NIS-3 (Sedlak & Broadhurst, 1996), nearly 50,000 children were victims of serious physical abuse in 1993; that is, they suffered from life-threatening injuries, long-term physical impairment, or required professional treatment to prevent long-term physical impairment. It has also been estimated that over 18,000 children become severely disabled each year as a result of severe child abuse (Baladerian, 1991). We could not locate epidemiological data on more specific physical outcomes, such as the rate of nonorganic failure to thrive (resulting from gross neglect) among infants, specific injuries such as broken bones, or pregnancies or sexually transmitted diseases resulting from sexual abuse.

The threat of injury often is devastating; however, we need more specific information on injury and disease for several reasons. First, physical safety is the foremost concern about family violence; thus, clear evidence of harm is essential for a number of purposes ranging from basic knowledge to justifying intervention. Second, specific physical injuries provide one relatively unambiguous (if limited) definition of family violence; thus, data would allow investigators to clearly chart the scope of the problem and changes over time. Third, clear information on the physical consequences of family violence should allow researchers to more adequately investigate the psychological consequences of physical abuse, especially given evidence for a dose-response relation, which we review shortly.

Psychological Consequences. All types of family violence are linked with diverse psychological problems ranging from aggression, to anxiety, to depression. Still evidence on more specific psychological outcomes may be clouded by the need to consider (a) risk factors correlated with family violence; (b) clusters of symptoms (e.g., disorders like posttraumatic stress disorder [PTSD]) in addition to specific symptoms; (c) subtle psychological consequences that are difficult to document empirically, particularly among children; and (d) psychological processes (not just psychological outcomes) set into motion or disrupted by the experience of family violence.

Correlated Risk Factors. Family violence is associated with a number of factors known to place children and adults at risk for psychological problems—for example, poverty, troubled family environments, genetic liability, and so on. These risk factors may account for the apparent relation between family violence and specific psychological outcomes, or their effects may interact with the consequences of abuse. Thus, for example, research comparing abused and nonabused children may actually reflect the psychological effects of anxious attachments, social isolation, or general family stress rather than the consequences of violence per se

(National Research Council, 1993). It is essential to consider this possibility both in research and in clinical interventions, which appropriately may focus on risk factors in addition to, or even instead of, the abuse, particularly when abuse is less serious.

Subtle Psychological Effects on Victims. The diagnoses of PTSD and acute stress disorder include symptoms such as reexperiencing and dissociation; thus, they offer further impetus for studying some of the more subtle psychological consequences of abuse, particularly among children. The vehement controversy over recovered memories of abuse provides another rationale for research on more subtle psychological reactions (Loftus, 1994). A review of the recovered memories controversy is beyond the scope of this article, but we do note that one longitudinal study found that documented sexual abuse among girls ages 10 months to 12 years was not reported by nearly 40% of the victims when they were asked about their history as young adults (Williams, 1994a). Our present concern is not how supportive or unsupportive these data are of the recovered memories controversy (Williams, 1994b) but the need to better understand the emotional and cognitive processes that may contribute to the forgetting of (or the dissociation from) abuse in the 17 years between victimization and reinterview.

PREVENTION AND INTERVENTION...

Child Abuse Reporting

By all accounts, child protective services are driven by—and overwhelmed with—the investigation of child abuse reports. As we have noted, over three million reports are made every year, but less than one-third are substantiated. Moreover, social service workers are so swamped with reports and investigations that about 40% of substantiated cases receive no services at all (McCurdy & Daro, 1993). Finally, only a small proportion of substantiated cases, less than 20%, involve any type of formal court action. Rather, child protective workers encourage the great majority of those involved in substantiated cases to enter treatment voluntarily (Melton et al., 1995).

In order to facilitate the support of families under stress, we would exempt mental health professionals from reporting less serious cases of abuse, whether known or suspected, when a family is actively engaged in treatment. We do not attempt to define these cases here, but we do note the need to define less serious cases (what we have called maltreatment) clearly in the law, as has been done in some experimental programs. Such a change would be a first step toward the broader goal of refocusing the child protection system on supporting rather than policing families under stress, while simultaneously pursuing more vigorous, coercive intervention with cases of serious family violence.

Supportive Interventions

A number of supportive interventions have been developed in an attempt to reduce violent behavior within families, including individual and group therapies for both victims and perpetrators, couples therapy for victims of domestic violence, parent-training and family therapy, and home-visiting programs for the prevention of child abuse. In general, the more serious and chronic the nature of the abuse, the less success these programs have in changing behaviors (National Research Council, 1993). With problems of mild to moderate abuse, however, multilevel programs, which combine behavioral methods, stress management, and relationship skills (parent–child or spousal), lower stress in families and may reduce the likelihood of continued aggression in both child and spouse abuse cases (National Research Council, 1993).

Some interventions show promise, but the need for early intervention and especially prevention is underscored by the difficulty of changing entrenched family violence and the stressful life circumstances that promote abuse (Garbarino & Kostelny, 1992). Home-visitor programs for new parents living in difficult circumstances are one

especially promising form of prevention. Home-visitor programs simultaneously assist with material needs (e.g., cribs, child care, transportation), psychological needs (e.g., parenting education and support), and educational needs (e.g., job skills) and many both improve general family well-being and reduce child maltreatment.

Coercive Intervention

As a final point, we note that distinguishing between levels of abuse also may help to improve intervention in cases of serious and extreme family violence. One benefit would be to free investigators and police to focus on the most serious cases. Shockingly, between 35% and 50% of all fatalities that are due to child abuse or neglect occur in cases that have already been brought to the attention of law enforcement and child protection agencies (Lung & Daro, 1996). Another benefit might be to help to clarify when coercive legal intervention is and is not appropriate. For example, debates have errupted over the once promising policy of mandatory arrest for violent partners (even over the objection of victims), because the reduced recidivism found in early research (Sherman & Berk, 1984) has not been replicated in subsequent studies. As another example, many commentators have questioned the overriding goal of family reunification following child abuse; especially in cases like these, the termination of parental rights and early adoption may be the appropriate intervention, especially given the many problems with the overwhelmed foster-care system (Tatara, 1994).

In both of these examples, distinguishing between levels and abuse should help to resolve controversy and thereby clarify the appropriate use of coercive legal intervention. The idea of termination of parental rights or of arrest is far more threatening when our definitions of abuse include relatively minor acts, as they currently do. Clearly, it is difficult to draw a line between cases that should or should not lead to arrest or termination of parental rights (Azar, Benjet, Fuhrmann, & Cavallero, 1995), but the challenge of distinguishing between levels of abuse should not deter us from the task. What we have called maltreatment may be on a continuum of what we have termed violence, and both acts of abuse may differ from normal family aggression only by a matter of degree. As we have argued throughout this article, however, drawing distinctions between levels of abuse seems consistent with the state of our knowledge about the prevalence of abuse, its development, its consequences, and appropriate intervention.

24

Historical Trends in Divorce in the United States

PAUL R. AMATO AND SHELLEY IRVING

When a group of people were asked what they believe was the divorce rate, they responded with the nice, round figure of 50%. In reality, the divorce rate is 24%. How is it that our opinion on the divorce rate is so off-tangent? Perhaps it is because those reporting the statistics do not understand statistics. Perhaps it is because items such as cohabitation, divorce, nonmarried families, and singles are more eye-catching than the fact that marriage is so popular that nearly 90% of all adults will marry at least once and most will remarry after a divorce. Perhaps it is because divorce has always been controversial. This article clears up some of these factors surrounding the divorce rate.

As you read this selection, consider the following questions as guides:

1. *Why has the divorce rate not declined in this age of love and affection as the basis for marriage?*
2. *Like divorce, marriage has been undergoing change. What are some of these changes? Why do you think they are occurring?*
3. *Why would you marry or not marry?*

...THE 20TH CENTURY

The Frequency of Divorce

During the first half of the 20th century, the divorce rate rose and fell in response to specific events and changing social circumstances. The divorce rate increased gradually during the early years of the 20th century, and then it surged in the years following World War I. Presumably, this occurred for the reasons noted earlier; that is, couples were separated for long periods, some married individuals engaged in adultery during the time of separation, and many veterans had difficulty fitting into civilian life after the war. In addition, there was a rapid increase in the marriage rate during the war years. Given that some of these marriages were contracted hastily (before men-left for military service), it is not surprising that many ill-considered matches were made. The divorce rate declined during the Great Depression of the 1930s—a time when many unhappy spouses literally could not afford to divorce. World War II pulled the United States out of the depression, and, consistent with earlier trends, divorce rates spiked sharply upward during the war. In the aftermath of the war, the divorce rate declined, and by the end of the 1950s, the rate of divorce was back to where it had been 20 years earlier.

Source: *In Handbook of Divorce and Relationship Dissolution.* Mark A. Fine & John H. Harvey (Eds). Lawrence Erlbaum Associates, 2006.

The 1950s represented a relatively "profamily" period in U.S. history. In addition to the stable divorce rate, the marriage rate was high and fertility increased dramatically—a period known as the baby boom. Several forces came together to create the stable, two-parent, child-oriented family of the l950s: a strong economy, increases in men's wages, veterans' taking advantage of the GI bill to obtain university educations, the growth of home construction in the suburbs, and a desire on the part of many people to move beyond the tumultuous war years and to concentrate on home and family life (Cherlin, 1992).

After being stable for more than a decade, the divorce rate increased sharply during the 1960s and 1970s. It reached an all-time peak in 1980 and then declined modestly. Currently, the rate of divorce is about 20(/1000), which means that about 2% of all marriages end in divorce every year [(20/1,000) × 100]. Although the 2% figure may seem low, it is based on a single year. By applying duration-specific probabilities of divorce across all the years of marriage, it is possible to project the percentage of marriages that will end in divorce. Using this method, demographers estimate that about one half of first marriages, and about 60% of second marriages, will end in divorce (Cherlin, 1992). These figures represent a historically high level of marital disruption in the United States. In comparison, about one eighth of all marriages ended in divorce in 1900, and about one fourth of all marriages ended in divorce in 1950 (Preston & McDonald, 1979).

The Legal Regulation of Divorce

...Throughout American history, divorces were granted only when one spouse demonstrated to the court's satisfaction that the other spouse was "guilty" of violating the marriage contract. This process required that courts gather evidence, including witness testimony, either from people familiar with the family or from the spouses themselves. Under this system, the "innocent" spouse often received a better deal than the guilty spouse with respect to the division of marital property, alimony, and child custody. In other words, the law punished the spouse who

was responsible for undermining the marriage (Katz, 1994).

Fault-based divorce did not allow for the possibility that two spouses simply might be unhappy with their marriage and wish to go their separate ways. Instead, by making divorce difficult, and by requiring one spouse to accept responsibility for violating the marriage contract, the law affirmed its commitment to the norm of marital permanence. Despite the difficulty and cost of proving fault in court, however, demand for divorce increased throughout the first half of the 20th century. To accommodate this growing demand, courts broadened the grounds for marital dissolution, and an increasing proportion of divorces were granted on the relatively vague grounds of mental cruelty. More specifically, in the mid-19th century, only 12% of divorces occurred for reasons of cruelty. This figure rose to 28% in 1920, 40% in 1930, 48% in 1940, and 59% in 1950 (DiFonzo, 1997).

In 1933, New Mexico was the first state to add "incompatibility" as a ground for divorce (DiFonzo, 1997). Twenty years later, Oklahoma instituted a form of no-fault divorce based on mutual consent (Vlosky & Monroe, 2002). Although fault-based divorces still were possible in New Mexico and Oklahoma, spouses could dissolve their marriages even if neither spouse had committed a serious marital offense. Several other states added no-fault options during the next decade, including Alaska in 1963 and New York in 1967 (Vlosky & Monroe). It is important to note, however, that fault-based divorce existed side by side with no-fault divorce in each of these states.

The most dramatic change in divorce law occurred in California in 1969, when the state legislature eliminated fault-based divorce entirely and replaced it with no-fault divorce. Under the new legislation, only one ground for divorce existed: the marriage was "irretrievably broken" as a result of "irreconcilable differences" (Glendon, 1989). Moreover, courts in California granted divorces even if one spouse wanted the divorce and the other did not—a system know as unilateral no-fault divorce. The assumption underlying unilateral no-fault divorce was that it takes two committed spouses to form a marriage. If one spouse no longer wished to remain in the relationship, then the marriage was not practicable. California also removed the notion of fault from the award of spousal support and the division of marital property (Parkman, 2000; but note that the notion of fault was retained to a certain extent, however, in decisions about child custody).

Other states quickly followed California's lead, and, by the mid-1980s, all 50 states had adopted no-fault divorce. Most states adopted versions of unilateral no-fault divorce. A few states introduced no-fault divorce, but only by mutual consent. In Pennsylvania, for example; if one spouse wants a divorce and the other does not, then a no-fault divorce can be obtained only after the spouses are separated for 2 years. Otherwise, the spouse who wants the divorce must prove fault. Despite the fact that some states require mutual consent, and despite the fact that some states have retained fault provision, the great majority of divorces in the United States today take place under unilateral no-fault divorce regimes (Katz, 1994).

States introduced no-fault divorce for several reasons. First, it was widely known that many spouses who wished to dissolve their marriages colluded to fabricate grounds for divorce. For example, spouses might claim that infidelity had occurred, even if it had not. The recognition that many couples were making a mockery of the law was one factor that led to divorce reformation (Katz, 1994). In addition, legal scholars increasingly accepted the proposition that spouses had a legal right to end their marriages if they were incompatible, had fallen out of love, or were no longer happy living together (Glendon, 1989). Finally, fault-based divorce is an inherently adversarial procedure. Many legislators believed that no-fault divorce would lessen the level of animosity between former spouses, and, hence, make it easier for them to cooperate in raising their children following marital dissolution (Glendon, 1989; Katz, 1994).

Some scholars believe that the liberalization of divorce laws stimulated further demand for divorce (for a review of this evidence, see Parkman, 2000). Other scholars disagree with this claim (e.g., Glenn,

1999). Nevertheless, even if legal changes encouraged some couples to divorce, this effect was modest. Most scholars believe that changes in divorce laws were more a consequence than a cause of marital breakdown (e.g., Cherlin, 1992).

Not everyone was happy with the shift to no-fault divorce, especially after the sharp increase in divorce during the 1970s. Some conservative reformers hoped to lower the frequency of marital dissolution by placing restrictions on unilateral no-fault divorce. For example, during the 1990s, legislators in nearly a dozen states that had unilateral divorce laws introduced bills that required the consent of both spouses for a no-fault divorce, provided that they had dependent children. In some attempted reforms, fault-based divorce would be available without mutual consent in cases of abuse, desertion, or adultery. Other bills attempted to lengthen the waiting period prior to divorce or to require marital counseling (with the goal of attempting a reconciliation) before divorce. Despite attracting a good deal of media attention, none of these bills was passed in any state (DiFonzo, 1997).

Divorce law reformers were more successful with the introduction of covenant marriage in three states. Louisiana was the first state to introduce this legislation in 1997, followed by Arizona and Arkansas. Under this system, couples choose between two types of marriage: a standard marriage or a covenant marriage. To obtain a covenant marriage, couples must attend premarital education classes and promise to seek marital counseling to preserve the marriage if problems arise later. Unilateral no-fault divorce is not an option for ending a covenant marriage. Instead, to terminate a covenant marriage, one spouse must prove fault, although a couple can also obtain a divorce after a 2-year separation. (For a detailed description of covenant marriage in Louisiana, see Thompson & Wyatt, 1999.) Although proponents of covenant marriage see it as a way to strengthen marriage and lower the divorce rate, only a small percentage of couples in Louisiana have chosen covenant marriages (Sanchez, Nock, Wright, Pardee, & Ionescu, 2001). For this reason, the introduction of covenant marriage has, so far, been unsuccessful in lowering the rate of divorce.

Public Attitudes Toward Divorce

As the divorce rate surged following World War I, divorce emerged again as a major social issue. During the 1920s and 1930s, articles on divorce appeared frequently in the media, including in magazines such as the *Saturday Evening Post, Vanity Fair, Ladies Home Journal,* and *Good Housekeeping.* Echoing the concerns of a half-century earlier, conservative writers felt that divorce posed a threat to the family and to the larger society, and that divorcing couples were placing their own needs ahead of the needs of their children. Many of these critics placed the blame for marital instability on increases in wives' employment, rising expectations, for personal satisfaction from marriage, and the declining stigma of divorce (DiFonzo, 1997).

Many early family scholars also took a dim view of divorce. Developmental theories of the time, such as Freudian theory, assumed that children need to grow up with two parents to develop normally. For this reason, most social scientists in the first half of the 20th century saw the rising level of marital disruption as a serious social problem. Sociologists were concerned that one of the fundamental institutions of society—the family—was being undermined. In the 1920s and 1930s, social scientists published books with titles such as *The Marriage Crisis* (Groves, 1928), *Family Disorganization* (Mowrer, 1927), and *Marriage at the Crossroads* (Stekel, 1931). Curiously, few studies focused on children, presumably because the idea that divorce was bad for children seemed self-evident. Instead, researchers focused primarily on factors that promote or erode marital happiness (e.g., Burgess & Cottrell, 1939; Terman, 1938).

The debate over divorce subsided during the 1950s and early 1960s. As we noted earlier, this was a profamily time in American history, with high marriage rates, a stable divorce rate, and rising levels of affluence. Nevertheless, many people continued to see divorce as being unrespectable. As an illustration, Adlai Stevenson (who was divorced in 1949) served as the Democratic Party's candidate for President in 1952.

During the campaign, Dwight Eisenhower (the Republican Party candidate) raised Stevenson's divorce as a campaign issue—an issue that resonated strongly among women voters (Rothstein, 2002). Partly for this reason, Stevenson lost the election.

Public attitudes toward divorce became more liberal during the 1960s and 1970s, For example, surveys indicate that *disagreement* with the statement, "When there are children in the family, parents should stay together even if they don't get along," increased from 51% in 1962 to 80% in 1977 (Thornton & Young-DeMarco, 2001). As noted earlier, Adlai Stevenson's divorce was a major campaign issue in the 1952 presidential election. In contrast, Ronald Reagan ran successfully for President in 1980, and the fact that his first marriage had ended in divorce was never raised as a campaign issue (Rothstein, 2002). By the end of the 1970s, the great majority of Americans viewed divorce as an unfortunate but common event, and the stigma of divorce, although still present, was considerably weaker than in earlier eras.

The growing acceptance of divorce was consistent with a larger cultural shift that occured during the 1960s and 1970s. During these decades, personal happiness and self-fulfillment came to be seen as the main goals of marriage (Bellah, Madsen, Sullivan, Swidler, & Tipton, 1985; Cherlin, 2004; Popenoe, 1996). During the same time, people's expectations for sexual satisfaction with marriage also increased (Seidman, 1991). With the satisfaction of personal needs becoming the main criterion by which people judged their marriages, spouses tended to seek divorces when they became dissatisfied with their relationships, even if their marriages did not include serious problems such as cruelty or adultery. Moreover, rather than condemning the decision to divorce, friends and family members tended to support spouses who left unsatisfying marriages.

The last two decades of the 20th century, however, saw a reaction to the liberal views of the 1960s and 1970s. Two national surveys, one carried out in 1980 (Booth, Johnson, White, & Edwards, 1991) and the other carried out in 2000 (Amato, Johnson, Booth, & Rogers, 2003), provide the latest information on this issue. These surveys revealed that agreement with the statement, "Couples are able to get divorced too easily today," increased from 33% in 1980 to 47% in 2000. Correspondingly, agreement with the statement, "The personal happiness of an individual is more important than putting up with a bad marriage," declined from 74% in 1980 to 64% in 2000. Overall, scores on a scale based on these and other items (coded to reflect support for the norm of lifelong marriage) increased by more than one fourth of a standard deviation (0.27) during this 20-year period. Moreover, growing support for the norm of lifelong marriage was apparent among wives as well as husbands. A similar trend can be observed in the General Social Survey. This annual survey of the American population revealed that the percentage of people who believe that divorce should be more difficult to obtain increased from 44% in 1974 to 51% in 2002 (see http://webapp.icpsr.umich.edu/GSS).

Although attempts to scale back no-fault divorce during the 1990s were unsuccessful (as we noted earlier), legislative efforts to strengthen marriage (rather than restrict access to divorce) met with some success. Although not widely recognized at the time, the 1996 federal welfare reform legislation referred to promoting marriage and encouraging the formation and maintenance of two-parent families as explicit policy goals (Ooms, 2002). Since that time, several state governments have enacted legislation and programs to strengthen marriage. For example, Oklahoma (as part of the Oklahoma Marriage Initiative) currently provides publicly funded premarital education classes to a wide range of couples, including poor, unmarried parents. Florida decreased the fee for a marriage license, along with the waiting period between obtaining a license and getting married, for couples that have taken a premarital education course. Florida also requires all high school students to take a course on marriage skills. In Arizona, Florida, Louisiana, and Utah, couples are given marriage materials (booklets or videos) that include information on how to build strong marriages, the effects of divorce on children, and available community resources (see Parke & Ooms, 2002, for details on these policies.)

Despite this recent shift in attitudes and policy, it would be a mistake to conclude that the American public is generally against divorce. Instead, people appear to be deeply ambivalent. For example, in the 2000 survey described earlier (Amato et al., 2003), only a minority of people (17%) agreed with the statement, "It's okay for people to get married thinking that, if it does not work out, they can always get a divorce." However, as also noted earlier, the majority of people continue to believe that personal happiness is more important than remaining in an unhappy marriage. Similarly, a poll by *Time Magazine* in 2000 found that 66% of people believed that children are better off in a divorced family than in "an unhappy marriage in which parents stay together mainly for the kids." At the same time, however, 64% of people believed that children always or frequently are harmed when parents get divorced (Kim, 2000). It appears that many people today hold contradictory, unresolved views on divorce.

Public confusion about divorce is reflected in (and may have been shaped partly by) debate among family scholars during the last two decades of the 20th century. Some scholars view the retreat from marriage and the corresponding spread of single-parent families as a cause for alarm (Glenn, 1996; Popenoe, 1996; Waite & Gallagher, 2000; Whitehead, 1993; Wilson, 2002). These scholars believe that American culture has become increasingly individualistic, and people have become inordinately preoccupied with the pursuit of personal happiness. Because people no longer wish to be hampered with obligations to others, commitment to traditional institutions that require these obligations, such as marriage, has eroded. Indeed, in a society that encourages the maximization of self-interest, people are more concerned about their own well-being than the well-being of their spouses or children. For example, Popenoe (1996) made this argument:

> Traditionally, marriage has been understood as a social obligation—an institution designed mainly for economic security and procreation. Today, marriage is understood mainly as a path toward self-fulfillment.... No longer comprising a set of norms and social obligations that are widely

enforced, marriage today is a voluntary relationship that individuals can make and break at will. (p. 533)

According to Popenoe, people are no longer willing to remain married through the difficult times, for better or for worse, until death do us part. Instead, marital commitment lasts only as long as people are happy and feel that their own needs are being met.

Other scholars reject the notion that our culture has shifted toward greater individualism and selfishness in recent decades (Coontz, 1988, Scanzoni, 2001; Skolnick, 1991; Stacey, 1996). These scholars also are skeptical of claims that the proportion of unsuccessful marriages has increased. According to this perspective, marriages were as likely to be troubled in the past as in the present, but because obtaining a divorce was time-consuming and expensive, and because divorced individuals were stigmatized, these troubled marriages remained "intact." Rather than view the rise in marital instability with alarm, advocates of this perspective point out that divorce provides a second chance at happiness for adults and an escape from a dysfunctional and aversive home environment for many children. Moreover, because children are adaptable and can develop successfully in a variety of family structures, the spread of alternatives to mandatory lifelong marriage poses few problems for the next generation. Coontz (2000) stated the following:

> The amount of diversity in U.S. families today is probably no larger than in most periods of the past.... Most of the contemporary debate over family forms and values is not occasioned by the existence of diversity but by its increasing legitimation. Historical studies of family life ... make it clear that families have always differed. Many different family forms and values have worked (or not worked) for various groups at different times. There is no reason to assume that family forms and practices that differ from those of the dominant ideal are necessarily destructive. (p. 28)

Feminist scholars, in particular, have argued that changes in family life during the past several decades have strengthened, rather than undermined, the quality of intimate relationships. For example, Stacey (1996, p. 9) stated that "changes in work, family, and sexual opportunities for women and men ... open the prospect of introducing greater democracy, equality and choice than ever before into our most intimate relationships, especially for women and members of sexual minorities."

GENERAL TRENDS AND EXPLANATIONS

Four General Trends

Looking at the broad sweep of American history leads to four general conclusions about divorce trends. First, despite the fact that the divorce rate in the United States increased during some periods and decreased during others, the overarching trend—starting in the 17th century and continuing through the 20th century—has been a gradual rise in the rate of divorce. The roots of our current high divorce rate are over 300 years old. The fact that half of all current first marriages end in divorce, therefore, cannot be a function of relatively recent social conditions such as the shift to no-fault divorce. Instead, the seeds of widespread marital instability were present in American society from its very beginning. Any theoretical explanation for the current high rate of divorce must take this fact into account.

Second, the conditions that courts, state legislatures, and the public view as justifying divorce have expanded continuously throughout American history. During the Colonial era, divorces were granted for a limited number of causes, such as desertion and adultery. These grounds expanded significantly during the 19th and 20th centuries. The ground of cruelty is particularly noteworthy, as it referred to relatively severe forms of physical abuse in the early 1800s and later incorporated acts of emotional cruelty and unkindness. A particularly dramatic shift occurred in the 1970s and early 1980s, when all states accepted versions of no-fault divorce. This change eliminated the necessity of proving that the marriage contract had been violated and made it possible for couples to divorce for virtually any reason. Moreover, unilateral no-fault divorce (available in most states) meant that one spouse could obtain a divorce, even if the other spouse wanted the marriage to continue. The gradual expansion of the grounds for divorce suggests that the law was responding to an ever-increasing demand from the American public, along with rising expectations for what constitutes a good marriage.

A third noteworthy trend is the tendency for wives to initiate divorce more often than husbands. This gender difference existed during the Colonial era (Phillips, 1991), the 19th century (Willcox, 1891), and the 20th century (Braver, 1998; Kitson, 1992; Maccoby & Mnookin, 1992). Why are wives more likely than husbands to seek a divorce? One explanation is that men are more likely to behave badly within marriage than are women. For example, men are more likely than are women to engage in serous forms of violence, both inside and outside of marriage (Felson, 2002). Men also are more likely than women to abuse alcohol and other substances, and husbands are more likely than wives to engage in infidelity. . . . Recent research shows that wives are especially likely to end their marriages because of their spouses' cruelty, adultery, or drunkenness (Amato & Previti, 2003)—a difference that appears to have been as true in the past as it is today.

In addition, in the Colonial era through the 19th century, husbands were considerably more likely than wives to desert their families, largely because men had more opportunities for economic independence. Given that desertion was one of the few grounds for divorce in many states, it is not surprising that wives were especially likely to file for divorce. However, with the industrialization of the American economy, women gained greater access to employment, and, correspondingly, the means to live independently from men (albeit at a lower standard of living). Economic autonomy made it possible for wives to leave their marriages, just as husbands had left (deserted) their marriages in earlier eras. Rather than deserting their husbands, however, wives were able to obtain divorces on the expanding grounds of cruelty, and later, on the basis of no-fault provisions. In other words, although wives have always been more likely than husbands to initiate divorce, their reasons for doing so shifted as social conditions changed.

The feminist perspective, which argues that marriage provides substantially greater benefits to husbands than to wives, provides another explanation for the tendency for wives to initiate divorce. This point was articulated by 19th-century feminist activists, such as Susan B. Anthony and Elizabeth Cady Stanton, and it has been echoed in the writings of more contemporary feminist scholars (e.g., Bernard, 1972; Stacey, 1996). For example, Bernard argued that "marriage introduces such profound discontinuities into the lives of women as to constitute genuine emotional health hazards" (p. 37). Some empirical work supports the notion that marriage improves the mental and physical health of husbands and lowers the mental and physical health of wives (e.g., Gove & Tudor, 1973). The great majority of recent longitudinal studies, however, indicate that marriage generally improves the mental and physical health of *both* sexes (e.g., Ross, 1995; Waite & Gallagher, 2000; Williams, 2003)—a finding that contradicts the feminist hypothesis.

Despite the lack of empirical support for the feminist hypothesis (that marriage benefits husbands more than wives), this notion is partly consistent with our claim that men are especially prone to exhibit behaviors that undermine the quality and practicability of marriage. As we noted earlier, men are more likely than women to abuse alcohol and other substances, to engage in severe violence, and (especially in the past) to desert their families. Nevertheless, only a minority of husbands engage in these behaviors. In most cases, therefore, marriage benefits wives as well as husbands. However, in a minority of marriages, wives suffer from their husbands' antisocial behavior. Consistent with this interpretation, if we eliminate badly behaved husbands from the married population, then wives and husbands are equally likely to initiate divorce (Amato & Previti, 2003).

Finally, our review indicates that divorce always has been controversial in American society. During the Colonial era, sharp differences existed between New England and the southern colonies about whether divorce should be allowed at all. In the latter part of the 18th century, divorce became a matter of widespread and contentious debate in the media and among policy-makers. After World War I, many observers (including many social scientists) believed that the family was in danger, with harmful consequences for children and society in general. During the last two decades of the 20th century, this debate emerged yet again among policymakers, the media, the general public, and family scholars. Curiously, many of the arguments that are advanced today (both in favor of and against divorce) are similar, if not identical, to arguments made in previous centuries.

The Deinstitutionalization of Marriage

We believe that these various historical trends—along with differences of opinion among contemporary family scholars, policymakers, and the general public about the meaning and implications of these trends—can be explained with reference to the *deinstitutionalization* of marriage. The view was originally put forward by Ernest Burgess, a sociologist who wrote on marriage and family life in the first part of the 20th century (Burgess & Cottrell, 1939; Burgess, Locke, & Thomes, 1963; Burgess & Wallin, 1953). Burgess argued that marriage was in a process of transition from a social institution to a private arrangement based on companionship. By *institution*, Burgess meant a fundamental unit of social organization—a formal status regulated by social norms, public opinion, law, and religion. In contrast, the emerging form of marriage was based primarily on emotional bonds between two autonomous individuals.

According to Burgess, the industrialization and urbanization of the United States weakened the institutional basis of marriage. Prior to the second half of the 19th century, the United States was predominantly a rural society, and most adults and children lived and worked on family farms. During this era, marriage was essential to the welfare not only of individual family members but also to the larger community. Before the development of specialized services and institutions, family members relied on each other to meet a wide range of needs, including child care, economic production, job training, and elder care. Because cohesive families were necessary for survival, the entire community had an interest in ensuring marital stability. Once married, spouses were expected not only to conform to traditional standards of behavior but also to sacrifice their personal goals, if necessary, for the sake of the marriage. Of course, because marriages were patriarchal, wives made more sacrifices than did husbands. Nevertheless, through marriage, men and women participated in an institution that was larger and more significant than themselves. For these reasons, divorce was frowned on and was allowed only if one partner had seriously violated the marriage contract, such as by engaging in physical abuse or desertion.

By 1900, the United States had become industrialized, and two-parent breadwinner–homemaker families replaced farm families as the dominant family form (Hernandez, 1993). Burgess believed that industrialization and urbanization provided individuals with more control over their marriages. Several factors were responsible for this change, including the greater geographical mobility of families (which freed spouses from the domination of parents and the larger kin group), the rise of democratic institutions (which increased the status and power of women), and a decline in religious control (which resulted in more freedom to adopt unconventional views and behaviors). As we noted earlier, the growth of employment opportunities for women also gave wives more economic independence from their husbands. As parents, religion, community expectations, and patriarchal traditions exerted less control over individuals, marriage was based increasingly on the mutual affection and individual preferences of spouses.

Burgess referred to the new model (and ideal) of marriage as *companionate marriage*. According to Burgess, companionate marriage is characterized by egalitarian rather than patriarchal relationships between spouses. Companionate marriage is held together not by bonds of social obligation but by ties of love, friendship, and common interest. Unlike institutional marriage, which emphasizes conformity to social norms, companionate marriage allows for an ample degree of self-expression and personal development. Of course, people in the United States have always expected marriage to be a source of love and emotional support. Furthermore, growing support for companionate marriage can be traced from the Revolutionary War through the end of the 19th century (Griswald, 1982). However, the notion that marriage should be based *primarily* on mutual satisfaction began to gain widespread public acceptance only during the early decades of the 20th century. Psychologists, educators, and social service providers applied these ideas in their professional practice, and it was in this context that marital counseling emerged as a discipline, with its goal being to help couples achieve emotional closeness and sexual satisfaction through improved communication and conflict management. By the 1950s, the great majority of Americans, irrespective of social class, accepted the companionate model of marriage as the cultural ideal (Mintz & Kellog, 1988).

The concept of deinstitutionalization helps to explain the historical trends described in this chapter. Because marriages held together by mutual satisfaction are intrinsically less stable than are marriages held together by community expectations, legal requirements, and religious restrictions, the gradual decline of institutional constraints made the long-term rise in divorce inevitable.

Moreover, as people's standards for a good marriage increased, so did the percentage of marriages that failed to live up to these standards. As a result, marriages that were seen as being tolerable in the 19th century came to be seen as intolerable in the 20th century. It seems likely that rising standards for a "good" marriage fueled public demand for divorce, which in turn placed pressure on the states and courts to liberalize the grounds for marital dissolution (see DiFonzo, 1997, for a historical argument along these lines).

Finally, it is not surprising that the gradual deinstitutionalization of marriage has been generating debate for three centuries. Public controversies usually revolve around strongly held values, and values related to marriage and family life are anchored deeply in the American psyche. Americans, it seems, have always had a love–hate relationship with divorce. On the one hand, people value the freedom to leave unhappy unions, correct earlier mistakes, and find greater happiness with new partners. On the other hand, people are deeply concerned about social stability, tradition, and the overall impact of high levels of marital instability on the well-being of children. The clash between these two concerns reflects a fundamental contradiction within marriage itself; that is, marriage is designed to promote both institutional and personal goals. Happy and stable marriages meet both of these goals without friction. In contrast, when unhappy spouses wrestle with the decision to end their marriages, they are caught between their desire to further their own personal happiness and their sense of obligation to others, including their spouses, their children, their churches, and their communities. Because tension between the values of freedom and obligation is inherent in marriage, controversies over marital stability will always be common in American society.

REFERENCES

Amato, P.R., & Previti, D. (2003). People's reasons for divorcing: Gender, social class, the life course, and adjustment. *Journal of Family Issues, 24,* 602–626.

Amato, P.R., Johnson, D., Booth, A., & Rogers, S. J. (2003). Continuity and change in marriage between 1980 and 2000. *Journal of Marriage and Family, 65,* 1–22.

Bellah, R. N., Madsen, R., Sullivan, W. M., Swidler, A., & Tipron, S. M. (1985). *Habits of the heart: Individualism and commitment in American life.* New York: Harper & Row.

Bernard, J. (1972). *The future of marriage.* New York: World Book.

Booth, A., Johnson, D. R., White, L. K., & Edwards, J. (1981). *Female labor force participation and marital instability: Methodology report.* Lincoln: University of Nebraska–Lincoln. Bureau of Sociological Research.

Braver, S. L. (1998). *Divorced dads.* New York: Putnam.

Burgess, E. W., & Cottrell, L. S. (1939). *Predicting success or failure in marriage.* New York: Prentice-Hall.

Burgess, E. W., & Wallin, P. (1953). *Engagement and marriage.* Chicago: Lippincott.

Burgess, E. W., Locke, H. J., & Thomes, M. (1963). *The family: From institution to companionship.* New York: American Book.

Cherlin, A. J. (1992). *Marriage, divorce, and remarriage.* Cambridge, MA: Harvard University Press.

Cherlin, A. J. (2004). The deinstitutionalization of marriage. *Journal of Marriage and Family, 66,* 848–862.

Coontz, S. (2000). Historical perspectives on family diversity. In D. H. Demo, K. R. Allen, & M. A. Fine (Eds.), *Handbook of family diversity* (pp. 15–31). New York: Oxford University Press.

DiFonzo, J. H. (1997). *Beneath the fault line: The popular and legal culture of divorce in twentieth-century America.* Charlottesville: University Press of Virginia.

Felson, R. B. (2002). *Violence and gender reexamined.* Washington, DC: American Psychological Association.

Glendon, M. A. (1989). *The transformation of family law: State, law, and family in the United States and Western Europe.* Chicago: University of Chicago Press.

Glenn, N. D. (1999). Further discussion of the effects of no-fault divorce on divorce rates. *Journal of Marriage and the Family, 61,* 800–802.

Gove, W. R., & Tudor, J. F. (1973). Adult sex roles and mental illness. *American Journal of Sociology, 78,* 812–835.

Groves, E. R. (1928). *The marriage crisis.* New York: Longman, Green.

Griswald, R. L. (1982). *Family and Divorce in California, 1850–1890.* Albany: State University of New York Press.

Katz, S. N. (1994). Historical perspective and current trends in the legal process of divorce. *Future of Children, 4,* 44–62.

Kim, W. (2000, September). Should you stay together for the kids? *Time Magazine, 156,* 75–82.

Kitson, G. C. (1992). *Portrait of divorce: Adjustment to marital breakdown.* New York: Guilford.

Maccoby, E. E., & Mnookin, R. H. (1992). *Dividing the child.* Cambridge, MA: Harvard University Press.

Mintz, S., & Kellog, S. (1989). *Domestic revolutions: A social history of American family life.* New York: The Free Press.

Mowrer, E. R. (1927). *Family disorganization.* Chicago: University of Chicago Press.

Parke, M., & Ooms, T. (2002, October). *More than a dating service? State activities designed to strengthen and promote marriage* (Policy Brief No. 2: Couples and Marriage Series). Washington, DC: Center for Law and Social Policy.

Parkman, A. M. (2000). *Good intentions gone awry: No-fault divorce and the American family.* Lanham, MD: Rowman and Littlefield.

Phillips, R. (1991). *Untying the knot: A short history of divorce.* New York: Cambridge University Press.

Popenoe, D. (1996). *Life without father: Compelling new evidence that fatherhood and marriage are indispensable for the good of children and society.* New York: Martin Kessler Books.

Preston, S. H., & McDonald, J. (1979). The incidence of divorce within cohorts of American marriages contracted since the civil war. *Demography, 16,* 1–25.

Sanchez, L., Nock, S. L., Wright, J. D., Pardee, J. W., & Ionescu, M. (2001). The implementation of covenant marriage in Louisiana. *The Virginia Journal of Social Policy and the Law, 23,* 192–223.

Seidman, S. (1991). *Romantic longings: Love in America, 1830–1980.* New York: Routledge.

Skolnick, A. S. (1991). *Embattled paradise: The American family in an age of uncertainty.* New York: Basic Books.

Stacey, J. (1996). *In the name of the family: Rethinking family values in the postmodern age.* Boston: Beacon Press.

Stekel, W. (1931). *Marriage at the crossroads.* New York: Godwin.

Thompson, R. A., & Wyatt, J. M. (1999). Values, policy, and research on divorce. In R. A. Thompson & P. R. Amato (Eds.), *The postdivorce family: Children, parenting, and society* (pp. 191–232). Thousand Oaks, CA: Sage.

Thornton, A., & Young-DeMarco, L. (2001). Four decades of trends in attitudes toward family issues in the United States: The 1960s through the 1990s. *Journal of Marriage and the Family, 63,* 1009–1037.

Vlosky, D. A., & Monroe, P. A. (2002). The effective dates of no-fault divorce laws in the 50 states. *Family Relations, 51,* 317–324.

Waite, L. J., & Gallagher, M. (2000). *The case for marriage: Why married people are happier, healthier, and better off financially.* New York: Doubleday.

Whitehead, B. (1993, April). "Dan Quayle Was Right." *The Atlantic Monthly,* pp. 47–84.

Willcox, W. F. (1891). *The divorce problem: A study in statistics.* New York: Columbia University Press.

Williams, K. (2003). Has the future of marriage arrived? A contemporary examination of gender, marriage, and psychological well-being. *Journal of Health and Social Behavior, 44,* 470–487.

Wilson, J. (2002). *The marriage problem: How our culture has weakened families.* New York: Harper Collins.

Chapter 7

Education/Religion

Institutions in the Crossfire

Much of our time up to the age of 18, and often well beyond, is spent learning the roles necessary for survival in society. All societies are concerned with socializing their young to develop the skills and knowledge deemed necessary for inculcating loyalty to the social system. In some societies, this education takes place informally, through the imitation of elders. In others, formal schools have become the key mechanism for the transmission of culture and knowledge. As can be seen, education is a basic institution that must meet a growing list of social needs. Thus a major controversy involving this institution concerns its success in carrying out these tasks.

A description of the American education system might include a decline in academic standards with grade inflation and a narrow emphasis on vocation, a diminished respect for teaching, overly programmed competitiveness, weakened salience of written texts, and inertia in the face of these critical challenges. The following trends seem to confirm the accusation that U.S. schools are rife with mediocrity. During the 1980s and 1990s, the American education system was accused at every level of demonstrating such mediocrity. That is, the education system was not functioning at the expected level. It is estimated that 3 million Americans cannot read or write at all. Another 30 million are functionally illiterate.

A 50-state graded assessment of schools' level of achievement found the following grades: no grade for achievement because the evidence was discouraging; B for standards and achievements; C for quality of teaching—millions of students have teachers who lack the minimum requirements set by the state to teach public schools; C for school climate; and C+ for spending sufficient funds (*Education Week*, 1997). The main explanation for these poor educational results lies in the fact that American students spend less than half as many hours studying academic subjects as do students in France, Germany, and Japan: studies of the students' social environment reveal that 13-year-olds in this country are more likely to spend a great deal of time watching television and ignoring their homework than are students in other countries.

A major suggestion geared toward solving the problems of education is to have more equal educational opportunities for everyone by implementing effective integration of schools via a metropolitan plan in which inner-city schools are merged with suburban schools and students are bused within each district. Also, magnet schools with unique educational programs have been proposed as possible solutions.

Another suggestion for solving the educational problem is to provide compensatory education via special programs and assistance. An example of this kind of program is Project Head Start, which provides preschool instruction to disadvantaged children. Another such program is Title I of the Elementary and Secondary Education Act, which provides federal funds to give extra help to disadvantaged students already in school.

There is also a proposal to improve U.S. education by raising the schools' requirements—for example, having more academic courses in place of electives, increasing the amount of time students spend in school to at least seven hours per day and between 200 and 220 days per year, and making the students work harder by requiring more homework.

Finally, there is a need to make improvements in the educational system by updating facilities and equipment, reducing class size to 15 or fewer students, and hiring teachers who are certified to teach in the areas they are hired to teach.

As can be seen, to improve the U.S. educational system, it will be necessary to change other institutions in the larger social structure. However, because this chapter focuses on education, the readings deal just with the education infrastructure. For most of us, formal education starts with kindergarten. In the first reading of this chapter, Harry Gracey discusses how children are initially socialized into the rules of society by learning the rules of the classroom—discipline, obedience, and routines. In this way, schools prepare children for their later roles in society.

In the next reading, Jonathan Kozol turns to one of the major problems noted earlier—the unequal opportunities that exist in our schools. In three excerpts he describes problems found in many schools: a school that closes periodically because of defects in the physical plant, a school with large classes because of a lack of facilities, and a poor school contrasted with one in a wealthy suburb.

In the third reading on education, Orfield et al. describe a major problem that persists in today's schools: segregation. These authors eloquently describe the difficulties in achieving equal opportunity in education.

The importance of religion in society is emphasized by Emile Durkheim in the next reading in this chapter. He notes that religion is a reflection of society and is, therefore, connected to and an influence on all other social institutions. For example, it is virtually impossible to discuss the role of education in society without mentioning the role of religion on education. Early twentieth-century conflicts over elementary and secondary education were shaped by religious divisions over morality and values in public schools: The mission of public schools to prepare citizens for democracy owes its roots to the mainline Protestant understanding of public life and the role of religion in it. At the same time, conservative Protestantism has argued over the legitimacy of secular public schools because of their inclusion of courses dealing with sex education and the science curriculum, which some see as hostile to moral and spiritual values. Thus there has been controversy over the teaching of evolution in the classroom, school prayer, and so-called liberal textbooks. This religious impact on education should not come as a surprise because religion is intimately involved in other social institutions, such as the economic institution. Adam Smith (*The Wealth of Nations*, 1981, pp. 788–814) indicated that self-interest motivates clergy just as it does secular producers; that market forces constrain churches just as they constrain secular firms; that the benefits of competition, the burden of monopoly, and the hazards of government regulation are as real for religion as for any other sector of the economy; and that religious institutions strongly influence politics despite the purported claim of a separation between church and state in the United States. Political functions performed by religious institutions include being an incubator for civic skills, being agents for mobilization, and acting as information providers (Lee, Pachon, & Baretto, 2002). In addition, church leaders endorse candidates, take stands on political issues, and allow their churches to be sites where political debates and mobilization occur. In this way, religious leaders

influence politics. As a consequence, school board candidates who support creationism in the classroom, school prayer, and school vouchers are more likely to be conservative Protestants. Add to these facts that the United States is clearly a religious society—a majority of adults believe in God (95%), have a religious affiliation (58%), read the Bible (69%), and regularly attend a place of worship (40%) (Dillon, 2003)—and you realize that the impact of religion on these institutions is not minor.

According to Durkheim, its influence on the other institutions in society flows from religion's ability to allow the individual to transcend human experience; thus religion represents a means of dealing with conditions of uncertainty. Religion helps people overcome their feelings of powerlessness and, in this manner, helps them to control the conditions of their lives and to cope with their unfulfilled economic and psychological needs. Considering the role religion plays in people's lives, it is not surprising that for believers, religious beliefs transcend national loyalties—see the article by James Aho on the extreme measures that are being wrought and the government's reluctance to intrude on the activities of religious organizations, political or otherwise. Another major political area involving religious conflict is the fight over abortion rights. Dallas Blanchard, writing in the final article in this chapter, examines this controversy.

As you read this chapter, consider the following questions: What are the consequences of inequality in education? Which solutions to this problem would you propose? Which changes would you make to help this country live up to its promise of a separation of church and state?

25

Learning the Student Role
Kindergarten as Academic Boot Camp

HARRY L. GRACEY

Schools are agents of socialization, preparing students for life in society. In preparing children, they teach expectations and demand conformity to society's norms. They help teach children the attitudes and behaviors appropriate to the society. Gracey points out that this process of teaching children to "fit in" begins early. His focus is the routines children in kindergarten are expected to follow.

As you read the article, think about the following questions:

1. *What are key elements of the learning experience of kindergarteners?*
2. *How does the kindergarten routine encourage conformity and help prepare children for later life?*
3. *What might happen in later schooling and life if children do not receive this early childhood socialization?*
4. *How did your schooling prepare you for what you are doing and plan to do?*

GLOSSARY **Educational institution/system** The structure in society that provides systematic socialization of young. **Student role** Behavior and attitudes regarded by educators as appropriate to children in schools.

Education must be considered one of the major institutions of social life today. Along with the family and organized religion, however, it is a "secondary institution," one in which people are prepared for life in society as it is presently organized. The main dimensions of modern life, that is, the nature of society as a whole, is determined principally by the "Primary institutions," which today are the economy, the political system, and the military establishment. Education has been defined by sociologists, classical and contemporary, as an institution which serves society by socializing people into it through a formalized, standardized procedure. At the beginning of this century Emile Durkheim told student teachers at the University of Paris that education "consists of a methodical socialization of the younger generation." He went on to add:

> It is the influence exercised by adult generations on those that are not ready for social life. Its object is to arouse and to

Source: In Dennis Wrong and Harry L. Gracey (eds.), *Readings in Introductory Sociology*. New York: Macmillan, 1967. Reprinted with permission.

develop in the child a certain number of physical, intellectual, and moral states that are demanded of him by the political society as a whole and by the special milieu for which he is specifically destined.... To the egotistic and asocial being that has just been born, [society] must, as rapidly as possible, add another, capable of leading a moral and social life. Such is the work of education.[1]

The education process, Durkheim said, "is above all the means by which society perpetually recreates the conditions of its very existence."[2] The contemporary educational sociologist, Wilbur Brookover, offers a similar formulation in his recent textbook definition of education:

> Actually, therefore, in the broadest sense education is synonymous with socialization. It includes any social behavior that assists in the induction of the child into membership in the society or any behavior by which the society perpetuates itself through the next generation.[3]

The educational institution is, then, one of the ways in which society is perpetuated through the systematic socialization of the young, while the nature of the society which is being perpetuated—its organization and operation, its values, beliefs, and ways of living—are determined by the primary institutions. The educational system, like other secondary institutions, *serves* the society which is *created* by the operation of the economy, the political system, and the military establishment.

Schools, the social organizations of the educational institution, are today for the most part large bureaucracies run by specially trained and certified people. There are few places left in modern societies where formal teaching and learning is carried on in small, isolated groups, like the rural, one-room schoolhouses of the last century. Schools are large, formal organizations which tend to be parts of larger organizations, local community School Districts. These School Districts are bureaucratically organized and their operations are supervised by

state and local governments. In this context, as Brookover says:

> The term education is used ... to refer to a system of schools, in which specifically designated persons are expected to teach children and youth certain types of acceptable behavior. The school system becomes a ... unit in the total social structure and is recognized by the members of the society as a separate social institution. Within this structure a portion of the total socialization process occurs.[4]

Education is the part of the socialization process which takes place in the schools; and these are, more and more today, bureaucracies within bureaucracies.

Kindergarten is generally conceived by educators as a year of preparation for school. It is thought of as a year in which small children, five or six years old, are prepared socially and emotionally for the academic learning which will take place over the next twelve years. It is expected that a foundation of behavior and attitudes will be laid in kindergarten on which the children can acquire the skills and knowledge they will be taught in the grades. A booklet prepared for parents by the staff of a suburban New York school system says that the kindergarten experience will stimulate the child's desire to learn and cultivate the skills he will need for learning in the rest of his school career. It claims that the child will find opportunities for physical growth, for satisfying his "need for self-expression," acquire some knowledge, and provide opportunities for creative activity. It concludes, "The most important benefit that your five-year-old will receive from kindergarten is the opportunity to live and grow happily and purposefully with others in a small society." The kindergarten teachers in one of the elementary schools in this community, one we shall call the Wilbur Wright School, said their goals were to see that the children "grew" in all ways: physically, of course, emotionally, socially, and academically. They said they wanted children to like school as a result of their kindergarten experiences and that they wanted them to learn to get along with others.

None of these goals, however, is unique to kindergarten; each of them is held to some extent by teachers in the other six grades at Wright School. And growth would occur, but differently, even if the child did not attend school. The children already know how to get along with others, in their families and their play groups. The unique job of the kindergarten in the educational division of labor seems rather to be teaching children the student role. The student role is the repertoire of behavior and attitudes regarded by educators as appropriate to children in school. Observation in the kindergartens of the Wilbur Wright School revealed a great variety of activities through which children are shown and then drilled in the behavior and attitudes defined as appropriate for school and thereby induced to learn the role of student. Observations of the kindergartens and interviews with the teachers both pointed to the teaching and learning of classroom routines as the main element of the student role. The teachers expended most of their efforts, for the first half of the year at least, in training the children to follow the routines which teachers created. The children were, in a very real sense, *drilled* in tasks and activities created by the teachers for their own purposes and beginning and ending quite arbitrarily (from the child's point of view) at the command of the teacher. One teacher remarked that she hated September, because during the first month "everything has to be done rigidly, and repeatedly, until they know exactly what they're supposed to do." However, "by January," she said, "they know exactly what to do [during the day] and I don't have to be after them all the time." Classroom routines were introduced gradually from the beginning of the year in all the kindergartens, and the children were drilled in them as long as was necessary to achieve regular compliance. By the end of the school year, the successful kindergarten teacher has a well-organized group of children. They follow classroom routines automatically, having learned all the command signals and the expected responses to them. They have, in our terms, learned the student role. The following observation shows one such classroom operating at optimum organization on an afternoon late in May. It is

the class of an experienced and respected kindergarten teacher.

AN AFTERNOON IN KINDERGARTEN

At about 12:20 in the afternoon on a day in the last week of May, Edith Kerr leaves the teachers' room where she has been having lunch and walks to her classroom at the far end of the primary wing of Wright School. A group of five- and six-year-olds peers at her through the glass doors leading from the hall cloakroom to the play area outside. Entering her room, she straightens some material in the "book corner" of the room, arranges music on the piano, takes colored paper from her closet and places it on one of the shelves under the window. Her room is divided into a number of activity areas through the arrangement of furniture and play equipment. Two easels and a paint table near the door create a kind of passageway inside the room. A wedge-shaped area just inside the front door is made into a teacher's area by the placing of "her" things there: her desk, file, and piano. To the left is the book corner, marked off from the rest of the room by a puppet stage and a movable chalkboard. In it are a display rack of picture books, a record player, and a stack of children's records. To the right of the entrance are the sink and clean-up area. Four large round tables with six chairs at each for the children are placed near the walls about halfway down the length of the room, two on each side, leaving a large open area in the center for group games, block building, and toy truck driving. Windows stretch down the length of both walls, starting about three feet from the floor and extending almost to the high ceilings. Under the windows are long shelves on which are kept all the toys, games, blocks, paper, paints, and other equipment of the kindergarten. The left rear corner of the room is a play store with shelves, merchandise, and cash register; the right rear corner is a play kitchen with stove, sink, ironing board, and bassinette with baby dolls in it. This area is partly shielded from the rest of the room by a large standing display rack for posters and children's art work. A sandbox is found against the back wall between these two areas. The room is light, brightly colored and filled with things adults feel five- and six-year-olds will find interesting and pleasing.

At 12:25 Edith opens the outside door and admits the waiting children. They hang their sweaters on hooks outside the door and then go to the center of the room and arrange themselves in a semi-circle on the floor, facing the teacher's chair, which she has placed in the center of the floor. Edith follows them in and sits in her chair checking attendance while waiting for the bell to ring. When she has finished attendance, which she takes by sight, she asks the children what the date is, what day and month it is, how many children are enrolled in the class, how many are present, and how many are absent.

The bell rings at 12:30 and the teacher puts away her attendance book. She introduces a visitor, who is sitting against the wall taking notes, as someone who wants to learn about schools and children. She then goes to the back of the room and takes down a large chart labeled "Helping Hands." Bringing it to the center of the room, she tells the children it is time to change jobs. Each child is assigned some task on the chart by placing his name, lettered on a paper "hand," next to a picture signifying the task—e.g., a broom, a blackboard, a milk bottle, a flag, and a Bible. She asks the children who wants each of the jobs and rearranges their "hands" accordingly. Returning to her chair, Edith announces, "One person should tell us what happened to Mark." A girl raises her hand, and when called on says, "Mark fell and hit his head and had to go to the hospital." The teacher adds that Mark's mother had written saying he was in the hospital.

During this time the children have been interacting among themselves, in their semi-circle. Children have whispered to their neighbors, poked one another, made general comments to the group, waved to friends on the other side of the circle. None of this has been disruptive, and the teacher has ignored it for the most part. The children seem to know just how much of each kind of interaction is permitted—they may greet in a soft voice someone who sits next to them, for example, but may not shout greetings to a friend who sits across the circle, so they confine themselves to waving and remain well within understood limits.

At 12:35 two children arrive. Edith asks them why they are late and then sends them to join the circle on the floor. The other children vie with each other to tell the newcomers what happened to Mark. When this leads to a general disorder Edith asks, "Who has serious time?" The children become quiet and a girl raises her hand. Edith nods and the child gets a Bible and hands it to Edith. She reads the Twenty-third Psalm while the children sit quietly. Edith helps the child in charge begin reciting the Lord's Prayer; the other children follow along for the first unit of sounds, and then trail off as Edith finishes for them. Everyone stands and faces the American flag hung to the right of the door. Edith leads the pledge to the flag, with the children again following the familiar sounds as far as they remember them. Edith then asks the girl in charge what song she wants and the child replies, "My Country." Edith goes to the piano and plays "America," singing as the children follow her words.

Edith returns to her chair in the center of the room and the children sit again in the semi-circle on the floor. It is 12:40 when she tells the children, "Let's have boys' sharing time first." She calls the name of the first boy sitting on the end of the circle, and he comes up to her with a toy helicopter. He turns and holds it up for the other children to see. He says, "It's a helicopter." Edith asks, "What is it used for?" and he replies, "For the army. Carry men. For the war." Other children join in, "For shooting submarines." "To bring back men from space when they are in the ocean." Edith sends the boy back to the circle and asks the next boy if he has something. He replies "No" and she passes on to the next. He says "Yes" and brings a bird's nest to her. He holds it for the class to see, and the teacher asks, "What kind of bird made the nest?" The boy replies, "My friend says a rain bird made it." Edith asks what the nest is made of and different children reply, "mud," "leaves," and "sticks." There is also a bit of moss woven into the nest, and Edith tries to describe it to the children. They, however, are more interested in seeing if anything is inside it, and Edith lets the boy carry it around the semi-circle showing the children its insides. Edith tells the children of some baby robins in a nest in her yard, and some of the children tell about baby birds they have seen. Some children are asking about a small object in the nest which they say looks like an egg, but all have seen the nest now and Edith calls on the next boy. A number of children say, "I know what Michael has, but I'm not telling." Michael brings a book to the teacher and then goes back to his place in the circle of children. Edith reads the last page of the book to the class. Some children tell of books which they have at home. Edith calls the next boy, and three children call out, "I know what David has." "He always has the same thing." "It's a bang-bang." David goes to his table and gets a box which he brings to Edith. He opens it and shows the teacher a scale-model of an old-fashioned dueling pistol. When David does not turn around to the class, Edith tells him, "Show it to the children" and he does. One child says, "Mr. Johnson [the principal] said no guns." Edith replies, "Yes, how many of you know that?" Most of the children in the circle raise their hands. She continues, "That you aren't supposed to bring guns to school?" She calls the next boy on the circle and he brings two large toy soldiers to her which the children enthusiastically identify as being from "Babes in Toyland." The next boy brings an American flag to Edith and shows it to the class. She asks him what the stars and stripes stand for and admonishes him to treat it carefully. "Why should you treat it carefully?" she asks the boy. "Because it's our flag," he replies. She congratulates him, saying, "That's right."

"Show and Tell" lasted twenty minutes and during the last ten one girl in particular announced that she knew what each child called upon had to show. Edith asked her to be quiet each time she spoke out, but she was not content, continuing to offer her comment at each "show." Four children from other classes had come into the room to bring

something from another teacher or to ask for something from Edith. Those with requests were asked to return later if the item wasn't readily available.

Edith now asks if any of the children told their mothers about their trip to the local zoo the previous day. Many children raise their hands. As Edith calls on them, they tell what they liked in the zoo. Some children cannot wait to be called on, and they call out things to the teacher, who asks them to be quiet. After a few of the animals are mentioned, one child says, "I liked the spooky house," and the others chime in to agree with him, some pantomiming fear and horror. Edith is puzzled, and asks what this was. When half the children try to tell her at once, she raises her hand for quiet, then calls on individual children. One says, "The house with nobody in it"; another, "The dark little house." Edith asks where it was in the zoo, but the children cannot describe its location in any way which she can understand. Edith makes some jokes but they involve adult abstractions which the children cannot grasp. The children have become quite noisy now, speaking out to make both relevant and irrelevant comments, and three little girls have become particularly assertive.

Edith gets up from her seat at 1:10 and goes to the book corner, where she puts a record on the player. As it begins a story about the trip to the zoo, she returns to the circle and asks the children to go sit at the tables. She divides them among the tables in such a way as to indicate that they don't have regular seats. When the children are all seated at the four tables, five or six to a table, the teacher asks, "Who wants to be the first one?" One of the noisy girls comes to the center of the room. The voice on the record is giving directions for imitating an ostrich and the girl follows them, walking around the center of the room holding her ankles with her hands. Edith replays the record, and all the children, table by table, imitate ostriches down the center of the room and back. Edith removes her shoes and shows that she can be an ostrich too. This is apparently a familiar game, for a number of children are calling out, "Can we have the crab?" Edith asks one of the children to do a crab "so we can all remember how," and then plays the part of the record with music for imitating crabs by. The children from the first table line up across the room, hands and feet on the floor and faces pointing toward the ceiling. After they have "walked" down the room and back in this posture they sit at their table and the children of the next table play "crab." The children love this; they run from their tables, dance about on the floor waiting for their turns and are generally exuberant. Children ask for the "inch worm," and the game is played again with the children squirming down the floor. As a conclusion Edith shows them a new animal imitation, the "lame dog." The children all hobble down the floor on three "legs," table by table to the accompaniment of the record.

At 1:30 Edith has the children line up in the center of the room: she says, "Table one, line up in front of me," and children ask, "What are we going to do?" Then she moves a few steps to the side and says, "Table two over here; line up next to table one," and more children ask, "What for?" She does this for table three and table four, and each time the children ask, "Why, what are we going to do?" When the children are lined up in four lines of five each, spaced so that they are not touching one another, Edith puts on a new record and leads the class in calisthenics, to the accompaniment of the record. The children just jump around every which way in their places instead of doing the exercises, and by the time the record is finished, Edith, the only one following it, seems exhausted. She is apparently adopting the President's new "Physical Fitness" program for her classroom.

At 1:35 Edith pulls her chair to the easels and calls the children to sit on the floor in front of her, table by table. When they are all seated she asks, "What are you going to do for worktime today?" Different children raise their hands and tell Edith what they are going to draw. Most are going to make pictures of animals they saw in the zoo. Edith asks if they want to make pictures to send to Mark in the hospital, and the children agree to this. Edith gives drawing paper to the children, calling them to her one by one. After getting a piece of paper, the children go to the crayon box on the righthand shelves, select a number of colors, and go to the tables, where they begin drawing. Edith is again trying to quiet the perpetually talking girls. She keeps two of them standing by her so they won't disrupt the others. She asks them, "Why do you feel you have to talk all the time?" and then scolds them for not listening to her. Then she sends them to their tables to draw.

Most of the children are drawing at their tables, sitting or kneeling in their chairs. They are all working very industriously and, engrossed in their work, very quietly. Three girls have chosen to paint at the easels, and having donned their smocks, they are busily mixing colors and intently applying them to their pictures. If the children at the tables are primitives and neo-realists in their animal depictions, these girls at the easels are the class abstract-expressionists, with their broad-stroked, colorful paintings.

Edith asks of the children generally, "What color should I make the cover of Mark's book?" Brown and green are suggested by some children "because Mark likes them." The other children are puzzled as to just what is going on and ask, "What book?" or "What does she mean?" Edith explains what she thought was clear to them already, that they are all going to put their pictures together in a "book" to be sent to Mark. She goes to a small table in the play-kitchen corner and tells the children to bring her their pictures when they are finished and she will write their message for Mark on them.

By 1:50 most children have finished their pictures and given them to Edith. She talks with some of them as she ties the bundle of pictures together—answering questions, listening, carrying on conversations. The children are playing in various parts of the room with toys, games, and blocks which they have taken off the shelves. They also move from table to table examining each other's pictures, offering compliments and suggestions. Three girls at the table are cutting up colored paper for a collage. Another girl is walking about the room in a pair of high heels with a woman's purse over her arm. Three boys are playing in the center of the room with the large block set, with which they are building walk-ways and walking on them.

Edith is very much concerned about their safety and comes over a number of times to fuss over them. Two or three other boys are pushing trucks around the center of the room, and mild altercations occur when they drive through the block constructions. Some boys and girls are playing at the toy store, two girls are serving "tea" in the play kitchen and one is washing a doll baby. Two boys have elected to clean the room, and with large sponges they wash the movable blackboard, the puppet stage, and then begin on the tables. They run into resistance from the children who are working with construction toys on the tables and do not want to dismantle their structures. The class is like a room full of bees, each intent on pursuing some activity, occasionally bumping into one another, but just veering off in another direction without serious altercation. At 2:05 the custodian arrives pushing a cart loaded with half-pint milk containers. He places a tray of cartons on the counter next to the sink, then leaves. His coming and going is unnoticed in the room (as, incidentally, is the presence of the observer, who is completely ignored by the children for the entire afternoon).

At 2:15 Edith walks to the entrance of the room, switches off the lights, and sits at the piano and plays. The children begin spontaneously singing the song, which is "Clean up, clean up. Everybody clean up." Edith walks around the room supervising the clean-up. Some children put their toys, the blocks, puzzles, games, and so on back on their shelves under the windows. The children making a collage keep right on working. A child from another class comes in to borrow the 45-rpm adapter for the record player. At more urging from Edith the rest of the children shelve their toys and work. The children are sitting around their tables now, and Edith asks, "What record would you like to hear while you have your milk?" There is some confusion and no general consensus, so Edith drops the subject and begins to call the children, table by table, to come get their milk. "Table one," she says, and the five children come to the sink, wash their hands and dry them, pick up a carton of milk and a straw, and take it back to their table. Two talking girls wander about the

room interfering with the children getting their milk and Edith calls out to them to "settle down." As the children sit, many of them call out to Edith the name of the record they want to hear. When all the children are seated at tables with milk, Edith plays one of these records called "Bozo and the Birds" and shows the children pictures in a book which go with the record. The record recites, and the book shows the adventures of a clown, Bozo, as he walks through a woods meeting many different kinds of birds who, of course, display the characteristics of many kinds of people or, more accurately, different stereotypes. As children finish their milk, they take blankets or pads from the shelves under the windows and lie on them in the center of the room, where Edith sits on her chair showing the pictures. By 2:30 half the class is lying on the floor on their blankets, the record is still playing, and the teacher is turning the pages of the book. The child who came in previously returns the 45-rpm adapter, and one of the kindergartners tells Edith what the boy's name is and where he lives.

The record ends at 2:40. Edith says, "Children, down on your blankets." All the class is lying on blankets now. Edith refuses to answer the various questions individual children put to her because, she tells them, "It's rest time now." Instead she talks very softly about what they will do tomorrow. They are going to work with clay, she says. The children lie quietly and listen. One of the boys raises his hand and when called on tells Edith, "The animals in the zoo looked so hungry yesterday." Edith asks the children what they think about this and a number try to volunteer opinions, but Edith accepts only those offered in a "rest-time tone," that is, softly and quietly. After a brief discussion of animal feeding, Edith calls the names of the two children on milk detail and has them collect empty milk cartons from the tables and return them to the tray. She asks the two children on clean-up detail to clean up the room. Then she gets up from her chair and goes to the door to turn on the lights. At this signal, the children all get up from the floor and return their blankets and pads to the shelf. It is raining (the reason for no outside play this afternoon) and cars driven by mothers clog the school drive and line up along the street. One of the talkative little girls comes over to Edith and pointing out the window says, "Mrs. Kerr, see my mother in the new Cadillac?"

At 2:50 Edith sits at the piano and plays. The children sit on the floor in the center of the room and sing. They have a repertoire of songs about animals, including one in which each child sings a refrain alone. They know these by heart and sing along through the ringing of the 2:55 bell. When the song is finished, Edith gets up and coming to the group says, "Okay, rhyming words to get your coats today." The children raise their hands and as Edith calls on them, they tell her two rhyming words, after which they are allowed to go into the hall to get their coats and sweaters. They return to the room with these and sit at their tables. At 2:59 Edith says, "When you have your coats on, you may line up at the door." Half of the children go to the door and stand in a long line. When the three o'clock bell rings, Edith returns to the piano and plays. The children sing a song called "Goodbye," after which Edith sends them out.

TRAINING FOR LEARNING AND FOR LIFE

The day in kindergarten at Wright School illustrates both the content of the student role as it has been learned by these children and the processes by which the teacher has brought about this learning, or "taught" them the student role. The children have learned to go through routines and to follow orders with unquestioning obedience, even when these make no sense to them. They have been disciplined to do as they are told by an authoritative person without significant protest. Edith has developed this discipline in the children by creating and enforcing a rigid social structure in the classroom through which she effectively controls the behavior of most of the children for most of the school day. The "living with others in a small society" which the school pamphlet tells parents is the

most important thing the children will learn in kindergarten can be seen now in its operational meaning, which is learning to live by the routines imposed by the school. This learning appears to be the principal content of the student role.

Children who submit to school-imposed discipline and come to identify with it, so that being a "good student" comes to be an important part of their developing identities, *become* the good students by the school's definitions. Those who submit to the routines of the school but do not come to identify with them will be adequate students who find the more important part of their identities elsewhere, such as in the play group outside school. Children who refuse to submit to the school routines are rebels, who become known as "bad students" and often "problem children" in the school, for they do not learn the academic curriculum and their behavior is often disruptive in the classroom. Today schools engage clinical psychologists in part to help teachers deal with such children.

In looking at Edith's kindergarten at Wright School, it is interesting to ask how the children learn this role of student—come to accept school-imposed routines—and what, exactly, it involves in terms of behavior and attitudes. The most prominent features of the classroom are its physical and social structures. The room is carefully furnished and arranged in ways adults feel will interest children. The play store and play kitchen in the back of the room, for example, imply that children are interested in mimicking these activities of the adult world. The only space left for the children to create something of their own is the empty center of the room, and the materials at their disposal are the blocks, whose use causes anxiety on the part of the teacher. The room, being carefully organized physically by the adults, leaves little room for the creation of physical organization on the part of the children.

The social structure created by Edith is a far more powerful and subtle force for fitting the children to the student role. This structure is established by the very rigid and tightly controlled set of rituals and routines through which the children are put during the day. There is first the rigid "locating procedure" in which the children are asked to find themselves in terms of the month, date, day of the week, and the number of the class who are present and absent. This puts them solidly in the real world as defined by adults. The day is then divided into six periods whose activities are for the most part determined by the teacher. In Edith's kindergarten the children went through Serious Time, which opens the school day, Sharing Time, Play Time (which in clear weather would be spent outside), Work Time, Clean-up Time, after which they have their milk, and Rest Time after which they go home. The teacher has programmed activities for each of these Times.

Occasionally the class is allowed limited discretion to choose between proffered activities, such as stories or records, but original ideas for activities are never solicited from them. Opportunity for free individual action is open only once in the day, during the part of Work Time left after the general class assignment has been completed (on the day reported the class assignment was drawing animal pictures for the absent Mark). Spontaneous interests or observations from the children are never developed by the teacher. It seems that her schedule just does not allow room for developing such unplanned events. During Sharing Time, for example, the child who brought a bird's nest told Edith, in reply to her question of what kind of bird made it, "My friend says it's a rain bird." Edith does not think to ask about this bird, probably because the answer is "childish," that is, not given in accepted adult categories of birds. The children then express great interest in an object in the nest, but the teacher ignores this interest, probably because the object is uninteresting to her. The soldiers from "Babes in Toyland" strike a responsive note in the children, but this is not used for a discussion of any kind. The soldiers are treated in the same way as objects which bring little interest from the children. Finally, at the end of Sharing Time the child-world of perception literally erupts in the class with the recollection of "the spooky house" at the zoo. Apparently this made more of an impression on the children than did any of the animals, but Edith is unable to make any sense of it for herself. The tightly imposed order of the class begins to

break down as the children discover a universe of discourse of their own and begin talking excitedly with one another. The teacher is effectively excluded from this child's world of perception and for a moment she fails to dominate the classroom situation. She reasserts control, however, by taking the children to the next activity she has planned for the day. It seems never to have occurred to Edith that there might be a meaningful learning experience for the children in re-creating the "spooky house" in the classroom. It seems fair to say that this would have offered an exercise in spontaneous self-expression and an opportunity for real creativity on the part of the children. Instead, they are taken through a canned animal imitation procedure, an activity which they apparently enjoy, but which is also imposed upon them rather than created by them.

While children's perceptions of the world and opportunities for genuine spontaneity and creativity are being systematically eliminated from the kindergarten, unquestioned obedience to authority and rote learning of meaningless material are being encouraged. When the children are called to line up in the center of the room they ask "Why?" and "What for?" as they are in the very process of complying. They have learned to go smoothly through a programmed day, regardless of whether parts of the program make any sense to them or not. Here the student role involves what might be called "doing what you're told and never mind why." Activities which might "make sense" to the children are effectively ruled out, and they are forced or induced to participate in activities which may be "senseless," such as calisthenics.

At the same time the children are being taught by rote meaningless sounds in the ritual oaths and songs, such as the Lord's Prayer, the Pledge to the Flag, and "America." As they go through the grades children learn more and more of the sounds of these ritual oaths, but the fact that they have often learned meaningless sounds rather than meaningful statements is shown when they are asked to write these out in the sixth grade; they write them as groups of sounds rather than as a series of words, according to the sixth grade teachers at Wright

School. Probably much learning in the elementary grades is of this character, that is, having no intrinsic meaning to the children, but rather being tasks inexplicably required of them by authoritative adults. Listening to sixth-grade children read social studies reports, for example, in which they have copied material from encyclopedias about a particular country, an observer often gets the feeling that he is watching an activity which has no intrinsic meaning for the child. The child who reads, "Switzerland grows wheat and cows and grass and makes a lot of cheese" knows the dictionary meaning of each of these words but may very well have no conception at all of this "thing" called Switzerland. He is simply carrying out a task assigned by the teacher *because* it is assigned, and this may be its only "meaning" for him.

Another type of learning which takes place in kindergarten is seen in children who take advantage of the "holes" in the adult social structure to create activities of their own, during Work Time or out-of-doors during Play Time. Here the children are learning to carve out a small world of their own within the world created by adults. They very quickly learn that if they keep within permissible limits of noise and action they can play much as they please. Small groups of children formed during the year in Edith's kindergarten who played together at these times, developing semi-independent little groups in which they created their own worlds in the interstices of the adult-imposed physical and social world. These groups remind the sociological observer very much of the so-called "informal groups" which adults develop in factories and offices of large bureaucracies.[5] Here, too, within authoritatively imposed social organizations people find "holes" to create little subworlds which support informal, friendly, unofficial behavior. Forming and participating in such groups seems to be as much part of the student role as it is of the role of bureaucrat.

The kindergarten has been conceived of here as the year in which children are prepared for their schooling by learning the role of student. In the classrooms of the rest of the school grades, the children will be asked to submit to systems and routines imposed by the teachers and the curriculum. The days will be much like those of kindergarten, except that academic subjects will be substituted for the activities of the kindergarten. Once out of the school system, young adults will more than likely find themselves working in large-scale bureaucratic organizations, perhaps on the assembly line in the factory, perhaps in the paper routines of the white collar occupations, where they will be required to submit to rigid routines imposed by "the company" which may make little sense to them. Those who can operate well in this situation will be successful bureaucratic functionaries. Kindergarten, therefore, can be seen as preparing children not only for participation in the bureaucratic organization of large modern school systems, but also for the large-scale occupational bureaucracies of modern society.

NOTES

1. Emile Durkheim, *Sociology and Education* (New York: The Free Press, 1956), pp. 71–72.

2. *Ibid.*, p. 123.

3. Wilbur Brookover, *The Sociology of Education* (New York: American Book Company, 1957), p. 4.

4. *Ibid.*, p. 6.

5. See, for example, Peter M. Blau, *Bureaucracy in Modern Society* (New York: Random House, 1956), Chapter 3.

26

Savage Inequalities
Children in America's Schools

JONATHAN KOZOL

In addition to differences in structure and curricula in schools, rich and poor neighborhoods have different schools. The following excerpts from Jonathan Kozol's Savage Inequalities *illustrate the gap between schools in three different communities.*

As you read, consider the following questions:

1. *In what ways do the schools in these three communities differ?*
2. *What is the impact of differences between schools on children in those schools?*
3. *What are some causes of differences between schools and what might be done about these differences?*
4. *Are you aware of differences between schools in your area?*

GLOSSARY **Savage inequalities** Differences between schools for children from poor families in inner cities or poor suburbs compared to children from rich communities.

The problems of the streets in urban areas, as teachers often note, frequently spill over into public schools. In the public schools of East St. Louis this is literally the case.

"Martin Luther King Junior High School," notes the *Post-Dispatch* in a story published in the early spring of 1989, "was evacuated Friday afternoon after sewage flowed into the kitchen.... The kitchen was closed and students were sent home." On Monday, the paper continued, "East St. Louis Senior High School was awash in sewage for the second time this year." The school had to be shut because of "fumes and backed-up toilets." Sewage flowed into the basement, through the floor, then up into the kitchen and the students' bathrooms. The backup, we read, "occurred in the food preparation areas."

School is resumed the following morning at the high school, but a few days later the overflow recurs. This time the entire system is affected, since the meals distributed to every student in the city are prepared in the two schools that have been flooded. School is called off for all 16,500 students in the district. The sewage backup, caused by the failure of two pumping stations, forces officials at the high school to shut down the furnaces.

At Martin Luther King, the parking lot and gym are also flooded. "It's a disaster," says a legislator.

Source: From *Savage Inequalities* by Jonathan Kozol, copyright 1991 by Jonathan Kozol. Used by permission of Crown Publishers, a division of Random House, Inc.

"The streets are underwater; gaseous fumes are being emitted from the pipes under the schools," she says, "making people ill."

In the same week, the schools announce the layoff of 280 teachers, 166 cooks and cafeteria workers, 25 teacher aides, 16 custodians, and 18 painters, electricians, engineers and plumbers. The president of the teachers' union says the cuts, which will bring the size of kindergarten and primary classes up to 30 students, and the size of fourth to twelfth grade classes up to 35, will have "an unimaginable impact" on the students. "If you have a high school teacher with five classes each day and between 150 and 175 students …, it's going to have a devastating effect." The school system, it is also noted, has been using more than 70 "permanent substitute teachers," who are paid only $10,000 yearly, as a way of saving money.

Governor Thompson, however, tells the press that he will not pour money into East St. Louis to solve long-term problems. East St. Louis residents, he says, must help themselves. "There is money in the community," the governor insists. "It's just not being spent for what it should be spent for."

The governor, while acknowledging that East St. Louis faces economic problems, nonetheless refers dismissively to those who live in East St. Louis. "What in the community," he asks, "is being done right?" He takes the opportunity of a visit to the area to announce a fiscal grant for sewer improvement to a relatively wealthy town nearby.

In East St. Louis, meanwhile, teachers are running out of chalk and paper, and their paychecks are arriving two weeks late. The city warns its teachers to expect a cut of half their pay until the fiscal crisis has been eased.

The threatened teacher layoffs are mandated by the Illinois Board of Education, which, because of the city's fiscal crisis, has been given supervisory control of the school budget. Two weeks later the state superintendent partially relents. In a tone very different from that of the governor, he notes that East St. Louis does not have the means to solve its education problems on its own. "There is no natural way," he says, that "East St. Louis can bring itself out of this situation." Several cuts will be required in any case—one quarter of the system's teachers, 75 teacher aides, and several dozen others will be given notice—but, the state board notes, sports and music programs will not be affected.

East St. Louis, says the chairman of the state board, "is simply the worst possible place I can imagine to have a child brought up.… The community is in desperate circumstances." Sports and music, he observes, are, for many children here, "the only avenues of success." Sadly enough, no matter how it ratifies the stereotype, this is the truth; and there is a poignant aspect to the fact that, even with class size soaring and one quarter of the system's teachers being given their dismissal, the state board of education demonstrates its genuine but skewed compassion by attempting to leave sports and music untouched by the overall austerity.

Even sports facilities, however, are degrading by comparison with those found and expected at most high schools in America. The football field at East St. Louis High is missing almost everything—including goalposts. There are a couple of metal pipes—no crossbar, just the pipes. Bob Shannon, the football coach, who has to use his personal funds to purchase footballs and has had to cut and rake the football field himself, has dreams of having goalposts someday. He'd also like to let his students have new uniforms. The ones they wear are nine years old and held together somehow by a patchwork of repairs. Keeping them clean is a problem, too. The school cannot afford a washing machine. The uniforms are carted to a corner laundromat with fifteen dollars' worth of quarters.

Other football teams that come to play, according to the coach, are shocked to see the field and locker rooms. They want to play without a halftime break and get away. The coach reports that he's been missing paychecks, but he's trying nonetheless to raise some money to help out a member of the team whose mother has just died of cancer.

"The days of the tight money have arrived," he says. "It don't look like Moses will be coming to this school."

He tells me he has been in East St. Louis 19 years and has been the football coach for 14 years. "I was born," he says, "in Natchez,

Mississippi. I stood on the courthouse steps of Natchez with Charles Evers. I was a teen-age boy when Michael Schwerner and the other boys were murdered. I've been in the struggle all along. In Mississippi, it was the fight for legal rights. This time, it's a struggle for survival.

"In certain ways," he says, "it's harder now because in those days it was a clear enemy you had to face, a man in a hood and not a statistician. No one could persuade you that you were to blame. Now the choices seem like they are left to you and, if you make the wrong choice, you are made to understand you are to blame...."

Night-time in this city, hot and smoky in the summer, there are dealers standin' out on every street. Of the kids I see here, maybe 55 percent will graduate from school. Of that number, maybe one in four will go to college. How many will stay? That is a bigger question.

"The basic essentials are simply missing here. When we go to wealthier schools I look at the faces of my boys. They don't say a lot. They have their faces to the windows, lookin' out. I can't tell what they are thinking. I am hopin' they are saying, 'This is something I will give my kids someday.'"

Tall and trim, his black hair graying slightly, he is 45 years old.

"No, my wife and I don't live here. We live in a town called Ferguson, Missouri. I was born in poverty and raised in poverty. I feel that I owe it to myself to live where they pick up the garbage."

In the visitors' locker room, he shows me lockers with no locks. The weight room stinks of sweat and water-rot. "See, this ceiling is in danger of collapsing. See, this room don't have no heat in winter. But we got to come here anyway. We wear our coats while working out. I tell the boys, 'We got to get it done. Our fans don't know that we do not have heat.'"

He tells me he arrives at school at 7:45 A.M. and leaves at 6:00 P.M.—except in football season, when he leaves at 8:00 P.M. "This is my life. It isn't all I dreamed of and I tell myself sometimes that I might have accomplished more. But growing up in poverty rules out some avenues. You do the best you can."...

On the following morning I visit P.S. 79, an elementary school in the same district. "We work under difficult circumstances," says the principal, James Carter, who is black. "The school was built to hold one thousand students. We have 1,550. We are badly overcrowded. We need smaller classes but, to do this, we would need more space. I can't add five teachers. I would have no place to put them."

Some experts, I observe, believe that class size isn't a real issue. He dismisses this abruptly. "It doesn't take a genius to discover that you learn more in a smaller class. I have to bus some 60 kindergarten children elsewhere, since I have no space for them. When they return next year, where do I put them?"

"I can't set up a computer lab. I have no room. I had to put a class into the library. I have no librarian. There are two gymnasiums upstairs but they cannot be used for sports. We hold more classes there. It's unfair to measure us against the suburbs. They have 17 to 20 children in a class. Average class size in this school is 30."

"The school is 29 percent black, 70 percent Hispanic. Few of these kids get Head Start. There is no space in the district. Of 200 kindergarten children, 50 maybe get some kind of preschool."

I ask him how much difference preschool makes.

"Those who get it do appreciably better. I can't overestimate its impact but, as I have said, we have no space."

The school tracks children by ability, he says. "There are five to seven levels in each grade. The highest level is equivalent to 'gifted' but it's not a full-scale gifted program. We don't have the funds. We have no science room. The science teachers carry their equipment with them."

We sit and talk within the nurse's room. The window is broken. There are two holes in the ceiling. About a quarter of the ceiling has been patched and covered with a plastic garbage bag.

"Ideal class size for these kids would be 15 to 20. Will these children ever get what white kids in the suburbs take for granted? I don't think so. If you ask me why, I'd have to speak of race and social class.

I don't think the powers that be in New York City understand, or want to understand, that if they do not give these children a sufficient education to lead healthy and productive lives, we will be their victims later on. We'll pay the price someday—in violence, in economic costs. I despair of making this appeal in any terms but these. You cannot issue an appeal to conscience in New York today. The fair-play argument won't be accepted. So you speak of violence and hope that it will scare the city into action."

While we talk, three children who look six or seven years old come to the door and ask to see the nurse, who isn't in the school today. One of the children, a Puerto Rican girl, looks haggard. "I have a pain in my tooth," she says. The principal says, "The nurse is out. Why don't you call your mother?" The child says, "My mother doesn't have a phone." The principal sighs. "Then go back to your class." When she leaves, the principal is angry. "It's amazing to me that these children ever make it with the obstacles they face. Many *do* care and they *do* try, but there's a feeling of despair. The parents of these children want the same things for their children that the parents in the suburbs want. Drugs are not the cause of this. They are the symptom. Nonetheless, they're used by people in the suburbs and rich people in Manhattan as another reason to keep children of poor people at a distance."

I ask him, "Will white children and black children ever go to school together in New York?"

"I don't see it," he replies. "I just don't think it's going to happen. It's a dream. I simply do not see white folks in Riverdale agreeing to cross-bus with kids like these. A few, maybe. Very few. I don't think I'll live to see it happen."

I ask him whether race is the decisive factor. Many experts, I observe, believe that wealth is more important in determining these inequalities.

"This," he says—and sweeps his hand around him at the room, the garbage bag, the ceiling—"would not happen to white children."

In a kindergarten class the children sit cross-legged on a carpet in a space between two walls of books. Their 26 faces are turned up to watch their teacher, an elderly black woman. A little boy who sits beside me is involved in trying to tie bows in his shoelaces. The children sing a song: "Lift Every Voice." On the wall are these handwritten words: "Beautiful, also, are the souls of my people."

In a very small room on the fourth floor, 52 people in two classes do their best to teach and learn. Both are first grade classes. One, I am informed, is "low ability." The other is bilingual.

"The room is barely large enough for one class," says the principal.

The room is 25 by 50 feet. There are 26 first graders and two adults on the left, 22 others and two adults on the right. On the wall there is the picture of a small white child, circled by a Valentine and a Gainsborough painting of a child in a formal dress....

Children who go to school in towns like Glencoe and Winnetka do not need to steal words from a dictionary. Most of them learn to read by second or third grade. By the time they get to sixth or seventh grade, many are reading at the level of the seniors in the best Chicago high schools. By the time they enter ninth grade at New Trier High, they are in a world of academic possibilities that far exceed the hopes and dreams of most schoolchildren in Chicago.

"Our goal is for students to be successful," says the New Trier principal. With 93 percent of seniors going on to four-year colleges—many to schools like Harvard, Princeton, Berkeley, Brown, and Yale—this goal is largely realized.

New Trier's physical setting might well make the students of Du Sable High School envious. The *Washington Post* describes a neighborhood of "circular driveways, chirping birds and white-columned homes." It is, says a student, "a maple land of beauty and civility." While Du Sable is sited on one crowded city block, New Trier students have the use of 27 acres. While Du Sable's science students have to settle for makeshift equipment, New Trier's students have superior labs and up-to-date technology. One wing of the school, a physical education center that includes three separate gyms, also contains a fencing room, a wrestling room, and studios for dance instruction. In all, the school has seven gyms as well as an Olympic pool.

The youngsters, according to a profile of the school in *Town and Country* magazine, "make good use of the huge, well-equipped building, which is immaculately maintained by a custodial staff of 48."

It is impossible to read this without thinking of a school like Goudy, where there are no science labs, no music or art classes and no playground—and where the two bathrooms, lacking toilet paper, fill the building with their stench.

"This is a school with a lot of choices," says one student at New Trier; and this hardly seems an over-statement if one studies the curriculum. Courses in music, art and drama are so varied and abundant that students can virtually major in these subjects in addition to their academic programs. The modern and classical language department offers Latin (four years) and six other foreign languages. Elective courses include the literature of Nobel winners, aeronautics, criminal justice, and computer languages. In a senior literature class, students are reading Nietzsche, Darwin, Plato, Freud, and Goethe. The school also operates a television station with a broadcast license from the FCC, which broadcasts on four channels to three counties.

Average class size is 24 children; classes for slower learners hold 15. This may be compared to Goudy—where a remedial class holds 39 children and a "gifted" class has 36.

Every freshman at New Trier is assigned a faculty adviser who remains assigned to him or her through graduation. Each of the faculty advisers—they are given a reduced class schedule to allow them time for this—gives counseling to about two dozen children....

The ambience among the students at New Trier, of whom only 1.3 percent are black, says *Town and Country*, is "wholesome and refreshing, a sort of throwback to the Fifties." It is, we are told, "a preppy kind of place." In a cheerful photo of the faculty and students, one cannot discern a single nonwhite face.

New Trier's "temperate climate" is "aided by the homogeneity of its students," *Town and Country* notes. "... Almost all are of European extraction and harbor similar values."

"Eighty to 90 percent of the kids here," says a counselor, "are good, healthy, red-blooded Americans."

The wealth of New Trier's geographical district provides $340,000 worth of taxable property for each child; Chicago's property wealth affords only one-fifth this much. Nonetheless, *Town and Country* gives New Trier's parents credit for a "willingness to pay enough ... in taxes" to make this one of the state's best-funded schools. New Trier, according to the magazine, is "a striking example of what is possible when citizens want to achieve the best for their children." Families move here "seeking the best," and their children "make good use" of what they're given.

27

Deepening Segregation in American Public Schools

GARY ORFIELD, MARK D. BACHMEIER, DAVID R. JAMES, AND TAMELA EITLE

This excerpt from a report by the Harvard Project on School Desegregation provides a brief history of desegregation in public schools, along with recent statistics and trends relating to the status of race segregation in public schools. The issue of resegregation of public school is of increasing concern in this country. This selection provides some background surrounding desegregation, including the court cases and legislation that formed its core. In addition, current statistics are included so you can see how far we have come.

Questions to consider as you read this selection:

1. *How did the desegregation of schools begin? What were the key decisions that made desegregation possible? Was it successful?*
2. *How does segregation affect the school experiences of students from different ethnic groups? Consider the article "Savage Inequalities" by Kozol. What might be the effects of resegregation?*
3. *What is the status of school segregation in your community. What changes, if any, are needed? Are taking place?*

GLOSSARY   ***Brown v. Board of Education*** U.S. Supreme Court decision judging separate school facilities to be unequal.   **De jure segregation** Separation required by law.   **Segregation** To separate groups, usually by racial or ethnic differences.

Decades of legal and political struggle were required to end the apartheid system of mandated segregation in the schools of 17 states and to transform the South from an area of absolute segregation for black students to the most integrated region of the country. We often celebrate this accomplishment as if it were a permanent reversal of a history of segregation and inequality.

From the 1950s through the late 1980s, African American students experienced declining segregation, particularly in the southern and border states.

The changes begun by the 1954 Supreme Court decision in *Brown v. Board of Education*, however, are now coming undone. The statistics analyzed for this article show that segregation is increasing for blacks, particularly in the states that once mandated racial separation. For Latinos, an even more severe level of segregation is intensifying across the nation.

The trends reported here are the first since the Supreme Court, in the 1990s, approved a return to segregated neighborhood schools under some conditions. A number of major cities have recently received court approval for such changes and others are in court. The segregation changes reported here are most striking in the southern and border states, but segregation is spreading across the nation, particularly affecting our rapidly growing Latino communities in the West. This report shows that the racial and ethnic segregation of African American and Latino students has produced a deepening isolation from middle-class students and from successful schools. It also highlights a little noticed but extremely important expansion of segregation to the suburbs, particularly in larger metropolitan areas. Expanding segregation is a mark of a polarizing society that lacks effective policies for building multiracial institutions.

Latino students, who will soon be the largest minority group in American public schools, were granted the right to desegregated education by the Supreme Court in 1973, but new data show they now are significantly more segregated than black students, with clear evidence of increasing isolation across the nation. Part of this trend is caused by the very rapid growth in the number of Latino students in several major states. Regardless of the reasons, Latino students now experience more isolation from whites and more concentration in high poverty schools than any other group of students.[1,2]

Desegregation is not just sitting next to someone of another race. Economic class and family and community educational background are also critically important for educational opportunity. School segregation effects go beyond racial separation. Segregated black and Latino schools are fundamentally different from segregated white schools in terms of the background of the children and many things that relate to educational quality. This report shows that only a twentieth of the nation's segregated white schools face conditions of concentrated poverty among their children, but more than 80% of segregated black and Latino schools do. A child moving from a segregated African American or Latino school to a white school will very likely exchange conditions of concentrated poverty for a middle-class school. Exactly the opposite is likely when a child is sent back from an interracial school to a segregated neighborhood school, as is happening under a number of recent court orders that end busing or desegregation choice plans.

The Supreme Court concluded in 1954 that intentionally segregated schools were "inherently unequal," and contemporary evidence indicates that this remains true today. Thus, it is very important to continuously monitor the extent to which the nation is realizing the promise of equal educational opportunity in schools that are now racially segregated. Education was vital to the success of the black tenth of the U.S. population when *de jure* segregation was declared unconstitutional—it is far more important today, when millions of good, low-education jobs have vanished, and when one-third of public school students are nonwhite.[3]

With the stakes for educational opportunity much higher today, this report shows that we are moving backward toward greater racial separation, rather than pressing gradually forward as we were between the 1950s and the mid-1980s. It shows a delayed impact of the Reagan administration campaign to reverse desegregation orders, which made no progress while Reagan was president, but now has had a substantial impact through appointments that transformed the federal courts. The 1991–95 period following the Supreme Court's first decision authorizing resegregation witnessed the continuation of the largest backward movement toward segregation for blacks in the 43 years since *Brown*.

During the 1980s, the courts rejected efforts to terminate school desegregation, and the level of desegregation actually increased, although the Reagan and Bush administrations advocated reversals. Congress rejected proposals for major steps to reverse desegregation, and there has been no trend toward increasing hostility to desegregation in public opinion. In fact, opinion is becoming more favorable.[4] The policy changes have come from the courts. The Supreme Court, in decisions from 1991 to 1995, has given lower courts discretion to approve resegregation on a large scale, and it is beginning to occur.

The statistics reported here show only the first phase of what is likely to be an accelerating trend. These statistics for the 1994–95 school year do not reflect post-1994 decisions that terminated desegregation plans in metropolitan Wilmington, Broward County (Florida), Denver, Buffalo, Mobile, Cleveland, and a number of other areas. Important cases in several other cities are pending in court now. These decisions are virtually certain to accelerate the trend toward increased racial and economic segregation of African American and Latino students. Thus, the trends reported today should be taken as portents of larger changes now under way.

BACKGROUND OF DESEGREGATION

In 1954 the Supreme Court began the process of desegregating American public education in its landmark decision, *Brown v. Board of Education*. Congress took its most powerful action for school desegregation with the passage of the 1964 Civil Rights Act. In 1971, the great national battle over urban desegregation began with the Supreme Court's decision in the Charlotte, North Carolina, busing case, *Swann v. Charlotte-Mecklenberg Board of Education*.[5] With *Swann*, there was a comprehensive set of pol   in place for ma: : desegregation i the South.

No similar body of law ever developed in the North and West. The Supreme Court first extended some desegregation requirements to the cities of the North and recognized the rights of Hispanic as well as black students from illegal segregation in 1973.[6] In the early 1970s Congress enacted legislation to help pay for the training and educational changes (but not the busing) needed to make desegregation more

effective. These last major initiatives intended to foster desegregation took place more than two decades ago.

Since 1974 almost all of the federal policy changes have been negative, even while the nation's non–white population has dramatically increased, particularly its school age children. In what is rapidly becoming a society dominated by suburbia, only a small fraction of white middle-class children are growing up in central cities. The key Supreme Court decision of *Milliken v. Bradley*[7] in 1974 reversed lower court plans to desegregate metropolitan Detroit and provided a drastic limitation on the possibility of substantial and lasting city-suburban school desegregation. That decision ended significant movement toward less segregated schools and made desegregation virtually impossible in many metropolitan areas when the non–white population was concentrated in central cities. (It is not, therefore, surprising that the state of Michigan ranks second in the nation in segregation of black students two decades after the Supreme Court confined desegregation efforts within the boundaries of a largely black and economically declining city.)[8]

The Supreme Court ruled that the courts could try to make segregated schools more equal in its second Detroit decision in 1977, *Milliken v. Bradley II*.[9] The Court authorized an order that the State of Michigan pay for some needed programs in Detroit which were aimed at repairing the harms inflicted by segregation in schools that would remain segregated because of the 1974 decision blocking city-suburban desegregation. Unfortunately, there was little serious follow-up by the courts on the educational remedies, and the Supreme Court severely limited such remedies in the 1995 *Missouri v. Jenkins*[10] decision.

The government turned actively against school desegregation in 1981 under the Reagan administration, with the Justice Department reversing policy on many pending cases and attacking urban desegregation orders. Congress accepted the administration's proposal to end the federal desegregation assistance program in the 1981 Omnibus Budget Reconciliation Act. Twelve years of active efforts to reverse desegregation orders and remake the federal courts followed. The Clinton administration in its first term defended some orders but developed no coherent policy and took no significant initiatives for desegregation.

By far the most important changes in policy have come from the Supreme Court. The appointment of Justice Clarence Thomas in 1991 consolidated a majority favoring cutting back civil rights remedies requiring court-ordered changes in racial patterns. In the 1991 *Board of Education of Oklahoma City v. Dowell*[11] decision, the Supreme Court ruled that a school district that had complied with its court order for several years could be allowed to return to segregated neighborhood schools if it met specific conditions. In the 1992 *Freeman v. Pitts*[12] decision, the Court made it easier to end student desegregation even when the other elements of a full desegregation order had never been accomplished. Finally, in its 1995 *Jenkins* decision, the Court's majority ruled that the court-ordered programs designed to make segregated schools more equal educationally and to increase the attractiveness of the schools to accomplish desegregation through voluntary choices were temporary and did not have to work before they could be discontinued.

In other words, desegregation was redefined from the goal of ending schools defined by race to a temporary and limited process that created no lasting rights and need not overcome the inequalities growing out of a segregated history. These decisions stimulated efforts in a number of cities to end the court orders, sometimes even over the objection of the school district involved.

RACIAL COMPOSITION OF
AMERICAN SCHOOLS

As the courts were cutting back on desegregation requirements, the proportion of minority students in public schools was growing rapidly and becoming far more diverse. In the fall of 1994, American public schools enrolled more than 43 million students, of whom 66% were white, 17% African

American, 13% Latino, 4% Asian, and 1% Indian and Alaskan. The proportion of Latinos in the United States was higher than that of blacks at the time desegregation began in 1954, and the proportion of whites was far lower. The two regions with the largest school enrollments, the South and the West, were 58% and 57% white, foreshadowing a near future in which large regions of the United States would have white minorities. Table 7.1 shows that there has been a huge growth (178%) in the number of Latino students during the 26 years since 1968, when data was first available nationally, to 1994. Meanwhile, the number of white (Anglo) students declined 9%, and the number of black students rose 14%.

On a regional level, African Americans remained the largest minority group in the schools of all regions except the West and Alaska and Hawaii. The proportion of black students in the South was, however, about twice the proportion in the Northeast and Midwest and more than four times the level in the West. Latinos, on the other hand, made up more than a fourth of the enrollment in the West but only about a 50th in the Border region and a 25th in the Midwest.

The dramatic changes in the composition of American school enrollment is most apparent in five states that already have a majority of non-white students statewide. These include the nation's two most populous states, California and Texas, which enroll 8.8 million students and are both moving rapidly toward a Latino majority in their school systems.

TABLE 7.1   Public School Enrollment Changes, 1968–94 (In Millions)

	1968	1980	1994	Change 1968–94
Hispanics	2.00	3.18	5.57	+3.57 (178%)
Anglos	34.70	29.16	28.46	−6.24 (−18%)
Blacks	6.28	6.42	7.13	+0.85 (14%)

SOURCE: DB5 Corp., 1982, 1987; Gary Orfield, Rosemary George, and Amy Orfield, "Racial Change in U.S. School Enrollments, 1968–1984," paper presented at National Conference on School Desegregation, University of Chicago, 1968. 1994–95 NCES Common Core of Data.

NATIONAL INCREASE IN SEGREGATION

In the fall of 1972, after the Supreme Court's 1971 busing decision that led to new court orders for scores of school districts, 63.6% of black students were in schools with less than half white enrollment. Fourteen years later, that percentage was virtually the same, but it rose to 67.1% by 1994–95 (Table 7.2). Desegregation remained at its high point until about 1988 but then began to fall significantly on this measure.

A second measure of segregation, calculated as the number of students experiencing intense isolation in schools with less than one-tenth whites (i.e., 90–100% minority enrollment), shows that the proportion of black students facing extreme isolation dropped sharply with the busing decisions, declining from 64.3% in 1968 to 38.7% in 1972 and continuing to decline slightly through the mid-1990s (Table 7.2). This isolation increased gradually from 1988 to 1991 but actually declined slightly from 1991 to 1994. This is the only measure that does not show increased black segregation.

The third measure of desegregation used in this study, the exposure index—which calculates the percentage of white students in a school attended

TABLE 7.2   **Percentage of U.S. Black and Latino Students in Predominantly Minority and 90–100% Minority Schools, 1968–94**

	50–100% Minority		90–100% Minority	
	Blacks	Latinos	Blacks	Latinos
1968–69	76.6	54.8	64.3	23.1
1972–73	63.6	56.6	38.7	23.3
1980–81	62.9	68.1	33.2	28.8
1986–87	63.3	71.5	32.5	32.2
1991–92	66.0	73.4	33.9	34.0
1994–95	67.1	74.0	33.6	34.8

SOURCE: U.S. Department of Education Office for Civil Rights data in Orfield, *Public School Desegregation in the United States,1968–1980,* tables 1 and 10; 1991–92 and 1994–95 NCES Common Core of Data.

TABLE 7.3   **Percentage of White Students in Schools Attended by Typical Black or Latino Students, 1970–94**

	Blacks	Latinos
1970	32.0	43.8
1980	36.2	35.5
1986	36.0	32.9
1991	34.4	31.2
1994	33.9	30.6

SOURCE: 1994–95 NCES Common Core of Data Public School Universe.

by typical black students—shows the level of contact almost as low as it was before the busing decisions in the early 1970s: 32%, down from its 1980 level of 36.2% (Table 7.3). Overall, the level of black segregation in U.S. schools is increasing slowly, continuing an historic reversal first apparent in the 1991 enrollment statistics....

TRENDS FOR LATINO STUDENTS

Latino segregation has become substantially more severe than African American segregation by each of the measures used in this study. In the Northeast, the West, and the South, more than three-fourths of all Latino students are in predominantly non-white schools, a level of isolation found for African American students only in the Northeast.... We have been reporting these trends continuously for two decades. They are clearly related to inferior education for Latino students.[13] Although data are limited, the surveys that have been done tend to show considerable interest in desegregated education among Latinos and substantial support for busing if there is no other way to achieve integration.[14]

All three measures of segregation reported in Tables 7.2 and 7.3 show a continuing gradual increase in segregation for Latino students nationally. The most significant change comes in the proportion of students in intensely segregated schools, which rose to 34.8% in 1994. In 1968, only 23.1% of Latino students were in these isolated and highly impoverished schools, compared to 64.3% of black students (Table 7.2). Now the percentage of Latino students in such schools is up by almost half and is slightly higher than the level of intense segregation for black students....

RACE AND POVERTY

The relationship between segregation by race and segregation by poverty in public schools across the nation is exceptionally strong. The correlation between the percentage of black and Latino enrollments and the percentage of students receiving free lunches is an extremely high .72. This means that racially segregated schools are very likely to be segregated by poverty as well.

There is strong and consistent evidence from national and state data from across the United States as well as from other nations that high poverty schools usually have much lower levels of educational performance on virtually all outcomes. This is not all caused by the school; family background is a more powerful influence. Schools with concentrations of low income children have less prepared children. Even better prepared children can be harmed academically if they are placed in a school with few other prepared students and, in some cases, in a social setting where academic achievement is not supported.

School achievement scores in many states and in the nation show a very strong relation between poverty concentrations and low achievement.[15] This is because high poverty schools are unequal in many ways that affect educational outcomes. The students' parents are far less educated—a very powerful influence—and the child is much more likely to be living in a single parent home that is struggling with multiple problems. Children are much more likely to have serious developmental and untreated health problems. Children move much more often, often in the middle of a school year, losing continuity and denying schools sufficient time to make an impact on their learning.

High poverty schools have to devote far more time and resource to family and health crises, security, children who come to school not speaking standard English, seriously disturbed children, children with no educational materials in their homes, and many children with very weak educational preparation. These schools tend to draw less qualified teachers and to hold them for shorter periods of time. They tend to have to invest much more heavily in remediation and much less adequately in advanced and gifted classes and demanding materials. The levels of competition and peer group support for educational achievement are much lower in high poverty schools. Such schools are viewed much more negatively in the community and by the schools and colleges at the next level of education as well as by potential employers. In those states that have implemented high stakes testing, which denies graduation or flunks students, the high poverty schools tend to have by far the highest rates of sanctions.[16]

None of this means that the relationship between poverty and educational achievement is inexorable, or that there are not exceptions. Many districts have one or a handful of high poverty schools that perform well above the normal pattern. Students from the same family background may perform at very different levels of achievement, and there are some highly successful students and teachers in virtually every school. The overall relationships, however, are very powerful. Students attending high poverty schools face a much lower level of competition regardless of their own interests and abilities.

This problem is intimately related to racial segregation ... [o]f the schools in the United States, 60.7% have less than one-fifth black and Latino students, while 9.2% ... have 80–100% black and Latino students. At the extremes, only 5.4% of the schools with 0–10% black and Latino students have more than half low income students; 70.5% ... of them have less than one-fourth poor students. Among schools that are 90–100% black and/or Latino, on the other hand, almost nine-tenths (87.7%) are predominantly poor, and only about 3% ... have less than one-fourth poor children.

A student in a segregated minority school is 16.3 times more likely to be in a concentrated poverty school than a student in a segregated white school.[17]

WHERE IS SEGREGATION CONCENTRATED? THE LONG-TERM EFFECTS OF THE SUPREME COURT'S DECISION AGAINST SUBURBAN DESEGREGATION

Blacks living in rural areas and in small and medium-sized towns or the suburbs of small metropolitan areas are far more likely to experience substantial school desegregation than those living in the nation's large cities; they attend schools with an average of about 50% white students. … In contrast, blacks in large cities attend schools that have an average of only 17% white students and those in smaller cities attend schools with an average of 38% white students. Black students in the suburbs of large cities attend schools with an average of 41% white students. … Considering the small proportion of minority students in many suburban rings, this level of segregation is a poor omen for the future of suburbs that will become more diverse.

The nation's non-white population is extremely concentrated in metropolitan areas. Outside the South, this concentration tends to be in the largest metropolitan areas with the largest ghettos and barrios. Many of the small cities and towns in Illinois and Michigan, for example, have few African American students, and the vast majority of the nation's white students live in suburbs divided into scores of separate school districts, all laid over extremely segregated metropolitan housing markets. This means that the central city school districts become extremely isolated by race and poverty and are critical only for non-white students. Since the minority communities are constantly expanding along their boundaries, and virtually all-white developments are continuously being constructed on the outer periphery of suburbia, the central cities have a continual increase in their proportion of black and Latino students.

The suburbs are now the dominant element of our society and our politics. As the nation's population changes dramatically in the coming decades, suburbs are destined to become much more diverse. What kind of access black and Latino children will have to mainstream suburban society will be affected by the racial characteristics of suburban schools. It raises serious concerns to realize that by 1994, blacks were in schools that averaged only 41% white students, and Latinos were in schools that averaged just 36% white students, in the suburbs of the largest cities. Whites in those suburban rings were in schools with an average of only 14% … combined black and Latino enrollments. Latino students, but not blacks, were almost as segregated in the suburbs of smaller metropolitan areas. … If these patterns intensify as the suburban African American and Latino population grows, we may be facing problems that are as serious as those that led to desegregation conflicts in many cities.

It would be profoundly ironic if the Supreme Court decision that meant to protect suburban boundary lines (*Milliken v. Bradley*) ended up making it impossible for suburban communities in the path of racial change to avoid rapid resegregation. Individual suburban school districts are often so small that they can go through racial change much more rapidly and irreversibly than a huge city. A suburb will often have only the enrollment of a single high school attendance area in a city and has little hope of stabilizing its enrollment, once a major racial change begins, without drawing on students from a broader geographic area. This means that in areas with many fragmented school districts, not only the city but also substantial portions of suburban rings may face high levels of segregation. Since non-white suburbanization began in earnest in the 1970s, the cities also have been losing many of their minority middle-class families, leaving the cities with an escalating concentration of poverty.

… These districts contain 18% black students, 23% Latino students, 13% Asian students, but only 2% of the whites. About a fifth of black and Latino students depend on districts that do not matter to

98% of white families. Most of these systems have faced recurrent fiscal and political crises for years and have low levels of educational achievement. Desegregation has become virtually impossible in some of these systems since the *Milliken* decision. The trends of metropolitan racial change since World War II suggest that segregation will become worse in the future.

Consequences of Smaller Districts

Different parts of the country traditionally have very different patterns of organizing school districts, depending in part on local traditions and, in part, on whether or not the districts were organized back when the horse and buggy meant that units of local government had to be very small. In much of the South, counties have traditionally been more important and municipalities less important. In New England the towns existed long before they became part of suburban rings in large metropolitan areas. School districts in the South were often county-wide; in the Northeast and the older parts of the Midwest, they were often defined by the structure of local town government set generations in the past. When the Supreme Court decided in 1974 to make it very difficult to desegregate across school district boundary lines, it virtually guaranteed that the regions with large districts would be far less segregated than those with small districts.

The nation's largest school districts are in Maryland, Florida, Louisiana, North Carolina, West Virginia, and Delaware, all states in the Southern and Border regions. Illinois, New York, New Jersey, and Michigan, consistently the most segregated states for black students, have much smaller districts, as do Texas and California, where most Latino students are concentrated. There has been no absolute relationship between district size and segregation, of course, because of the widely varying proportions of black and Latino students within various states and because of different types of desegregation plans in place in different areas. Nonetheless, the most segregated states tend to be fragmented into a great many small districts, and the most integrated states tend to have very different

patterns, although there are significant exceptions to this pattern. The Maryland and Louisiana statistics show, for example, that it is possible to have very large districts with very high levels of segregation, while the Indiana and Ohio statistics show that small districts in states with multiple desegregation orders and a small black population are compatible with a lower level of intense segregation.

No state with small districts and a substantial African American population has come anywhere near the level of desegregation achieved in the most successful states with large systems. North Carolina and Tennessee, which already have relatively large districts, have seen city and suburban districts consolidated in recent years to create more countywide systems....

Segregation of Whites at the State Level

In a nation where whites are destined to become one of several minorities in the schools if the existing trends continue, it is important not only to consider the isolation of non-white students from whites but also the isolation of whites from the growing parts of the population.

Except in the historic *de jure* states for blacks and the states taken from Mexico in the war in the 1840s, most white students have not yet experienced substantial desegregation. Although they are growing up in a society where the U.S. Census Bureau predicts that more than half of schoolage children will be non-white in a third of a century, many are being educated in overwhelmingly white schools with little contact with black or Latino students.[18] Those students may be ill prepared as the American workforce changes and skills in race relations become increasingly valuable in many jobs....

CONCLUSION

In American race relations, the bridge from the twentieth century may be leading back into the nineteenth century. We may be deciding to bet the future of the country once more on "separate but equal." There is no evidence that separate but equal today works any more than it did a century ago.

The debate that has been stimulated by recent Supreme Court decisions is a debate about how and when to end desegregation plans. The most basic need now is for a serious national examination of the cost of resegregation and the alternative solutions to problems with existing desegregation plans. Very few Americans prefer segregation, and most believe that desegregation has had considerable value, but most whites are still opposed to plans that involve mandatory transportation of students. During the last 15 years, plans have been evolving to include more educational reforms and choice mechanisms to try to achieve desegregation and educational gains simultaneously. A stronger fair housing law, a number of settlements of housing segregation cases, and federal initiatives to change the operation of subsidized housing—as well as the very rapid creation of brand-new communities in the sunbelt—all offer opportunities to try to change the pattern of segregated housing that underlies school segregation. Policies that would help move the country toward a less polarized society include:

1. Resumption of serious enforcement of desegregation by the Justice Department and serious investigation of the degree to which districts have complied with all Supreme Court requirements by the Department of Education. Such requirements could be appropriately specified in a federal regulation.

2. Creation of a new federal education program to train students, teachers, and administrators in human relations, conflict resolution, and multi-ethnic education techniques and to help districts devise appropriate plans and curricula for successful multiracial schools.

3. Serious federal research on multiracial schools and the comparative success of segregated and desegregated schools.

4. A major campaign to increase non-white teachers and administrators through a combination of employment discrimination enforcement and resources for recruitment and education of potential teachers.

5. Incorporation of successful desegregation into the national educational goals.

6. Federal and state efforts to expand the use of integrated two-way bilingual programs from the demonstration stage to a major technique for improving second language acquisition for both English speakers and other language speakers and for building successful ethnic relationships.

7. Additional Title IV resources to expand state education department staffs working on desegregation and racial equity in the schools.

8. Federal, state, and local plans to coordinate housing policy with school desegregation policy.

9. Examination of choice and charter school plans to assure that they are not increasing segregation and to reinforce their potential contribution to desegregation.

10. Examination of high stakes state testing programs to ensure that they are not punishing the minority students who must attend inferior segregated school under existing state and local policies.

NOTES

1. Distribution of Latinos by ethnicity and state is reported in M. Beatriz Arias, "The Context of Education for Hispanic Students: An Overview," *American Journal of Education* 95(1) (November 1986): 26–57.

2. In this report "white" means non-Hispanic whites. Hispanic or Latino is treated as part of the non-white population although many Latinos define themselves as whites in racial terms. These definitions are used to avoid the awkward and confusing language that would otherwise be necessary and is not an attempt to define Latinos as a race.

3. U.S. Bureau of Census Projections, and Steven A. Holmes, "Census Sees a Profound Ethnic Shift in

U.S.," *New York Times*, March 14, 1996; *Education Week*, March 27, 1996, p. 3.

4. Gary Orfield, "Public Opinion and School Desegregation," *Teachers College Record 96*(4) (Summer 1995); Gallup Poll in *USA Today*, May 12, 1994; Gallup Poll in *Phi Delta Kappan*, September 1996. The 1996 survey reported that "the percentages who say integration has improved the quality of education for blacks and for whites have been increasing steadily since these questions were first asked in 1971" (p. 48). The report also showed that 83% of the public believed that interracial schools were desirable.

5. 402 U.S. 1 (1971).

6. *Keyes v. Denver School District No. 1*, 413 U.S. 189 (1973).

7. 418 U.S. 717 (1974).

8. Calculations of metropolitan segregation from 1992 Common Core data.

9. 433 U.S. 267 (1977).

10. 115 S.Ct. 2038 (1995).

11. 498 U.S. 237 (1991).

12. 503 U.S. 467 (1992).

13. Ruben Espinosa and Alberto Ochoa, "Concentration of California Hispanic Students in Schools with Low Achievement: A Research Note," *American Journal of Education 95*(1), (1986): 77–95; Ruben Donato, Martha Menchaca, and Richard R. Valencia, "Segregation, Desegregation, and the Integration of Chicano Students: Problems and Prospects" in Richard R. Valencia, ed., *Chicano School Failure and Success: Research and Policy Agendas for the 1990s* (London: Falmer Press, 1991): 27–63.

14. A national survey by the *Boston Globe* in 1992 reported that 82% of Latinos said that they favored busing if there was no other way to achieve integration and that Latinos expressing an opinion said that they would be "willing to have your own children go to school by bus so the schools would be integrated" ("Poll Shows Wide Support Across U.S. for Integration," *Boston Globe*, January 5, 1992: 15); for evidence of more closely divided opinion earlier see Gary Orfield, "Hispanic Education: Challenges, Research, and Policies," *American Journal of Education 95*(1)(1986): 11–12.

15. See, for example, discussion of the relationship of disadvantaged school status and educational opportunity in Jeannie Oakes, *Multiplying Inequalities: The Effects of Race, Social Class, and Tracking on Opportunities to Learn Mathematics and Science* (Santa Monica: RAND, 1990), figure 2.3.; 1988 NAEP data reported in Educational Testing Service, *The State of Inequality*; Peter Scheirer, "Poverty Not Bureaucracy," working paper, Metropolitan Opportunity Project, Univ. of Chicago, 1989; Samuel S. Peng, Margaret C. Wang, and Herbert J. Walberg, "Demographic Disparities of Inner-City Eighth Graders," *Urban Education 26*(4) (January 1992): 441–459; David M. Cutler and Edward L. Glaeser, "Are Ghettos Good or Bad?" (Cambridge: National Bureau of Economic Research, 1995); Gary G. Wehlage, "Social Capital and the Rebuilding of Communities," Center on Organization and Restructuring of Schools, Issue Report No. 5, Fall 1993: 3–5; Raymond Hernandez, "New York City Students Lagging in Mastery of Reading and Math," *New York Times*, January 3, 1997: Al, B4; a 1997 study in the suburban Washington district of Prince George's County reported that "with each 10 percent increase in the number of students who qualified for free or reduced-price lunches, a school's score on the state test dropped by an average of 1.8 percentage points in reading, 4.1 percentage points in math and 2.2 percentage points in science" (*Washington Post*, May 15, 1997: A17).

16. A study of 1992 test scores in greater Cleveland showed that district poverty level differences "explained as much as 39 percent of the differences in school district passing rates" on the state proficiency test (*Cleveland Plain Dealer*, October 8, 1995: 3-C).

17. Asian students, who are having far greater success in U.S. schools, attend schools that average only about one-fourth African American and Latino students and they are far less likely to be in high poverty schools, factors that help account for their mobility (*Asian Students and Multiethnic Desegregation*, Harvard Project on School Desegregation, October 1994).

18. A recent Gallup Poll showed that 93% of the public and 96% of school parents thought it was very important to teach students "acceptance of people of different races and ethnic backgrounds" (*Phi Delta Kappan*, October 1993: 139, 145).

28

The Elementary Forms of Religious Life

EMILE DURKHEIM

The author claims that religious beliefs and rituals are real and reflect the societies in which they exist. Note his reasons for this claim. If you agree with his thesis, ask yourself what effect it has on your personal religious beliefs.

As you read, ask yourself the following questions:

1. *Why does the author say that religious beliefs and rituals are real and reflect society?*
2. *How do your religious beliefs and rituals reflect American society?*
3. *Considering religious impact, should religious institutions continue to receive tax breaks? Why?*

GLOSSARY   **Profane** Not devoted to religious purposes; secular.   ***Sui generis*** In itself.

The theorists who have undertaken to explain religion in rational terms have generally seen in it before all else a system of ideas, corresponding to some determined object. This object has been conceived in a multitude of ways: nature, the infinite, the unknowable, the ideal, etc.; but these differences matter but little. In any case, it was the conceptions and beliefs which were considered as the essential elements of religion. As for the rites, from this point of view they appear to be only an external translation, contingent and material, of these internal states which alone pass as having any intrinsic value. This conception is so commonly held that generally the disputes of which religion is the theme turn about the question whether it can conciliate itself with science or not, that is to say, whether or not there is a place beside our scientific knowledge for another form of thought which would be specifically religious.

But the believers, the men who lead the religious life and have a direct sensation of what it really is, object to this way of regarding it, saying that it does not correspond to their daily experience. In fact, they feel that the real function of religion is not to make us think, to enrich our knowledge, nor to add to the conceptions which we owe to science others of another origin and another character, but rather, it is to make us act,

Source: From *The Elementary Forms of Religious Life: A Study in Religious Sociology,* by Emile Durkheim, Routledge, 1915. Translated from the French by Joseph Ward Swain. London: Allen & Urwin, 1915. Reprinted with the permission of Simon and Schuster.

to aid us to live. The believer who has communicated with his god is not merely a man who sees new truths of which the unbeliever is ignorant; he is a man who is *stronger*. He feels within him more force, either to endure the trials of existence, or to conquer them. It is as though he were raised above the miseries of the world, because he is raised above his condition as a mere man; he believes that he is saved from evil, under whatever form he may conceive this evil. The first article in every creed is the belief in salvation by faith. But it is hard to see how a mere idea could have this efficacy. An idea is in reality only a part of ourselves; then how could it confer upon us powers superior to those which we have of our own nature? Howsoever rich it might be in affective virtues, it could add nothing to our natural vitality; for it could only release the motive powers which are within us, neither creating them nor increasing them. From the mere fact that we consider an object worthy of being loved and sought after, it does not follow that we feel ourselves stronger afterwards; it is also necessary that this object set free energies superior to these which we ordinarily have at our command and also that we have some means of making these enter into us and unite themselves to our interior lives. Now for that, it is not enough that we think of them; it is also indispensable that we place ourselves within their sphere of action, and that we set ourselves where we may best feel their influence; in a word, it is necessary that we act, and that we repeat the acts thus necessary every time we feel the need of renewing their effects. From this point of view, it is readily seen how that group of regularly repeated acts which form the cult get their importance. In fact, whoever has really practiced a religion knows very well that it is the cult which gives rise to these impressions of joy, of interior peace, of serenity, of enthusiasm which are, for the believer, an experimental proof of his beliefs. The cult is not simply a system of signs by which the faith is outwardly translated; it is a collection of the means by which this is created and recreated periodically. Whether it consists in material acts or mental operations, it is always this which is efficacious.

Our entire study rests upon this postulate that the unanimous sentiment of the believers of all times cannot be purely illusory.... We admit that these religious beliefs rest upon a specific experience whose demonstrative value is, in one sense, not one bit inferior to that of scientific experiments, though different from them. We, too, think that "a tree is known by its fruits," and that fertility is the best proof of what the roots are worth. But from the fact that a "religious experience," if we choose to call it this, does exist and that it has a certain foundation—and, by the way, is there any experience which has none?—it does not follow that the reality which is its foundation conforms objectively to the idea which believers have of it. The very fact that the fashion in which it has been conceived has varied infinitely in different times is enough to prove that none of these conceptions express it adequately. If a scientist states it as an axiom that the sensations of heat and light which we feel correspond to some objective cause, he does not conclude that this is what it appears to the senses to be. Likewise, even if the impressions which the faithful feel are not imaginary, still they are in no way privileged intuitions; there is no reason for believing that they inform us better upon the nature of their object than do ordinary sensations upon the nature of bodies and their properties. In order to discover what this object consists of, we must submit them to an examination and elaboration analogous to that which has substituted for the sensuous idea of the world another which is scientific and conceptual.

This is precisely what we have tried to do, and we have seen that this reality, which mythologies have represented under so many different forms, but which is the universal and eternal objective cause of these sensations *sui generis* out of which religious experience is made, is society. We have shown what moral forces it develops and how it awakens this sentiment of a refuge, of a shield, and of a guardian support which attaches the believer to his cult. It is that which raises him outside himself; it is even that which made him. For that which makes a man is the totality of the intellectual property which constitutes civilization, and civilization is the work of society. Thus is explained the preponderating role of the cult in all religions, whichever they may be. This is because society cannot make its influence felt unless it is in action, and it is not in action unless the individuals who compose it are assembled together and act in common. It is by common action that it takes consciousness of itself and realizes its position; it is before all else an active cooperation. The collective ideas and sentiments are even possible only owing to these exterior movements which symbolize them, as we have established. Then it is action which dominates the religious life, because of the mere fact that it is society which is its source....

Religious forces are therefore human forces, moral forces. It is true that since collective sentiments can become conscious of themselves only by fixing themselves upon external objects, they have not been able to take form without adopting some of their characteristics from other things: They have thus acquired a sort of physical nature; in this way they have come to mix themselves with the life of the material world, and then have considered themselves capable of explaining what passes there. But when they are considered only from this point of view and in this role, only their most superficial aspect is seen. In reality, the essential elements of which these collective sentiments are made have been borrowed by the understanding. It ordinarily seems that they should have a human character only when they are conceived under human forms,[1] but even the most impersonal and the most anonymous are nothing else than objectified statements....

Some reply that men have a natural faculty for idealizing, that is to say, of substituting for the real world another different one, to which they transport themselves by thought. But that is merely changing the terms of the problem; it is not resolving it or even advancing it. This systematic idealization is an essential characteristic of religions. Explaining them by an innate power of idealization is simply replacing one word by another which is the equivalent of the first; it is as if they said that men have made religions because they have a religious nature. Animals know only one world, the one which they perceive by experience, internal

as well as external. Men alone have the faculty of conceiving the ideal, of adding something to the real. Now where does this singular privilege come from? Before making it an initial fact or a mysterious virtue which escapes science, we must be sure that it does not depend upon empirically determinable conditions.

The explanation of religion which we have proposed has precisely this advantage, that it gives an answer to this question. For our definition of the sacred is that it is something added to and above the real: Now the ideal answers to this same definition; we cannot explain one without explaining the other. In fact, we have seen that if collective life awakens religious thought on reaching a certain degree of intensity, it is because it brings about a state of effervescence which changes the conditions of psychic activity. Vital energies are overexcited, passions more active, sensations stronger; there are even some which are produced only at this moment. A man does not recognize himself; he feels himself transformed and consequently he transforms the environment which surrounds him. In order to account for the very particular impressions which he receives, he attributes to the things with which he is in most direct contact properties which they have not, exceptional powers and virtues which the objects of everyday experience do not possess. In a word, above the real world where his profane life passes he has placed another which, in one sense, does not exist except in thought, but to which he attributes a higher sort of dignity than to the first. Thus, from a double point of view it is an ideal world.

The formation of the ideal world is therefore not an irreducible fact which escapes science; it depends upon conditions which observation can touch; it is a natural product of social life. For a society to become conscious of itself and maintain at the necessary degree of intensity the sentiments which it thus attains, it must assemble and concentrate itself. Now this concentration brings about an exaltation of the mental life which takes form in a group of ideal conceptions where is portrayed the new life thus awakened; they correspond to this new set of psychical forces which is added to those which we have at our disposition for the daily tasks of existence. A society can neither create itself nor recreate itself without at the same time creating an ideal. This creation is not a sort of work of supererogation for it, by which it would complete itself, being already formed; it is the act by which it is periodically made and remade. Therefore when some oppose the ideal society to the real society, like two antagonists which would lead us in opposite directions, they materialize and oppose abstractions. The ideal society is not outside of the real society; it is a part of it. Far from being divided between them as between two poles which mutually repel each other, we cannot hold to one without holding to the other. For a society is not made up merely of the mass of individuals who compose it, the ground which they occupy, the things which they use and the movements which they perform, but above all is the idea which it forms of itself. It is undoubtedly true that it hesitates over the manner in which it ought to conceive itself; it feels itself drawn in divergent directions. But these conflicts which break forth are not between the ideal and reality, but between two different ideals, that of yesterday and that of today, that which has the authority of tradition and that which has the hope of the future. There is surely a place for investigating whence these ideals evolve; but whatever solution may be given to this problem, it still remains that all passes in the world of the ideal.

Thus the collective ideal which religion expresses is far from being due to a vague innate power of the individual, but it is rather at the school of collective life that the individual has learned to idealize. It is in assimilating the ideals elaborated by society that he has become capable of conceiving the ideal. It is society which, by leading him within its sphere of action, has made him acquire the need of raising himself above the world of experience and has at the same time furnished him with the means of conceiving another. For society has constructed this new world in constructing itself, since it is society which this expresses. Thus both with the individual and in the group, the faculty of idealizing has nothing mysterious about it. It is not a sort of luxury which a man could get along

without, but a condition of his very existence. He could not be a social being, that is to say, he could not be a man, if he had not acquired it. It is true that in incarnating themselves in individuals, collective ideals tend to individualize themselves. Each understands them after his own fashion and marks them with his own stamp; he suppresses certain elements and adds others. Thus the personal ideal disengages itself from the social ideal in proportion as the individual personality develops itself and becomes an autonomous source of action. But if we wish to understand this aptitude, so singular in appearance, of living outside of reality, it is enough to connect it with the social conditions upon which it depends.

Therefore it is necessary to avoid seeing in this theory of religion a simple restatement of historical materialism: That would be misunderstanding our thought to an extreme degree. In showing that religion is something essentially social, we do not mean to say that it confines itself to translating into another language the material forms of society and its immediate vital necessities. It is true that we take it as evident that social life depends upon its material foundation and bears its mark, just as the mental life of an individual depends upon his nervous system and in fact his whole organism. But collective consciousness is something more than a mere epiphenomenon of its morphological basis, just as individual consciousness is something more than a simple efflorescence of the nervous system. In order that the former may appear, a synthesis *sui generis* of particular consciousnesses is required. Now this synthesis has the effect of disengaging a whole world of sentiments, ideas, and images which, once born, obey laws all their own. They attract each other, repel each other, unite, divide themselves, and multiply, though these combinations are not commanded and necessitated by the conditions of the underlying reality. The life thus brought into being even enjoys so great an independence that it sometimes indulges in manifestations with no purpose or utility of any sort, for the mere pleasure of affirming itself. We have shown that this is often precisely the case with ritual activity and mythological thought....

That is what the conflict between science and religion really amounts to. It is said that science denies religion in principle. But religion exists; it is a system of given facts; in a word, it is a reality. How could science deny this reality? Also, insofar as religion is action, and insofar as it is a means of making men live, science could not take its place, for even if this expresses life, it does not create it; it may well seek to explain the faith, but by that very act it presupposes it. Thus there is no conflict except upon one limited point. Of the two functions which religion originally fulfilled, there is one, and only one, which tends to escape it more and more: That is its speculative function. That which science refuses to grant to religion is not its right to exist, but its right to dogmatize upon the nature of things and the special competence which it claims for itself for knowing man and the world. As a matter of fact, it does not know itself. It does not even know what it is made of, nor to what need it answers. It is itself a subject for science, so far is it from being able to make the law for science! And from another point of view, since there is no proper subject for religious speculation outside that reality to which scientific reflection is applied, it is evident that this former cannot play the same role in the future that it has played in the past.

However, it seems destined to transform itself rather than to disappear....

NOTE

1.   It is for this reason that Frazer and even Preuss set impersonal religious forces outside of, or at least on the threshold of religion, to attach them to magic.

29

Popular Christianity and Political Extremism in the United States

JAMES AHO

Many believe that American political extremism is something that took place in the past and that it is unrelated to religion. According to the author of this article, neither of these beliefs is true.

As you read this selection, consider the following questions as guides:

1. *According to the author, extremists differ from others only in their biographies and in the fact that they belong to an extremist organization. Does this claim seem illogical to you? Why or why not?*
2. *Why do you believe or not believe in the role of political extremism in Christianity?*
3. *Think about a number of current movements in the United States and consider whether they are extremist. What would you do about them? Would you eliminate their tax-free status?*

EXTREMISM DEFINED

The word "extremism" is used rhetorically in everyday political discourse to disparage and undermine one's opponents. In this sense, it refers essentially to anyone who disagrees with me politically. In this chapter, however, "extremism" will refer exclusively to particular kinds of behaviors, namely, to non-democratic actions, regardless of their ideology—that is, regardless of whether we agree with the ideas behind them or not (Lipset and Raab 1970: 4–17). Thus, extremism includes: (1) efforts to deny civil rights to certain people, including their right to express unpopular views, their right to due process at law, to own property, etc.; (2) thwarting attempts by others to organize in opposition to us, to run for office, or vote; (3) not playing according to legal constitutional rules of political fairness: using personal smears like "Communist Jew-fag" and "nigger lover" in place of rational discussion; and above all, settling differences by vandalizing or destroying the property or life of one's opponents. The test is not the end as such, but the means employed to achieve it.

Source: From *Disruptive Religion: The Force of Faith in Social Movement Activism*, Christian Smith, ed. New York: Routledge, 1996, pp. 189–204.

CYCLES OF AMERICAN RIGHT-WING EXTREMISM . . .

American political history has long been acquainted with Christian-oriented rightist extremism. As early as the 1790s, for example, Federalist Party activists, inspired partly by Presbyterian and Congregationalist preachers, took up arms against a mythical anti-Christian cabal known as the Illuminati— Illuminati = bringers of light = Lucifer, the devil.

The most notable result of anti–Illuminatism was what became popularly known as the "Reign of Terror": passage of the Alien and Sedition acts (1798). These required federal registration of recent immigrants to America from Ireland and France, reputed to be the homes of Illuminatism lengthened the time of naturalization to become a citizen from five to fourteen years, restricted "subversive" speech and newspapers—that is, outlets advocating liberal Jeffersonian or what were known then as "republican" sentiments—and permitted the deportation of "alien enemies" without trial.

The alleged designs of the Illuminati were detailed in a three hundred-page book entitled *Proofs of a Conspiracy Against All the Religions and Governments of Europe Carried on in the Secret Meetings of … Illuminati* (Robison 1967 [1798]). Over two hundred years later *Proofs of a Conspiracy* continues to serve as a source-book for right-wing extremist commentary on American social issues. Its basic themes are: (1) *manichaenism:* that the world is divided into the warring principles of absolute good and evil; (2) *populism:* that the citizenry naturally would be inclined to ally with the powers of good, but have become indolent, immoral, and uninformed of the present danger to themselves; (3) *conspiracy:* that this is because the forces of evil have enacted a scheme using educators, newspapers, music, and intoxicants to weaken the people's will and intelligence; (4) *anti-modernism:* that the results of the conspiracy are the very laws and institutions celebrated by the unthinking masses as "progressive": representative government, the separation of church and State, the extension of suffrage to the property-less, free public education, public-health measures,

etc.; and (5) *apocalypticism:* that the results of what liberals call social progress are increased crime rates, insubordination to "natural" authorities (such as royal families and property-owning Anglo-Saxon males), loss of faith, and the decline of common decency—in short, the end of the world.

Approximately every thirty years America has experienced decade-long popular resurrections of these five themes. While the titles of the alleged evil-doers in each era have been adjusted to meet changing circumstances, their program is said to have remained the same. They constitute a diabolic *Plot Against Christianity* (Dilling 1952). In the 1830s, the cabal was said to be comprised of the leaders of Masonic lodges: in the 1890s, they were accused of being Papists and Jesuits; in the 1920s, they were the Hidden Hand; in the 1950s, the Insiders or Force X; and today they are known as Rockefellerian "one-world" Trilateralists or Bilderbergers.

Several parallels are observable in these periods of American right-wing resurgence. First, while occasionally they have evolved into democratically organized political parties holding conventions that nominate slates of candidates to run for office—the American Party, the Anti-Masonic Party, the People's Party, the Prohibition Party—more often, they have become secret societies in their own right, with arcane passwords, handshakes, and vestments, plotting campaigns of counter-resistance behind closed doors. That is, they come to mirror the fantasies against which they have taken up arms. Indeed, it is this ironic fact that typically occasions the public ridicule and undoing of these groups. The most notable examples are the Know Nothings, so-called because under interrogation they were directed to deny knowledge of the organization; the Ku Klux Klan, which during the 1920s had several million members; the Order of the Star Spangled Banner, which flourished during the 1890s; the Black Legion of Michigan, circa 1930; the Minutemen of the late 1960s; and most recently, the *Bruders Schweigen*, Secret Brotherhood, or as it is more widely known, The Order.

Secondly, the thirty-year cycle noted above evidently has no connection with economic booms and busts. While the hysteria of the 1890s took place during a nation-wide depression, McCarthyism exploded on the scene during the most prosperous era in American history. On close view, American right-wing extremism is more often associated with economic good times than with bad, the 1920s, the 1830s, and the 1980s being prime examples. On the contrary, the cycle seems to have more to do with the length of a modern generation than with any other factor.

Third, and most important for our purposes, Christian preachers have played pivotal roles in all American right-wing hysterias. The presence of Dan Gayman, James Ellison, and Bertrand Comparet spear-heading movements to preserve America from decline today continues a tradition going back to Jedidiah Morse nearly two centuries ago, continuing through Samuel D. Burchard, Billy Sunday, G. L. K. Smith, and Fred Schwarz's Christian Anti-Communist Crusade.

In the nineteenth century, the honorary title "Christian patriot" was restricted to white males with Protestant credentials. By the 1930s, however, Catholic ideologues, like the anti-Semitic radio priest Father Coughlin, had come to assume leadership positions in the movement. Today, somewhat uneasily, Mormons are included in the fold. The Ku Klux Klan, once rabidly anti-Catholic and misogynist, now encourages Catholic recruits and even allows females into its regular organization, instead of requiring them to form auxiliary groups.

CHRISTIANITY: A CAUSE OF POLITICAL EXTREMISM?

The upper Rocky Mountain region is the heartland of American right-wing extremism in our time. Montana, Idaho, Oregon, and Washington have the highest per capita rates of extremist groups of any area in the entire country (Aho 1994: 152–153). Research on the members of these groups show that they are virtually identical to the surrounding population in all respects but one (Aho 1991: 135–163)—they are not less formally educated

than the surrounding population. Furthermore, as indicated by their rates of geographic mobility, marital stability, occupational choice, and conventional political participation, they are no more estranged from their local communities than those with whom they live. And finally, their social status seems no more threatened than that of their more moderate neighbors. Indeed, there exists anecdotal evidence that American right-wing extremists today are drawn from the more favored, upwardly mobile sectors of society. They are college-educated, professional suburbanites residing in the rapidly growing, prosperous Western states (Simpson 1983).

In other words, the standard sociological theories of right-wing extremism—theories holding, respectively, that extremists are typically undereducated, if not stupid, transient and alienated from ordinary channels of belonging, and suffer inordinately from status insecurity—find little empirical support. Additionally, the popular psychological notion that right-wing extremists are more neurotic than the general population, perhaps paranoid to the point of psychosis, cannot be confirmed. None of the right-wing political murderers whose psychiatric records this author has accessed have been medically certified as insane (Aho 1991: 68–82; Aho 1994: 46–49). If this is true for right-wing murderers, it probably also holds for extremists who have not taken the lives of others.

The single way in which right-wing extremists *do* differ from their immediate neighbors is seen in their religious biographies. Those with Christian backgrounds generally, and Presbyterians, Baptists and members of independent fundamentalist Protestant groups specifically, all are overrepresented among intermountain radical patriots (Aho 1991: 164–182). Although it concerns a somewhat different population, this finding is consistent with surveys of the religious affiliations of Americans with conservative voting and attitudinal patterns (Lipset and Raab 1970: 229–232, 359–361, 387–392, 433–437, 448–452; Shupe and Stacey 1983; Wilcox 1992).

Correlations do not prove causality. Merely because American extremists are members of certain denominations and sects does not permit the conclusion that these religious groups compel their members to extremism. In the first place, the vast majority of independent fundamentalists, Baptists, and Presbyterians are not political extremists, even if they are inclined generally to support conservative causes. Secondly, it is conceivable that violently predisposed individuals are attracted to particular religions because of what they hear from the pulpit; and what they hear channels their *already* violent inclinations in political directions....

The point is not that every extremist is a violent personality searching to legitimize criminality with religion. Instead, the example illustrates the subtle ways in which religious belief, practice, and organization all play upon individual psychology to produce persons prepared to violate others in the name of principle. Let us look at each of these factors separately, understanding that in reality they intermesh in complicated, sometimes contradictory ways that can only be touched upon here.

BELIEF

American right-wing politics has appropriated from popular Christianity several tenets: the concept of unredeemable human depravity, the idea of America as a specially chosen people, covenant theology and the right to revolt, the belief in a national mission, millennialism, and anti-Semitism. Each of these in its own way has inspired rightist extremism.

The New Israel

The notion of America as the new Israel, for example, is the primary axiom of a fast-growing religiously-based form of radical politics known as Identity Christianity. Idaho's Aryan Nations Church is simply the most well-known Identity congregation. The adjective "identity" refers to its insistence that Anglo-Saxons are in truth the Israelites. They are "Isaac's-sons"—the Saxons—and hence the Bible is *their* historical record, not that of the Jews (Barkun 1994). The idea is that after its exile to what today is northern Iran around seven hundred B.C., the Israelites migrated over the Caucasus mountains— hence their racial type, "Caucasian"—and settled in

various European countries. Several of these allegedly still contain mementos of their origins: the nation of Denmark is said to be comprised of descendants from the tribe of Dan; the German-speaking Jutland, from the tribe of Judah; Catalonia, Scotland, from the tribe of Gad.

Covenant Theology

Identity Christianity is not orthodox Christianity. Nevertheless, the notion of America as an especially favored people, or as Ronald Reagan once said, quoting Puritan founders, a "city on a hill," the New Jerusalem, is widely shared by Americans. Reagan and most conservatives, of course, consider the linkage between America and Israel largely symbolic. Many right-wing extremists, however, view the relationship literally as an historical fact and for them, just as the ancient Israelites entered into a covenant with the Lord, America has done the same. According to radical patriots America's covenant is what they call the "organic Constitution." This refers to the original articles of the Constitution plus the first ten amendments, the Bill of Rights. Other amendments, especially the 16th establishing a federal income tax, are considered to have questionable legal status because allegedly they were not passed according to constitutional strictures.

The most extreme patriots deny the constitutionality of the 13th, 14th, and 15th amendments— those outlawing slavery and guaranteeing free men civil and political rights as full American citizens. Their argument is that the organic Constitution was written by white men exclusively for themselves and their blood descendents (Preamble 1986). Non-Caucasians residing in America are considered "guest peoples" with no constitutional rights. Their continued residency in this country is entirely contingent upon the pleasure of their hosts, the Anglo-Saxon citizenry. According to some, it is now time for the property of these guests to be confiscated and they themselves exiled to their places of origin (Pace 1985).

All right-wing extremists insist that if America adheres to the edicts of the organic Constitution, she, like Israel before her, shall be favored among the world's nations. Her harvests shall be bountiful, her communities secure, her children obedient to the voices of their parents, and her armies undefeated. But if she falters in her faith, behaving in ways that contravene the sacred compact, then calamities, both natural and human-made, shall follow. This is the explanation for the widespread conviction among extremists today for America's decline in the world. In short, the federal government has established agencies and laws contrary to America's divine compact: these include the Internal Revenue Service; the Federal Reserve System; the Bureau of Alcohol, Tobacco and Firearms; the Forest Service; the Bureau of Land Management; Social Security; Medicare and Medicaid; the Environmental Protection Agency; Housing and Urban Development; and the official apparatus enforcing civil rights for "so-called" minorities.

Essentially, American right-wing extremists view the entire executive branch of the United States government as little more than "jack-booted Nazi thugs," to borrow a phrase from the National Rifle Association fund-raising letter: a threat to freedom of religion, the right to carry weapons, freedom of speech, and the right to have one's property secure from illegal search and seizure.

Clumsy federal-agency assaults, first on the Weaver family in northern Idaho in 1992, then on the Branch Davidian sect in Waco, Texas, in 1993, followed by passage of the assault weapons ban in 1994, are viewed as indicators that the organic Constitution presently is imperiled. This has been the immediate impetus for the appearance throughout rural and Western America of armed militias since the summer of 1994. The terrorists who bombed a federal building in Oklahoma City in the spring of 1995, killing one hundred sixty-eight, were associated with militias headquartered in Michigan and Arizona. One month after the bombing, the national director of the United States Militia Association warned that after the current government falls, homosexuals, abortionists, rapists, "unfaithful politicians," and any criminal not rehabilitated in seven years will be executed. Tax evaders will no longer be treated as felons;

instead they will lose their library privileges (Sherwood 1995).

Millennialism

Leading to both the Waco and Weaver incidents was a belief on the victims' parts that world apocalypse is imminent. The Branch Davidians split from the Seventh-Day Adventists in 1935 but share with the mother church its own millenarian convictions. The Weavers received their apocalypticism from *The Late Great Planet Earth* by fundamentalist lay preacher Hal Lindsey (1970), a book that has enjoyed a wide reading on the Christian right.

Both the Davidians and the Weavers were imbued with the idea that the thousand-year-reign of Christ would be preceded by a final battle between the forces of light and darkness. To this end both had deployed elaborate arsenals to protect themselves from the anticipated invasion of "Babylonish troops." These, they feared, would be comprised of agents from the various federal bureaucracies mentioned above, together with UN troops stationed on America's borders awaiting orders from Trilateralists. Ever alert to "signs" of the impending invasion, both fired at federal officers who had come upon their property; and both ended up precipitating their own martyrdom. Far from quelling millenarian fervor, however, the two tragedies were immediately seized upon by extremists as further evidence of the approaching End Times.

Millenarianism is not unique to Christianity, nor to Western religions; furthermore, millenarianism culminating in violence is not new—in part because one psychological effect of end-time prophesying is a devaluation of worldly things, including property, honors, and human life. At the end of the first Christian millennium (A.D. 1000) as itinerant prophets were announcing the Second Coming, their followers were taking up arms to prepare the way, and uncounted numbers died (Cohn 1967). It should not surprise observers if, as the second millennium draws to a close and promises of Christ's imminent return increase in frequency, more and more armed cults flee to the mountains, there to prepare for the final conflagration.

Anti-Semitism

Many post-Holocaust Christian and Jewish scholars alike recognize that a pervasive anti-Judaism can be read from the pages of the New Testament, especially in focusing on the role attributed to Jews in Jesus' crucifixion. Rosemary Ruether, for example, argues that anti-Judaism constitutes the "left-hand of Christianity," its archetypal negation (Ruether 1979). Although pre-Christian Greece and Rome were also critical of Jews for alleged disloyalty, anti-Semitism reached unparalleled heights in Christian theology, sometimes relegating Jews to the status of Satan's spawn, the human embodiments of Evil itself.

During the Roman Catholic era, this association became embellished with frightening myths and images. Jews—pictured as feces-eating swine and rats—were accused of murdering Christian children on high feast days, using their blood to make unleavened bread, and poisoning wells. Added to these legends were charges during the capitalist era that Jews control international banking and by means of usury have brought simple, kind-hearted Christians into financial ruin (Hay 1981 [1950]). All of this was incorporated into popular Protestant culture through, among other vehicles, Martin Luther's diatribe, *On the Jews and Their Lies*, a pamphlet that still experiences brisk sales from patriotic bookstores. This is one possible reason for a survey finding by Charles Glock and Rodney Stark that created a minor scandal in the late 1960s. Rigidly orthodox American Christians, they found, displayed far higher levels of Jew-hatred than other Christians, regardless of their education, occupation, race, or income (Glock and Stark 1966).

In the last thirty years there has been "a sharp decline" in anti-Semitic prejudice in America, according to Glock (1993: 68). Mainline churches have played some role in this decline by facilitating Christian–Jewish dialogue, de-emphasizing offensive scriptural passages, and ending missions directed at Jews. Nevertheless, ancient anti-Jewish calumnies continue to be raised by leaders of the groups that are the focus of interest in this reading. Far from being a product of neurotic syndromes like the so-called Authoritarian (or fascist) Personality, the Jew-hatred of many right-wing extremists today is directly traceable to what they have absorbed from these preachments, sometimes as children.

Human Depravity ...

One of the fundamentals of Calvinist theology, appropriated into popular American Christianity, is this: a transcendent and sovereign God resides in the heavens, relative to whom the earth and its human inhabitants are utterly, hopelessly fallen. True, Calvin only developed a line of thought already anticipated in Genesis and amplified repeatedly over the centuries. However, with a lawyer's penetrating logic, Calvin brought this tradition to its most stark, pessimistic articulation. It is this belief that accompanied the Pilgrims in their venture across the Atlantic, eventually rooting itself in the American psyche.

From its beginnings, a particular version of the doctrine of human depravity has figured prominently in American right-wing extremist discourse. It has served as the basis of its perennial misogyny, shared by both men and women. The female, being supposedly less rational and more passive, is said to be closer to earth's evil. Too, the theology of world devaluation is the likely inspiration for the right-wing's gossipy preoccupation with the body's appetites and the "perilous eroticism of emotion," for its prudish fulminations against music, dance, drink, and dress, and for its homophobia. Here, too, is found legitimation for the right-wing's vitriol against Satanist ouiji boards, "Dungeons and Dragons," and New Age witchcrafters with their horoscopes and aroma-therapies, and most recently, against "pagan-earth-worshippers" and "tree hugging idolaters" (environmentalists). In standing tall to "Satan's Kids" and their cravenness, certain neo-Calvinists in Baptist, Presbyterian, and fundamentalist clothing accomplish their own purity and sanctification.

Conspiratorialism

According to Calvin, earthquakes, pestilence, famine, and plague should pose no challenge to faith in God. We petty, self-absorbed creatures have no right to question sovereign reason. But even in Calvin's time, and more frequently later, many Christians have persisted in asking: if God is truly all-powerful, all-knowing, and all-good, then how is evil possible? Why do innocents suffer? One perennial, quasi-theological response is conspiratorialism. In short, there are AIDS epidemics, murderous holocausts, rampant poverty, and floods because counter-poised to God there exists a second hidden force of nearly equal power and omniscience: the Devil and His human consorers—Jews, Jesuits, Hidden Hands, Insiders, Masons, and Bilderbergers.

By conspiratorialism, we are not referring to documented cases of people secretly scheming to destroy co-workers, steal elections, or run competitors out of business. Conspiracies are a common feature of group life. Instead, we mean the attempt to explain the entirety of human history by means of a cosmic Conspiracy, such as that promulgated in the infamous *Protocols of the Learned Elders of Zion*. This purports to account for all modern institutions by attributing them to the designs of twelve or thirteen—one representing each of the tribes of Israel—Jewish elders (Aho 1994, 68–82). *The Protocols* enjoys immense and endless popularity on the right; and has generated numerous spin-offs: *The International Jew, None Dare Call It Conspiracy*, and the *Mystery of [Jewish] Iniquity*, to name three.

To posit the existence of an evil divinity is heresy in orthodox Christianity. But, theological objections aside, it is difficult indeed for some believers to resist the temptation of intellectual certitude conspiratorialism affords. This certainty derives from the fact that conspiratorialism in the cosmic sense cannot be falsified. Every historical event can be, and often is, taken as further verification of conspiracies. If newspapers report a case of government corruption, this is evidence of government conspiracy; if they do not, this is evidence of news media complicity in the conspiracy. If the media deny involvement in a cover-up, this is still further proof of their guilt; if they admit to having sat on the story, this is surely an admission of what is already known.

PRACTICE

Christianity means more than adhering to a particular doctrine. To be Christian is to live righteously. God-fearing righteousness may either be understood as a *sign* of one's salvation, as in orthodox Christianity or, as in Mormonism, a way to *earn* eternal life in the celestial heavens.

Nor is it sufficient for the faithful merely to display righteousness in their personal lives and businesses, by being honest, hardworking, and reliable. Many Christians also are obligated to witness to, or labor toward, salvation in the political arena; to work with others to remake this charnel-house world after the will of God; to help establish God's kingdom on earth. Occasionally this means becoming involved in liberal causes—abolitionism, civil rights, the peace and ecological movements; often it has entailed supporting causes on the right. In either case it may require that one publicly stand up to evil. For, as Saint Paul said, to love God is to hate what is contrary to God.

Such a mentality may lead to "holy war," the organized effort to eliminate human fetishes of evil (Aho 1994: 23–34). For some, in cleansing the world of putrefaction their identity as Christian is recognized, it is re-known. This is not to argue that holy war is unique to Christianity, or that all Christians participate in holy wars. Most Christians are satisfied to renew their faith through the rites of Christmas, Easter, baptism, marriage, or mass. Furthermore, those who *do* speak of holy war often use it metaphorically to describe a private spiritual battle against temptation, as in "I am a soldier of Christ, therefore I am not permitted to fight" (Sandford 1966). Lastly, even holy war in the political sense does not necessarily imply the use of violence. Although they sometimes have danced tantalizingly close to extremism (in the sense defined earlier), neither Pat Robertson nor Jerry Falwell, for example, has advocated non-democratic means in their "wars" to avert America's decline.

Let us examine the notion of Christian holy war more closely. The sixteenth-century father of Protestant reform, Martin Luther, repudiated the concept of holy war, arguing that there exist two realms: holiness, which is the responsibility of the Church, and warfare, which falls under the State's authority (Luther 1974). Mixing these realms, he says, perverts the former while unnecessarily hamstringing the latter. This does not mean that Christians may forswear warfare, according to Luther. In his infinite wisdom, God has ordained princes to quell civil unrest and protect nations from invasion.... To this day, Lutherans generally are less responsive to calls for holy wars than many other Christians.

John Calvin, on the other hand, rejected Luther's proposal to separate church from State. Instead, his goal was to establish a Christocracy in Geneva along Roman Catholic lines, and to attain this goal through force, if need be, as Catholicism had done. Calvin says that not only is violence to establish God's rule on earth permitted, it is commanded. "Good brother, we must bend unto all means that give furtherance to the holy cause" (Walzer 1965: 17, 38, 68–87, 90–91, 100–109; see Troeltsch 1960: 599–601, 651–652, 921–922 n. 399).... And it was the Calvinist ethic, not that of Luther, that was imported to America by the Puritans, informing the politics of Presbyterians and Congregationalists—the immediate heirs of Calvinism—as well as some Methodists and many Baptists. Hence, it is not surprising that those raised in these denominations are often overrepresented in samples of "saints" on armed crusades to save the world for Christ.

Seminal to the so-called pedagogic or educational function of holy war are two requirements. First, the enemy against whom the saint fights must be portrayed in terms appropriate to his status as a fetish of evil. Second, the campaign against him must be equal to his diabolism. It must be terrifying, bloodthirsty, uncompromising.

"Prepare War!" was issued by the now defunct Covenant, Sword and the Arm of the Lord, a fundamentalist Christian paramilitary commune headquartered in Missouri. A raid on the compound in the late 1980s uncovered one of the largest private arms caches ever in American history.... When the Lord God has delivered these enemies into our hands, warns the pamphlet quoting the Old Testament, "thou shalt save alive nothing that breatheth: but thou shalt utterly destroy them" (CSA n.d.: 20; see Deuteronomy 20: 10–18).

The 1990s saw a series of state-level initiatives seeking to deny homosexuals civil rights. Although most of these failed by narrow margins, one in Colorado was passed (later to be adjudged unconstitutional), due largely to the efforts of a consortium of fundamentalist Christian churches.... Acknowledging that the title of their pamphlet "Death Penalty for Homosexuals" would bring upon them the wrath of liberals, its authors insist that "such slanderous tactics" will not deter the anti-homosexual campaign. "For truth will ultimately prevail, no matter how many truth-bearers are stoned." And what precisely is this truth? It is that the Lord Himself has declared that "if a man also lie with mankind, as he lieth with a woman, both of them have committed an abomination: they shall surely be put to death; their blood shall be upon them" (Peters 1992: i; see Leviticus 20:13)....

What should Christians do in the face of this looming specter, asks the pamphlet? "We, today, can and should have God's Law concerning Homosexuality and its judgment of the death penalty." For "they which commit such things," says the apostle Paul, "are worthy of death" (CSA n.d.: 15; see Romans 1:27–32). Extremism fans the flames of extremism.

ORGANIZATION

Contrary to popular thinking, people rarely join right-wing groups because they have a prior belief in doctrines such as those enumerated above. Rather, they come to believe because they have first joined. That is, people first affiliate with right-wing activists and only then begin altering their intellectual outlooks to sustain and strengthen these ties. The original ties may develop from their jobs, among neighbors, among prison acquaintances, or through romantic relationships....

The point of this story is the sociological truth that the way in which some people become right-wing extremists is indistinguishable from the way others become vegetarians, peace activists, or members of mainline churches (Lofland and Stark 1965; Aho 1991: 185–211). *Their affiliations are mediated by significant others already in the movement.* It is from these others that they first learn of the cause; sometimes it is through the loaning of a pamphlet or videotape; occasionally it takes the form of an invitation to a meeting or workshop. As the relationship with the other tightens, the recruit's viewpoint begins to change. At this stage old friends, family members, and cohorts, observing the recruit spending inordinate time with "those new people," begin their interrogations: "What's up with you, man?" In answer, the new recruit typically voices shocking things: bizarre theologies, conspiracy theories, manichaeistic worldviews. Either because of conscious "disowning" or unconscious avoidance, the recruit finds the old ties loosening, and as they unbind, the "stupidity" and "backwardness" of prior acquaintances become increasingly evident.

Pushed away from old relationships and simultaneously pulled into the waiting arms of new friends, lovers, and comrades, the recruit is absorbed into the movement. Announcements of full conversion to extremism follow. To display commitment to the cause, further steps may be deemed necessary: pulling one's children out of public schools where "secular humanism" is taught; working for radical political candidates to stop America's "moral decline"; refusing to support ZOG with taxes; renouncing one's citizenship and throwing away Social Security card and driver's license; moving to a rugged wilderness to await the End Times. Occasionally it means donning camouflage, taking up high-powered weaponry, and confronting the "forces of Satan" themselves.

There are two implications to this sociology of recruitment. First and most obviously, involvement in social networks is crucial to being mobilized into right-wing activism. Hence, contrary to the claims of the estrangement theory of extremism mentioned above, those who are truly isolated from their local communities are the last and least likely to become extremists themselves. My research (Aho 1991, 1994) suggests that among the most important of these community ties

is membership in independent fundamentalist, Baptist, or Presbyterian congregations.

Secondly, being situated in particular networks is largely a matter of chance. None of us choose our parents. Few choose their co-workers, fellow congregants, or neighbors, and even friendships and marriages are restricted to those available to us by the happenstance of our geography and times. What this means is that almost any person could find themselves in a Christian patriot communications network that would position them for recruitment into right-wing extremism.

As we have already pointed out, American right-wing extremists are neither educationally nor psychologically different from the general population. Nor are they any more status insecure than other Americans. What makes them different is how they are socially positioned. This positioning includes their religious affiliation. Some people find themselves in churches that expose them to the right-wing world. This increases the likelihood of their becoming right-wingers.

CONCLUSION

Throughout American history, a particular style of Christianity has nurtured right-wing extremism. Espousing doctrines like human depravity, white America as God's elect people, conspiratorialism, Jews as Christ killers, covenant theology and the right to revolt, and millennialism, this brand of Christianity is partly rooted in orthodox Calvinism and in the theologically questionable fantasies of popular imagination. Whatever its source, repeatedly during the last two centuries, its doctrines have served to prepare believers cognitively to assume hostile attitudes toward "un-Christian"—hence un-American—individuals, groups, and institutional practices.

This style of Christianity has also given impetus to hatred and violence through its advocacy of armed crusades against evil. Most of all, however, the cults, sects, and denominations wherein this style flourishes have served as mobilization centers for recruitment into right-wing causes. From the time of America's inception, right-wing political leaders in search of supporters have successfully enlisted clergymen who preach these principles to bring their congregations into the fold in "wars" to save America for Christ.

It is a mistake to think that modern Americans are more bigoted and racist than their ancestors were. Every American generation has experienced right-wing extremism, even that occasionally erupting into vigilante violence of the sort witnessed daily on the news today. What is different in our time is the sophistication and availability of communications and weapons technology. Today, mobilizations to right-wing causes has been infinitely enhanced by the availability of personal computer systems capable of storing and retrieving information on millions of potential recruits. Mobilization has also been facilitated by cheap shortwave radio and cable-television access, the telephone tree, desktop publishing, and readily available studio-quality recorders. Small

coteries of extremists can now activate supporters across immense distances at the touch of a button. Add to this the modern instrumentality for maiming and killing available to the average American citizen: military-style assault weaponry easily convertible into fully automatic machine guns, powerful explosives manufactured from substances like diesel oil and fertilizer, harmless in themselves, hence purchasable over-the-counter. Anti-tank and aircraft weapons, together with assault vehicles, have also been uncovered recently in private arms caches in the Western states.

Because of these technological changes, religious and political leaders today have a greater responsibility to speak and write with care regarding those with whom they disagree. Specifically, they must control the temptation to demonize their opponents, lest, in their declarations of war they bring unforeseen destruction not only on their enemies, but on themselves.

REFERENCES

Aho, J. 1991. *The Politics of Righteousness: Idaho Christian Patriotism.* Seattle: University of Washington Press.

———. 1994. *This Thing of Darkness: A Sociology of the Enemy.* Seattle: University of Washington Press.

Barkun, M. 1994. *Religion and the Racist Right: The Origins of the Christian Identity Movement.* Chapel Hill: North Carolina University Press.

Cohn, N. 1967. *The Pursuit of the Millennium.* New York: Oxford University Press.

Dilling, E. 1952. *The Plot Against Christianity,* n.p.

Glock, C. 1993. "The Churches and Social Change in Twentieth–Century America." *Annals of the American Academy of Political and Social Science.* 527: 67–83.

Glock, C., and R. Stark. 1966. *Christian Beliefs and Anti-Semitism.* New York: Harper & Row.

Hay, M. 1981 (1950). *The Roots of Christian Anti-Semitism.* New York: Anti-Defamation League of B'nai B'rith.

Lindsey, H. 1970. *The Late Great Planet Earth.* Grand Rapids: Zondervan.

Lipset, S. M., and E. Raab. 1970. *The Politics of Unreason: Right-Wing Extremism in America, 1790–1970.* New York: Harper & Row.

Lofland, J., and R. Stark. 1965. "Becoming a World-Saver: A Theory of Conversation to a Deviant Perspective." *American Sociological Review.* 30: 862–875.

Luther, M. 1974. *Luther: Selected Political Writings,* J. M. Porter, ed. Philadelphia: Fortress Press.

Mannheim, K. 1952. "The Problem of Generations," in *Essays in the Sociology of Knowledge.* London: Routledge and Kegan Paul.

Nisbet, R. 1953. *The Quest for Community.* New York: Harper and Brothers.

Pace, J. O. 1985. *Amendment to the Constitution.* Los Angeles: Johnson, Pace, Simmons and Fennel.

Peters, P. 1992. *Death Penalty for Homosexuals.* LaPorte, Colorado: Scriptures for America.

Preamble. 1986. "Preamble to the United States Constitution: Who Are the Posterity?" Oregon City, Oregon: Republic vs. Democracy Redress.

Robison, J. 1967 (1798). *Proofs of a Conspiracy....* Los Angeles: Western Islands.

Ruether, R. 1979. *Faith and Fratricide: The Theological Roots of Anti-Semitism.* New York: Seabury.

Sandford, F. W. 1966. *The Art of War for the Christian Soldier.* Amherst, New Hampshire: Kingdom Press.

Schlesinger, A. 1986. *The Cycles of American History.* Boston: Houghton Mifflin.

Sherwood, "Commander" S. 1995. Quoted in *Idaho State Journal.* May 21.

Shupe, A., and W. Stacey. 1983. "The Moral Majority Constituency," in *The New Christian Right,* R. Liebman and R. Wuthnow, eds. New York: Aldine.

Simpson, J. 1983. "Moral Issues and Status Politics," in *The New Christian Right,* R. Liebman and R. Wuthnow, eds. New York: Aldine.

Solt, L. 1971. *Saints in Arms: Puritanism and Democracy in Cromwell's Army.* New York: AMS Press.

Stark, R., and William Bainbridge. 1985. *The Future of Religion: Secularization, Revival and Cult Formation.* Berkeley: University of California Press.

Stouffer, S. A. 1966. *Communism, Conformity and Civil Liberties.* New York: John Wiley.

Troeltsch, E. 1960. *Social Teachings of the Christian Churches.* Trans. by O. Wyon. New York: Harper & Row.

Watzer, M. 1965. *The Revolution of the Saints.* Cambridge, MA: Harvard University Press.

Wilcox, C. 1992. *God's Warriors: The Christian Right in Twentieth Century America.* Baltimore, MD: Johns Hopkins University Press.

30

Motivation and Ideology

What Drives the Anti-Abortion Movement

DALLAS A. BLANCHARD

Blanchard evaluates the anti-abortion movement, including who joins it and why, and which belief systems drive this and other fundamentalist causes. The anti-abortion movement involves many belief systems and activities—and is even a way of life for some participants.

As you read, ask yourself the following questions:

1. *Who are those most likely to join the anti-abortion movement, and why?*
2. *Which religious and social factors stimulate people to join movements?*
3. *What do fundamentalists from different religions have in common?*
4. *Have you ever joined a social movement? What motivated you to join?*

GLOSSARY **Social movement** Organized activities to encourage or oppose some aspect of change (in this case, oppose abortion). **Fundamentalism** Adherence to traditional religious or cultural norms such as respect for authority.

WHY AND HOW PEOPLE JOIN THE ANTI-ABORTION MOVEMENT

Researchers have posited a variety of explanations for what motivates people to join the anti-abortion movement. As with any other social movement, the anti-abortion movement has within it various subgroups, or organizations, each of which attracts different kinds of participants and expects different levels of participation. It might in fact be more appropriate to speak of anti-abortion *movements*.

Those opposing abortion are not unified. Some organizations have a single-issue orientation, opposing abortion alone, while others take what they consider to be a "pro-life" stance on many issues, opposing abortion as well as euthanasia, capital punishment, and the use of nuclear and chemical arms....

Source: From *The Anti-Abortion Movement and the Rise of the Religious Right: From Polite to Fiery Protest* by Dallas A. Blanchard. New York: Twayne, 1994. Reprinted by permission of the Gale Group.

Researchers have identified a number of pathways for joining the anti-abortion movement. Luker, in her 1984 study of the early California movement, found that activists in the initial stages of the movement found their way to it through professional associations. The earliest opponents of abortion liberalization were primarily physicians and attorneys who disagreed with their professional associations' endorsement of abortion reform. It is my hypothesis that membership in organizations that concentrate on the education of the public or religious constituencies and on political lobbying is orchestrated primarily through professional networks. With the passage of the California reform bill and the increase in abortion rates several years later, many recruits to the movement fell into the category Luker refers to as "self-selected"; that is, they were not recruited through existing networks but sought out or sometimes formed organizations through which to express their opposition.

Himmelstein (1984), in summarizing the research on the anti-abortion movement available in the 10 years following *Roe v. Wade*, concluded that religious networks were the primary source of recruitment. Religious networks appear to be more crucial in the recruitment of persons into high-profile and/or violence prone groups (Blanchard and Prewitt 1993)—of which Operation Rescue is an example—than into the earlier, milder activist groups (although such networks are generally important throughout the movement). Such networks were also important, apparently, in recruitment into local Right to Life Committees, sponsored by the National Right to Life Committee and the Catholic church. The National Right to Life Committee, for example, is 72 percent Roman Catholic (Granberg 1981). It appears that the earliest anti-abortion organizations were essentially Catholic and dependent on church networks for their members; the recruitment of Protestants later on has also been dependent on religious networks (Cuneo 1989, Maxwell 1992).

Other avenues for participation in the antiabortion movement opened up through association with other issues. Feminists for Life, for example, was founded by women involved in the feminist movement. Sojourners, a socially conscious evangelical group concerned with issues such as poverty and racism, has an anti-abortion position. Some anti-nuclear and anti-death-penalty groups have also been the basis for the organization of anti abortion efforts.

Clearly, preexisting networks and organizational memberships are crucial in initial enlistment into the movement. Hall (1993) maintains that individual mobilization into a social movement requires the conditions of attitudinal, network, and biographical availability. My conclusions regarding the anti-abortion movement support this contention. Indeed, biographical availability—the interaction of social class, occupation, familial status, sex, and age—is particularly related to the type of organization with which and the level of activism at which an individual will engage.

General social movement theory places the motivation to join the anti-abortion movement into four basic categories: status defense; anti-feminism; moral commitment; and cultural fundamentalism, or defense.

The earliest explanation for the movement was that participants were members of the working class attempting to shore up, or defend, their declining social status. Clarke, in his 1987 study of English anti-abortionists finds this explanation to be inadequate, as do Wood and Hughes in their 1984 investigation of an anti-pornography movement group.

Petchesky (1984) concludes that the movement is basically anti-feminist—against the changing status of women. From this position, the primary goal of the movement is to "keep women in their place" and, in particular, to make them suffer for sexual "libertinism." Statements by some anti-abortion activists support this theory. Cuneo (1989), for example, finds what he calls "sexual puritans" on the fringe of the anti-abortion movement in Toronto. Abortion opponent and longtime right-wing activist Phyllis Schlaffley states this position: "It's very healthy for a young girl to be deterred from promiscuity by fear of contracting a painful, incurable disease, or cervical cancer, or sterility, or the likelihood of giving birth to a dead, blind, or brain-damaged baby (even ten years later when she may be happily married)" (Planned Parenthood pamphlet, no title, n.d. [1990])…. A number of researchers have concluded that sexual moralism is the strongest predictor of anti-abortion attitudes.

The theory of moral commitment proposes that movement participants are motivated by concern for the human status of the fetus. It is probably as close as any explanation comes to "pure altruism." Although there is a growing body of research on altruism, researchers on the abortion issue have tended to ignore this as a possible draw to the movement, while movement participants almost exclusively claim this position: that since the fetus is incapable of defending itself, they must act on its behalf.

In examining and categorizing the motivations of participants in the anti-abortion movement in Toronto, Cuneo (1989:85ff.) found only one category—civil rights—that might be considered altruistic. The people in this category tend to be nonreligious and embarrassed by the activities of religious activists; they feel that fetuses have a right to exist but cannot speak for themselves. Cuneo's other primary categories of motivation are characterized by concerns related to the "traditional" family, the status of women in the family, and religion. He also finds an activist fringe composed of what he calls religious seekers; sexual therapeutics, "plagued by guilt and fear of female sexual power" (115); and punitive puritans, who want to punish women for sexual transgressions. All of Cuneo's categories of participant, with the exception of the civil rights category, seek to maintain traditional male/female hierarchies and statuses.

The theory of cultural fundamentalism, or defense, proposes that the anti-abortion movement is largely an expression of the desire to return to what its proponents perceive to be "traditional culture." This theory incorporates elements of the status defense and anti-feminist theories.

It is important to note that a number of researchers at different points in time (Cuneo 1989; Ginsburg 1990; Luker 1984; Maxwell 1991, 1992) have indicated that (1) there have been changes over time in who gets recruited into the movement and why, (2) different motivations tend to bring different kinds of people into different types of activism, and (3) even particular movement organizations draw different kinds of people with quite different motivations. At this point in the history of the anti-abortion movement, the dominant motivation, particularly in the more activist organizations such as Operation Rescue, appears to be cultural fundamentalism. Closely informing cultural fundamentalism are the tenets of religious fundamentalism, usually associated with certain Protestant denominations but also evident in the Catholic and Mormon faiths….

RELIGIOUS AND CULTURAL FUNDAMENTALISM DEFINED

Cultural fundamentalism is in large part a protest against cultural change: against the rising status of women; against the greater acceptance of "deviant" life-styles such as homosexuality; against the loss of prayer and Bible reading in the schools; and against the increase in sexual openness and freedom. Wood and Hughes (1984) describe cultural fundamentalism as "adherence to traditional norms, respect for family and religious authority, asceticism and control of impulse. Above all, it is an unflinching and thoroughgoing moralistic outlook on the world; moralism provides a common orientation and common discourse for concerns with the use of alcohol and pornography, the rights of homosexuals, 'pro-family' and 'decency' issues." The theologies of Protestants and Catholics active in the anti-abortion movement—many of whom could also be termed fundamentalists—reflect these concerns….

There are at least six basic commonalities to what can be called Protestant, Catholic, and Mormon fundamentalisms: (1) an attitude of certitude—that one may know the final truth, which includes antagonism to ambiguity; (2) an external source for that certitude—the Bible or church dogma; (3) a belief system that is at root dualistic; (4) an ethic based on the "traditional" family; (5) a justification for violence; and, therefore, (6) a rejection of modernism (secularization)….

Taking those six commonalities point by point:

1. The certitude of fundamentalism rests on dependence on an external authority. That attitude correlates with authoritarianism, which includes obedience to an external authority, and, on that basis, the willingness to assert authority over others.

2. While the Protestant fundamentalists accept their particular interpretation of the King James Version of the Bible as the authoritative source, Mormon and Catholic fundamentalists tend to view church dogma as authoritative.

3. The dualism of Catholics, Protestants, and Mormons includes those of body/soul, body/mind, physical/spiritual. More basically, they see a distinction between God and Satan, the forces of good and evil. In the fundamentalist worldview, Satan is limited and finite; he can be in only one place at one time. He has servants, however, demons who are constantly working his will, trying to deceive believers. A most important gift of the Spirit is the ability to distinguish between the activities of God and those of Satan and his demons.

4. The "traditional" family in the fundamentalist view of things has the father as head of the household, making the basic decisions, with the wife and children subject to his wishes. Obedience is stressed for both wives and children. Physical punishment is generally approved for use against both wives and children.

     This "traditional" family with the father as breadwinner and the mother as homemaker, together rearing a large family, is really not all that traditional. It arose on the family farm, prior to 1900, where large numbers of children were an economic asset. Even then, women were essential in the work of the farm.... In the urban environment, the "traditional" family structure was an option primarily for the middle and upper classes, and they limited their family size even prior to the development of efficient birth control methods. Throughout human history women have usually been breadwinners themselves, and the "traditional" family structure was not an option.

5. The justification for violence lies in the substitutionary theory of the atonement theology of both Protestants and Catholics. In this theory, the justice of God demands punishment for human sin. This God also supervises a literal hell, the images of which come more from Dante's *Inferno* than from the pages of the Bible. Fundamentalism, then, worships a violent God and offers a rationale for human violence (such as Old Testament demands for death when adultery, murder, and other sins are committed). The fundamentalist mindset espouses physical punishment of children, the death penalty, and the use of nuclear weapons; fundamentalists are more frequently wife abusers, committers of incest, and child abusers.

6. Modernism entails a general acceptance of ambiguity contingency, probability (versus certitude), and a unitary view of the universe—that is, the view that there is no separation between body and soul, physical and spiritual, body and mind (when the body dies, the self is thought to die with it). Rejection of modernism and postmodernism is inherent in the rejection of a unitary worldview in favor of a dualistic worldview. The classic fundamentalist position embraces a return to religion as the central social institution, with education, the family, economics, and politics serving religious ends, fashioned after the social structure characteristic of medieval times.

Also characteristic of Catholic, Protestant, and Mormon fundamentalists are beliefs in individualism (which supports a naive capitalism); pietism; a chauvinistic Americanism (among some fundamentalists) that sees the United States as the New Israel and its inhabitants as God's new chosen people; and a general opposition to intellectualism, modern science, the tenets of the Social Gospel, and communism. (Some liberal Christians may share some of these views.) Amid this complex of beliefs and alongside the opposition to evolution, interestingly, is an underlying espousal of social Darwinism, the "survival of the fittest" ethos that presumes American society to be truly civilized, the pinnacle of social progress. This nineteenth-century American neo-colonialism dominates the contemporary political views held by the religious right. It is also inherent in their belief in individualism and opposition to social welfare programs.

Particular personality characteristics also correlate with the fundamentalist syndrome: authoritarianism, self-righteousness, prejudice against minorities, moral absolutism (a refusal to compromise on perceived moral issues), and anti-analytical, anti-critical thinking. Many fundamentalists refuse to accept ambiguity as a given in moral decision making and tend to arrive at simplistic solutions to complex problems. For example, many hold that the solution to changes in the contemporary family can be answered by fathers' reasserting their primacy, by forcing their children and wives into blind obedience. Or, they say, premarital sex can be prevented by promoting abstinence. One popular spokesman, Tim LaHaye, asserts that the antidote to sexual desire, especially on the part of teenagers, lies in censoring reading materials (LaHaye 1980). Strict parental discipline automatically engenders self-discipline in children, he asserts. The implication is that enforced other-directedness by parents produces inner-directed children, while the evidence indicates that they are more likely to exchange parental authoritarianism for that of another parental figure. To develop inner-direction under such circumstances requires, as a first step, rebellion against the rejection of parental authority—the opposite of parental intent.

One aspect of fundamentalism, particularly the Protestant variety, is its insistence on the subservient role of women. The wife is expected to be subject to the direction of her husband, children to their father. While Luker (1984) found that anti-abortionists in California supported this position and that proponents of choice generally favored equal status for women, recent research has shown that reasons for involvement in the anti-abortion movement vary by denomination. That is, some Catholics tend to be involved in the movement

more from a "right to life" position, while Protestants and other Catholics are more concerned with sexual morality. The broader right to life position is consistent with the official Catholic position against the death penalty and nuclear arms, while Protestant fundamentalists generally support the death penalty and a strong military. Thus, Protestant fundamentalists, and some Catholic activists, appear to be more concerned with premarital sexual behavior than with the life of the fetus.

Protestants and Catholics (especially traditional, ethnic Catholics) however, are both concerned with the "proper," or subordinate, role of women and the dominant role of men. Wives should obey their husbands, and unmarried women should refrain from sexual intercourse. Abortion, for the Protestants in particular, is an indication of sexual licentiousness (see, for example, LaHaye 1980). Therefore, the total abolition of abortion would be a strong deterrent to such behavior helping to reestablish traditional morality in women. Contemporary, more liberal views of sexual morality cast the virgin female as deviant. The male virgin has long been regarded as deviant. The fundamentalist ethic appears to accept this traditional double standard with its relative silence on male virginity.

Another aspect of this gender role ethic lies in the home-related roles of females. Women are expected to remain at home, to bear children, and to care for them, while also serving the needs of their husbands. Again, this is also related to social class and the social role expectations of the lower and working classes, who tend to expect women to "stay in their place."

Luker's (1984) research reveals that some women in the anti-abortion movement are motivated by concern for maintaining their ability to rely on men (husbands) to support their social roles as mothers, while pro-choice women tend to want to maintain their independent status. Some of the men involved in anti-abortion violence are clearly acting out of a desire to maintain the dependent status of women and the dominant roles of men. Some of those violent males reveal an inability to establish "normal" relationships with women, which indicates that their violence may arise from a basic insecurity with the performance of normal male roles in relationships with women. This does not mean that these men do not have relationships with women. Indeed, it is in the context of relationships with women that dominance-related tendencies become more manifest. It is likely that insecurity-driven behaviors are characteristic of violent males generally, but psychiatric data are not available to confirm this, even for the population in question.

THE COMPLEX OF FUNDAMENTALIST ISSUES

The values and beliefs inherent to religious and cultural fundamentalism are expressed in a number of issues other than abortion. Those issues bear some discussion here, particularly as they relate to the abortion question.

1. *Contraception.* Fundamentalists, Catholic, Protestant, and Mormon, generally oppose the use of contraceptives since they limit family size and the intentions of God in sexuality. They especially oppose sex education in the schools and the availability of contraceptives to minors without the approval of their parents. (See *Nightline,* 21 July 1989.) This is because control of women and sexuality are intertwined. If a girl has knowledge of birth control, she is potentially freed of the threat of pregnancy if she becomes sexually active. This frees her from parental control and discovery of illegitimate sexual intercourse.

2. *Prenatal testing, pregnancies from rape or incest, or those endangering a woman's life.* Since every pregnancy is divinely intended, opposition to prenatal testing arises from its use to abort severely defective fetuses and, in some cases, for sex selection. Abortion is wrong regardless of the origins of the pregnancy or the consequences of it.

3. *In vitro fertilization, artificial fertilization, surrogate motherhood.* These are opposed because they interfere with the "natural" fertilization process and because they may mean the destruction of some fertilized embryos.

4. *Homosexuality.* Homophobia is characteristic of fundamentalism, because homosexual behavior is viewed as being "unnatural" and is prohibited in the Bible.

5. *Uses of fetal tissue.* The use of fetal tissue in research and in the treatment of medical conditions such as Parkinson's disease is opposed, because it is thought to encourage abortion. (See *New York Times,* 16 August 1987; *Good Morning America,* 25 July 1991; *Face to Face with Connie Chung,* 25 November 1989; and *Nightline,* 6 January 1988.)

6. *Foreign relations issues.* Fundamentalists generally support aid to Israel and military funding (Diamond 1989). Indeed, as previously mentioned, they commonly view the United States as the New Israel. Protestant fundamentalists tend to be premillennialists, who maintain that biblical prophecies ordain that the reestablishment of the State of Israel will precede the Second Coming of Christ. Thus, they support aid to Israel to hasten the Second Coming, which actually, then, has an element of anti-Semitism to it, since Jews will not be among the saved.

7. *Euthanasia.* So-called right to life groups have frequently intervened in cases where relatives have sought to remove a patient from life-support systems. Most see a connection with abortion in that both abortion and removal of life-support interfere with God's decision as to when life should begin and end.

The most radical expression of cultural fundamentalism is that of Christian Reconstructionism.... The adherents of Christian Reconstructionism, while a distinct minority, have some congregations of up to 12,000 members and count among their number Methodists, Presbyterians, Lutherans, Baptists, Catholics, and former Jews. They are unabashed theocratists. They believe every area of life—law, politics, the arts, education, medicine, the media, business, and especially morality—should be governed in accordance with the tenets of Christian Reconstructionism. Some, such as Gary North, a prominent reconstructionist and son-in-law of Rousas John Rushdoony, considered the father of reconstructionism, would deny religious liberty—the freedom of religious expression—to "the enemies of God," whom the reconstructionists, of course, would identify.

The reconstructionists want to establish a "God-centered government," a Kingdom of God on Earth, instituting the Old Testament as the Law of the Land. The goal of reconstructionism is to reestablish biblical, erusalemic society. Their program is quite specific. Those criminals which the Old Testament condemned to death would be executed, including homosexuals, sodomites, rapists, adulterers, and "incorrigible" youths. Jails would become primarily holding tanks for those awaiting execution or assignment as servants indentured to those whom they wronged as one form of restitution. The media would be censored extensively to reflect the views of the church. Public education and welfare would be abolished (only those who work should eat), and taxes would be limited to the tithe, 10 percent of income, regardless of income level, most of it paid to the church. Property, Social Security, and inheritance taxes would be eliminated. Church elders would serve as judges in courts overseeing moral issues, while "civil" courts would handle other issues. The country would return to the gold standard. Debts, including, for example, 30-year mortgages, would be limited to six years. In short, Christian Reconstructionists see democracy as being opposed to Christianity, as placing the rule of man above the rule of God. They also believe that "true" Christianity has its earthly rewards. They see it as the road to economic prosperity, with God blessing the faithful.

REFERENCES

Editors' Note: The original chapter from which this selection was taken has extensive footnotes and references that could not be listed in their entirety here. For more explanation and documentation of sources, please see the source note on page 234.

Blanchard, Dallas A., and Terry J. Prewitt. 1993. *Religious Violence Abortion: The Gideon Project.* Gainesville: University Press of Florida.

Clarke, Alan. 1987. "Collective Action against Abortion Represents a Display of, and Concern for, Cultural Values, Rather than an Expression of Status Discontent." *British Journal of Sociology* 38:235–53.

Cuneo, Michael. 1989. *Catholics against the Church: Anti-Abortion Protest in Toronto, 1969–1985.* Toronto: University of Toronto Press.

Diamond, Sara. 1989. *Spiritual Warfare: The Politics of the Christian Right.* Boston: South End Press.

Ginsburg, Faye. 1990. *Contested Lives: The Abortion Debate in an American Community.* Berkeley: University of California Press.

Granberg, Donald. 1981. "The Abortion Activists." *Family Planning Perspectives* 18:158–61.

Hall, Charles. 1993. "Social Networks and Availability Factors: Mobilizing Adherents for Social Movement Participation" (Ph.D. dissertation, Purdue University).

Himmelstein, Jerome L. 1984. "The Social Basis of Anti-Feminism: Religious Networks and Culture." *Journal for the Scientific Study of Religion* 25:1–25.

Chapter 8

Economics and Politics

Implications for Survival

E̲ach society must have a system for the production and distribution of goods and services. The economic institution fulfills these functions. Like other social institutions, the economic institution intertwines with the other major social institutions but seemingly more so with the political institution. An examination of various political systems reveals this close relationship. When the strong central government of the ancient Roman Empire came to an end with the empire's fall, its monetary system and civic safeguards also came to an end. The people living under this empire in Europe were forced to return to farming to grow their own food instead of being able to purchase it and were forced to submit to marauding ex-soldiers for protection. Eventually, these ex-soldiers confiscated the land and set up a system of government with themselves in power and a strong individual from their ranks elected as a leader—their king. This economic–political system was called feudalism.

As the feudal European states grew and merged (willingly or not), their leaders became strong enough to journey overseas and impose their political will on less developed but desirable landholdings. Thus began the economic–political exploitation system of colonialism. The indigenous people of those areas were unwilling to work the large plantations and mines because they had their own land on which to live, grow food, and hunt. Thus it was necessary for the conquerors to develop a political system to come to the aid of the economic system: The best native lands were seized, the natives were forced to pay in cash rather than in goods for necessities, and many natives were arrested on trumped-up charges and sentenced to work on the plantations and mines.

After the middle of the nineteenth century, the developed world began a rapid transference from an agricultural society to an industrial one. In most industrialized nations, this development occurred under the economic system of capitalism—a system in which there is private (rather than state) ownership of wealth and control of the production and distribution of goods. An important aspect of this system is the global economy—a significant increase in the amount of international investment and trade. As technology has driven down the cost of international communication and travel, there has been an increase in the flow of capital across borders and growth in trade.

The development of capitalism led to the development of large corporations as individual companies merged to operate the developing vast industries. A side effect of these mergers was the development of corporations that were large enough to dominate various industries and to stifle the benefits of competition. The ideology accompanying a capitalistic economic system tends to be conservative regarding the role of government. Conservatives tend to emphasize a free-market, laissez-faire ideology that wants to restrict the role of government in regulating the economy because it is believed that this scheme would avoid the inefficiency that such regulation produces. The alternative belief is that government should be *more* involved in economic planning as a means of fostering greater economic competition and building a social safety net for corporations via subsidies and the twins accompanying employment—namely, unemployment and underemployment. At any one time, unemployment may seem low. Because it affects different people for different lengths of time, however, its overall effect is a lot higher than the official figures. Underemployment occurs when people work part-time when they desire full-time work, work at jobs that are below their skill level, or work full-time for poverty wages.

With increased development, social change became more rapid. With this evolution came changes in the economy, the nature of work, and the underlying workforce. Because a major goal of all corporate organizations is to increase profits, they work at both increasing profits and reducing costs. This has led to corporate downsizing and a lower number of workers, an increase in the use of temporary workers, and a resulting decline in income for many families. In the first reading in this chapter, Jeremy Rifkin notes that these corporate goals have produced a decline in the global labor force and the birth of a post-market era. He sees the age-old dream of abundance and leisure becoming even more distant at the dawn of this Information Age. In short, Rifkin sees a future in which work is no longer central to our lives.

If Rifkin's prediction is correct, then measures must be taken by both economic and political institutions to deal with the resulting system of fewer overall jobs and fewer jobs that pay a livable wage. Both outcomes would likely lower consumers' purchasing power—that is, their ability to buy corporate products.

These economic trends are seen in the changing nature of the distribution of U.S. income. From the end of World War II until around 1973, a strong economy meant a trend toward greater income equalization. Since that time, however, there has been a vast reversal of this trend. The most affluent fifth of the country has seen their share of the nation's wealth grow to almost half of this wealth (48%) while the bottom fifth receives only 3.5%. So rapid has been the growth of income for the most affluent that the top 5% earned almost 21% of all income and the top 1% earned almost 17% of this income. The result is that the disparities in income and wealth in the United States have grown faster and become sharper than in any other advanced industrial Western democracy. Actually, overall wealth is far more unequally distributed than income, given that wealthy people tend to have wealthy parents who leave them substantial estates. The result is that the wealthiest 20% of Americans own almost 75% of all the wealth in the United States. In fact, the top 1% of Americans own almost 40% of all U.S. wealth. Meanwhile, the bottom 20% tend to have zero net worth because their debts exceed their assets.

Seemingly, the only means for helping this situation for the poor at the bottom is to distribute income more equally, provide more and better jobs, improve welfare, and organize the poor to demand more. These events are unlikely to occur, however, because the people who accumulate more money, power, and prestige will be more likely to create ideologies and laws that maintain and increase their benefits. The more inequality in a social system, the less opportunity people at the bottom of the system will have to change the system.

Many people tend to believe that welfare supports lazy cheaters (see the Zuckerman article). Nevertheless, the poor perform many valuable functions. (see Gans 40)

As you read about the economic institution in this chapter, think of some current economic issues and consider how they might be addressed. What can be done about such current issues as private and public debts, unemployment, and wage stagnation? Some ideas include these:

- Decrease taxes on the poor
- Guarantee an increased minimum income
- Improve employment programs
- Improve housing subsidies
- Increase the earned income tax credit

- Provide adequate health care for all
- Raise the minimum wage

In considering these ideas, keep in mind that the people who accumulate money, power, and prestige will develop vested interests in maintaining and increasing their benefits.

Perhaps the first question that needs answering when examining the political institution is why is there a need for government. An examination of this institution reveals the following functions of government:

- Distributing power equitably so that each citizen has equal opportunities and is provided with a safety net—a minimum standard of living;
- Handling international relations and protecting the citizenry from foreign aggressors and terrorism
- Protecting civil liberties such as legal protection
- Taking responsibility for overall planning and direction of society and thereby dealing with needs unmet by other institutions
- Securing order by enforcing society's norms and being the final arbiter in disputes

A glance at these functions of government reveals that they are all not performed equably for all citizens—there are obvious inequalities in power among social groups, in economic control, in political voice, and in swaying public opinion. This is why in studying the political institution, sociologists concern themselves with the groups involved in the political process and the conditions that tend to generate political involvement or apathy. In turn, this leads to two related interests: the problems of a democracy and the influence on the political structure exerted by the economic structure.

Democracy has become an important area of inquiry since the creation of a number of new states following World War II. The process is continuing as more ethnic groups seek their independence from other dominating ethnic groups. Many of these new countries express the desire to become democracies but, as the many political clashes have shown, this desire is difficult to attain and to maintain. Thus many countries that claim to be democratic are, in reality, not what they seem.

This situation is perhaps best seen in the United States—a country that holds itself up as a model for democracy despite having some of the lowest rates of voter registration and voter participation among all Western industrial nations. Registration requirements tend to have a negative impact on lower-income groups and younger people, so more of these people are denied the vote—despite the fact that voter registration is actually unnecessary because most people are registered numerous times via property taxes, auto license taxes, and other official functions. Eleven states have shown that voter registration is unnecessary by allowing registration at the same time as voting, with far higher rates of participation and without fraud. Voting turnout for national elections is usually reported as 50%, but this represent only registered voters. If the denied non-registered voters were included, then the voter turnout would be only about 30%—a rate achieved only in national presidential elections. This figure is the lowest in the Western industrialized world and one of the lowest in the entire world of voting countries.

An apt question to ask at this point is why? The answer appears to lie in the day provided for voting. The United States is only one of three industrialized countries that votes on a workday. Most other countries vote on a Saturday, a Sunday, or both days. Our low voter turnout appears to reflect the fact that this country makes it more difficult to vote, despite our claims to be a model of democracy. Why do you think this condition has not been corrected? To answer this question, you must ask who benefits from the current electoral system.

Another aspect of the political voting system in America is the way in which campaigns are financed. With political campaigns now reduced to media-run endeavors, candidates need lots of funds to compete. This need has provided the economic corporate world an opening into political power via the development of political action committees (PACs). Thus about 1% of the wealthiest Americans contribute the bulk of political donations. Of course, the politicians claim that the contributions do not affect their voting behavior—it simply allows the major contributors easier access to their office to be heard. This claim, of course, is also saying that hearing about only one side of a proposal has no effect in decision making or future contributions from the major donors. What these political donations do show is that politicians have two types of constituents—those who receive bailouts, subsidies, and tax breaks and those who get cuts in their welfare benefits (see the article by Zuckerrman). It appears that the main point for the politician is getting reelected, which requires lots and lots of money in today's voting system. Incumbents have an 8-to-1 advantage in PAC contributions and so outspend their opponents by more than four fold. Incumbents are also allowed to roll over unspent political contributions, and their accumulated war chest can be and often is enough to scare away legitimate contenders. In almost all elections, 98% of the House of Representative incumbents and about 85% of the Senate incumbents are reelected. What do you think should be done about this situation? What *can* be done about this situation? Several of these inequalities of political voice are covered in the next article by Schlozman and her associates.

Although political terrorism is not a new phenomenon, the seemingly large increase in these acts has important implication for new and existing democracies. We may wonder why the believers are willing to kill so easily. In the reading by Meyer, we learn that many are willing to kill strangers when ordered to do so by voices of authority.

At this point in our reading about democracy, it seems time to return to its fundamentals by examining the conditions necessary for a state to achieve and maintain democracy. The article by Lipset reminds us that change is a normative feature of society but that a number of factors can delay, encourage, or prevent change: the media, those with large economic assets, those with political authority, and our own socialized beliefs. Thus we do not know whether the low voter turnout for elections is a result of artificial barriers to voting such as registration requirements and workday elections or whether it is due to our belief that it makes no difference who wins because both parties are more alike than different in representing the people. What do you think?

31

The End of Work

JEREMY RIFKIN

The author notes that we are at the dawn of the informational age and that it is already having a worldwide impact in the loss of employment.

As you read this selection, consider the following questions as guides:

1. *The author paints a grim picture of the future of employment. Do you agree with his assessment? Why?*
2. *Can you think of areas of employment that will escape huge layoffs?*
3. *Can you think of possible new areas for future employment?*
4. *What, if anything, is the government doing about current and future unemployment?*

From the beginning, civilization has been structured, in large part, around the concept of work. From the Paleolithic hunter/gatherer and Neolithic farmer to the medieval craftsman and assembly line worker of the current century, work has been an integral part of daily existence. Now, for the first time, human labor is being systematically eliminated from the production process. Within less than a century, "mass" work in the market sector is likely to be phased out in virtually all of the industrialized nations of the world. A new generation of sophisticated information and communication technologies is being hurried into a wide variety of work situations. Intelligent machines are replacing human beings in countless tasks, forcing millions of blue and white collar workers into unemployment lines, or worse still, breadlines.

Our corporate leaders and mainstream economists tell us that the rising unemployment figures represent short-term "adjustments" to powerful market-driven forces that are speeding the global economy into a Third Industrial Revolution. They hold out the promise of an exciting new world of high-tech automated production, booming global commerce, and unprecedented material abundance.

Millions of working people remain skeptical. Every week more employees learn they are being let go. In offices and factories around the world, people wait, in fear, hoping to be spared one more day. Like a deadly epidemic inexorably working its way through the marketplace, the strange, seemingly inexplicable new economic disease spreads, destroying lives and destabilizing whole communities in its wake. In the United States, corporations are eliminating more than 2 million jobs annually.[1] In Los Angeles, the First Interstate Bankcorp, the nation's thirteenth-largest bank holding company, recently restructured its operations, eliminating 9,000 jobs, more than 25 percent of its workforce. In Columbus, Indiana, Arvin

Source: From *The End of Work* by Jeremy Rifkin. New York: G. P. Putnam's Sons, 1995, pp. XV–XVIII, 3–19.

Industries streamlined its automotive components factory and gave out pink slips to nearly 10 percent of its employees. In Danbury, Connecticut, Union Carbide re-engineered its production, administration, and distribution systems to trim excess fat and save $575 million in costs by 1995. In the process, more than 13,900 workers, nearly 22 percent of its labor force, were cut from the company payroll. The company is expected to cut an additional 25 percent of its employees before it finishes "re-inventing" itself in the next two years.[2]

Hundreds of other companies have also announced layoffs. GTE recently cut 17,000 employees. NYNEX Corp said it was eliminating 16,800 workers. Pacific Telesis has riffed more than 10,000. "Most of the cuts," reports *The Wall Street Journal*, "are facilitated, one way or another, by new software programs, better computer networks and more powerful hardware" that allow companies to do more work with fewer workers.[3]

While some new jobs are being created in the U.S. economy, they are in the low-paying sectors and generally temporary employment. In April of 1994, two thirds of the new jobs created in the country were at the bottom of the wage pyramid. Meanwhile, the outplacement firm of Challenger, Gray and Christmas reported that in the first quarter of 1994, layoffs from big corporations were running 13 percent over 1993, with industry analysts predicting even steeper cuts in payrolls in the coming months and years.[4]

The loss of well-paying jobs is not unique to the American economy. In Germany, Siemens, the electronics and engineering giant, has flattened its corporate management structure, cut costs by 20 to 30 percent in just three years, and eliminated more than 16,000 employees around the world. In Sweden, the $7.9 billion Stockholm-based food cooperative, ICA, re-engineered its operations, installing a state-of-the-art computer inventory system. The new labor-saving technology allowed the food company to shut down a third of its warehouses and distribution centers, cutting its overall costs in half. In the process, ICA was able to eliminate more than 5,000 employees, or 30 percent of its wholesale workforce, in just three years, while revenues grew by more than 15 percent. In Japan, the telecommunications company NTT announced its intentions to cut 10,000 employees in 1993, and said that, as part of its restructuring program, staff would eventually be cut by 30,000—15 percent of its workforce.[5]

The ranks of the unemployed and underemployed are growing daily in North America, Europe, and Japan. Even developing nations are facing increasing technological unemployment as transnational companies build state-of-the-art high-tech production facilities all over the world, letting go millions of laborers who can no longer compete with the cost efficiency, quality control, and speed of delivery achieved by automated manufacturing. In more and more countries the news is filled with talk about lean production, re-engineering, total quality management, post-Fordism, decruiting, and downsizing. Everywhere men and women are worried about their future. The young are beginning to vent their frustration and rage in increasing antisocial behavior. Older workers, caught between a prosperous past and a bleak future, seem resigned, feeling increasingly trapped by social forces over which they have little or no control. Throughout the world there is a sense of momentous change taking place—change so vast in scale that we are barely able to fathom its ultimate impact. Life as we know it is being altered in fundamental ways.

SUBSTITUTING SOFTWARE FOR EMPLOYEES

While earlier industrial technologies replaced the physical power of human labor, substituting machines for body and brawn, the new computer-based technologies promise a replacement of the human mind itself, substituting thinking machines for human beings across the entire gamut of economic activity. The implications are profound and far-reaching. To begin with, more than 75 percent of the labor force in most industrial nations engage in work that is little more than simple repetitive tasks. Automated machinery, robots, and increasingly sophisticated computers can perform many if not most of these jobs. In the United States alone, that means that in the years ahead more than 90 million jobs in a labor force of 124 million are potentially vulnerable to replacement by machines. With current surveys showing that less than 5 percent of companies around the world have even begun to make the transition to the new machine culture, massive unemployment of a kind never before experienced seems all but inevitable in the coming decades.[6] Reflecting on the significance of the transition taking place, the distinguished Nobel laureate economist Wassily Leontief has warned that with the introduction of increasingly sophisticated computers, "the role of humans as the most important factor of production is bound to diminish in the same way that the role of horses in agricultural production was first diminished and then eliminated by the introduction of tractors."[7]

Caught in the throes of increasing global competition and rising costs of labor, multinational corporations seem determined to hasten the transition from human workers to machine surrogates. Their revolutionary ardor has been fanned, of late, by compelling bottom-line considerations. In Europe, where rising labor costs are blamed for a stagnating economy and a loss of competitiveness in world markets, companies are hurrying to replace their workforce with the new information and telecommunication technologies. In the United States, labor costs in the past eight years have more than tripled relative to the cost of capital equipment. (Although real wages have failed to keep up with inflation and in fact have been dropping, employment benefits, especially health care costs, have been rising sharply.) Anxious to cut costs and improve profit margins, companies have been substituting machines for human labor at an accelerating rate. Typical is Lincoln Electric, a manufacturer of industrial motors in Cleveland, which announced plans to increase its capital expenditures in 1993 by 30 percent over its 1992 level. Lincoln's assistant to the CEO, Richard Sobow, reflects the thinking of many others in the business community when he

says, "We tend to make a capital investment before hiring a new worker."[8]

Although corporations spent more than a trillion dollars in the 1980s on computers, robots, and other automated equipment, it has been only in the past few years that these massive expenditures have begun to pay off in terms of increased productivity, reduced labor costs, and greater profits. As long as management attempted to graft the new technologies onto traditional organizational structures and processes, the state-of-the-art computer and information tools were stymied, unable to perform effectively and to their full capacity. Recently, however, corporations have begun to restructure the workplace to make it compatible with the high-tech machine culture.

RE-ENGINEERING

"Re-engineering" is sweeping through the corporate community, making true believers out of even the most recalcitrant CEOs. Companies are quickly restructuring their organizations to make them computer friendly. In the process, they are eliminating layers of traditional management, compressing job categories, creating work teams, training employees in multilevel skills, shortening and simplifying production and distribution processes, and streamlining administration. The results have been impressive. In the United States, overall productivity jumped 2.8 percent in 1992, the largest rise in two decades.[9] The giant strides in productivity have meant wholesale reductions in the workforce. Michael Hammer, a former MIT professor and prime mover in the restructuring of the workplace, says that re-engineering typically results in the loss of more than 40 percent of the jobs in a company and can lead to as much as a 75 percent reduction in a given company's workforce. Middle management is particularly vulnerable to job loss from re-engineering. Hammer estimates that up to 80 percent of those engaged in middle-management tasks are susceptible to elimination.[10]

Across the entire U.S. economy, corporate re-engineering could eliminate between 1 million and 2.5 million jobs a year "for the foreseeable future," according to *The Wall Street Journal*.[11] By the time the first stage of re-engineering runs its course, some studies predict a loss of up to 25 million jobs in a private sector labor force that currently totals around 90 million workers. In Europe and Asia, where corporate restructuring and technology displacement is beginning to have an equally profound impact, industry analysts expect comparable job losses in the years ahead. Business consultants like John C. Skerritt worry about the economic and social consequences of re-engineering. "We can see many, many ways that jobs can be destroyed," says Skerritt, "but we can't see where they will be created." Others, like John Sculley, formerly of Apple Computer, believe that the "reorganization of work" could be as massive and destabilizing as the advent of the Industrial Revolution. "This may be the biggest social issue of the next 20 years," says Sculley.[12] Hans Olaf Henkel, the CEO of IBM Deutschland, warns, "There is a revolution under way."[13]

Nowhere is the effect of the computer revolution and re-engineering of the workplace more pronounced than in the manufacturing sector. One hundred and forty-seven years after Karl Marx urged the workers of the world to unite, Jacques Attali, a French minister and technology consultant to socialist president François Mitterrand, confidently proclaimed the end of the era of the working man and woman. "Machines are the new proletariat," proclaimed Attali. "The working class is being given its walking papers."[14]

The quickening pace of automation is fast moving the global economy to the day of the workerless factory. Between 1981 and 1991, more than 1.8 million manufacturing jobs disappeared in the United States.[15] In Germany, manufacturers have been shedding workers even faster, eliminating more than 500,000 jobs in a single twelve-month period between early 1992 and 1993.[16] The decline in manufacturing jobs is part of a long-term trend that has seen the increasing replacement of human beings by machines at the workplace. In the 1950s, 33 percent of all U.S. workers were employed in manufacturing. By the 1960s, the number of manufacturing jobs had

dropped to 30 percent, and by the 1980s to 20 percent. Today, less than 17 percent of the workforce is engaged in blue collar work. Management consultant Peter Drucker estimates that employment in manufacturing is going to continue dropping to less than 12 percent of the U.S. workforce in the next decade.[17]

For most of the 1980s it was fashionable to blame the loss of manufacturing jobs in the United States on foreign competition and cheap labor markets abroad. Recently, however, economists have begun to revise their views in light of new in-depth studies of the U.S. manufacturing sector. Noted economists Paul R. Krugman of MIT and Robert L. Lawrence of Harvard University suggest, on the basis of extensive data, that "the concern, widely voiced during the 1950s and 1960s, that industrial workers would lose their jobs because of automation, is closer to the truth than the current preoccupation with a presumed loss of manufacturing jobs because of foreign competition."[18]

Although the number of blue: collar workers continues to decline, manufacturing productivity is soaring. In the United States, annual productivity, which was growing at slightly over 1 percent per year in the early 1980s, has climbed to over 3 percent in the wake of the new advances in computer automation and the restructuring of the workplace. From 1979 to 1992, productivity increased by 35 percent in the manufacturing sector while the workforce shrank by 15 percent.[19]

William Winpisinger, past president of the International Association of Machinists, a union whose membership has shrunk nearly in half as a result of advances in automation, cites a study by the International Metalworkers Federation in Geneva forecasting that within thirty years, as little as 2 percent of the world's current labor force "will be needed to produce all the goods necessary for total demand."[20] Yoneji Masuda, a principal architect of the Japanese plan to become the first fully computerized information: based society, says that "in the near future, complete automation of entire plants will come into being, and during the next twenty to thirty years there will probably emerge... factories that require no manual labor at all."[21]

While the industrial worker is being phased out of the economic process, many economists and elected officials continue to hold out hope that the service sector and white: collar work will be able to absorb the millions of unemployed laborers in search of work. Their hopes are likely to be dashed. Automation and re-engineering are already replacing human labor across a wide swath of service related fields. The new "thinking machines" are capable of performing many of the mental tasks now performed by human beings, and at greater speeds. Andersen Consulting Company, one of the world's largest corporate restructuring firms, estimates that in just one service industry, commercial banking and thrift institutions, re-engineering will mean a loss of 30 to 40 percent of the jobs over the next seven years. That translates into nearly 700,000 jobs eliminated.[22]

Over the past ten years more than 3 million white: collar jobs were eliminated in the United States. Some of these losses, no doubt, were casualties of increased international competition. But as David Churbuck and Jeffrey Young observed in *Forbes,* "Technology helped in a big way to make them redundant." Even as the economy rebounded in 1992 with a respectable 2.6 percent growth rate, more than 500,000 additional clerical and technical jobs simply disappeared.[23] Rapid advances in computer technology, including parallel processing and artificial intelligence, are likely to make large numbers of white: collar workers redundant by the early decades of the next century.

Many policy analysts acknowledge that large businesses are shedding record numbers of workers but argue that small companies are taking up the slack by hiring on more people. David Birch, a research associate at MIT, was among the first to suggest that new economic growth in the high-tech era is being led by very small firms—companies with under 100 employees. At one point Birch opined that more than 88 percent of all the new job creation was taking place in small businesses, many of whom were on the cutting edge of the new technology revolution. His data was cited by conservative economists during the Reagan-Bush era as proof positive that new technology innovations were creating as many jobs as were being lost to technological

displacement. More recent studies, however, have exploded the myth that small businesses are powerful engines of job growth in the high-tech era. Political economist Bennett Harrison, of the H. J. Heinz III School of Public Policy and Management at Carnegie-Mellon University, using statistics garnered from a wide variety of sources, including the International Labor Organization of the United Nations and the U.S. Bureau of the Census, says that in the United States "the proportion of Americans working for small companies and for individual establishments...has barely changed at all since at least the early 1960s." The same holds true, according to Harrison, for both Japan and West Germany, the other two major economic superpowers.[24]

The fact is that while less than 1 percent of all U.S. companies employ 500 or more workers, these big firms still employed more than 41 percent of all the workers in the private sector at the end of the last decade. And it is these corporate giants that are re-engineering their operations and letting go a record number of employees.[25]

The current wave of job cuts takes on even greater political significance in light of the tendency among economists continually to revise upward the notion of what is an "acceptable" level of unemployment. As with so many other things in life, we often adjust our expectations for the future, on the basis of the shifting present circumstances we find ourselves in. In the case of jobs, economists have come to play a dangerous game of accommodation with steadily rising unemployment figures, sweeping under the rug the implications of a historical curve that is leading inexorably to a world with fewer and fewer workers.

A survey of the past half-century of economic activity discloses a disturbing trend. In the 1950s the average unemployment for the decade stood at 4.5 percent. In the 1960s unemployment rose to an average of 4.8 percent. In the 1970s it rose again to 6.2 percent, and in the 1980s it increased again, averaging 7.3 percent for the decade. In the first three years of the 1990s, unemployment has averaged 6.6 percent.[26]

As the percentage of unemployed workers edged ever higher over the postwar period, economists have changed their assumptions of what constitutes full employment. In the 1950s, 3 percent unemployment was widely regarded as full employment. By the 1960s, the Kennedy and Johnson administrations were touting 4 percent as a full employment goal. In the 1980s, many mainstream economists considered 5 or even 5.5 percent unemployment[27] as near full employment. Now, in the mid-1990s, a growing number of economists and business leaders are once again revising their ideas on what they regard as "natural levels" of unemployment. While they are reluctant to use the term "full employment," many Wall Street analysts argue that unemployment levels should not dip below 6 percent, lest the economy risk a new era of inflation.[28]

The steady upward climb in unemployment, in each decade, becomes even more troubling when we add the growing number of part-time workers who are in search of full-time employment and the number of discouraged workers who are no longer looking for a job. In 1993, more than 8.7 million people were unemployed, 6.1 million were working part-time but wanted full-time employment, and more than a million were so discouraged they stopped looking for a job altogether. In total, nearly 16 million American workers, or 13 percent of the labor force, were unemployed or underemployed in 1993.[29]

The point that needs to be emphasized is that, even allowing for short-term dips in the unemployment rate, the long-term trend is toward ever higher rates of unemployment. The introduction of more sophisticated technologies, with the accompanying gains in productivity, means that the global economy can produce more and more goods and services employing an ever smaller percentage of the available workforce.

A WORLD WITHOUT WORKERS...

When the first wave of automation hit the industrial sector in the late 1950s and early 1960s, labor leaders, civil rights activists, and a chorus of social critics were quick to sound the alarm. Their concerns, however, were little shared by business leaders at the time who continued to believe that increases in productivity brought about by the new automated technology would only enhance economic growth and promote increased employment and purchasing power. Today, however, a small but growing number of business executives are beginning to worry about where the new high-technology revolution is leading us. Percy Barnevik is the chief executive officer of Asea Brown Boveri, a 29-billion-dollar-a-year Swiss Swedish builder of electric generators and transportation systems, and one of the largest engineering firms in the world. Like other global companies, ABB has recently re-engineered its operations, cutting nearly 50,000 workers from the payroll, while increasing turnover 60 percent in the same time period. Barnevik asks, "Where will all these [unemployed] people go?" He predicts that the proportion of Europe's labor force employed in manufacturing and business services will decline from 35 percent today to 25 percent in ten years from now, with a further decline to 15 percent twenty years down the road. Barnevik is deeply pessimistic about Europe's future: "If anybody tells me, wait two or three years and there will be a hell of a demand for labor, I say, tell me where? What jobs? In what cities? Which companies? When I add it all together, I find a clear risk that the 10 percent unemployed or underemployed today could easily become 20 to 25 percent."[30]

Peter Drucker, whose many books and articles over the years have helped facilitate the new economic reality, says quite bluntly that "the disappearance of labor as a key factor of production" is going to emerge as the critical "unfinished business of capitalist society."[31]

For some, particularly the scientists, engineers, and employers, a world without work will signal the beginning of a new era in history in which human beings are liberated, at long last, from a life of back-breaking toil and mindless repetitive tasks. For others, the workerless society conjures up the notion of a grim future of mass unemployment and global destitution, punctuated by increasing social unrest and upheaval. On one point virtually all of the contending parties agree. We are, indeed, entering into a new period in history—one in which machines increasingly replace human beings in the process of making and moving goods and providing services. This realization led the editors of *Newsweek* to ponder the unthinkable in a recent issue dedicated to technological unemployment. "What if there were really no more jobs?" asked *Newsweek.*[32] The idea of a society not based on work is so utterly alien to any notion we have about how to organize large numbers of people into a social whole, that we are faced with the prospect of having to rethink the very basis of the social contract.

Most workers feel completely unprepared to cope with the enormity of the transition taking place. The rash of current technological breakthroughs and economic restructuring initiatives seem to have descended on us with little warning. Suddenly, all over the world, men and women are asking if there is a role for them in the new future unfolding across the global economy. Workers with years of education, skills, and experience face the very real prospect of being made redundant by the new forces of automation and information. What just a few short years ago was a rather esoteric debate among intellectuals and a small number of social writers around the role of technology in society is now the topic of heated conversation among millions of working people. They wonder if they will be the next to be replaced by the new thinking machines. In a 1994 survey conducted by *The New York Times,* two out of every five American workers expressed worry that they might be laid off, required to work reduced hours, or be forced to take pay cuts during the next two years. Seventy-seven percent of the respondents said they personally knew of someone who had lost his or her job in the last few years, while 67 percent said that joblessness was having a substantial effect on their communities.[33]

In Europe, fear over rising unemployment is leading to widespread social unrest and the emergence of neo-fascist political movements. Frightened, angry voters have expressed their frustration at the ballot box, boosting the electoral fortunes of extreme-right-wing parties in Germany, Italy, and Russia. In Japan, rising concern over unemployment is forcing the major political parties to address the jobs issue for the first time in decades.

We are being swept up into a powerful new technology revolution that offers the promise of a great social transformation, unlike any in history. The new high-technology revolution could mean fewer hours of work and greater benefits for millions. For the first time in modern history, large numbers of human beings could be liberated from long hours of labor in the formal marketplace, to be free to pursue leisure-time activities. The same technological forces could, however, as easily lead to growing unemployment and a global depression. Whether a utopian or dystopian future awaits us depends, to a great measure, on how the productivity gains of the Information Age are distributed. A fair and equitable distribution of the productivity gains would require a shortening of the workweek around the world and a concerted effort by central governments to provide alternative employment in the third sector—the social economy—for those whose labor is no longer required in the marketplace. If, however, the dramatic productivity gains of the high-tech revolution are not shared, but rather used primarily to enhance corporate profit, to the exclusive benefit of stockholders, top corporate managers, and the emerging elite of high-tech knowledge workers, chances are that the growing gap between the haves and the have-nots will lead to social and political upheaval on a global scale.

All around us today, we see the introduction of breathtaking new technologies capable of extraordinary feats. We have been led to believe that the marvels of modern technology would be our salvation. Millions placed their hopes for a better tomorrow on the liberating potential of the computer revolution. Yet the economic fortunes of most working people continue to deteriorate amid the embarrassment of technological riches. In every industrial country, people are beginning to ask why the age-old dream of abundance and leisure, so anticipated by generations of hardworking human beings, seems further away now, at the dawn of the Information Age, than at any time in the past half century. The answer lies in understanding a little-known but important economic concept that has long dominated the thinking of both business and government leaders around the world.

32

Welfare Reform in America
A Clash of Politics and Research

DIANA M. ZUCKERMAN

All institutions are interrelated. Zuckerman deals with the impact of politics on welfare reform, illustrating the close relationship of the political and economic institutions.

As you read this article, ask yourself the following questions:

1. *Many people see welfare as "something for nothing." Do you agree that welfare is necessary? Why?*
2. *If welfare is "something for nothing," would not subsidies to various industries also fit this description? Why?*
3. *How would you change the welfare system?*

The 1996 Personal Responsibility and Work Opportunity Reconciliation Act (PRWORA) radically changed welfare as we knew it, and data on its impact on the most vulnerable Americans are just becoming available....

The controversial passage of PRWORA was the culmination of many years of debate as well as concerns expressed across the political spectrum about the extent to which the welfare system should be considered a failure or an essential safety net. On the right, there were many years of anecdotes about "welfare queens" driving fancy cars, buying steaks with their food stamps, and teaching their children that welfare made working unnecessary.... On the left, there was an assumption that welfare saved innocent lives but also a growing concern about the deteriorating conditions in the inner city, where welfare dependence was sometimes a way of life being handed down from generation to generation, often accompanied by drug abuse, violence, teen pregnancy, and other social ills.

Research should have provided essential information to help determine how the welfare system should be changed, but instead, research was used as an ideological weapon to support conflicting points of view.... They could be used to prove both that most families on welfare were on it for short periods of time and that there was a hard-core group of families that stay on welfare for many years (Pavetti, 1996). Similarly, the statistics could be used to show that teen pregnancies were statistically significant predictors of long-term welfare and poverty (GAO, 1994, pp. 94–115) or that many welfare

Source: A longer version of this article was published in the *Journal of Social Issues*, Winter 2000, Vol. 56, No. 4.

Dr. Zuckerman is president of the National Center for Policy Research (CPR) for Women & Families and can be reached at dz@center4policy.org.

recipients were the victims of a crisis and just needed help for a few months before they were able to support themselves (Greenberg, 1993)....

This article describes how political and public pressures and compelling anecdotes overpowered the efforts of progressive public policy organizations and researchers, resulting in legislation focused on getting families off welfare rather than getting families out of poverty....

WELFARE REFORM AS A PRESIDENTIAL ISSUE

As a presidential candidate, Bill Clinton made it clear that he would not defend the welfare system but instead would work to change it.... As President, he quickly appointed experts to work on welfare reform, with particular focus on toughening child support laws and requiring welfare recipients to prepare for self-sufficiency (Koppelman, 1993). Although the public and policymakers were ready for major changes in the welfare system, many did not realize that a law passed in 1988, the Family Support Act, already had strengthened child support collection and ordered states to require able-bodied recipients to enter remedial education or job training projects. President Clinton had been instrumental in negotiating that legislation as an officer of the National Governors' Association, and the bill was expected to help single mothers and to encourage more women to move from welfare to work. However, the weakened economy of the late 1980s led to a surge in the welfare rolls instead, with the size of those rolls increasing 25% from 1989 to 1992. As a result of the recession, states were unable to provide matching funds to claim their full share of the federal funding for Job Opportunities and Basic Skills (JOBS), a training program. The law did not seem to be working, and the pressure to "do something about welfare" grew.

As a candidate, Clinton had promised to reform health care as well as welfare, and as President, he decided to focus on health care reform legislation first. This was a logical choice, because the lack of affordable health care was a major disincentive for single mothers who wanted to move from welfare to work. The Clinton administration strategy was to first pass a law that would make health care affordable for the working poor, so that it would be easier to make the other legislative changes necessary to reduce the welfare rolls. Improving access to health care would help those families that would lose Medicaid when they left welfare for low-level jobs that did not offer health insurance as a benefit.

While health care reform took center stage in the public eye, a welfare reform plan was being quietly developed by Clinton administration officials. As Assistant Secretary for Children and Families at HHS, Mary Jo Bane wanted to refocus the welfare system on work and to give states more flexibility regarding welfare policies (Bane, 1997). States were already submitting welfare reform proposals to the federal government to request waivers from the federal requirements....

Meanwhile, polls were showing tremendous public support for requiring all "able-bodied" welfare recipients, including mothers with young children, to get education or training for up to 2 years and then to work (Ellwood, 1996). Assistant HHS Secretaries Bane and David Ellwood wanted the federal law to set consistent national criteria and restrictions, in order to avoid a race to the bottom by the states. According to Ellwood, their proposal to reform welfare had four major goals:

1. *Make work pay.* Low-income workers need a living wage, health care, and child care to make working make sense instead of welfare.

2. *Two-year limits.* Transform the welfare system from a handout to a hand up, with clear requirements to work after 2 years of training or education.

3. *Child support enforcement.* Require absent parents to pay, whether or not they were married.

4. *Fight teen pregnancy.* The plan offered grants to high-risk schools that proposed innovative initiatives to lower rates of teen pregnancy and also supported a national clearinghouse and a few intensive demonstration projects.

The public seemed ready for these kinds of changes, but the Clinton administration was concerned about the cost. In order to move single mothers from welfare to work, it would be necessary to provide job training, child care, and other services for many of them. The cost of those programs and services would initially be high, and it was expected that the savings as families moved off welfare would not be great enough to make up for those extra costs. Since President Clinton had also promised to cut the country's enormous budget deficit, and since a major goal of welfare reform was to save money and make the federal government smaller, it did not seem politically feasible to expect taxpayers to pay more for welfare reform than they were paying for the existing welfare system, even in the short term. To save money, the Clinton plan included a slow phase-in of the program, which caused some conservative critics to accuse the administration of not being serious about reform (Ellwood, 1996).

CONGRESS AND WELFARE REFORM

For a variety of reasons, President Clinton had difficulty obtaining the support of his own party on welfare, health care, and other issues. As a moderate "New Democrat," his positions often seemed too conservative for the more liberal Democrats and too liberal for the more conservative ones, most of whom were southern and concerned about re-election as the Republican Party gained support in the South. Conservative Democrats developed their own welfare reform plan with a faster phase-in. In addition, conservatives from both parties were concerned that the Clinton plan did not sufficiently address out-of-wedlock childbearing and provided too many federal requirements instead of giving the states the autonomy to decide about welfare reform (Ellwood, 1996).

Why did the Clinton plan fail? Ellwood (1996) speculates that they should have tried to pass welfare reform before health care reform instead of

afterward.... As someone involved in health care reform legislation in the Senate at the time, I believe that the problem with the timing was not lack of attention but rather the weakened credibility of the Clinton administration because of the barrage of criticism aimed at its health care proposal. Ellwood also speculates that the President's promise to "end welfare as we know it" was a potent sound bite but did not address the concerns of many Democrats about whether the new system would be better than the old. Ellwood believes that the phrase "2 years and you're off" was even more destructive, because it implied no help at all after 2 years, which he says is "never what was intended." I agree that these were problems, and in addition, the tension and lack of trust between the congressional Democrats and the Clinton administration contributed to the view of many Democrats that the Clinton welfare plan was too controversial and would make them politically vulnerable....

The 1994 election, which resulted in the Republican takeover of the House and Senate, changed the political dynamics. The result was that every congressional committee was chaired by a Republican instead of a Democrat and composed primarily of Republicans rather than Democrats. Instead of being chaired by the most liberal Democrats, the committees that would vote on welfare reform were now chaired by conservative Republicans. Although the Republicans had only a small majority in the House, they had the control of committee chairmanships and disproportionate membership on the committees. This meant that the Republican leadership had tremendous power to control the welfare reform bill.

Even more important than the leadership of the congressional committees was the leadership of the House of Representatives, which changed from a liberal Democrat to Newt Gingrich, an outspoken critic of "big government," especially for social programs. Gingrich had been a major architect of the Contract with America, which specified that welfare should be available for only 2 years and that benefits should not be available to minor mothers or for children born to mothers on welfare. The shellacking that the Democrats experienced in the election was perceived as a mandate for the Contract with America (Merida, 1994), including a more punitive welfare reform plan than the one the Clinton administration had proposed.

Meanwhile, state governors demanded more say about how money in the new welfare program would be spent, resulting in a bill that looked more like a block grant than a social program. This had the political benefit of getting Congress off the hook: It would not have to make the difficult political decisions about restrictions and instead could put those decisions in the hands of the state governments.

The media were also influential and tended to focus on the shortcomings of the welfare program that was in place, Aid for Families with Dependent Children (AFDC)....

As the ideological battle continued around it, Congress was under tremendous pressure to pass a welfare reform bill....

Throughout 1995 and 1996, the Republican majority controlled the policy agenda and welfare reform was their cause. In that political climate, it would have been very difficult for progressive organizations to succeed in their public education and lobbying efforts, regardless of research results. It was especially difficult for liberal legislators to suggest that welfare mothers should be able to stay home and care for their children when nationwide 55% of mothers with children under the age of 3 were employed, most of them full time (Pavetti, 1997). To make matters worse, the progressive advocates did not have solid research findings to support their opposition to the Republican leadership's welfare reform bill, other than frightening but questionable statistics describing the number of children who would fall into poverty if their mothers were thrown off welfare. This left the most progressive members of Congress with little ammunition to use against the most conservative proposals. The slim Republican majority in Congress was joined by enough Democrats to create a substantial majority for the welfare reform bill that passed in 1996.

The bill that Congress eventually passed is lengthy and complicated, but the introductory section includes "findings" that explicitly show the ideology behind the legislation. There are 10 findings, and the first three set the tone:

1. Marriage is the foundation of a successful society.

2. Marriage is an essential institution of a successful society which promotes the interests of children.

3. Promotion of responsible fatherhood and motherhood is integral to successful child rearing and the well-being of children.
(U.S. Congress, 1996, p. 6)

The bill's focus on marriage and responsible parenthood reflects the Republican Party's ties to the Christian right as well as the growing disenchantment with government programs throughout the country in the early and mid-1990s. The rest of the findings, however, use research data to support this focus, and many of the statistics are ones that have been employed by experts across the ideological spectrum to support public policies aimed at reducing teen pregnancy, reducing poverty, and other goals that are as popular among liberal Democrats as conservative Republicans, as well as everything in between. For example, the next finding pointed out that "only 54 percent of single-parent families with children had a child support order established and, of that 54 percent, only about one-half received the full amount due" (U.S. Congress, 1996, p. 6).

Another major concern was the growth of welfare. The fifth finding points out that the number of individuals on welfare had more than tripled since 1965 and that more than two-thirds were children. The number of children on welfare every month increased from 3.3 million in 1965 to 9.3 million in 1992, although the number of children in the United States declined during those years. The legislation also points out that 89% of the children receiving AFDC were living in homes without fathers and that the percentage of unmarried women nearly tripled between 1970 and 1991.

The next finding includes details regarding pregnancies among unmarried teens and concludes "if the current trend continues, 50 percent of all births by the year 2015 will be out-of-wedlock" (U.S. Congress, 1996, p. 7). The findings recommended that strategies to combat teenage pregnancy "must address the issue of male responsibility, including statutory rape culpability and prevention" and points out that most teen mothers "have histories of sexual and physical abuse, primarily with older adult men."

Despite the clear ideological underpinnings of the bill, some of the findings could be embraced by both political parties. For example, the bill correctly points out that unmarried teenage mothers are more likely to go on welfare and to spend more years on welfare. It also points out that babies of unwed mothers are at risk for very low or moderately low birth weight, for growing up to have lower cognitive attainment and lower educational aspirations, and for child abuse and neglect. They are more likely to grow up to be teen parents and to go on welfare and less likely to have an intact marriage.

The problems of single parenting were also described, showing how they set in motion a cycle of welfare and poverty:

- Mothers under 20 years of age are at the greatest risk of bearing low-birth-weight babies.

- The younger the single-parent mother, the less likely she is to finish high school.

- Young women who have children before finishing high school are more likely to receive welfare assistance for a longer period of time.

- Children of teenage single parents have lower cognitive scores, lower educational aspirations, and a greater likelihood of becoming teenage parents themselves.

- Children of single-parent homes are three times more likely to fail and repeat a year in grade school and almost four times more likely to be expelled or suspended from school than are children from intact two-parent families.

- Of those youth held for criminal offenses within the State juvenile justice system, only 30% lived primarily in a home with both parents, compared to 74% of the general population of children.

The costs of teen parenting were delineated, with an estimate that "between 1985 and 1990, the public cost of births to teenage mothers under

the Aid to Families with Dependent Children program, the food stamp program, and the Medicaid program" (U.S. Congress, 1996, p. 8) was $120 billion. In a climate of deficit reduction and support for smaller government, that estimate was extremely compelling.

The welfare reform bill includes nine sections, referred to as "titles," and several are more generous than would have been expected given the Republican control of Congress and the Republicans' opposition to "big government." For example, Title VI on child care authorizes increased federal money so that child care is more available and affordable. President Clinton, however, expressed considerable concerns about Title IV, which banned most legal immigrants from most federal benefit programs, and Title VIII, which cut the food stamp program across the board and also restricted food stamps to unemployed adults without disabilities or dependents to 3 months out of every 36.

The welfare reform law had a great deal of support from governors around the country, because it gave them enormous flexibility regarding the spending of federal funds. The rationale was that states could experiment with new approaches, under the assumption that what works in a rural state, for example, might not work in a more urban environment. Heavy subsidies for day care or job training might be useful in some areas, for example, but unnecessary in others.

WHY NOT WAIT FOR DATA?

Although data were quoted in the welfare reform bill, there were no convincing data to predict what would actually happen if the bill passed. A "natural experiment" was taking place, however, that could have answered those questions. Between January 1987 and August 1996, 46 states had received approval for waivers to experiment with AFDC and welfare-to-work programs (GAO, 1997). Since welfare reform represented such a dramatic change in policies, with many lives at stake, it would have been logical to delay a federal welfare reform law until the data were analyzed from those programs. For example, by May 1997, the General Accounting Office, which is a research branch of Congress, had published a report entiled *Welfare Reform: States' Early Experiences with Benefits Termination*, based on a study conducted at the request of Senator Pat Moynihan. The study found that the benefits of 18,000 families were terminated under waivers through December 1996, most of them in Iowa, Massachusetts, and Wisconsin. More than 99% of these families failed to comply with program requirements; for example, some wanted to stay home with their children or were unwilling to do community service or work for low wages (GAO, 1997). These findings could have been used to design a welfare reform process that protected some of these families, but the bill was passed before the data were available.

Instead, the little research that was already completed was used to push welfare reform forward. For example, Vermont had applied for waivers to the welfare requirements, and its welfare restructuring project was the nation's first statewide demonstration of time-limited welfare (Zengerle, 1997). In 2 years, Vermont raised the number of welfare parents with jobs from 20% to 26% and increased their average monthly earnings from $373 to $437. Of course, it was also important to note that Vermont increased its social services budget by 50%, in part to create new jobs. Welfare recipients who were unable to find jobs in the private sector were eligible for 10 months' employment in public jobs or working for nonprofit organizations. It also would have been logical to note that Vermont has relatively few welfare recipients and that its experience is likely to be different from that of most other states. Nevertheless, this study was used to show the success of welfare reform....

THE REALITY OF WELFARE REFORM

Although the welfare reform bill that passed in 1996 was radical and potentially devastating to poor families, the booming economy during the next four years and political compromises resulted in a law that no longer seemed as extreme or partisan as it once did. For example, in the short term, almost every state received more federal funds under welfare reform that it did prior to reform (Nightingale & Brennan, 1998). Perhaps most important, the Clinton administration was able to influence the bill through regulations. The final regulations, announced in April 1999, contain exceptions to the rules that make the bill less rigid. For example, states may continue to provide welfare benefits for longer than 60 months to up to 20% of the welfare caseload based on hardship or domestic violence, and the 20% limit can be exceeded if there are federally recognized "good cause domestic violence waivers" (Schott, Lazere, Goldberg, & Sweeney, 1999). In addition, states will be penalized if they sanction a single parent caring for a child under age 6 if the parent can demonstrate his or her inability to obtain child care.

Research results are finally coming in from across the country. The Urban Institute (Loprest, 1999) reports that most adults leaving welfare between 1995 and 1997 got a new job or increased earnings. They also report that former welfare recipients work more hours than employed near-poor mothers who were not previously receiving welfare. Nevertheless, many families are doing poorly since welfare reform was implemented, and the number of children in extreme poverty (less than half the poverty level) has increased (Sherman, 1999). Between 1995 and 1997, the average income of the poorest 20% of female-headed households fell an average of $580 per family, primarily due to the loss of food stamps and other government benefits (Primus, Rawlings, Larin, & Porter, 1999).

Research had little impact on the passage of welfare reform in 1996, although statistics were used by both sides in the national debate. Is it possible that new research on the impact of welfare reform will influence welfare policies in the future? The studies in this issue provide very useful information about the barriers to success for welfare mothers who attempt to move into the workforce on a permanent basis, with important implications for how policies can maximize success and minimize tragedy for vulnerable families. Now that the costs of the welfare program have been drastically reduced and the national annual budget deficit has turned into a surplus, these studies may manage to attract the attention of policymakers. Unfortunately, welfare reform has made national policy changes in this area even more difficult, because decisions are now made in 50 different states, rather than by the federal government.

REFERENCES

Bane, M. J. (1997). Welfare as we might know it. *American Prospect*, 30, 47–53.

Elwood, D. T. (1996). Welfare reform as I knew it: When bad things happen to good policies. *American Prospect*, 26, 22–29.

General Accounting Office (GAO). (1994). *Teenage mothers least likely to become self-sufficient.* GAO/HEHS-94-115. Washington, DC: GAO.

General Accounting Office (GAO). (1997). *Welfare reform: States' early experience with benefit termination.* GAO/HEHS-97-74. Washington, DC: GAO.

Greenberg, M. (1993). *What state AFDC studies on length of stay tell us about welfare as a "way of life."* Washington, DC: Center for Law and Social Policy.

Koppelman, J. (1993). Helping AFDC children escape the cycle of poverty: Can welfare reform be used to achieve this goal? *National Health Policy Forum, 627*, 2–9.

Krauthammer, C. (1994, November 11). Republican mandate. *Washington Post*, p. A31.

Loprest, P. (1999). *How families that left welfare are doing: A national picture.* Washington, DC: Urban Institute.

Merida, K. (1994, December 28). Last rites for liberalism? Democrats' legacy now symbolizes their woes. *Washington Post*, p. A1.

Nightingale, D. S., & Brennan, K. (1998). *The welfare-to-work grants program: A new link in the welfare reform chain.* Washington, DC: Urban institute.

Pavetti, L. (1996, May 23). *Time on welfare and welfare dependency.* Testimony before the House of Representatives Committee on Ways and Means, 104th Congress.

Pavetti, L. (1997), *How much more can they work? Setting realistic expectations for welfare mothers.* Washington, DC: Urban Institute.

Primus, W., Rawlings, L., Larin, K., & Porter, K. (1999). *The initial impacts of welfare reform on the incomes of single-mother families.* Washington, DC: Center on Budget and Policy Priorities.

Schott, L., Lazere, E., Goldberg, H., & Sweeney, E. (1999). *Highlights of the final TANF regulations.* Washington, DC: Center on Budget and Policy Priorities.

Sherman, A. (1999). *Extreme child poverty rises sharply in 1997.* Washington, DC: Children's Defense Fund.

U.S. Congress. H.R. 3734: Personal Responsibility and Work Opportunity Reconciliation Act of 1996 (Enrolled Bill Sent to President) (1996) [On-line]. Available: http://thomas.loc.gov/cgi-bin/query/D?cl04:1:./temp/~c104aiBNdl:e19928

Zengerle, J. S. (1997). Welfare as Vermont knows it. *American Prospect*, 30, 54–55.

33

The Credit Card
Private Troubles and Public Issues

GEORGE RITZER

The credit card is seen as a personal part of the individual's economy. Ritzer reveals that anything this ubiquitous has many ramifications for the society in general.

As you read, ask yourself the following questions:

1. *According to the author, what are some good and bad things that can be said about credit cards?*
2. *How does your personal experience tie in with the author's claims about credit cards?*
3. *Which restrictions would you put on credit card companies? Why?*

GLOSSARY **Machinations** Crafty schemes.

The credit card has become an American icon. It is treasured, even worshipped, in the United States and, increasingly, throughout the rest of the world. The title of the book [from which this selection was taken], *Expressing America*, therefore has a double meaning: The credit card expresses something about the essence of modern American society and, like an express train, is speeding across the world's landscape delivering American (and more generally consumer) culture....

The credit card is not the first symbol of American culture to play such a role, nor will it be the last. Other important contemporary American icons include Coca-Cola, Levi's, Marlboro, Disney, and McDonald's. What they have in common is that, like credit cards, they are products at the very heart of American society, and they are highly valued by, and have had a profound effect on, many other societies throughout the world. However, the credit card is distinctive because it is a means that can be used to obtain those other icons, as well as virtually anything else available in the world's marketplaces. It is because of this greater versatility that the credit card may prove to be the most important American icon of all. If nothing else, it is likely to continue to exist long after other icons have become footnotes in the history of American culture. When the United States has an entirely new set of icons, the credit card will remain an important means for obtaining them....

Source: From George Ritzer, *Expressing America: A Critique of the Global Credit Card Society.* Copyright 1995 by Pine Forge Press. Reprinted by permission of Sage Publications, Inc.

THE ADVANTAGES OF CREDIT CARDS

…The most notable advantage of credit cards, at least at the societal level, is that they permit people to spend more than they have. Credit cards thereby allow the economy to function at a much higher (and faster) level than it might if it relied solely on cash and cash-based instruments.

Credit cards also have a number of specific advantages to consumers, especially in comparison to using cash for transactions:

- Credit cards increase our spending power, thereby allowing us to enjoy a more expansive, even luxurious lifestyle.

- Credit cards save us money by permitting us to take advantage of sales, something that might not be possible if we had to rely on cash on hand.

- Credit cards are convenient. They can be used 24 hours a day to charge expenditures by phone, mail, or home computer. Thus, we need no longer be inconvenienced by the fact that most shops and malls close overnight. Those whose mobility is limited or who are housebound can also still shop.

- Credit cards can be used virtually anywhere in the world, whereas cash (and certainly checks) cannot so easily cross national borders. For example, we are able to travel from Paris to Rome on the spur of the moment in the middle of the night without worrying about whether we have, or will be able to obtain on arrival, Italian lire. [Italy adopted the euro as its currency after this selection was written.]

- Credit cards smooth out consumption by allowing us to make purchases even when our incomes are low. If we happen to be laid off, we can continue to live the same lifestyle, at least for a time, with the anticipation that we will pay off our credit card balances when we are called back to work. We can make emergency purchases (of medicine, for example) even though we may have no cash on hand.

- Credit cards allow us to do a better job of organizing our finances, because we are provided each month with a clear accounting of expenditures and of money due.

- Credit cards may yield itemized invoices of tax-deductible expenses, giving us systematic records at tax time.

- Credit cards allow us to refuse to pay a disputed bill while the credit card company investigates the transaction. Credit card receipts also help us in disputes with merchants.

- Credit cards give us the option of paying our bills all at once or of stretching payments out over a length of time.

- Credit cards are safer to carry than cash is and thus help to reduce cash-based crime….

A KEY PROBLEM WITH CREDIT CARDS

In the course of the twentieth century, the United States has gone from a nation that cherished savings to one that reveres spending, even spending beyond one's means….

At the level of the national government, our addiction to spending is manifest in a once-unimaginable level of national debt, the enormous growth rate of that debt, and the widespread belief that the national debt cannot be significantly reduced, let alone eliminated. As a percentage of gross national product (GNP),[*] the federal debt declined rather steadily after World War II, reaching 33.3% in 1981. However, it then rose dramatically, reaching almost 73% of GNP in 1992. In dollar terms, the federal debt was just under $1 trillion in 1981, but by September 1993, it had more than quadrupled, to over $4.4 trillion. There is widespread fear that a huge and growing federal debt may bankrupt

---

[*]While the term *GNP* is still used for historical purpose it should be noted that the term *GDP* (gross domestic product) is now preferred.

the nation and a near consensus that it will adversely affect future generations.

Our addiction to spending is also apparent among the aggregate of American citizens. Total personal savings was less in 1991 than in 1984, in spite of the fact that the population was much larger in 1991. Savings fell again in the early 1990s from about 5.2% of disposable income in late 1992 to approximately 4% in early 1994. A far smaller percentage of families (43.5%) had savings accounts in 1989 than did in 1983 (61.7%). And the citizens of many other nations have a far higher savings rate. At the same time, our indebtedness to banks, mortgage companies, credit card firms, and so on is increasing far more dramatically than similar indebtedness in other nations....

WHO IS TO BLAME?

The Individual

In a society that is inclined to "psychologize" all problems, we are likely to blame individuals for not saving enough, for spending too much, and for not putting sufficient pressure on officials to restrain government expenditures. We also tend to "medicalize" these problems, blaming them on conditions that are thought to exist within the individual.... Although there are elements of truth to psychologistic and medicalistic perspectives, there is also a strong element of what sociologists term "blaming the victim." That is, although individuals bear some of the responsibility for not saving, for accumulating mounting debt, and for permitting their elected officials to spend far more than the government takes in, in the main individuals have been victimized by a social and financial system that discourages saving and encourages indebtedness.

Why are we so inclined to psychologize and medicalize problems like indebtedness? For one thing, American culture strongly emphasizes individualism. We tend to trace both success and failure to individual efforts, not larger social conditions. For another, large social and financial systems expend a great deal of time, energy, and money seeking, often successfully, to convince us that they are

not responsible for society's problems. Individuals lack the ability and the resources to similarly "pass the buck." Of perhaps greatest importance, however, is the fact that individual, especially medical, problems appear to be amenable to treatment and even seem curable. In contrast, large-scale social problems (pollution, for example) seem far more intractable. It is for these, as well as many other reasons, that American society has a strong tendency to blame individuals for social problems.

The Government

... Since the federal debt binge began in 1981, the government has also been responsible for creating a climate in which financial imprudence seems acceptable. After all, the public is led to feel, if it is acceptable for the government to live beyond its means, why can't individual citizens do the same? If the government can seemingly go on borrowing without facing the consequences of its debt, why can't individuals?

If the federal government truly wanted to address society's problems, it could clearly do far more both to encourage individual savings and to discourage individual debt. For example, the government could lower the taxes on income from savings accounts or even make such income tax-free. Or it could levy higher taxes on organizations and agencies that encourage individual indebtedness. The government could also do more to control and restrain the debt-creating and debt-increasing activities of the credit card industry.

Business

Although some of the blame for society's debt and savings problem must be placed on the federal government, the bulk of the responsibility belongs with those organizations and agencies associated with our consumer society that do all they can to get people to spend not only all of their income but also to plunge into debt in as many ways, and as deeply, as possible. We can begin with American business.

Those in manufacturing, retailing, advertising, and marketing (among others) devote their working hours and a large portion of their energies to figuring out ways of getting people to buy things.... One example is the dramatic proliferation of seductive catalogs that are mailed to our homes. Another is the advent and remarkable growth in popularity of the television home shopping networks. What these two developments have in common is their ability to allow us to spend our money quickly and efficiently without ever leaving our homes. Because the credit card is the preferred way to pay for goods purchased through these outlets, catalogs and home shopping networks also help us increase our level of indebtedness.

Banks and Other Financial Institutions

The historical mission of banks was to encourage savings and discourage debt. Today, however, banks and other financial institutions lead us away from savings and in the direction of debt. Saving is discouraged by, most importantly, the low interest rates paid by banks. It seems foolish to people to put their money in the bank at an interest rate of, say, 2.5% and then to pay taxes on the interest, thereby lowering the real rate of return to 2% or even less. This practice seems especially asinine when the inflation rate is, for example, 3% or 4%. Under such conditions, the saver's money is declining in value with each passing year. It seems obvious to most people that they are better off spending the money before it has a chance to lose any more value.

While banks are discouraging savings, they are in various ways encouraging debt. One good example is the high level of competition among the banks (and other financial institutions) to offer home equity lines of credit to consumers. As the name suggests, such lines of credit allow people to borrow against the equity they have built up in their homes.... Banks eagerly lend people money against this equity. Leaving the equity in the house is a kind of savings that appreciates with the value of the real estate, but borrowing against it allows people to buy more goods and services. In the process, however, they acquire a large new debt. And the house itself could be lost if one is unable to pay either the original mortgage or the home equity loan.

The credit card is yet another invention of the banks and other financial institutions to get people to save less and spend more. In the past, only the relatively well-to-do were able to get credit, and getting credit was a very cumbersome process (involving letters of credit, for example). Credit cards democratized credit, making it possible for the masses to obtain at least a minimal amount. Credit cards are also far easier to use than predecessors like letters of credit. Credit cards may thus be seen as convenient mechanisms whereby banks and other financial institutions can lend large numbers of people what collectively amounts to an enormous amount of money.

Normally, no collateral is needed to apply for a credit card. The money advanced by the credit card firms can be seen as borrowing against future earnings. However, because there is no collateral in the conventional sense, the credit card companies usually feel free to charge usurious interest rates.

Credit cards certainly allow the people who hold them to spend more than they otherwise would.... Some...overwhelmed by credit card debt, take out home equity lines of credit to pay it off. Then, with a clean slate, at least in the eyes of the credit card companies, such people are ready to begin charging again on their credit cards. Very soon many of them find themselves deeply in debt both to the bank that holds the home equity loan and to the credit card companies.

A representative of the credit card industry might say that no one forces people to take out home equity lines of credit or to obtain credit cards; people do so of their own volition and therefore are responsible for their financial predicament. Although this is certainly true at one level, at another level it is possible to view people as the victims of a financial (and economic) system that depends on them to go deeply into debt and itself grows wealthy as a result of that indebtedness. The newspapers, magazines, and broadcast media are full of advertisements offering various inducements to apply for a particular credit card or home equity loan rather than the ones offered by competitors. Many people are bombarded with mail offering all sorts of attractive benefits to those who sign up for yet another card or loan. More generally, one is made to feel foolish, even out of step, if one refuses to be an active part of the debtor society. Furthermore, it has become increasingly difficult to function in our society without a credit card. For example, people who do not have a record of credit card debt and payment find it difficult to get other kinds of credit, like home equity loans, car loans, or even mortgage loans.

AN INDICTMENT OF THE FINANCIAL SYSTEM

The major blame for our society's lack of savings and our increasing indebtedness must be placed on the doorstep of large institutions. We focus on one of those institutions—the financial system, which is responsible for making credit card debt so easy and attractive that many of us have become deeply and perpetually indebted to the credit card firms....

CASE IN POINT: GETTING THEM HOOKED WHILE THEY'RE YOUNG

Before moving on to a more specific discussion of the sociological perspective on the problems associated with credit cards, one more example of the way the credit card industry has created problems for people would be useful: the increasing effort by credit card firms to lure students into possessing their own credit cards. The over 9 million college students (of which 5.6 million are in school on a full-time basis) represent a huge and lucrative market for credit card companies. According to one estimate, about 82% of full-time college students now have credit cards. The number of undergraduates with credit cards increased by 37% between 1988 and 1990. The credit card companies have been aggressively targeting this population not only because of the immediate increase in business it offers but also because of the long-term income

possibilities as the students move on to full-time jobs after graduation. To recruit college students, credit card firms are advertising heavily on campus, using on-campus booths to make their case and even hiring students to lure their peers into the credit card world. In addition, students have been offered a variety of inducements. I have in front of me a flyer aimed at a college-age audience. It proclaims that the cards have no annual fee, offer a comparatively low interest rate, and offer "special student benefits," including a 20% discount at retailers like MusicLand and Gold's Gym and a 5% discount on travel.

The credit card firms claim that the cards help teach students to be responsible with money (one professor calls it a "training-wheels operation"). The critics claim that the cards teach students to spend, often beyond their means, instead of saving....

In running up credit card debt, it can be argued, college students are learning to live a lie. They are living at a level that they cannot afford at the time or perhaps even in the future. They may establish a pattern of consistently living beyond their means. However, they are merely postponing the day when they have to pay their debts....

Not satisfied with the invasion of college campuses, credit card companies have been devoting increasing attention to high schools. One survey found that as of 1993, 32% of the country's high school students had their own credit cards and others had access to an adult's card. Strong efforts are underway to greatly increase that percentage. The president of a marketing firm noted, "It used to be that college was the big free-for-all for new customers.... But now, the big push is to get them between 16 and 18." Although adult approval is required for a person under 18 years of age to obtain a credit card, card companies have been pushing more aggressively to gain greater acceptance in this age group....

The motivation behind all these programs is the industry view that about two-thirds of all people remain loyal to their first brand of card for 15 or more years. Thus the credit card companies are trying to get high school and college students accustomed to using their card instead of a competitor's. The larger fear is that the credit card companies are getting young people accustomed to buying on credit, thereby creating a whole new generation of debtors.

A SOCIOLOGY OF CREDIT CARDS

…Sociologists have grown increasingly dissatisfied with having to choose between large-scale, macroscopic theories and small-scale, microscopic theories. Thus, there has been a growing interest in theories that integrate micro and macro concerns. In Europe, expanding interest in what is known there as agency-structure integration parallels the increasing American preoccupation with micro-macro integration.

Mills: Personal Troubles, Public Issues

Of more direct importance here is the now-famous distinction made by Mills in his 1959 work, *The Sociological Imagination*, between micro-level personal troubles and macro-level public issues. Personal troubles tend to be problems that affect an individual and those immediately around him or her. For example, a father who commits incest with his daughter is creating personal troubles for the daughter, other members of the family, and perhaps himself. However, that single father's actions are not going to create a public issue; that is, they are not likely to lead to a public outcry that society ought to abandon the family as a social institution. Public issues, in comparison, tend to be problems that affect large numbers of people and perhaps society as a whole. The disintegration of the nuclear family would be such a public issue.…

A useful parallel can be drawn between the credit card and cigarette industries. The practices of the cigarette industry create a variety of personal troubles, especially illness and early death. Furthermore, those practices have created a number of public issues (the cost to society of death and illness traceable to cigarette smoke), and thus many people have come to see cigarette industry practices themselves as public issues.

Examples of industry practices that have become public issues are the aggressive marketing of cigarettes overseas.… Similarly, the practices of the credit card industry help to create personal problems (such as indebtedness) and public issues (such as the relatively low national savings rate). Furthermore, some industry practices—such as the aggressive marketing of credit cards to teenagers—have themselves become public issues.

One of our premises is that we need to begin adopting the same kind of critical outlook toward the credit card industry that we use in scrutinizing the cigarette industry.…

Mills's ideas give us remarkably contemporary theoretical tools for undertaking a critical analysis of the credit card industry and the problems it generates.…

Marx: Capitalist Exploitation

In addition to Mills's general approach, there is the work of the German social theorist Karl Marx (1818–1881), especially his ideas on the exploitation that he saw as endemic to capitalist society.…

There have been many changes in the capitalist system, and a variety of issues have come to the fore that did not exist in Marx's day. As a result, a variety of neo-Marxian theories have arisen to deal with these capitalist realities. One that concerns us here is the increasing importance to capitalists of the market for goods and services. According to neo-Marxians, exploitation of the worker continues in the labor market, but capitalists also devote increasing attention to getting consumers to buy more goods and services. Higher profits can come from both cutting costs and selling more products.

The credit card industry plays a role by encouraging consumers to spend more money, in many cases far beyond their available cash, on the capitalists' goods and services. In a sense, the credit card companies have helped the capitalists to exploit consumers. Indeed, one could argue that modern capitalism has come to depend on a high level of consumer indebtedness. Capitalism could have progressed only so far by extracting cash from the consumers. It had to find a way to go further.…

Simmel: The Money Economy

… Simmel pointed to many problems associated with a money economy, but three are of special concern:

- The first problem…is the "temptation to imprudence" associated with a money economy. Simmel argued that money, in comparison to its predecessors, such as barter, tends to tempt people into spending more and going into debt. My view is that credit cards are even more likely than money to make people imprudent. People using credit cards are not only likely to spend more but are also more likely to go deeply into debt….

- Second, Simmel believed that money makes possible many types of "mean machinations" that were not possible, or were more difficult, in earlier economies. For example, bribes for political influence or payments for assassinations are more easily made with money than with barter…. Although bribes or assassinations are generally less likely to be paid for with a credit card than with cash, other types of mean machinations become more likely with credit cards. For example, some organizations associated with the credit card industry engage in fraudulent or deceptive practices in order to maximize their income from credit card users….

- The third problem with a money economy that concerned Simmel was the issue of secrecy, especially the fact that a money economy makes payments of bribes and other types of secret transactions more possible. However, our main concern is the increasing lack of secrecy and the invasion of privacy associated with the growth of the credit card industry….

Weber: Rationalization

…Weber defined rationalization as the process by which the modern world has come to be increasingly dominated by structures devoted to efficiency, predictability, calculability, and technological control. Those rational structures (for example, the capitalist marketplace and the bureaucracy) have had a progressively negative effect on individuals. Weber described a process by which more and more of us would come to be locked in an "iron cage of rationalization." …The credit card industry has also been an integral part of the rationalization process. By rationalizing the process by which consumer loans are made, the credit card industry has contributed to our society's dehumanization….

Globalization and Americanization

A sociology of credit cards requires a look at the relationship among the credit card industry, personal troubles, and public issues on a global scale. It is not just the United States, but also much of the rest of the world, that is being affected by the credit card industry and the social problems it helps create. To some degree, this development is a result of globalization, a process that is at least partially autonomous of any single nation and that involves the reciprocal impact of many economies. In the main, however, American credit card companies dominate the global market….

The central point is that, in many countries around the world, Americanization is a public issue that is causing personal troubles for their citizens. The credit card industry Americanizes and homogenizes life around the world, with the attendant loss of cultural and individual differences….

OTHER REASONS FOR EXAMINING CREDIT CARDS

Something New in the History of Money

Money in all its forms, especially in its cash form, is part of a historical process. It may seem hard to believe from today's vantage point, but at one time there was no money. Furthermore, some predict that there will come a time in which money, at least in the form of currency, will become less important if not disappear altogether, with the emergence of a "cashless society."…

More important for our purposes, money in the form of currency is being increasingly supplanted by the credit card. Instead of plunking down cash or even writing a check, more of us are saying "Charge it!" This apparently modest act is, in fact, a truly revolutionary development in the history of money. Furthermore, it is having a revolutionary impact on the nature of consumption, the economy, and the social world more generally. In fact, rather than simply being yet another step in the development of money, I am inclined to agree with the contention that credit cards are "an entirely new idea in value exchange." A variety of arguments can be marshaled in support of the idea that in credit cards we are seeing something entirely new in the history of economic exchange, especially relative to cash:

- Credit card companies are performing a function formerly limited to the federal government. That is, they create money…the Federal Reserve is no longer alone in this ability. The issuing of a new credit card with a $1,000 limit can be seen as creating $1,000. Thus, the credit industry is creating many billions of dollars each year and, among other things, creating inflationary pressures in the process.

- Credit cards do not have a cash or currency form. In fact, they are not even backed by money until a charge is actually made.

- With cash we are restricted to the amount on hand or in the bank, but with credit cards our ceiling is less clear. We are restricted only by the ever-changing limits of each of our credit cards as well as by the aggregate of the limits of all those cards.

- Although we can use our cash anytime we wish, the use of our credit card requires the authorization of another party.

- Unlike cash, which allows for total anonymity, one's name is printed on the front of the credit card and written on the back; a credit card may even have one's picture on it. Furthermore, credit card companies have a great deal of computerized information on us that is drawn on to approve transactions.

- Although cash is simple to produce and use over and over, credit cards require the backing of a complex, huge, and growing web of technologies.

- There is no direct cost to the consumer for using cash, but fees and interest may well accrue with credit card use.

- Although everyone, at least theoretically, has access to cash, some groups (the poor, the homeless, the unemployed) may be denied access to credit cards. Such restrictions sometimes occur unethically or illegally through the "redlining" of certain types of consumers or geographic areas.

- Because of their accordionlike limits, credit cards are more likely than cash to lead to consumerism, overspending, and indebtedness.…

A Growing Industry

Another reason for focusing on credit cards is their astounding growth in recent years, which reflects their increasing importance in the social and economic worlds. There are now more than a billion credit cards of all types in the United States. Receivables for the industry as a whole in 1993 were up by almost 16% from the preceding year and by over 400% in a decade. The staggering proliferation of credit cards is also reflected in other indicators of use in the United States:

- Sixty-one percent of Americans now have at least one credit card.

- The average cardholder carries nine different cards.

- In 1992, consumers used the cards to make 5 billion transactions, with a total value of $420 billion.

There has been, among other things, growth in the number of people who have credit cards, the average number of credit cards held by each person, the amount of consumer debt attributable to credit card purchases, the number of facilities accepting credit cards, and the number of organizations issuing cards.... The average outstanding balance owed to Visa and MasterCard increased from less than $400 in the early 1980s to $970 in 1989 and to $1,096 in 1993. The amount of high-interest credit card debt owed by American consumers rose from $2.7 billion in 1969 to $50 billion in 1980 and was approaching $300 billion in 1994....

A Symbol of American Values

A strong case can be made that the credit card is one of the leading symbols of 20th-century America or, as mentioned earlier, that the credit card is an American icon. Indeed, one observer calls the credit card "the twentieth century's symbol par excellence." Among other things, the credit card is emblematic of affluence, mobility, and the capacity to overcome obstacles in the pursuit of one's goals. Thus, those hundreds of millions of people who carry credit cards are also carrying with them these important symbols. And when they use a credit card, they are turning the symbols into material reality....

DEBUNKING CREDIT CARD MYTHS

To most of us, credit cards appear to have near-magical powers, giving us greater access to a cornucopia of goods and services. They also seem to give us something for nothing. That is, without laying out any cash, we can leave the mall with an armload of purchases. Most of us like what we can acquire with credit cards, but some like credit cards so much that they accumulate as many as they can. Lots of credit cards, with higher and higher spending limits, are important symbols of success. That most people adopt a highly positive view of credit cards is borne out by the proliferation of the cards throughout the United States and the world....

A debunking sociology is aimed at revealing the spuriousness of various ideologies. As Berger puts it, "In such analyses the ideas by which men explain their actions are unmasked as self-deception, sales talk, the kind of sincerity ... of a man who habitually believes his own propaganda." From this perspective, the credit card companies can be seen as purveyors of self-deceptive ideologies. They are, after all, in the business of selling their wares to the public, and they will say whatever is necessary to accomplish their goal....

34

Inequalities of Political Voice

KAY LEHMAN SCHLOZMAN, BENJAMIN I. PAGE, SIDNEY VERBA, AND MORRIS P. FIORINA

Most people want to believe that, because we live in a democracy, we have equality in the political voice of our government. On further consideration, we also realize that there are large gaps in economic equality and that those with more wealth have more political power.
As you read this selection, consider the following questions as guides:

1. *Make a list of all known economic inequalities and compare them with the ones mentioned in the article.*
2. *Make a list of all known political inequalities and compare them with the ones mentioned in the article.*
3. *Which laws would you enact to lessen economic inequality? Political inequality?*
4. *Do you think these laws will ever be enacted? Why? Why not?*

WHAT AMERICANS THINK ABOUT INEQUALITY...

Inequalities Among Social Groups

One of the great stories of the past century in the United States has been the gradual rejection of inequalities based on such social characteristics as race, gender, ethnicity, disability, or sexual orientation. More and more, Americans have come to oppose any government-enforced discrimination on such grounds. Most now say they oppose private prejudice and discrimination as well. The chief remaining areas of controversy concern whether and how laws and regulations should prohibit, or counteract the effects of, private discrimination.

The change in public attitudes about African Americans has been particularly dramatic. Between 1942 and 1985, the proportion of Americans saying that black and white children should go to the same schools (rather than "separate schools") rose in a steady, linear fashion, from a meager 31 percent to an overwhelming 93 percent. Similarly, the proportion of whites who opposed "separate sections for Negroes" on streetcars and buses grew from 46 percent in 1942 to 88 percent in 1970. The 45 percent minority who said in 1944 that "Negroes should have as good a chance as white people to get any kind of job" grew to a near-unanimous, 97 percent majority in 1972. Opinion also turned against the segregation of public accommodations and housing. Between 1963 and 1990, opposition to "laws against marriages between blacks and whites" rose from 38 percent to 79 percent.[1]

Still, even now, public support for federal government policies that enforce pro-integration

Source: L. R. Jacobs and T. Skocpol, eds. 2005. *Inequality and American Democracy: What We Know and What We Need to Learn.* New York: Russell Sage.

principles is mixed. Only a minority of Americans—a minority that declined in the 1970s and 1980s—has said that the government in Washington should "see to it" that schools are integrated. Large majorities oppose school busing or job quotas. Reactions to affirmative action depend on the details. Support for open housing laws has increased, reaching a small majority (57 percent) in favor by 1990. But only with respect to public accommodations, where racial interactions are transient and casual, have substantial and growing majorities of whites favored government action.[2]

Does this opposition to concrete measures for achieving integration and remedying past discrimination result from continuing racism? Or does it reflect broader, "race-neutral" principles, like individualism and color-blind application of the law? This question continues to provoke scholarly controversy. Some scholars tend to blame "symbolic racism" or "racial resentment." Others point toward ideological principles.[3]

With respect to gender discrimination, public opinion has also moved strongly in favor of legal equality for women. In 1937, only 18 percent of Americans said they approved of a woman "earning money in business or industry" if she had a "husband capable of supporting her." This figure rose markedly over the years to 82 percent in 1990. Large majorities oppose job discrimination and favor equal pay for equal work. The proposed Equal Rights Amendment regularly won 60 percent support, and fully 77 percent when the text was read to respondents.[4]

By 1982, affirmative action programs in *industry* ("provided there are no rigid quotas") won 72 percent support or higher with respect to women, "Spanish-Americans," blacks, and especially the physically handicapped. Support for "affirmative action programs in *higher education*" was a bit stronger. After hitting a plateau in the 1980s, support for such programs remains high. Even in the case of homosexuality, which many Americans still call "wrong," majorities do not think it should be illegal and oppose job discrimination—with some distinctions among occupations (Page and Shapiro 1992, 97–100). Moreover, Americans' attitudes toward homo-

sexuality have become substantially more tolerant over the past two decades (Fiorina 2005, chap 5).

Economic Inequalities

Americans' views of inequalities in the economic domain are multidimensional. Most Americans say that all people are "created equal" and strongly favor equality of opportunity. However, there is considerably more tolerance of inequality of economic results—especially when people perceive extensive opportunities to get ahead, or when economic inequality can be plausibly justified as providing incentives to work and invest in ways that may benefit everyone. At the same time, most Americans oppose what they see as unfair economic disparities that do not reflect merit or effort, and most favor a number of government programs with egalitarian effects.

Large majorities of Americans approve of private property and free enterprise. They believe that hard work and ambition are rewarded, that their children will be better off than they are, and that it benefits the country to have a class of rich people. But most Americans also think that the rich have too much political power and do not pay enough taxes. They consider lawyers, CEOs, doctors, investment bankers, and various celebrities to be "overpaid," while restaurant workers, school teachers, secretaries, policemen, nurses and factory workers "underpaid." Most feel that "money and wealth in this country should be more evenly distributed."[5] These attitudes appear to vary somewhat with the business cycle and with international events. Economic downturns tend to produce more egalitarian sentiments, and extra sacrifices are sought from the affluent during major wars.

Government Action Against Economic Inequality. According to opinion surveys, although Americans are not necessarily more likely to embrace economic inequalities than are citizens of other advanced democracies, they are much more likely than Europeans to blame individuals rather than government for disparities of income and wealth. Perhaps reflecting a general skepticism about government, they also express little enthusiasm—less than people elsewhere—for redistribution of income or wealth by the government, at least when the issue is posed in the abstract.[6] For example, very few favor "a law limiting the amount of money and individual is allowed to earn in a year," and most agree that "people should be allowed to accumulate as much wealth as they can, even if some make millions while others live in poverty." On a 7-point scale, responses tilt only slightly more toward the sentiment that "Washington ought to reduce income differences between rich and poor," than toward the feeling that the "government should not concern itself" with such matters.[7]

At the same time, there has long been a difference between Americans' "ideological conservatism" and their "operational liberalism."[8] Considerable majorities favor a number of concrete policies that would have, or actually do have, substantial redistributive effects. Large numbers of Americans support having moderately—though not highly—progressive taxes. Most favor closing tax "loopholes" used by the wealthy. While most people call their own taxes "too high," antitax fervor among ordinary Americans is considerably weaker than politicians sometimes imply. Substantial majorities of the public are willing to pay the taxes needed to fund popular spending programs. In 2001 and 2003, for example, when large, regressive cuts in income and estate taxes were enacted, polls showed that substantial majorities of Americans would have preferred to keep the money and bolster the Social Security system or reduce the budget deficit. In 2004 as well, most said that they preferred budget balancing and more progressive taxes.[9]

It is noteworthy that Social Security—the largest single program in the federal budget—is extremely popular. The primary effect of Social Security is to smooth out middle-class people's incomes over their lifetimes, but it also redistributes incomes among individuals by providing substantial benefits even for those who had low incomes during their working lives. For that reason, poverty experts have called Social Security "crucial" in reducing poverty among the elderly (Blank 1997,

228; Page and Simmons 2000, chap 4). Year after year, the overwhelming majority of Americans have said that they want to keep Social Security at its present level or spend more on it; only a tiny minority wants to cut back. There has been considerable resistance to any "reforms" that would reduce guaranteed Social Security benefits in any way, even through reduced cost-of-living increases or stretched-out retirement ages. What appears to be high public support for partial privatization of the program (for example, allowing individuals to invest part of their payroll taxes in personal retirement accounts) drops sharply when it is made clear that such a change would imply cuts in guaranteed benefits. This pattern remained unchanged in early 2005, as President Bush began his serious push for private accounts.[10]

The American public also generally favors universalistic government help with education (including day care and pre-schooling), which can be a major equalizing force in society. Most Americans want government help with jobs and employment, including a surprising (though seldom surveyed) level of support for job creation through public works programs. Large majorities want to help the uninsured get health care, and favor a variety of possible methods including the expansion of Medicare to younger people, providing catastrophic health insurance coverage to everyone, expanding community health clinics, and subsidizing private health insurance. The term "welfare" is despised, but a wide range of programs for the "deserving" poor win public approval, so long as the able-bodied are willing to work. Public support has been high, for example, for Supplemental Social Security Income, Unemployment Insurance, and even the former Aid for Families with Dependent Children. Well into the George W. Bush administration, large majorities of Americans said they favored universal, government-run health insurance.[11]

Political Inequalities

In contrast to the economic realm, substantial majorities of Americans endorse a high degree of equality in the realm of politics—in terms of both abstract principles and concrete legal arrangements. For example, in an early study with a local sample, as many as 95 percent of Americans endorsed the idea that "every citizen should have an equal chance to influence government policy"; and 91 percent said that everyone should have an equal right to hold public office. True, roughly half of Americans have said that government should pay most attention to "people of intelligence and character" or to "the people who really know something about the subject." But large majorities have said that elected officials would "badly misuse their power" if they were not watched by voters, and that elections are one of the best ways to keep officials on their toes. A large majority of Americans have said that all adult citizens should be allowed to vote, "regardless of how ignorant they may be" (McClosky and Zaller 1984, 74, 75, 79).

When it comes to specific institutional arrangements, most Americans favor a variety of reforms aimed at reducing political inequalities in the control of government. Even before the controversies surrounding the 2000 presidential election, solid majorities of Americans (73 percent or 86 percent in different polls) favored abolishing the Electoral College and electing the president by popular vote. Equally large majorities of 70 percent or 80 percent have also favored choosing presidential candidates in a nationwide primary rather than through party conventions. Before the 1971 ratification of the Twenty-sixth Amendment, 60 percent to 70 percent of Americans favored granting the vote to eighteen-year-olds. Concerned about the power of money in politics, substantial majorities of Americans favor limits on private campaign contributions and on campaign spending—not, however, public financing of elections (Page and Shapiro 1992, 166–67).

UNEQUAL VOICES IN SURVEYS WEIGHING PUBLIC OPTION

These opinions are mostly those expressed in national public opinion surveys—which, by interviewing random samples of adult Americans, try to ascertain what all Americans think. In such surveys, everyone is supposed to have an equal voice. Yet there is evidence that surveys do not in fact report the opinions of all citizens equally, and that the results are biased in the direction of making Americans' opinions seem less egalitarian than they actually are.

The problem is that in forming political opinions, and expressing them to survey interviewers, is easier for those who have abundant resources in skills, income, and, especially, education. Less advantaged people—who also have real political wants and needs—are more likely to be uncertain or confused or to say "don't know" when interviewers ask their opinions. Natural supporters of egalitarian social welfare policies are also the least likely to register their opinions. By contrast, the very people who have abundant resources and make their voices most heard in surveys are the same people who are least concerned about gaps between the rich and the poor.

As a result, survey data appear to be subject to a consistent "exclusion bias" that tends to make Americans look, on average, somewhat more conservative and antiegalitarian (less likely, for example, to say that government should "reduce income differences between the rich and the poor") than they actually are. This bias has been estimated to be rather small—smaller, for example, than the resource-related biases… in voting and nearly every other form of individual or collective political input. Still, the exclusion bias appears to reinforce other political barriers faced by disadvantaged.[12] The survey results presented here reveal substantial proegalitarian sentiments despite this bias. Data corrected for the bias would probably move further in the same direction, especially on issues of economic inequality and redistribution.

Public Opinion as Effect Rather than Cause

In democratic political systems it is natural to think of public opinion as a cause, or at least a possible cause, of what governments do. We tend to judge how well democracy works partly in terms of how well government policies correspond with what citizens say they want. But this way of thinking neglects the possibility that public opinion may be an effect as well as a cause of political processes. Charles E. Lindblom referred to this as the problem of "circularity."[13] If political leaders, organized interest groups, large corporations, or others can manipulate the opinions of ordinary citizens, democracy will be compromised even though the government responds perfectly to those opinions. If public opinion can be manipulated, and if the tools of opinion manipulation are most available to the wealthy and powerful—who tend to occupy "bully pulpits" and to have the rhetorical skills or money needed to persuade others—the result may be a subtle, indirect, but pervasive kind of inequality in political influence.

It is fairly well established that the political contents of the mass media—especially the reported views of ostensibly nonpartisan commentators and "experts"—tend to affect the priorities and policy preferences of the public (Iyengar and Kinder 1987; Page, Shapiro, and Dempsey 1987). It is also well known that what appears in the media is heavily influenced by public officials, who are major sources of political news, and that business corporations spend a great deal of money funding the think-tanks, universities, and foundations that produce and publicize "expert" opinions. Corporations and others have also spent large amounts of money on issue advocacy advertisements, often trying to turn the public against egalitarian policies (Saloma 1984; Stefancic and Delgado 1996; West and Loomis 1998). Still, it is very difficult to ascertain the net impact, if any, of this activity on public opinion. And even when there are measurable effects, the interpretation is often disputed. If ordinary citizens change their opinions in response to persuasion by public officials or other policy elites, how are we to discern whether the process is one of "education" or "manipulation"? At minimum, it seems important to bear in mind the possibility that the wealthier and more powerful members of American society have been able to influence the opinions of the less affluent, reducing public support for policies that would combat economic inequality and adding to inequalities of political voice.

Similar problems arise in attempting to assess the possibility that recent increases in economic inequality have themselves affected public opinion, perhaps by making people angry cynical, and distrustful of government, which they may see as not doing much about—or, even, as exacerbating—their troubles. Seymour Martin Lipset and William Schneider (1983) demonstrate that much of the sharp decline in Americans' trust and confidence in government and other institutions occurred very early—in the late 1960s and early 1970s—before economic inequality began to increase. The causes of these declines, then, may have had more to do with political disenchantment over Vietnam and Watergate than with economic trends. But the long subsequent history of negative political attitudes, punctuated by protest candidates and rejections of incumbent officials, suggests that economic stagnation and increased gaps between rich and poor may indeed have contributed to the souring of the political views of many Americans or, at a minimum, reinforced the distrust that was precipitated by Vietnam and Watergate. We cannot be sure.

Public Opinion and Egalitarian Policies

This broad framework of opinion should be taken into account as we think about the reasons for egalitarian and antiegalitarian public policies. To recapitulate, although most Americans support a high level of equality among social groups and favor equality of opportunity, they appear to be less concerned about inequality in economic outcomes. For example, there is little public support for a massive redistribution of income or wealth. At the same time, however, there can be little doubt that large majorities of Americans prefer a democracy with a high level of political equality among citizens. Moreover, most support a number of concrete government policies that have, or would have, substantially egalitarian economic effects.

…[If] we wish to judge the extent to which political equality among citizens does or does not prevail in policymaking, it can be useful to compare policies that are actually enacted and implemented with policies that majorities of Americans say they favor. Evidence indicates that most Americans favor a number of policies that would tax the wealthy and upper-income people at higher levels, and spend more money to help middle- and lower-income people, than is currently the case. In other words, there are indications of something other than perfect political equality in U.S. policymaking. It is not the case that every citizen has one effective vote, one equal amount of influence upon political outcomes.[14] If this leads to a tilt against egalitarian public policies, political inequality in the United States may tend to reinforce economic inequalities....

REFERENCES

Blank, Rebecca M. 1997. *It Takes a Nation: A New Agenda for Fighting Poverty*. Princeton, N.J.: Princeton University Press.

Fiorina, 2005. *Culture War: The Myth of a Polarized America*. New York: Pearson Longman.

Iyengar, Shanto, and Donald R. Kinder. 1987. *News that Matters*. Chicago: University of Chicago Press.

Lipset, Seymour Martin, and William Schneider. 1983. *The Confidence Gap: Business, Labor, and Government in the Public Mind*. New York: Macmillan.

McClosky, Herbert, and John R. Zaller. 1984. *The American Ethos: Public Attitudes Toward Capitalism and Democracy*. Cambridge, Mass.: Harvard University Press.

Page, Benjamin I. and Robert Y. Shapiro. 1992. *The Rational Public: Fifty Years of Trends in American's Policy Preferences*. Chicago: University of Chicago Press.

Page, Benjamin I. and Robert Y. Shapiro, and Glenn R. Dempsey. 1987. "What Moves Public Opinion?" *America Political Science Review* 81(1): 23–43.

Page, Benjamin I., and James R. Simmons. 2000. *What Government Can Do: Dealing with Poverty and Inequality*. Chicago: University of Chicago Press.

Saloma, John S. III. 1984. *Ominous Politics: The New Conservative Labyrinth*. New York: Hill and Wang.

Stefancic, Jean, and Richard Delgado. 1996. *No Mercy: How Conservative Think Thanks and Foundations Changed America's Social Agenda*. Philadelphia: Temple University Press.

West, Darell M., and Burdett A. Loomis. 1998. *The Sound of Money: How Political Interests Get What They Want*. New York: W. W. Norton.

NOTES

1. On this point, which is confirmed by more recent surveys, see Schuman, Steeh, and Bobo (1985, 74–76); and Page and Shapiro (1992, 63, 68–71). These and other generalizations about public opinion are based not only on the published sources cited but also on the most recent data available (early 2005) on such web sites as Polling Report.com, ropercenter.uconn.edu, and those of particular news and polling organizations.

2. See Schuman, Steeh, and Bobo (1985, esp. 88–90); and Page and Shapiro (1992, 71–75).

3. For the former perspective, see, for example, Sears (1988); Kinder and Sanders (1996); and Gilens (1999); for the latter, see, for example, Sniderman and Piazza (1993). Sears, Sidanius, and Bobo (2000) bring together a variety of positions.

4. On public attitudes toward the ERA, see Mansbridge (1986); and Page and Shapiro (1992, 64, 100–2, 105–10).

5. On Americans' attitudes toward economic equality, see Honchschild (1981); McClosky and Zaller (1984, 108, 116, 133, 140); and Ladd and Bowman (1998, esp. 17, 18, 20–21, 97–98, 110).

6. For example, data from the International Social Survey Programme indicate that the average American's position on whether it should be "the government's responsibility to reduce income

differences" between rich and poor is less favorable to redistribution than is that of citizens of any of the thirteen other countries studied, even though there is little indication of American exceptionalism concerning desired income discrepancies across occupations (Osberg and Smeeding, 2003, 31, 37, 41). See also, Kleugel, Mason, and Wegener (1995); Weakliem, Andersen, and Health (2003, 47–48); and Mehrtens (2004).

7. See, Lipset and Schneider (1983); McClosky and Zaller (1984, 120, 141, 1430; and Ladd and Bowman (1998, 108–9, 111).

8. On the distinction between "ideological" and "operational" liberalism, see Free and Cantril (1968).

9. A June 2003 Harris survey found a solid 65 percent majority saying that that year's tax cuts were generally "unfair" in incidence among income groups. An AP/Ipsos survey at the time of the 2004 election (November 3 through 5, 2004) found much more support for balancing the budget (66 percent) than for cutting taxes (31 percent...).

10. A CNN/USA Today/Gallup poll (January 1 through 5, 2005) asked, "As you may know, one idea to address concerns with the Social Security system would allow people who retire in future decades to invest some of their Social Security in the stock market and bonds, but would reduce the guaranteed benefits they get when they retire. Do you think this is a good idea or a bad idea?" Forty percent responded that it is a good idea, 55 percent that it is a bad idea. To a similar question in a PSRA poll for Pew (January 1 through 5, 2005) that emphasized choice (noting possible rises or falls in investment values) versus guaranteed monthly benefits linked to past earnings, 65 percent preferred guaranteed benefits and only 29 percent favored younger workers deciding for themselves how to invest part of their Social Security taxes.

11. For example, an October 2003 ABC/WP survey found that a large majority of Americans (79 percent to 17 percent) said that it was important to get health coverage for everyone, even if it meant raising taxes. A majority (62 percent to 33 percent) preferred universal, government-provided health insurance over the current provision by employers or no one, and similar majorities favored it even if this limited their own choice of doctors or meant waiting lists for some nonemergency treatments. On public attitudes toward programs of government income support, see also Cook (1979); McClosky and Zaller (1984, 272–76); Page and Shapiro (1992, chap. 4); Cook and Barrett (1992, esp. 62); and Page (2001).

12. Berinsky (2002, 285) estimates that in 1996 the exclusion bias led to a .11 point overestimation of the average respondent's opinion on the 7 point "reduce income differences between the rich and the poor" scale. See also Brehm (1993); Bartels (1996); Althaus (1998); and Berinsky (2004).

13. See Lindblom (1977, chap.15) for a discussion of the "circularity" problem.

35

If Hitler Asked You to Electrocute a Stranger, Would You? Probably

PHILIP MEYER

Many have wondered how a former corporal (Hitler) could manage to influence so many people and get them to commit atrocities such as those that occurred in the concentration camps. This reading reveals the roles of authority and charisma in those decisions.

As you read this selection, ask yourself the following questions:

1. *What were two of the findings of the Milgram study that were surprising?*
2. *In the position of the testees, how would you have reacted? Why?*
3. *Do the findings explain the violence in the world? Some of it? Why?*

GLOSSARY **Pathological** Disorder in behavior. **Macabre** Gruesome or horrible. **Sadistic** Deriving pleasure from inflicting pain.

In the beginning, Stanley Milgram was worried about the Nazi problem. He doesn't worry much about the Nazis anymore. He worries about you and me, and, perhaps, himself a little bit too.

Stanley Milgram is a social psychologist, and when he began his career at Yale University in 1960 he had a plan to prove, scientifically, that Germans are different. The Germans-are-different hypothesis has been used by historians, such as William L. Shirer, to explain the systematic destruction of the Jews by the Third Reich. One madman could decide to destroy the Jews and even create a master plan for getting it done. But to implement it on the scale that Hitler did meant that thousands of other people had to go along with the scheme and help to do the work. The Shirer thesis, which Milgram set out to test, is that Germans have a basic character flaw which explains the whole thing, and this flaw is a readiness to obey authority without question, no matter what outrageous acts the authority commands.

The appealing thing about this theory is that it makes those of us who are not Germans feel better about the whole business. Obviously, you and I are not Hitler, and it seems equally obvious that we would never do Hitler's dirty work for him. But now, because of Stanley Milgram, we are compelled to wonder. Milgram developed a laboratory experiment which provided a systematic way to measure obedience. His plan was to try it out in New Haven on Americans and then go to Germany and try it out on Germans. He was strongly motivated by scientific curiosity, but there was also some moral content in his decision to pursue this line of research, which was, in turn, colored by his own Jewish background. If he could show that Germans are more obedient than Americans, he could then vary the conditions of the experiment and try to find out just what it is that makes some people more obedient than others. With this understanding, the world might, conceivably, be just a little bit better.

But he never took his experiment to Germany. He never took it any farther than Bridgeport. The first finding, also the most unexpected and disturbing finding, was that we Americans are an obedient people: not blindly obedient, and not blissfully obedient, just obedient. "I found so much obedience," says Milgram softly, a little sadly, "I hardly saw the need for taking the experiment to Germany."

There is something of the theater director in Milgram, and his technique, which he learned from one of the old masters in experimental psychology, Solomon Asch, is to stage a play with every line rehearsed, every prop carefully selected, and everybody an actor except one person. That one person is the subject of the experiment. The subject, of course, does not know he is in a play. He thinks he is in real life. The value of this technique is that the experimenter, as though he were God, can change a prop here, vary a line there, and see how the subject responds. Milgram eventually had to change a lot of the script just to get people to stop obeying. They were obeying so much, the experiment wasn't working—it was like trying to measure oven temperature with a freezer thermometer.

The experiment worked like this: If you were an innocent subject in Milgram's melodrama, you read an ad in the newspaper or received one in the mail asking for volunteers for an educational experiment. The job would take about an hour and pay $4.50. So you make an appointment and go to an old Romanesque stone structure on High Street with the imposing name of The Yale Interaction Laboratory. It looks something like a broadcasting studio. Inside, you meet a young, crew-cut man in a laboratory coat who says he is Jack Williams, the experimenter. There is another citizen, fiftyish, Irish face, an accountant, a little overweight, and very mild and harmless-looking. This other citizen seems nervous and plays with his hat while the two of you sit in chairs side by side and are told that the $4.50 checks are yours no matter what happens. Then you listen to Jack Williams explain the experiment.

It is about learning, says Jack Williams in a quiet, knowledgeable way. Science does not know much about the conditions under which people learn and this experiment is to find out about negative reinforcement. Negative reinforcement is getting punished when you do something wrong, as opposed to positive reinforcement which is getting rewarded when you do something right. The negative reinforcement in this case is electric shock. You notice a book on the table titled, *The Teaching-Learning Process*, and you assume that this has something to do with the experiment.

Then Jack Williams takes two pieces of paper, puts them in a hat, and shakes them up. One piece of paper is supposed to say, "Teacher" and the other, "Learner." Draw one and you will see which you will be. The mild-looking accountant draws one, holds it close to his vest like a poker player, looks at it, and says, "Learner." You look at yours. It says, "Teacher." You do not know that the drawing is rigged, and both slips say "Teacher." The experimenter beckons to the mild-mannered "learner."

"Want to step right in here and have a seat, please?" he says. "You can leave your coat on the back of that chair ... roll up your right sleeve, please. Now what I want to do is strap down your arms to avoid excessive movement on your part during the experiment. This electrode is connected to the shock generator in the next room.

"And this electrode paste," he says, squeezing some stuff out of a plastic bottle and putting it on the man's arm, "is to provide a good contact and to avoid a blister or burn. Are there any questions now before we go into the next room?"

You don't have any, but the strapped-in "learner" does.

"I do think I should say this," says the learner. "About two years ago, I was at the veterans' hospital ... they detected a heart condition. Nothing serious, but as long as I'm having these shocks, how strong are they—how dangerous are they?"

Williams, the experimenter, shakes his head casually. "Oh, no," he says. "Although they may be painful, they're not dangerous. Anything else?"

Nothing else. And so you play the game. The game is for you to read a series of word pairs: for example, blue girl, nice day, fat neck. When you finish the list, you read just the first word in each pair and then a multiple-choice list of four other words, including the second word of the pair. The learner, from his remote, strapped-in position, pushes one of four switches to indicate which of the four answers he thinks is the right one. If he gets it right, nothing happens and you go on to the next one. If he gets it wrong, you push a switch that buzzes and gives him an electric shock. And then you go to the next word. You start with 15 volts and increase the number of volts by 15 for each wrong answer. The control board goes from 15 volts on one end to 450 volts on the other. So that you know what you are doing, you get a test shock yourself, at 45 volts. It hurts. To further keep you aware of what you are doing to that man in there, the board has verbal descriptions of the shock levels, ranging from "Slight Shock" at the left-hand side, through "Intense Shock" in the middle, to "Danger: Severe Shock" toward the far right. Finally, at the very end, under 435- and 450-volt switches, there are three ambiguous X's. If, at any point, you hesitate, Mr. Williams calmly tells you to go on. If you still hesitate, he tells you again.

Except for some terrifying details, which will be explained in a moment, this is the experiment. The object is to find the shock level at which you disobey the experimenter and refuse to pull the switch.

When Stanley Milgram first wrote this script, he took it to fourteen Yale psychology majors and asked them what they thought would happen. He put it this way: Out of one hundred persons in the teacher's predicament, how would their break-off points be distributed along the 15-to-450 volt scale? They thought a few would break off very early; most would quit someplace in the middle, and a few would go all the way to the end. The highest estimate of the number out of one hundred who would go all the way to the end was three. Milgram

then informally polled some of his fellow scholars in the psychology department. They agreed that a very few would go to the end. Milgram thought so too.

"I'll tell you quite frankly," he says, "before I began this experiment, before any shock generator was built, I thought that most people would break off at 'Strong Shock' or 'Very Strong Shock.' You would get only a very, very small proportion of people going out to the end of the shock generator, and they would constitute a pathological fringe."

In his pilot experiments, Milgram used Yale students as subjects. Each of them pushed the shock switches one by one, all the way to the end of the board.

So he rewrote the script to include some protests from the learner. At first, they were mild, gentlemanly, Yalie protests, but "it didn't seem to have as much effect as I thought it would or should," Milgram recalls. "So we had more violent protestations on the part of the person getting the shock. All of the time, of course, what we were trying to do was not to create a macabre situation, but simply to generate disobedience. And that was one of the first findings. This was not only a technical deficiency of the experiment, that we didn't get disobedience. It really was the finding: that obedience would be much greater than we had assumed it would be and disobedience would be much more difficult than we had assumed."

As it turned out, the situation did become rather macabre. The only meaningful way to generate disobedience was to have the victim protest with great anguish, noise, and vehemence. The protests were tape-recorded so that all the teachers ordinarily would hear the same sounds and nuances, and they started with a grunt at 75 volts, proceeded through a "Hey, that really hurts," at 125 volts, got desperate with, "I can't stand the pain, don't do that," at 180 volts, reached complaints of heart trouble at 195, an agonized scream at 285, a refusal to answer at 315, and only heartrending, ominous silence after that.

Still, 65 percent of the subjects, twenty- to fifty-year-old American males, everyday, ordinary people, like you and me, obediently kept pushing those levers in the belief that they were shocking the mild-mannered learner, whose name was Mr. Wallace, and who was chosen for the role because of his innocent appearance, all the way up to 450 volts.

Milgram was now getting enough disobedience so that he had something he could measure. The next step was to vary the circumstances to see what would encourage or discourage obedience. There seemed very little left in the way of discouragement. The victim was already screaming at the top of his lungs and feigning a heart attack. So whatever new impediment to obedience reached the brain of the subject had to travel by some route other than the ear. Milligan thought of one.

He put the learner in the same room with the teacher. He stopped strapping the learner's hand down. He rewrote the script so that at 150 volts the learner took his hand off the shock plate and declared that he wanted out of the experiment. He rewrote the script some more so that the experimenter then told the teacher to grasp the learner's hand and physically force it down on the plate to give Mr. Wallace his unwanted electric shock.

"I had the feeling that very few people would go on at that point, if any," Milgram says. "I thought that would be the limit of obedience that you find in the laboratory."

It wasn't.

Although seven years have now gone by, Milgram still remembers the first person to walk into the laboratory in the newly rewritten script. He was a construction worker, a very short man. "He was so small," says Milgram, "that when he sat on the chair in front of the shock generator, his feet didn't reach the floor. When the experimenter told him to push the victim's hand down and give the shock, he turned to the experimenter, and he turned to the victim, his elbow went up, he fell down on the hand of the victim, his feet kind of tugged to one side, and he said, 'Like this, boss?' ZZUMPH!"

The experiment was played out to its bitter end. Milgram tried it with forty different subjects. And 30 percent of them obeyed the experimenter and kept on obeying.

"The protests of the victim were strong and vehement, he was screaming his guts out, he refused to participate, and you had to physically struggle with him in order to get his hand down on the shock generator," Milgram remembers. But twelve out of forty did it.

Milgram took his experiment out of New Haven. Not to Germany, just twenty miles down the road to Bridgeport. Maybe, he reasoned, the people obeyed because of the prestigious setting of Yale University. If they couldn't trust a center of learning that had been there for two centuries, whom could they trust? So he moved the experiment to an untrustworthy setting.

The new setting was a suite of three rooms in a run-down office building in Bridgeport. The only identification was a sign with a fictitious name: "Research Associates of Bridgeport." Questions about professional connections got only vague answers about "research for industry."

Obedience was less in Bridgeport. Forty-eight percent of the subjects stayed for the maximum shock, compared to 65 percent at Yale. But this was enough to prove that far more than Yale's prestige was behind the obedient behavior.

For more than seven years now, Stanley Milgram had been trying to figure out what makes ordinary American citizens so obedient. The most obvious answer—that people are mean, nasty, brutish, and sadistic—won't do. The subjects who gave the shocks to Mr. Wallace to the end of the board did not enjoy it. They groaned, protested, fidgeted, argued, and in some cases, were seized by fits of nervous, agitated giggling.

"They even try to get out of it," says Milgram, "but they are somehow engaged in something from which they cannot liberate themselves. They are locked into a structure, and they do not have the skills or inner resources to disengage themselves."

Milgram, because he mistakenly had assumed that he would have trouble getting people to obey the orders to shock Mr. Wallace, went to a lot of trouble to create a realistic situation.

There was crew-cut Jack Williams and his grey laboratory coat. Not white, which might denote a medical technician, but ambiguously authoritative grey. Then there was the book on the table, and the other appurtenances of the laboratory which emitted the silent message that things were being performed here in the name of science, and were therefore great and good.

But the nicest touch of all was the shock generator. When Milgram started out, he had only a $300 grant from the Higgins Fund of Yale University. Later he got more ample support from the National Science Foundation, but in the beginning he had to create this authentic-looking machine with very scarce resources except for his own imagination. So he went to New York and roamed around the electronic shops until he found some little black switches at Lafayette Radio for a dollar apiece. He bought thirty of them. The generator was a metal box, about the size of a small footlocker, and he drilled the thirty holes for the thirty switches himself in a Yale machine shop. But the fine detail was left to professional industrial engravers. So he ended up with a splendid-looking control panel dominated by the row of switches, each labeled with its voltage, and each having its own red light that flashed on when the switch was pulled. Other things happened when a switch was pushed. Besides the ZZUMPH-ing noise, a blue light labeled "voltage energizer" went on, and a needle on a dial labeled "voltage" flicked from left to right. Relays inside the box clicked. Finally, in the upper left-hand corner of the control panel was this inscription, engraved in precise block letters:

SHOCK GENERATOR TYPE ZLB
DYSON INSTRUMENT COMPANY
WALTHAM, MASS.
OUTPUT: 15 VOLTS—450 VOLTS

One day a man from the Lehigh Valley Electronics Company of Pennsylvania was passing through the laboratory, and he stopped to admire the shock generator.

"This is a very fine shock generator," he said. "But who is this Dyson Instrument Company?" Milgram felt proud at that, since Dyson Instrument Company existed only in the recesses of his imagination.

When you consider the seeming authenticity of the situation, you can appreciate the agony some of the subjects went through. It was pure conflict. As Milgram explains to his students, "When a parent says, 'Don't strike old ladies,' you are learning two things: the content, and, also, to obey authority. This experiment creates conflicts between the two elements."

Subjects in the experiment were not asked to give the 450-volt shock more than three times. By that time, it seemed evident that they would go on indefinitely. "No one," says Milgram, "who got within five shocks of the end ever broke off. By that point, he had resolved the conflict."

Why do so many people resolve the conflict in favor of obedience?

Milgram's theory assumes that people behave in two different operating modes as different as ice and water. He does not rely on Freud or sex or toilet-training hang-ups for this theory. All he says is that ordinarily we operate in a state of autonomy, which means we pretty much have and assert control over what we do. But in certain circumstances, we operate under what Milgram calls a state of agency (after agent, n....one who acts for or in the place of another by authority from him; a substitute; a deputy.—*Webster's Collegiate Dictionary*). A state of agency, to Milgram, is nothing more than a frame of mind.

"There's nothing bad about it, there's nothing good about it," he says. "It's a natural circumstance of living with other people....I think of a state of agency as a real transformation of a person; if a person has different properties when he's in that state, just as water can turn to ice under certain conditions of temperature, a person can move to the state of mind that I call agency...the critical thing is that you see yourself as the instrument of the execution of another person's wishes. You do not see yourself as acting on your own. And there's a real transformation, a real change of properties of the person."

To achieve this change, you have to be in a situation where there seems to be a ruling authority whose commands are relevant to some legitimate purpose; the authority's power is not unlimited.

But situations can be and have been structured to make people do unusual things, and not just in Milgram's laboratory. The reason, says Milgram, is that no action, in and of itself, contains meaning.

"The meaning always depends on your definition of the situation. Take an action like killing another person. It sounds bad."

"But then we say the other person was about to destroy a hundred children, and the only way to stop him was to kill him. Well, that sounds good."

"Or, you take destroying your own life. It sounds very bad. Yet, in the Second World War, thousands of persons thought it was a good thing to destroy your own life. It was set in the proper context. You sipped some saki from a whistling cup, recited a few haiku. You said, 'May my death be as clean and as quick as the shattering of crystal.' And it almost seemed like a good, noble thing to do, to crash your kamikaze plane into an aircraft carrier. But the main thing was, the definition of what a kamikaze pilot was doing had been determined by the relevant authority. Now, once you are in a state of agency, you allow the authority to determine, to define what the situation is. The meaning of your actions is altered."

So, for most subjects in Milgram's laboratory experiments, the act of giving Mr. Wallace his painful shock was necessary, even though unpleasant, and besides they were doing it on behalf of somebody else and it was for science. There was still strain and conflict, of course. Most people resolved it by grimly sticking to their task and obeying. But some broke out. Milgram tried varying the conditions of the experiment to see what would help break people out of their state of agency.

"The results, as seen and felt in the laboratory," he has written, "are disturbing. They raise the possibility that human nature, or more specifically the kind of character produced in American democratic society, cannot be counted on to insulate its citizens from brutality and inhumane treatment at the direction of malevolent authority. A substantial proportion of people do what they are told to do, irrespective of the content of the act and without limitations of conscience, so long as they perceive that the command comes from a legitimate authority.

If in this study, an anonymous experimenter can successfully command adults to subdue a fifty-year-old man and force on him painful electric shocks against his protest, one can only wonder what government, with its vastly greater authority and prestige, can command of its subjects."

This is a nice statement, but it falls short of summing up the full meaning of Milgram's work. It leaves some questions still unanswered.

The first question is this: Should we really be surprised and alarmed that people obey? Wouldn't it be even more alarming if they all refused to obey? Without obedience to a relevant ruling authority there could not be a civil society. And without a civil society, as Thomas Hobbes pointed out in the seventeenth century, we would live in a condition of war, "of every man against every other man," and life would be "solitary, poor, nasty, brutish, and short."

In the middle of one of Stanley Milgram's lectures at CUNY recently, some mini-skirted undergraduates started whispering and giggling in the back of the room. He told them to cut it out. Since he was the relevant authority in that time and place, they obeyed, and most people in the room were glad that they obeyed.

This was not, of course, a conflict situation. Nothing in the coeds' social upbringing made it a matter of conscience for them to whisper and giggle. But a case can be made that in a conflict situation it is all the more important to obey. Take the case of war, for example. Would we really want a situation in which every participant in a war, direct, or indirect—from front-line soldiers to the people who sell coffee and cigarettes to employees at the Concertina barbed-wire factory in Kansas—stops and consults his conscience before each action? It is asking for an awful lot of mental strain and anguish from an awful lot of people. The value of having civil order is that one can do his duty, or whatever interests him, or whatever seems to benefit him at the moment, and leave the agonizing to others. When Francis Gary Powers was being tried by a Soviet military tribunal after his U-2 spy plane was shot down, the presiding judge asked if he had thought about the possibility that his flight might have provoked a war. Powers replied with Hobbesian clarity: "The people who sent me should think of these things. My job was to carry out orders. I do not think it was my responsibility to make such decisions."

It was not his responsibility. And it is quite possible that if everyone felt responsible for each of the ultimate consequences of his own tiny contributions to complex chains of events, then society simply would not work. Milgram, fully conscious of the moral and social implications of his research, believes that people should feel responsible for their actions. If someone else had invented the experiment, and if he had been the naive subject, he feels certain that he would have been among the disobedient minority.

"There is no very good solution to this," he admits, thoughtfully. "To simply and categorically say that you won't obey authority may resolve your personal conflict, but it creates more problems for society which may be more serious in the long run. But I have no doubt that to disobey is the proper thing to do in this [the laboratory] situation. It is the only reasonable value judgment to make."

The conflict between the need to obey the relevant ruling authority and the need to follow your conscience becomes sharpest if you insist on living by an ethical system based on a rigid code—a code that seeks to answer all questions in advance of their being raised. Code ethics cannot solve the obedience problem. Stanley Milgram seems to be a situation ethicist, and situation ethics does offer a way out: When you feel conflict, you examine the situation and then make a choice among the competing evils. You may act with a presumption in favor of obedience, but reserve the possibility that you will disobey whenever obedience demands a flagrant and outrageous affront to conscience. This, by the way, is the philosophical position of many who resist the draft. In World War II, they would have fought. Vietnam is a different, an outrageously different, situation.

Life can be difficult for the situation ethicist, because he does not see the world in straight lines, while the social system too often assumes such a God-given, squared-off structure. If your moral code includes an injunction against all war, you may be deferred as a conscientious objector. If you merely oppose this particular war, you may not be deferred.

Stanley Milgram has his problems, too. He believes that in the laboratory situation he would not have shocked Mr. Wallace. His professional critics reply that in his real-life situation he has done the equivalent. He has placed innocent and naive subjects under great emotional strain and pressure in selfish obedience to his quest for knowledge. When you raise this issue with Milgram, he has an answer ready. There is, he explains patiently, a critical difference between his naive subjects and the man in the electric chair. The man in the electric chair (in the mind of the naive subject) is helpless, strapped in. But the naive subject is free to go at any time.

Immediately after he offers this distinction, Milgram anticipates the objection.

"It's quite true," he says, "that this is almost a philosophic position, because we have learned that some people are psychologically incapable of disengaging themselves. But that doesn't relieve them of the moral responsibility."

The parallel is exquisite. "The tension problem was unexpected," says Milgram in his defense. But he went on anyway. The naive subjects didn't expect the screaming protests from the strapped-in learner. But that went on.

"I had to make a judgment," says Milgram. "I had to ask myself, was this harming the person or not? My judgment is that it was not. Even in the extreme cases, I wouldn't say that permanent damage results."

Sound familiar? "The shocks may be painful," the experimenter kept saying, "but they're not dangerous."

After the series of experiments was completed, Milgram sent a report of the results to his subjects and a questionnaire, asking whether they were glad or sorry to have been in the experiment. Eighty-three and seven-tenths percent said they were glad and only 1.3 percent were sorry; 15 percent were neither sorry nor glad. However, Milgram could not be sure at the time of the experiment that only 1.3 percent would be sorry.

Kurt Vonnegut Jr. put one paragraph in the preface to *Mother Night*, in 1966, which pretty much says it for the people with their fingers on the shock-generator switches, for you and me, and maybe even for Milgram. "If I'd been born in Germany," Vonnegut says, "I suppose I would have *been* a Nazi, bopping Jews and gypsies and Poles around, leaving boots sticking out of snowbanks, warming myself with my sweetly virtuous insides. So it goes."

Just so. One thing that happened to Milgram back in New Haven during the days of the experiment was that he kept running into people he'd watched from behind the one-way glass. It gave him a funny feeling, seeing those people going about their everyday business in New Haven and knowing what they would do to Mr. Wallace if ordered to. Now that his research results are in and you've thought about it, you can get this funny feeling too. You don't need one-way glass. A glance in your own mirror may serve just as well.

36

The Social Requisites of Democracy Revisited

SEYMOUR MARTIN LIPSET

*Since the end of World War II, there has been a movement toward the creation of new
states. In recent years there has also been a movement toward more democratic states.
By looking at the conditions for a democracy, Lipset indicates what these states need if they
are truly to achieve democracy. Considering these ideas, what do you see as the future of
world affairs?*

As you read this selection, ask yourself the following questions:

1. *What are some of the requisites needed to maintain democracy?*
2. *Based on the requisites needed to maintain democracy, how is
    democracy doing? Why?*
3. *Does the United States have a democracy? Why, or why not?*

GLOSSARY **Efficacy** Producing desired results. **Bourgeois** Member of the
middle class. **Facade** Superficial appearance. **Meritocratic** Hiring on the
basic of ability rather than patronage. **Totalitarian** Absolute control by
the state.

The recent expansion of democracy ... began in
the mid-1970s in Southern Europe. Then,
in the early and mid-1980s, it spread to Latin
America and to Asian countries like Korea,
Thailand, and the Philippines, and then in the late
1980s and early 1990s to Eastern Europe, the Soviet
Union, and parts of sub-Saharan Africa. Not long
ago, the overwhelming majority of the members of
the United Nations had authoritarian systems. As of
the end of 1993, over half, 107 out of 186 countries, have competitive elections and various guar-
antees of political and individual rights—that is
more than twice the number of two decades earlier
in 1970 (Karatnycky 1994: 6; *Freedom Review*
1993:3–4, 10). The move toward democracy is

not a simple one.... Countries that previously
have had authoritarian regimes may find it difficult
to set up a legitimate democratic system, since their
traditions and beliefs may be incompatible with the
workings of democracy.

In his classic work *Capitalism, Socialism, and
Democracy*, Schumpeter (1950) defined democracy
as "that institutional arrangement for arriving at
political decisions in which individuals acquire the
power to decide by means of a competitive struggle
for the people's vote" (p. 250). This definition is
quite broad and my discussion here cannot hope
to investigate it exhaustively. Instead, I focus here
on ... the factors and processes affecting the prospects for the institutionalization of democracy.

Source: From the *American Sociological Review*, 1994, Vol. 59 (February), pp. 1–22.

HOW DOES DEMOCRACY ARISE?

Politics in Impoverished Countries

In discussing democracy, I want to clarify my biases and assumptions at the outset. I agree with the basic concerns of the founding fathers of the United States—that government, a powerful state, is to be feared (or suspected, to use the lawyer's term), and that it is necessary to find means to control governments through checks and balances. In our time, as economists have documented, this has been particularly evident in low-income nations. The "Kuznets curve" (Kuznets 1955; 1963; 1976), although still debated, indicates that when a less developed nation starts to grow and urbanize, income distribution worsens, but then becomes more equitable as the economy industrializes.* ... Before development, the class income structure resembles an elongated pyramid, very fat at the bottom, narrowing or thin toward the middle and top (Lipset 1981: 51). Under such conditions, the state is a major, usually *the* most important, source of capital, income, power, and status. This is particularly true in statist systems, but also characterizes many so-called free market economies. For a person or governing body to be willing to give up control because of an election outcome is astonishing behavior, not normal, not on the surface a "rational choice," particularly in new, less stable, less legitimate politics.

Marx frequently noted that intense inequality is associated with scarcity, and therefore that socialism, which he believed would be an egalitarian and democratic system with a politically weak state, could only occur under conditions of abundance (Marx 1958: 8–9). To try to move toward socialism under conditions of material scarcity would result in sociological abortions and in repression. The Communists proved him correct. Weffort (1992), a Brazilian scholar of democracy, has argued strongly that, although "the political equality of citizens, ... is ... possible in societies marked by a high degree of [economic] inequality," the contradiction between political and economic inequality "opens the field for tensions, institutional distortions, instability, and recurrent violence ... [and may prevent] the consolidation of democracy" (p. 22). Contemporary social scientists find that greater affluence and higher rates of well-being have been correlated with the presence of democratic institutions (Lipset, Seong, and Torres 1993:156–58; see also Diamond 1993a). Beyond the impact of national wealth and economic stratification, contemporary social scientists also agree with Tocqueville's analysis, that social equality, perceived as equality of status and respect for individuals regardless of economic condition, is highly conducive for democracy (Tocqueville 1976: vol. 2, 162–216).... Weffort (1992) emphasized, "such a 'minimal' social condition is absent from many new democracies,... [which can] help to explain these countries' typical democratic instability" (p. 18).

The Economy and the Polity

In the nineteenth century, many political theorists noted the relationship between a market economy and democracy (Lipset 1992:2). As Glassman (1991) has documented, "Marxists, classical capitalist economists, even monarchists accepted the link between industrial capitalism and parliamentary democracy" (p. 65). Such an economy, including a substantial independent peasantry, produces a middle class that can stand up against the state and provide the resources for independent groups.... Schumpeter (1950) held that, "modern democracy is a product of the capitalist process" (p. 297). Moore (1966), noting his agreement with the Marxists, concluded, "No bourgeois, no democracy" (p. 418)....

Waisman (1992:140–55), seeking to explain why some capitalist societies, particularly in Latin America, have not been democratic, has suggested that ... a free market needs democracy and vice versa.

But while the movement toward a market economy and the growth of an independent middle-class have weakened state power and enlarged human

---

*These generalizations do not apply to the East Asian NICS, South Korea, Taiwan, and Singapore.

rights and the rule of law, it has been the working class, particularly in the West, that has demanded the expansion of suffrage and the rights of parties. As John Stephens (1993) noted, "Capitalist development is associated with the rise of democracy in part because it is associated with the transformation of the class structure strengthening the working class" (p. 438)....

Therefore, a competitive market economy can be justified sociologically and politically as the best way to reduce the impact of nepotistic networks. The wider the scope of market forces, the less room there will be for rent-seeking by elites with privileged access to state power and resources. Beyond limiting the power of the state, however, standards of propriety should be increased in new and poor regimes, and explicit objective standards should be applied in allocating aid, loans, and other sources of capital from outside the state. Doing this, of course, would be facilitated by an efficient civil service selected by meritocratic standards. It took many decades for civil service reforms to take hold in Britain, the United States, and various European countries. To change the norms and rules in contemporary impoverished countries will not be achieved easily....

The Centrality of Political Culture

Democracy requires a supportive culture, the acceptance by the citizenry and political elites of principles underlying freedom of speech, media, assembly, religion, of the rights of opposition parties, of the rule of law, of human rights, and the like.... Such norms do not evolve overnight.... "Only four of the seventeen countries that adopted democratic institutions between 1915 and 1931 maintained them throughout the 1920s and 1930s. ...[O]ne-third of the 32 working democracies in the world in 1958 had become authoritarian by the mid-1970s" (Huntington 1991:17–21).

These experiences do not bode well for the current efforts in the former Communist states of Eastern Europe or in Latin America and Africa. And the most recent report by Freedom House concludes: "As 1993 draws to a close, freedom around

the world is in retreat while violence, repression, and state control are on the increase. The trend marks the first increase in five years ..." (Karatnycky 1994:4). A "reverse wave" in the making is most apparent in sub-Saharan Africa, where "9 countries showed improvement while 18 registered a decline" (p. 6). And in Russia, a proto-fascist movement led all other parties, albeit with 24 percent of the vote, in the December 1993 elections, while the Communists and their allies secured over 15 percent.

Almost everywhere that the institutionalization of democracy has occurred, the process has been a gradual one in which opposition and individual rights have emerged in the give and take of politics....

As a result, democratic systems developed gradually, at first with suffrage, limited by and linked to property and/or literacy. Elites yielded slowly in admitting the masses to the franchise and in tolerating and institutionalizing opposition rights. ... As Dahl (1971:36–37) has emphasized, parties such as the Liberals and Conservatives in nineteenth-century Europe, formed for the purpose of securing a parliamentary majority rather than to win the support of a mass electorate, were not pressed to engage in populist demagoguery.

Comparative politics suggest that the more the sources of power, status, and wealth are concentrated in the state, the harder it is to institutionalize democracy. Under such conditions the political struggle tends to approach a zero-sum game in which the defeated lose all. The greater the importance of the central state as a source of prestige and advantage, the less likely it is that those in power—or the forces of opposition—will accept rules of the game that institutionalize party conflict and could result in the turnover of those in office. Hence, once again it may be noted, the chances for democracy are greatest where, as in the early United States and to a lesser degree in other Western nations, the interaction between politics and economy is limited and segmented. In Northern Europe, democratization let the monarchy and the aristocracy retain their elite status, even though their powers were curtailed. In the United States, the central state was not a major source of privilege for the first half-century or more, and those at the center thus could yield office easily.

Democracy has never developed anywhere by plan, except when it was imposed by a democratic conqueror, as in post–World War II Germany and Japan. From the United States to Northern Europe, freedom, suffrage, and the rule of law grew in a piecemeal, not in a planned, fashion. To legitimate themselves, governmental parties, even though they did not like it, ultimately had to recognize the right of oppositions to exist and compete freely. Almost all the heads of young democracies, from John Adams and Thomas Jefferson to Indira Gandhi, attempted to suppress their opponents.... Democratic successes have reflected the varying strengths of minority political groups and lucky constellations, as much or more than commitments by new office holders to the democratic process.

Cross-national historical evaluations of the correlates of democracy have found that cultural factors appear even more important than economic ones (Lipset et al. 1993:168–70).... Dahl (1970:6), Kennan (1977:441–43), and Lewis (1993:93–94) have emphasized that the first group of countries that became democratic in the nineteenth century (about 20 or so) were Northwest European or settled by Northwest Europeans. "The evidence has yet to be produced that it is the natural form of rule for peoples outside these narrow perimeters" (Kennan 1977:41–43).[*]...

More particularly, recent statistical analyses of the aggregate correlates of political regimes have indicated that having once been a British colony is the variable most highly correlated with democracy (Lipset et al. 1993:168). ... The factors underlying this relationship are not simple (Smith 1978). In the British/non-British comparison, many former British colonies, such as those in North America before the revolution or India and Nigeria in more recent times, had elections, parties, and the rule of law before they became independent. In contrast, the Spanish, Portuguese, French, Dutch,

and Belgian colonies, and former Soviet-controlled countries did not allow for the gradual incorporation of "out/groups" into the polity. Hence democratization was much more gradual and successful in the ex-British colonies than elsewhere; their pre-independence experiences were important as a kind of socialization process and helped to ease the transition to freedom.

Religious Tradition

Religious tradition has been a major differentiating factor in transformations to democracy (Huntington 1993:25–29). Historically, there have been negative relationships between democracy and Catholicism, Orthodox Christianity, Islam, and Confucianism; conversely Protestantism and democracy have been positively interlinked. These differences have been explained by (1) the much greater emphasis on individualism in Protestantism and (2) the traditionally close links between religion and the state in the other four religions. Tocqueville (1975) and Bryce (1901) emphasized that democracy is furthered by a separation of religious and political beliefs, so that political stands are not required to meet absolute standards set down by the church....

Protestants, particularly the non-state-related sects, have been less authoritarian, more congregational, participatory, and individualistic. Catholic countries, however, have contributed significantly to the third wave of democratization during the 1970s and 1980s, reflecting "the major changes in the doctrine, appeal, and social and political commitments of the Catholic Church that occurred ... in the 1960s and 1970s" (Huntington 1991:281, 77–85)....

Conversely, Moslem (particularly Arab) states have not taken part in the third wave of democratization. Almost all remain authoritarian. Growth of democracy in the near future in most of these countries is doubtful because "notions of political freedom are not held in common...; they are alien to Islam." As Wright (1992) has stated, Islam "offers

---

[*]That evidence, of course, has emerged in recent years in South and East Asia, Latin America, and various countries descended from Southern Europe.

not only a set of spiritual beliefs, but a set of rules by which to govern society" (p. 133)....

Kazancigil (1991) has offered parallel explanations of the weakness of democracy in Islam with those for Orthodox Christian lands as flowing from their failures "to dissociate the religious from the political spheres" (p. 345)....

... It is significant that ... both Confucianism and Maoism in ideological content have explicitly stressed the problems of authority and order" (Pye 1968:16). Though somewhat less pessimistic, He Baogang's (1992) evaluation of cultural factors in mainland China concluded that "evidence reveals that the antidemocratic culture is currently stronger than the factors related to a democratic one" (p. 134). Only Japan, the most diluted Confucian country, "had sustained experience with democratic government prior to 1990,... [although its] democracy was the product of an American presence" (Huntington 1991:15). The others—Korea, Vietnam, Singapore, and Taiwan—were autocratic.... The situation, of course, has changed in recent years in response to rapid economic growth, reflecting the ways in which economic changes can impact on the political system undermining autocracy.

But India, a Hindu country that became democratic prior to industrialization, is different:

> The most salient feature of Indian civilization, from the point of view of our discussion, is that it is probably the only complete, highly differentiated civilization which throughout history has maintained its cultural identity without being tied to a given political framework.... [T]o a much greater degree than in many other historical imperial civilizations politics were conceived in secular forms.... Because of the relative dissociation between the cultural and the political order, the process of modernization could get underway in India without being hampered by too specific a traditional-cultural orientation toward the political sphere. (Eisenstadt 1968:32)

These generalizations about culture do not auger well for the future of the third wave of democracy in the former Communist countries. The Catholic Church played a substantial role in Poland's move away from Soviet Communism. But as noted previously, historically deeply religious Catholic areas have not been among the most amenable to democratic ideas. Poland is now troubled by conflicts flowing from increasing Church efforts to affect politics in Eastern Europe even as it relaxes its policies in Western Europe and most of the Americas. Orthodox Christianity is hegemonic in Russia and Belarus. The Ukraine is dominated by both the Catholic and Orthodox Churches. And fascists and Communists are strong in Russia and the Ukraine. Moslems are a significant group in the Central Asian parts of the former Soviet Union, the majority in some—these areas are among the consistently least democratic of the successor Soviet states. Led by the Orthodox Serbians, but helped by Catholic Croats and Bosnian Moslems, the former Yugoslavia is being torn apart along ethnic and religious lines with no peaceful, much less democratic, end in sight. We are fooling ourselves if we ignore the continuing dysfunctional effects of a number of cultural values and the institutions linked to them.

But belief systems change; and the rise of capitalism, a large middle class, an organized working class, increased wealth, and education are associated with secularism and the institutions of civil society which help create autonomy for the state and facilitate other preconditions for democracy. In recent years, nowhere has this been more apparent than in the economically successful Confucian states of East Asia—states once thought of as nearly hopeless candidates for both development and democracy. Tu (1993) noted their totally "unprecedented dynamism in democratization and marketization. Singapore, South Korea, and Taiwan all successfully conducted national elections in 1992, clearly indicating that democracy in Confucian societies is not only possible but also practical" (p. viii). Nathan and Shi (1993), reporting on "the first scientifically valid national sample survey done in China on political behavior and attitudes," stated: "When compared to residents of some of the most stable, long-established democracies in the world, the Chinese population scored lower on the variables we looked at, but not so low as to justify the conclusion that democracy is out of reach" (p. 116). Surveys which have been done in Russia offer similar positive conclusions (Gibson and Duch 1993), but the December 1993 election in which racist nationalists and pro-Communists did well indicate much more is needed. Democracy is not taking root in much of the former Soviet Union, the less industrialized Moslem states, nor many nations in Africa. The end is not in sight for many of the efforts at new democracies; the requisite cultural changes are clearly not established enough to justify the conclusion that the "third wave" will not be reversed. According to the Freedom House survey, during 1993 there were "42 countries registering a decline in their level of freedom [political rights and civil liberties] and 19 recording gains" (Karatnycky 1994:5).*

INSTITUTIONALIZATION

New democracies must be institutionalized, consolidated, and become legitimate. They face many problems....

Legitimacy

Political stability in democratic systems cannot rely on force. The alternative to force is legitimacy, an accepted systemic "title to rule...."

Weber (1946), the fountainhead of legitimacy theory, named three ways by which an authority may gain legitimacy. These may be summarized:

1. *Traditional*—through "always" having possessed the authority, the best example being the title held in monarchical societies.

2. *Rational-legal*—when authority is obeyed because of a popular acceptance of the appropriateness of the system of rules under which they have won and held office. In the United States, the Constitution is the basis of all authority.

3. *Charismatic*—when authority rests upon faith in a leader who is believed to be endowed with great personal worth, either from God, as in the case of a religious prophet or simply from the display of extraordinary talents. The "cult of personality" surrounding many leaders is an illustration of this (pp. 78–79).

Legitimacy is best gained by prolonged effectiveness, effectiveness being the actual performance of the government and the extent to which it satisfies the basic needs of most of the population and key power groups (such as the military and economic leaders) (Lipset 1979:16–23; Linz 1988:79–85). This generalization, however, is of no help to new systems for which the best immediate institutional advice is to separate the source and the agent of authority.

The importance of this separation cannot be underestimated. The agent of authority may be strongly opposed by the electorate and may be changed by the will of the voters, but the essence of the rules, the symbol of authority, must remain respected and unchallenged. Hence, citizens obey the laws and rules, even while disliking those who enforce them....

Rational-legal legitimacy is weak in most new democratic systems, since the law had previously operated in the interests of a foreign exploiter or domestic dictator. Efforts to construct rational-legal legitimacy necessarily involve extending the rule of the law and the prestige of the courts, which should be as independent from the rest of the polity as possible. As Ackerman (1992:60–62) and Weingast (1993) note, in new democracies, these requirements imply the need to draw up a "liberal" constitution *as soon as possible*. The constitution can provide a basis for legitimacy, for limitations on state power, and for political and economic rights....

To reiterate, if democratic governments which lack traditional legitimacy are to survive, they must

---

*In the Freedom House survey, a country may move up or down with respect to measures of freedom without changing its status as a democratic or authoritarian system.

be effective, or as in the example of some new Latin American and post-communist democracies, may have acquired a kind of negative legitimacy—an inoculation against authoritarianism because of the viciousness of the previous dictatorial regimes. Newly independent countries that are post-revolutionary, post-coup, or post-authoritarian regimes are inherently low in legitimacy. Thus most of the democracies established in Europe after World War I as a result of the overthrow of the Austro-Hungarian, German, and Czarist Russian empires did not last....

All other things being equal, an assumption rarely achieved, nontraditional authoritarian regimes are more brittle than democratic ones. By definition, they are less legitimate; they rely on force rather than belief to retain power. Hence, it may be assumed that as systems they are prone to be disliked and rejected by major segments of the population....

The record, as in the case of the Soviet Union, seems to contradict this, since that regime remained in power for three-quarters of a century. However, a brittle, unpopular system need not collapse. Repressive police authority, a powerful army, and a willingness by rulers to use brute force may maintain a regime's power almost indefinitely. The breakdown of such a system may require a major catalytic event, a defeat in war, a drastic economic decline, or a break in the unity of the government elite....

In contrast to autocracies, democratic systems rely on and seek to activate popular support and constantly compete for such backing. Government ineffectiveness need not spill into other parts of the society and economy. Opposition actually serves as a communication mechanism, focusing attention on societal and governmental problems. Freedom of opposition encourages a free flow of information about the economy as well as about the polity....

Non-traditional authoritarian regimes seek to gain legitimacy through cults of personality (e.g., Napoleon, Toussaint, Diaz, Mussolini, Hitler). New autocrats lack the means to establish legal-rational legitimacy through the rule of law. Communist governments, whose Marxist ideology explicitly denied the importance of "great men" in history and stressed the role of materialist forces and "the people," were forced to resort to charismatic legitimacy. Their efforts produced the cults of Lenin, Stalin, Mao, Tito, Castro, Ho, Kim, and others....

But charismatic legitimacy is inherently unstable. As mentioned earlier, a political system operates best when the source of authority is clearly separated from the agent of authority. If the ruler and his or her policies are seen as oppressive or exploitive, the regime and its rules will also be rejected. People will not feel obligated to conform or to be honest; force alone cannot convey a "title to rule."

EXECUTIVE AND ELECTORAL SYSTEMS ...

Executive Systems

In considering the relation of government structure to legitimacy it has been suggested that republics with powerful presidents will, all other things being equal, be more unstable than parliamentary ones in which powerless royalty or elected heads of state try to act out the role of a constitutional monarch. In the former, where the executive is chief of state, symbolic authority and effective power are combined in one person, while in the latter they are divided. With a single top office, it is difficult for the public to separate feelings about the regime from those held toward the policy makers. The difficulties in institutionalizing democracy in the many Latin American presidential regimes over the last century and a half may reflect this problem. The United States presents a special case, in which, despite combining the symbolic authority and power into the presidency, the Constitution has been so hallowed by ideology and prolonged effectiveness for over 200 years, that it, rather than those who occupy the offices it specifies, has become the accepted ultimate source of authority....

Evaluation of the relative worth of presidential and parliamentary systems must also consider the nature of each type. In presidential regimes, the power to enact legislation, pass budgets and appropriations, and make high-level appointments are divided among the president and (usually two) legislative Houses; parliamentary regimes are unitary regimes, in which the prime minister and cabinet can have their way legislatively. A prime minister with a parliamentary majority, as usually occurs in most Commonwealth nations and a number of countries in Europe, is much more powerful and less constrained than a constitutional president who can only propose while Congress disposes (Lijphart 1984:4–20). The weak, divided-authority system has worked in the United States, although it has produced much frustration and alienation at times. But, as noted, the system has repeatedly broken down in Latin America, although one could argue that this is explained not by the constitutional arrangements, but by cultural legacies and lower levels of productivity. Many parliamentary systems have failed to produce stable governments because they lack operating legislative majorities.... There is no consensus among political scientists as to which system, presidential or parliamentary, is superior, since it is possible to point to many failures for both types.

Electoral Systems

The procedures for choosing and changing administrations also affect legitimacy. Elections that offer the voters an effective way to change the government and vote the incumbents out will provide more stability; electoral decisions will be more readily accepted in those systems in which electoral rules, distribution of forces, or varying party strengths make change more difficult.

Electoral systems that emphasize single-member districts, such as those in the United States and in much of the Commonwealth, press the electorate to choose between two major parties....

In systems with proportional representation, the electorate may not be able to determine the composition of the government. In this type, a representation is assigned to parties which corresponds to their proportions of the vote.... Where no party has a majority, alliances may be formed out of diverse forces.... Small, opportunistic, or special interest parties may hold the balance of power and determine the shape and policies of post-election coalitions. The tendency toward instability and lack of choice in proportional systems can be reduced by setting up a minimum vote for representation, such as the five percent cut-off that exists in Germany and Russia. In any case, electoral systems, whether based on single-member districts or proportional representation, cannot guarantee particular types of partisan results (Gladdish 1993).

CIVIL SOCIETY AND POLITICAL PARTIES

Civil Society as a Political Base

More important than electoral rules in encouraging a stable system is a strong civil society—the presence of myriad "mediating institutions," including "groups, media, and networks" (Diamond 1993b:4), that operate independently between individuals and the state. These constitute "subunits, capable of opposing and countervailing the state."...

Citizen groups must become the bases of—the sources of support for—the institutionalized political parties which are a necessary condition for ... a modern democracy....

A fully operative civil society is likely to also be a participant one. Organizations stimulate interests and activity in the larger polity; they can be consulted by political institutions about projects that affect them and their members, and they can transfer this information to the citizenry. Civil organizations reduce resistance to unanticipated changes because they prevent the isolation of political institutions from the polity and can smooth over, or at least recognize, interest differences early on....

Totalitarian systems, however, do not have effective civil societies. Instead, they either seek to eliminate groups mediating between the individual and the state or to control these groups so there is no competition....

The countries of Eastern Europe and the former Soviet Union, however, are faced with the consequences of the absence of modern civil society, a lack that makes it difficult to institutionalize democratic polities.... Instead, they have had to create parties "from scratch."... "Instead of consolidation, there is fragmentation: 67 parties fought Poland's most recent general election, 74 Romania's" (*Economist* 1993a:4). As a result, the former Communists (now "socialists") have either been voted in as the majority party in parliament, as in Lithuania, or have become the largest party heading up a coalition cabinet, as in Poland. In January 1992, the Communist-backed candidate for president in Bulgaria garnered 43 percent of the vote (Malia 1992:73). These situations are, of course, exacerbated by the fact that replacing command economies by market processes is difficult, and frequently conditions worsen before they begin to improve.

Recent surveys indicate other continuing effects of 45 to 75 years of Communist rule. An overwhelming majority (about 70 percent) of the population in nearly all of the countries in Eastern Europe agree that "the state should provide a place of work, as well as a national health service, housing, education, and other services" (*Economist* 1993a:5)....

Political Parties as Mediators

Political parties themselves must be viewed as the most important mediating institutions between the citizenry and the state (Lipset 1993). And a crucial condition for a stable democracy is that major parties exist that have an almost permanent significant base of support. That support must be able to survive clear-cut policy failures by the parties. If this commitment does not exist, parties may be totally wiped out, thus eliminating effective opposition....

If, as in new democracies, parties do not command such allegiance, they can be easily eliminated. The Hamiltonian Federalist party, which competed in the early years of the American Republic with the Jeffersonian Democratic-Republicans, declined sharply after losing the presidency in 1800 and soon died out (Dauer 1953).... It may be argued then, that having at least two parties with an uncritically loyal mass base comes close to being a necessary

condition for a stable democracy. Democracy requires strong parties that can offer alternative policies and criticize each other....

Sources of Political Party Support

... These four sources of conflict, *center–periphery, state–church, land–industry,* and *capitalist–worker,* have continued to some extent in the contemporary world, and have provided a framework for the party systems of the democratic polities, particularly in Europe. Class became the most salient source of conflict and voting, particularly after the extension of the suffrage to all adult males (Lipset and Rokkan 1967). Both Tocqueville (1976:vol. 2, 89–93), in the early nineteenth century, and Bryce (1901:335), at the end of it, noted that at the bottom of the American political party conflict lay the struggle between aristocratic and democratic interests and sentiments.... Given all the transformations in Western society over the first half of the twentieth century, it is noteworthy how little the formal party systems changed. Essentially the conflicts had become institutionalized—the Western party systems of the 1990s resemble those of pre–World War II....

Beginning in the mid-1960s, the Western world appears to have entered a new political phase. It is characterized by the rise of so-called "post-materialistic issues, a clean environment, use of nuclear power, a better culture, equal status for women and minorities, the quality of education, international relations, greater democratization, and a more permissive morality, particularly as affecting familial and sexual issues" (Lipset 1981: 503–21). These have been perceived by some social analysts as the social consequences of an emerging third "revolution," the Post-Industrial Revolution, which is introducing new bases of social and political conflict. Inglehart (1990) and others have pointed to new cross-cutting lines of conflict—an *industrial-ecology* conflict—between the adherents of the industrial society's emphasis on production (who also hold conservative positions on social issues) and those who espouse the post-industrial emphasis on the quality-of-life and liberal social views when dealing with ecology, feminism, and nuclear energy. Quality-of-life concerns are difficult to formulate as party issues, but groups such as the Green parties and the New Left or New Politics—all educated middle-class groups—have sought to foster them....

The one traditional basis of party differentiation that seems clearly to be emerging in Russia is the center–periphery conflict, the first one that developed in Western society. The second, church–state (or church–secular), is also taking shape to varying degrees. Land–industry (or rural–urban) tension is somewhat apparent. Ironically, the capitalist–worker conflict is as yet the weakest, perhaps because a capitalist class and an independently organized working-class do not yet exist. Unless stable parties can be formed, competitive democratic politics is not likely to last in many of the new Eastern European and Central Asian polities....

THE RULE OF LAW AND ECONOMIC ORDER

Finally, order and predictability are important for the economy, polity, and society. The Canadian Fathers of Confederation, who drew up the newly unified country's first constitution in 1867, described the Constitution's objective as "peace, order, and good government" (Lipset 1990b:xiii). Basically, they were talking about the need for the "rule of law," for establishing rules of "due process," and an independent judiciary. Where power is arbitrary, personal, and unpredictable, the citizenry will not know how to behave; it will fear that any action could produce an unforeseen risk. Essentially, the rule of law means: (1) that people and institutions will be treated equally by the institutions administering the law—the courts, the police, and the civil service; and (2) that people and institutions can predict with reasonable certainty the consequences of their actions, at least as far as the state is concerned....

In discussing "the social requisites of democracy," I have repeatedly stressed the relationship between the level of economic development and the presence of democratic government....

Clearly, socioeconomic correlations are merely associational, and do not necessarily indicate cause.

Other variables, such as the force of historical incidents in domestic politics, cultural factors, events in neighboring countries, diffusion effects from elsewhere, leadership, and movement behavior can also affect the nature of the polity. Thus, the outcome of the Spanish Civil War, determined in part by other European states, placed Spain in an authoritarian mold, much as the allocation of Eastern Europe to the Soviet Union after World War II determined the political future of that area and that Western nations would seek to prevent the electoral victories of Communist-aligned forces. Currently, international agencies and foreign governments are more likely to endorse pluralistic regimes....

CONCLUSION

Democracy is an international cause. A host of democratic governments and parties, as well as various non-governmental organizations (NGOs) dedicated to human rights, are working and providing funds to create and sustain democratic forces in newly liberalized governments and to press autocratic ones to change (*Economist* 1993c:46). Various international agencies and units, like the European Community, NATO, the World Bank, and the International Monetary Fund (IMF), are requiring a democratic system as a condition for membership or aid. A diffusion, a contagion, or demonstration effect seems operative, as many have noted, one that encourages democracies to press for change and authoritarian rulers to give in. It is becoming both uncouth and unprofitable to avoid free elections, particularly in Latin America, East Asia, Eastern Europe, and to some extent in Africa (Ake 1991:33). Yet the proclamation of elections does not ensure their integrity. The outside world can help, but the basis for institutionalized opposition, for interest and value articulation, must come from within.

Results of research suggest that we be cautious about the long-term stability of democracy in many of the newer systems given their low level of legitimacy. As the Brazilian scholar Francisco Weffort (1992) has reminded us, "In the 1980s, the age of new democracies, the processes of political democratization occurred at the same moment in which those countries suffered the experience of a profound and prolonged economic crisis that resulted in social exclusion and massive poverty.... Some of those countries are building a political democracy on top of a minefield of social apartheid ..." (p. 20). Such conditions could easily lead to breakdowns of democracy as have already occurred in Algeria, Haiti, Nigeria, and Peru, and to the deterioration of democratic functioning in countries like Brazil, Egypt, Kenya, the Philippines, and the former Yugoslavia, and some of the trans-Ural republics or "facade democracies," as well as the revival of anti-democratic movements on the right and left in Russia and in other formerly Communist states.

What new democracies need, above all, to attain legitimacy is efficacy—particularly in the economic arena, but also in the polity....

REFERENCES

Eidtors's Note: The original chapter from which this selection was taken has extensive references that could not be listed here. For more documentation see the source note on page 283.

Ackerman, Bruce. 1992. *The Future of Liberal Revolution.* New Haven, CT: Yale University.

Bryce, James. 1901. *Study in History and Jurisprudence.* New York: Oxford University.

Dahl, Robert. 1970. *After the Revolution: Authority in a Good Society.* New Haven, CT: Yale University.

———. 1971. *Polyarchy: Participation and Opposition.* New Haven, CT: Yale University.

Gibson James L., and Raymond M. Duch. 1993. "Emerging Democratic Values in Soviet Political Culture." Pp. 69–94 in *Public Opinion and Regime Change,* A. A. Miller, W. M. Reisinger, and V. Hesli. Boulder, CO: Westview.

Gladdish, Ken. 1993. "The Primacy of the Particular." *Journal of Democracy* 4(1):53–65.

Glassman, Ronald. 1991. *China in Transition: Communism, Capitalism and Democracy.* Westport, CT: Praeger.

He Baogang. 1992. "Democratization: Antidemocratic and Democratic Elements in the Political Culture of China." *Australian Journal of Political Science* 27:120–36.

Huntington, Samuel. 1968. *Political Order in Changing Societies.* New Haven, CT: Yale University.

Kohli, Atul. 1992. "Indian Democracy: Stress and Resilience." *Journal of Democracy* 3(1):52–64.

Kuznets, Simon. 1955. "Economic Growth and Income Inequality." *American Economic Review* 45:1–28.

———. 1963. "Quantitative Aspects of the Economic Growth of Nations: VIII, The Distribution of Income by Size." *Economic Development and Cultural Change* 11:1–80.

———. 1976. *Modern Economic Growth: Rate, Structure and Spread.* New Haven, CT: Yale University.

Lewis, Bernard. 1993. "Islam and Liberal Democracy." *Atlantic Monthly.* 271(2):89–98.

Linz, Juan J. 1988. "Legitimacy of Democracy and the Socioeconomic System." Pp. 65–97 in *Comparing Pluralist Democracies: Strains on Legitimacy,* M. Dogan. Boulder, CO: Westview.

Lipset, Seymour Martin, and Stein Rokkan. 1967. "Cleavage Structures, Party Systems and Voter Alignments." Pp. 1–64 in *Party Systems and Voter Alignments,* S. M. Lipset and S. Rokkan. New York: Free Press.

Lipset, Seymour Martin, Kyoung–Ryung Seong, and John Charles Torres. 1993. "A Comparative Analysis of the Social Requisites of Democracy." *International Social Science Journal* 45:155–75.

Malia, Martin. 1992. "Leninist Endgame." *Daedalus* 121(2):57–75.

Weffort, Francisco C. 1992. "New Democracies, Which Democracies?" (Working Paper #198). Washington, DC: The Woodrow Wilson Center, Latin American Program.

Weingast, Barry. 1993. "The Political Foundations of Democracy and the Rule of Law." Stanford, CA: Hoover Institution, unpublished manuscript.

Wright, Robin. 1992. "Islam and Democracy." *Foreign Affairs* 71(3):131–45.

PART IV

Inequalities Between Groups

T he world can be kind or cruel to its inhabitants. Some people live privileged lives with such modern comforts as access to health care, communication systems, nice homes, plenty of food, and transportation facilities; others
struggle for survival due to disease, scarcities of food, and war. For example,
we are here in a university while most of our fellow young people in the
world will be lucky to get an elementary education. The chapters in this part
of the text examine in detail two major issues that separate the privileged from
those not so fortunate.

If asked, most of us would identify ourselves as middle class because it would
be elitist to think of ourselves as upper class and negative to think of ourselves as
lower class. The truth, however, is that we rarely think of the people in this
country as being categorized into different financial, gender, and racial classes.
Yet the chapter on gender revealed the wide disparities in our treatment of
men and women—a difference in treatment noted in our Declaration of
Independence: "all men are created equal." Our founding "fathers" obviously
considered women inferior to men and would deny them the vote for almost
100 years. Similarly, the last chapter on the economic and political institutions
revealed vast inequalities in wealth and political power. In short, people are
layered into various segments of privilege in society—they are *stratified* in the
system. Chapter 9 examines the causes and consequences of the inequalities of
stratification.

Another form of stratification is one based on the different ethnic and racial
groups in society. The intermingling of the world's peoples has produced group
relations that have led to discrimination, prejudice, and stereotyping based on
their cultural and physical differences. Chapter 10 examines the inequalities suffered by these ethnic and racial groups as well as their efforts to overcome them.

Members of both lower groups—the poor and the various ethnic/racial groups—are considered social problems because they are believed to be the primary cause of deviance and poverty in society. In short, most citizens see these social problems as individual problems caused by "bad" people. For this reason, it is necessary to understand the difference between individual and social problems.

C. Wright Mills (1958, pp. 8–9) summed up the difference between a social problem and an individual's personal problem by noting that social problems are "public issues of social structure," whereas personal problems are "personal troubles of milieu." A social problem, then, is a situation whose causes and solutions lie *outside* the individual and his or her immediate environment, whereas a personal problem is one whose causes and solutions lie *within* the individual and his or her immediate environment. For example, poverty is a problem experienced personally by numerous families; at the same time, it is a part of the larger social pattern of such problems as health care and unemployment. The problem cannot be resolved by dealing with only the one family's motivations. The question is no longer what is wrong with the individual family, but rather what is wrong with the economic institution? In the first case, the problem is blamed on the individual families involved: the "victim is being blamed as both the source and the solution to the problem."

Identifying the situation as a social problem, in contrast, leads to different actions.

The difference between a personal problem and a social problem leads to a final question in this regard: How does what is seen as a personal problem become translated into what is seen as a social problem? Obviously, some problems, such as war, have always been considered a social problem. Other problems, such as alcoholism, seem to fluctuate from being an individual problem to being a social problem needing social reform—Prohibition—back to being a personal problem. Still other problems originally seen as a personal problem—racial discrimination—are now labeled as a social problem.

For the latter change to occur, it required a social movement, the civil rights movement. A social movement occurs when groups of people band together to promote a particular cause. Surprisingly, regardless of the problem being developed, similar steps are involved:

1. People with a common interests and values in the issue begin to talk to one another about the issue.

2. Some individuals step forward to lead the developing movement.

3. The political power of the movement grows as it attracts members.

4. It becomes more organized.

5. Success takes time because there is the need for funds and there are other groups opposed to the aims of the movement.

Thus, to understand the interrelationship between personal and social problems within a society, it is necessary to understand two important factors. First, change is constant and is increasingly becoming more rapid. For example, a salary of a penny per day, doubling each day for 30 days may not sound like a very good wage—but at the end of those 30 days one would have earned over $10 million. Second, the values of the society determine how and in what manner that change is accepted. For example, for most of history the wife in a marriage was considered the husband's property, so divorce initiated by a wife was not possible. Similarly, child abuse was not considered a public problem for many years, so it was treated under a law to prevent cruelty to small animals. In short, the definition of a problem determines its treatment.

The inequalities discussed in Part IV are important to understand because they are the basis for the differences in experiences and opportunities encountered by individuals and groups around the world. It is also important to understand that problems for individuals can add up to social problems for society and thus need the aid of the social institutions of that society (Part III) to help relieve or resolve that individual problem.

Chapter 9

Stratification
Some Are More Equal
Than Others

We may deny it and attempt to cover it up with beliefs to the contrary, but most of us are aware that the United States is stratified by class. As in most societies, this class ranking is based on several categories that usually merge together: wealth, occupation, authority, age, and gender. The presence of a class system in a society that claims that "all men are created equal" raises two important questions: What are the effects of a class system on the various segments of the population, and why do the citizens tolerate such a system?

In answer to the first question, sociological research has revealed that social class affects almost all facets of our lives including our birth: Class affects our chances for survival during the birth process as well as our continuing health, health care, and educational attainment. Almost-unnoticed effects of class are also seen in the supposed guarantee of our society of equal protection and justice under the law: Our courts are more likely to find people of the lower class guilty of crimes and to send them to prison for far longer terms than members of the elite class who commit white-collar crimes. Another unseen aspect of unequal justice became visible when this country had a military draft: You were more likely to be drafted and to be wounded or killed if you were from the lower rankings of the military—a greater possibility if you belonged to a lower class. These findings amply illustrate how class status determines your control over your life options. In the first reading of this chapter, Andrew Hacker confirms the existence of a class system in the United States and notes the unequal balance in the number of people with money and influence and the number of people who are less fortunate.

The answer to the first question leads us to the second question: Why does a country that believes and claims equality for all tolerate an unequal system? Many reasons explain this contradiction, so we must limit our answers to just a few. One explanation lies in the meaning of the term *doublethink*. In his futuristic novel *1984*, George Orwell defined doublethink as holding two contradictory ideas at the same time and believing in both of them. This kind of thinking can be done only unconsciously. Its existence simply means that we have not examined our beliefs closely enough to note that they are contradictory. For example, Americans believe in equality and yet know that wealth affects such issues as justice.

Another form of doublethink relies on the dubious belief that it is possible to climb the social class ladder—if not in reality, then through imitation. This imitation of the wealthy gives the illusion of success and its resulting pleasure, as Schor describes. Another reason for the maintenance of a social class system lies in the fact that some people benefit from the system as it is and have the power to maintain it as it is. Fallers labels this effect as a "trickle effect." Finally, Herbert Gans claims a functional reason for the existence of class—the lower classes exist because they perform various important tasks for society. In short, the class structure may not be good for some, but it is good for getting many undesirable tasks accomplished.

As you read this chapter on stratification, ask yourself about your class position. Be realistic about your answer by adding an upper and lower place to each class position. Also ask yourself what you believe your future class position will be. Is this a realistic assessment? Besides marrying a rich person, what are things you could do to make your beliefs come true?

37

Money and the World We Want

ANDREW HACKER

Two issues are covered in this reading: First, is there a class structure, and what is its breakdown? Second, what are the consequences of this class structure?

As you read this selection, ask yourself the following questions:

1. *Which problems result from the way the United States distributes income?*
2. *Which changes would you make to help all people to be fully Americans?*
3. *Do you believe that Americans are becoming more equal? Why or why not?*

GLOSSARY **Corollary** A proposition following from one already proved. **Esoteric** Understood by those with special knowledge. **Skepticism** Doubt, disbelief.

The three decades spanning 1940 to 1970 were the nation's most prosperous years. Indeed, so far as can happen in America's kind of economy, a semblance of redistribution was taking place. During this generation, the share of national income that went to the bottom fifth of families rose by some fractions of a point to an all-time high, while that received by the richest fifth fell to its lowest level. This is the very period of shared well-being that many people would like to re-create in this country.

Yet in no way is this possible. It was an atypical era, in which America won an adventitious primacy because of a war it delayed entering and its geographic isolation which spared it the ravages that the other combatants suffered.

America's postwar upsurge lasted barely three decades. The tide began to turn in the early 1970s. Between 1970 and 1980, family income rose less than 7 percent. And that small increase resulted entirely from the fact that additional family members were joining the workforce. Indeed, the most vivid evidence of decline is found in the unremitting drop in men's earnings since 1970. Averages and medians, however, conceal important variations, such as that some American households have done quite well for themselves during the closing decades of the century. Those with incomes of $1 million or more have reached an all-time high, as are families and individuals making over $100,000. By all outward appearances, there is still plenty of money around, but it is landing in fewer hands. Yet it is by no means apparent that people are being paid these generous salaries and options and fees because their work is adding much of substance to the nation's output. Indeed, the coming century will test whether an economy can flourish by exporting most notably action movies and flavored water.

The term *upper class* is not commonly used to describe the people in America's top income tier

Source: Reprinted with permission of Scribner, a division of Simon & Schuster, Inc., from *Money: Who Has How Much and Why*, by Andrew Hacker. Copyright 1997 by Andrew Hacker.

How Families Fare: Three Tiers

Tier	Income Range	Number of Families/Percentage	
Comfortable	Over $75,000	12,961,000	18.6%
Coping	$25,000–$75,000	36,872,000	53.0%
Deprived	Under $25,000	19,764,000	28.4%

since it connotes a hereditary echelon that passes on its holdings from generation to generation. Only a few American families have remained at the very top for more than two or three generations, and even when they do, as have the du Ponts and Rockefellers, successive descendants slice the original pie into smaller and smaller pieces. So if America does not have a "class" at its apex, what should we call the people who have the most money? The answer is to refer to them as we usually do, as being "wealthy" or "rich."

The rich, members of the 68,064 households who in 1995 filed federal tax returns that declared a 1994 income of $1 million or more, have varied sources of income.... For present purposes, wealth may be considered holdings that would yield you an income of $1 million a year without your having to put in a day's work. Assuming a 7 percent return, it would take income-producing assets of some $15 million to ensure that comfort level.

Unfortunately, we have no official count of how many Americans possess that kind of wealth. It is measurably less than 68,064, since those households report that the largest segment of their incomes come from salaries. Indeed, among the chairmen of the one hundred largest firms, the median stock holding is only $8.4 million. A liberal estimate of the number of wealthy Americans would be about thirty thousand households, one-thirtieth of one percent of the national total.

This still puts almost 90 percent of all Americans between the very poor and the wealthy and the rich. One way to begin to define this majority is by creating a more realistic bottom tier than merely those people who fall below the official poverty line. Since Americans deserve more than subsistence, we may set $25,000 a year for a family of three as a minimum for necessities without frills. And even this is a pretty bare floor. Indeed, only

about 45 percent of the people questioned in the Roper-Starch poll felt that their households could "get by" on $25,000 a year. In 1995, almost 20 million families—28.4 percent of the total—were living below that spartan standard. This stratum is a varied group.... For over a third, all of their income comes from sources other than employment, most typically Social Security and public assistance. Over 43.3 percent of the households have only one earner, who at that income level is usually a woman and the family's sole source of support. The remainder consists of families where two or more members have had jobs of some kind, which suggests sporadic employment at close to the minimum wage. Whatever designation we give to these households—poor or just getting by—their incomes leave them deprived of even the more modest acquisitions and enjoyments available to the great majority of Americans.

The question of who belongs to America's middle class requires deciding where to place its upper and lower boundaries. While it is difficult to set precise boundaries for this stratum, it is accurate to say that, by one measure or another, most Americans fit into a middle class.

While it is meaningless to classify the households in the middle of America's income distribution because this stratum is so substantial and it encompasses such a wide range of incomes, meaningful divisions can be drawn in our country's overall income distribution.

This reading has sought to make clear that the prominent place of the rich tells us a great deal about the kind of country we are, as does the growing group of men and women with $100,000 salaries and households with $100,000 incomes. The same stricture applies to the poor, who, while not necessarily increasing in number, are too often permanently mired at the bottom. This noted, a three-tier division can be proposed, with the caveat that how you fare on a given income can depend on local costs and social expectations.

…

That the rich have become richer would seem to bear out Karl Marx's well-known prediction. The nation's greatest fortunes are substantially larger than those of a generation ago. Households with incomes exceeding $1 million a year are also netting more in real purchasing power. At a more mundane level, the top 5 percent of all households in 1975 averaged $122,651 a year; by 1995, in inflation-adjusted dollars, their average annual income had ascended to $188,962.

But Marx did not foresee that the number of rich families and individuals would actually increase over time. Between 1979 and 1994, the number of households declaring incomes of $1 million or more rose from 13,505 to 68,064, again adjusting for inflation. In 1996, *Forbes's* 400 richest Americans were all worth at least $400 million. In 1982, the year of the magazine's first list, only 110 people in the 400 had holdings equivalent to the 1996 cut-off figure. In other words, almost three-quarters of 1982's wealthiest Americans would not have made the 1996 list. And families with incomes over $100,000—once more, in constant purchasing power—increased almost threefold between 1970 and 1995, rising from 3.4 percent to 9 percent of the total. During the same period, the group of men making more than $50,000 rose from 12 percent of the total to 17 percent.

It is one thing for the rich to get richer when everyone is sharing in overall economic growth, and it is quite another for the better off to prosper while others are losing ground or standing still. But this is what has been happening. Thus 1995 found fewer men earning enough to place them in the $25,000 to $50,000 tier compared with twenty-five years earlier. And the proportion in the bottom bracket has remained essentially unchanged. But this is not necessarily a cause for cheer. In more halcyon times, it was assumed that each year would bring a measure of upward movement for people at the bottom of the income ladder and a diminution of poverty.

The wage gap of our time reflects both the declining fortunes of many Americans and the rise of individuals and households who have profited from recent trends. Economists generally agree on what brought about static wages and lowered living standards. In part, well-paying jobs are scarcer because goods that were once produced here, at American wage scales, are now made abroad and then shipped here for sale. A corollary cause has been the erosion of labor unions, which once safeguarded generous wages for their members. Between 1970 and 1996, the portion of the work-force represented by unions fell from 27 percent to 15 percent. Today, the most highly organized occupations are on public payrolls, notably teachers and postal employees. Only 11 percent of workers in the private sector belong to unions. For most of the other 89 percent this means that their current paychecks are smaller than they were in the past.

Analysts tend to differ on the extent to which the paltriness of the minimum wage has lowered living standards. Despite the 1996 increase, the minimum wage produces an income that is still below the poverty line. Even more contentious is the issue of to what degree immigrants and aliens have undercut wages and taken jobs once held by people who were born here. We can all cite chores that Americans are unwilling to do, at least at the wages customarily paid for those jobs. Scouring pots, laundering clothes, herding cattle, and caring for other people's children are examples of such tasks. At the same time, employers frequently use immigrants to replace better-paid workers, albeit by an indirect route. The most common practice is to remove certain jobs from the firm's payroll and then to hire outside contractors, who bring in their own staffs, which are almost always lower paid and often recently arrived in this country.

Most economists agree that the primary cause of diverging earnings among American workers has been the introduction of new technologies. These new machines and processes are so esoteric and complex that they require sophisticated skills that call for premium pay. In this view, the expanded stratum of Americans earning over $50,000 is made up largely of men and women who are adept at current techniques for producing goods, organizing information, and administering personnel. New technologies have reduced the number of people who are needed as telephone operators, tool and die makers, and aircraft mechanics. Between 1992 and 1996, Delta Airlines was scheduling the same number of flights each year, even though it was discharging a quarter of its employees. Closer to the top, there is a strong demand for individuals who are skilled at pruning payrolls. And this cadre has been doing its work well.... [I]n 1973, the five hundred largest industrial firms employed some 15.5 million men and women. By 1993, these firms employed only 11.5 million people. But this reduction in the industrial workforce amounted to more than a loss of 4 million positions. Given the increase in production that took place over this twenty-year period, it meant a comparable output could be achieved in 1993 with half as many American workers as were needed in 1973. In fact, American workers account for an even smaller share of the output, since many of the top five hundred industrial firms are having more of their production performed by overseas contractors and subsidiaries.

So what special skills do more highly educated workers have that make them eligible for rising salaries? In fact, such talents as they may display have only marginal ties to technological expertise. The years at college and graduate school pay off because they burnish students' personalities. The time spent on a campus imparts cues and clues on how to conduct oneself in corporate cultures and professional settings. This demeanor makes for successful interviews and enables a person to sense what is expected of him during the initial months on a job.

Does America's way of allocating money make any sense at all? Any answer to this question requires establishing a rationale. The most common explanation posits that the amounts people get are set in an open market. Thus, in 1995, employers offered some 14.3 million jobs that paid between $20,000 and $25,000, and were able to find 14.3 million men and women who were willing to take them. The same principle applied to the 1.7 million positions pegged at $75,000 to $85,000, and to the

dozen or so corporate chairmen who asked for or were given more than $10 million. By the same token, it can be argued that market forces operate at the low end of the scale. Wal-Mart and Pepsico's Pizza Hut cannot force people to work for $6.50 an hour, but those companies and others like them seem to attract the workers they need by paying that wage.

The last half century gave many groups a chance to shield themselves from the labor auction.

But many, if not most, of these protections are no longer being renewed. The up-and-coming generations of physicians, professors, and automobile workers are already finding that they must settle for lower pay and fewer safeguards and benefits. The most graphic exception to this new rule has been in the corporate world, where boards of directors still award huge salaries to executives, without determining whether such compensation is needed to keep their top people from leaving or for any other reason. They simply act as members of an inbred club who look after one another. Only rarely do outside pressures upset these arrangements, which is why they persist.

A market rationale also presumes that those receiving higher offers will have superior talents or some other qualities that put them in demand. Some of the reasons why one person makes more than another make sense by this standard. Of course, a law firm will pay some of its members more if they bring in new business or satisfy existing clients. Two roommates have just received master's degrees with distinction, one in education and the other in business administration. The former's first job is teaching second-graders and will pay $23,000. The latter, at an investment firm, will start at $93,000. About all that can be said with certainty is that we are unlikely to arrive at a consensus on which roommate will be contributing more to the commonwealth.

Would America be a better place to live, and would Americans be a happier people, if incomes were more evenly distributed? Even as the question is being posed, the answers can be anticipated.

One side will respond with a resounding "Yes!" After which will come a discourse on how poverty subverts the promise of democracy, while allowing wealth in so few hands attests to our rewarding greed and selfishness. There would be far less guilt and fear if the rich were not so rich and no Americans were poor. But the goal, we will be told, is not simply to take money from some people and give it to others. Rather, our goal should be to create a moral culture where citizens feel it is right to have no serious disparities in living standards. Other countries that also have capitalist economies have shown that this is possible.

Those who exclaim "No way!" in response to the same question will be just as vehement. To exact taxes and redistribute the proceeds is an immoral use of official power since it punishes the productive and rewards the indolent. And do we want the government telling private enterprises what wages they can offer? The dream of economic equality has always been a radical's fantasy. Apart from some primitive societies, such a system has never worked. If you want efficiency and prosperity, and almost everyone does, then variations in incomes are part of the equation.

These different responses arise in part from disparate theories of human nature. Since the earliest days of recorded history, philosophers have disagreed over whether our species is inherently competitive or cooperative.

Also at issue is whether greater economic equality can only be achieved by giving oppressive powers to the state, either to limit incomes through heavy taxes or by setting levels of earnings.

Of course, the price that America pays for economic inequality is its persisting poverty.

In one way or another all Americans will pay the high costs of poverty. California now spends more on its prison system than it does on higher education, and other states will soon be following suit. Bolstering police forces is hardly cheap: upward of $75,000 per officer, when overtime and benefits are added in. Being poor means a higher chance of being sick or being shot, or bearing low-birth-weight babies, all of which consume medical resources and have helped to make Medicaid one of the costliest public programs. In addition, poor Americans now represent the fastest-growing group

of AIDS victims: mainly drug users and the women and children they infect. Generally, $100,000 worth of medical treatment is spent on each person dying of AIDS. The poor also have more of their children consigned to "special education" classes, which most never leave and where the tab can reach three times the figure for regular pupils.

Additional expenses are incurred by families who put as much distance as possible between themselves and the poor. Doing so often entails the upkeep of gated communities, security systems, and privately supplied guards. Yet these are expenses better-off Americans readily bear. They are willing to foot the bills for more prisons and police, as well as the guns they keep in their bedrooms and the alarms for their cars. Indeed, their chief objection is to money given to non-married mothers who want to be at home with their pre-school-age children. Virtually every such penny is begrudged, followed by the demand that these mothers take jobs even before their children go to kindergarten. While the imposition of work may make some taxpayers cheerier, it will not do much to close the income gap.

How disparities in income affect the nation's well-being has long been debated by economists. Much of the argument centers on what people do with their money. The poor and those just trying to cope devote virtually all of what they have to necessities plus the few extras they can afford. If their incomes were raised, they would obviously spend more, which would create more demand and generate more jobs. It should not be forgotten that the year 1929, which was noted for a severe imbalance in incomes, gave us a devastating economic depression that lasted a decade.

The traditional reply to critics of economic inequality has been that we need not only the rich, but also a comfortably off class, who are able to put some of their incomes into investments. In other words, disparities give some people more than they "need," which allows them to underwrite the new enterprises that benefit everyone. While there is obvious validity to this argument, it should be added that much of this outlay now goes to paper contrivances, which have only a remote connection with anything productive, if any at all. Nor are the rich as necessary as they may once have been, since institutions now supply most invested capital. Metropolitan Life, Merrill Lynch, Bank of America, and the California Public Employees Pension Fund put substantially more into new production than do the 68,064 families with $1 million incomes.

In one sphere, the income gap comes closer to home. Most young Americans will not live as well as their parents did. Indeed, in many instances this is already occurring. A generation ago, many men had well-paid blue-collar jobs; now their sons are finding that those jobs are no longer available. In that earlier era, college graduates could enter a growing managerial stratum; today, firms view their payrolls as bloated and are ending the security of corporate careers....

The patterns of decline prevail if the entire nation is viewed as the equivalent of family. Federal programs now award nine times as much to retirees as they do to the nation's children, so senior citizens as a group fare better than younger Americans. Twice as many children as Americans over the age of sixty live in households below the poverty line. (And as death approaches, the government is more generous: almost 30 percent of the total Medicare budget is spent on the terminal year of elderly patients' lives.) Many retired persons have come to view a comfortable life as their entitlement, and have concluded that they no longer have obligations to repay. Grandparents tend to support campaigns for the rights of the elderly, not for school bonds and bigger education budgets.

In the end, the issue may be simply stated: what would be required for all Americans—or at least as many as possible—to make the most of their lives?...

Poverty takes its greatest toll in the raising of children. With a few exceptions, being poor consigns them to schools and surroundings that do little to widen their horizons. The stark fact is that we have in our midst millions of bright and talented children whose lives are fated to be a fraction of what they might be. And by any moderate standard, deprivation extends above the official poverty line. In most of the United States, families with incomes of less than $25,000 face real limits to the opportunities open to their children. Less than one child in ten from these households now enters and graduates from college. The statistics are apparent to outside observers, and to the children themselves, who very early on become aware of the barriers they face. And from this realization results much of the behavior that the rest of the society deplores. The principal response from solvent Americans has been to lecture the poor on improving their ways.

No one defends poverty, but ideologies differ on what can or should be done to alleviate it. Conservatives generally feel it is up to the individual: those at the bottom should take any jobs they can find and work hard to pull themselves up. Hence the opposition to public assistance, which is seen as eroding character and moral fiber. Indeed, conservatives suggest that people will display character and moral fiber if they are made to manage on their own.

Not many voting Americans favor public disbursements for the poor or even for single working mothers who cannot make ends meet. Most American voters have grown weary of hearing about the problems of low-income people. Yet even those who are unsettled by the persistence of income imbalances no longer feel that government officials and experts know how to reduce the disparities.

Of course, huge redistributions occur everyday. Funds for Social Security are supplied by Americans who are currently employed, providing their elders with pensions that now end up averaging $250,000 above what their own contributions would have warranted. Agricultural subsidies give farmers enough extra cash to ensure that they will have middle-class comforts. The same subventions furnish farms owned by corporations with generous profit margins....

In contrast, there is scant evidence that public programs have done much for the bottom tiers of American society. Despite the New Deal and the Great Society, including public works and public assistance, since 1935, the share of income going to the poorest fifth of America's households has remained between 3.3 percent and 4.3 percent. Thus, if many elderly Americans have been raised from poverty, it is clear that younger people are now taking their places....

All parts of the population except the richest fifth have smaller shares of the nation's income than they did twenty years ago. The gulf between the best-off and the rest shows no signs of diminishing, and by some political readings this should mean increased tensions between the favored fifth and everyone else. But declines in living standards have not been so severe or precipitous as to lead many people to question the equity of the economic system. The economy has ensured that a majority of Americans remain in moderate comfort and feel able to count their lives a reasonable success. Airline reservationists making $14,000 do not consider themselves "poor," and no one tells them that they are. Thus a majority of Americans still see themselves as middle class, and feel few ties or obligations to the minority with incomes less than their own.

Given this purview, why should the way America distributes its income be considered a problem? At this moment, certainly, there is scant sentiment for imposing further taxes on the well-to-do and doing more for the poor. As has been observed, there is little resentment felt toward the rich; if anything, greater animus is directed toward families receiving public assistance. Nor is it regarded as untoward if the well-off use their money to accumulate luxuries while public schools must cope with outdated textbooks and leaking roofs. Although this reading is about money, about why some have more and others less, it should not be read as a plea for income redistribution. The reason is straightforward: if people are disinclined to share what they have, they will not be persuaded by a reproachful tone. Rather, its aim is to enhance our understanding of ourselves, of the forces that propel us, and the shape we are giving to the nation of which we are a part.

How a nation allocates its resources tells us how it wishes to be judged in the ledgers of history and morality. America's chosen emphasis has been on offering opportunities to the ambitious, to those with the desire and the drive to surpass. America has more self-made millionaires and more men and women who have attained $100,000 than any other country.

But because of the upward flow of funds, which has accelerated in recent years, less is left for those who lack the opportunities or the temperament to succeed in the competition. The United States now has a greater percentage of its citizens in prison or on the streets, and more neglected children, than any of the nations with which it is appropriately compared. Severe disparities—excess alongside deprivations—under the society and subvert common aims. With the legacy we are now creating, millions of men, women, children are prevented from being fully American, while others pride themselves on how much they can amass.

38

Keeping Up with the Trumps

JULIET B. SCHOR

This reading deals with the sociological concept of the reference group. According to this concept, the group that we choose to compare ourselves to determines whether we feel positive or negative about ourselves in financial or other ways. Schor notes the effect of class beliefs on our own assumptions about class and spending habits.

As you read this selection, consider the following questions as guides:

1. *Do the television programs you watch affect the amount of money you spend on consumer goods or types of products?*
2. *Based on the material in this reading, what do you believe is your class ranking? Why?*
3. *Many people are overloaded with debt on their credit cards and are able to pay only the minimum amount due each month. Why do you think this is so? What can be done about it?*

For most of us, social space begins with relatives, friends, and co-workers. These are the people whose spending patterns we know and care most about. They are the people against whom we judge our own material lifestyles, and with whom we try and keep up. The comparisons we make between ourselves and them matter deeply to us. And we act upon that deep feeling. According to the research I have done, how people stack up financially against the group with which they most often compare themselves, or their reference group, has an enormous impact on their overall spending. I began my study by asking all the respondents at the company where I was conducting my research to identify their primary reference group. I then asked them, "How does your financial status compare to that of most of the members of the reference group you have chosen?"…

I elicited this information in order to figure out whether Americans really do keep up with others. I reasoned that a person who is trying to associate or identify with a group above himself or herself will spend more, all other things being equal, than someone who has chosen a comparison group of people with less money.…

The idea is that where you stand relative to those with whom you compare yourself has a significant impact on your spending.…

What is that evidence? I estimated statistical equations that explained the amount of saving and spending each person did in the year of the survey. I included a wide range of factors likely to affect

Source: From *The Overspent American: Upscaling, Downshifting, and the New Consumer.* Copyright © 1998 by Juliet B. Schor. Reprinted by permission of Basic Books.

spending, such as the respondent's age, number of dependents, household income in that year, long-term expected income (or what economists call permanent income), and so on. These are the standard variables that economists typically use to explain variations in spending propensities across the population. Then I added my own comparative variable—how the respondent stacked up financially compared to his or her reference group.

As it turns out, this variable has a very large impact. In the savings equation, each step a respondent moved down the scale (from much better off than the reference group to better off) reduced the amount saved by $2,953 a year. Moving down two steps reduced saving by twice that. The sheer magnitude of this effect can be appreciated when we remember that the average employee at the company where I conducted my research saved only $10,450 per year, including retirement savings. According to these estimates, disaster ensues as a person slides down the reference group scale. Moving from the top to the bottom would lead you to save $15,000 less each year—or more likely, to take on some of that amount in debt.

SCREEN WITH ENVY

While television has long been suspected as a promoter of consumer desire, there has been little hard evidence to support that view, at least for adult spending. After all, there's not an obvious connection. Many of the products advertised on television are everyday low-cost items such as aspirin, laundry detergent, and deodorant. Those TV ads are hardly a spur to excessive consumerism. Leaving aside other kinds of ads for the moment (for cars, diamonds, perfume) there's another counter to the argument that television causes consumerism: TV is a substitute for spending. One of the few remaining free activities, TV is a popular alternative to costly recreational spending such as movies, concerts, and restaurants. If it causes us to spend, that effect must be powerful enough to overcome its propensity to save us money.

Apparently it is. My research shows that the more TV a person watches, the more he or she spends. The likely explanation for the link between television and spending is that what we see on TV inflates our sense of what's normal. The lifestyles depicted on television are far different from the average American's: With a few exceptions, TV characters are upper-middle class, or even rich.

Studies by the consumer researchers Thomas O'Guinn and L. J. Shrum confirm this upward situation. The more people watch television, the more they think American households have tennis courts, private planes, convertibles, car telephones, maids, and swimming pools. Heavy watchers also overestimate the portion of the population who are millionaires, have had cosmetic surgery, and belong to a private gym, as well as those suffering from dandruff, bladder control problems, gingivitis, athlete's foot, and hemorrhoids (the effect of all those ads for everyday products). What one watches also matters. Dramatic shows—both daytime soap operas and prime time—have a stronger impact on viewer perceptions than other kinds of programs (say news, sports, or weather).

Heavy watchers are not the only ones, however, who tend to overestimate standards of living. Almost everyone does. (And almost everyone watches TV.) In one study, ownership rates for 22 of 27 consumer products were generally overstated. Your own financial position also matters. Television inflates standards for lower-, average-, and above-average-income students, but it does the reverse for really wealthy ones. (Among those raised in a financially rarefied atmosphere, TV is almost a reality check.) Social theories of consumption hold that the inflated sense of consumer norms promulgated by the media raises people's aspirations and leads them to buy more....

Television also affects norms by giving us real information about how other people live and what they have. It allows us to be voyeurs, opening the door to the "private world" inside the homes and lives of others.... As O'Guinn and Shrum note, television has replaced personal contact as our source of information about "what members of other social classes have and how they consume, even behind their closed doors."

Another piece of evidence for the TV–spending link is the apparent correlation between debt and excessive TV viewing. In the Merck Family Fund poll, the fraction responding that they "watch too much TV" rose steadily with indebtedness. More than half (56 percent) of all those who reported themselves "heavily" in debt also said they watched too much TV.

It is partly because of television that the top 20 percent of the income distribution, and even the top 5 percent within it, has become so important in setting and escalating consumption standards for more than just the people immediately below them. Television lets everyone see what these folks have and allows viewers to want it in concrete, product-specific ways. Let's not forget that television programming and movies are increasingly filled with product placements—the use of identifiable brands by characters. TV shows and movies are more and more like running ads. We've become so inured to this practice that it's hard to remember that a can of soda in TV show was once labeled "soda" rather than "Coke" or "Pepsi."…

Part of what keeps the see-want-borrow-and-buy sequence going is lack of attention. Americans live with high levels of denial about their spending patterns. We spend more than we realize, hold more debt than we admit to, and ignore many of the moral conflicts surrounding our acquisitions.… What is not well understood is that the spending of many normal consumers is also predicated on denial.…

Nowhere is denial so evident as with credit cards. Contrary to economists' usual portrayal of credit card debtors as fully rational consumers who use the cards to smooth out temporary shortfalls in income, the finding of the University of Maryland economist Larry Ausubel was that people greatly underestimate the amount of debt they hold on their cards—1992's actual $182 billion in debt was thought to be a mere $70 billion. Furthermore, most people do not expect to use their cards to borrow, but, of course, they do. Eighty percent end up paying finance charges within any given year, with just under half (47 percent) always holding unpaid balances.

Not paying attention to what we spend is also very common.…

FEAR OF FALLING

Fifty years ago, most people just wanted to secure their place in the American middle class, doing whatever it took to stay there. At one time, that was acquiring a houseful of "decencies," the status symbols of the middle class, situated between the necessities of the poor and the luxuries of the rich. Today, in a world where being middle class is not good enough for many people and indeed that social category seems like an endangered species, securing a place means going upscale. But when everyone's doing it, upscaling can mean simply keeping up. Even when we are aiming high, there's a strong defensive component to our comparisons. We don't want to fall behind or lose the place we've carved out for ourselves. We don't want to get stuck in the "wrong" lifestyle cluster. How we spend has become a crucial part of our self-image, personal identity, and social network.

The historical record highlights the fact that beneath—indeed driving—our system of competitive consumption are deep class inequalities. The classless-society and end-of-ideology literature of 25 years ago turns out to have been wishful thinking. Ironically, inequality began to arise soon after these ideas appeared. The dirty little secret of American society is that not everyone did become middle class. We still have rich and poor and gradations in between. Class background and income level affect not only the obvious—if and where you go to college, the quality of your children's elementary school, the kind of job you get—but also your likelihood of getting heart disease, the way you talk, and how respectfully you're treated by others. At all levels of the class structure, we are motivated, as Barbara Ehrenreich put it some years ago, by "fear of falling".… At all levels, a structure of inequality injects insecurity and fear into our psyches. The penalties of dropping down are perhaps the most powerful psychological hooks that keep us keeping up, even as the heights get dizzying.

However rational it may be for individuals to keep up with the upscaling of consumer standards, it can be deeply irrational for society as a whole.…

The more our consumer satisfaction is tied into so-cial comparisons—whether upscaling, just keeping up, or not falling too far behind—the less we achieve when consumption grows, because the people we compare ourselves to are also experienc-ing rising consumption. Our relative position does not change. Jones' delight at being able to afford the Honda Accord is dampened when he sees Smith's new Camry. Both must put in long hours to make the payments, suffer with congested high-ways and dirty air, and have less in the bank at the end of the day. And both remain frustrated when they think about the Land Cruiser down the street.

Of course, relative positions do change. Some people get promotions or pay raises that place them higher up in the hierarchy. Others fall behind. But these random changes cancel each other out. Of more interest is how the broad social groupings that make up the major comparison groups fare. From the end of the Second World War until the mid-1970s, growth was relatively equally distrib-uted. The rough doubling in living standards was experienced by most Americans, including the poor. In fact, the income distribution was even com-pressed, as people at the bottom gained some ground relative to those at the top. Since then, however, and particularly since the 1980s, the income groups have diverged.

Middle-class Americans began to experience themselves falling behind as their slow-growing wages and salaries lagged behind those of the groups above them. Their anxiety grew, and it became commonplace that it was no longer possible to achieve a middle-class standard of living on one salary. At the same time, increasing numbers began to lose completely the respectability that defined their class. Below them, a segment of downwardly mobile working people found that their reduced job prospects and declining wages had placed them in the ranks of the working poor. And the nonemployed poor fell even further as their num-bers grew and their average income fell.

Thus, relative position has worsened for most people, making it increasingly difficult to keep up. The excitement, convenience, or joy that house-holds may have experienced through the billions in additional spending between 1979 and the pres-ent seems to have been overshadowed by feelings of deprivation. Among the upper echelons, all those personal computers, steam showers, Caribbean va-cations, and piano lessons have not been sufficient to offset the anxieties inherent in a rapidly upscaling society.

The current mood has led to nostalgia about the older, simpler version of the American dream. There is a palpable sense of unease, a yearning for the less expansive, and less expensive, aspirations of our parents. In the words of one young man, "My dream is to build my own house. When my parents grew up, they weren't so much 'I want this, I've got to have that.' They just wanted to be comfort-able. Now we're more—I know I am—'I need this.' And it's not really a need."

39

A Note on the "Trickle Effect"

Lloyd A. Fallers

The author of this selection notes how items such as styles "trickle" down from the socio-economic elites and thereby provide lower-socioeconomic groups with a sense of equality. He claims that this is also true for occupational titles.

As you read this article, consider the following questions:

1. *Name changes in titles of several occupations. How do they reflect changes in the occupation?*
2. *Do higher incomes usually follow the change in title? Why?*
3. *Is it possible for occupational titles to reach some sort of peak? Why?*
4. *Which occupation would you like to enter? What do you believe its title will be when you enter that occupation? In 20 years?*

Much has been written—and much more spoken in informal social scientific shop talk—about the so-called "trickle effect"—the tendency in U.S. society (and perhaps to a lesser extent in Western societies generally) for new styles or fashions in consumption goods to be introduced via the socio-economic elite and then to pass down through the status hierarchy, often in the form of inexpensive, mass-produced copies.

In a recent paper, Barber and Lobel have analyzed this phenomenon in the field of women's clothes.[1] They point out that women's dress fashions are not simply irrational shifts in taste, but that they have definite functions in the U.S. status system. Most Americans, they say, are oriented toward status mobility. Goods and services consumed are symbolic of social status. In the family division of labor, the husband and father "achieves" in the occupational system and thus provides the family with monetary income. Women, as wives and daughters, have the task of allocating this income so as to maximize its status-symbolic value. Since women's clothing permits much subtlety of expression in selection and display, it becomes of great significance as a status-mobility symbol.[2] The ideology of the "open class" system, however, stresses broad "equality" as well as differential status. The tendency of women's dress fashions to "trickle down" fairly rapidly via inexpensive reproductions of originals crated at fashion centers helps to resolve this seeming inconsistency by preventing the development of rigid status distinctions.[3]

In the widest sense, of course, the "trickle effect" applies not only to women's dress but also to consumption goods of many other kinds. Most similar to women's dress fashions are styles in household

Source: "A Note on the 'Trickle Effect,'" by Lloyd A. Fallers, pp. 314–322, Fall, 1954, *Public Opinion Quarterly*. Reprinted with permission of the publisher.

furnishings. A colleague has pointed out to me that venetian blinds have had a similar status career—being introduced at relatively high levels in the status hierarchy and within a few years passing down to relatively low levels. Like women's dress styles, styles in household furnishings are to a substantial degree matters of taste and their adoption a matter of "learning" by lower-status persons that they are status relevant. The trickling down of other types of consumption goods is to a greater degree influenced by long-term increases in purchasing power at lower socio-economic levels. Such consumers' durables as refrigerators and automobiles, being products of heavy industry and hence highly standardized over relatively long periods and throughout the industries which produce them, are much less subject to considerations of taste. They do, however, trickle down over the long term and their possession is clearly status-relevant.

The dominant tendency among social scientists has been to regard the trickle effect mainly as a "battle of wits" between upper-status persons who attempt to guard their symbolic treasure and lower-status persons (in league with mass-production industries) who attempt to devalue the status-symbolic currency. There is much truth in this view. Latterly we have observed a drama of this sort being played out in the automotive field. Sheer ownership of an automobile is no longer symbolic of high status and neither is frequent trading-in. Not even the "big car" manufacturers can keep their products out of the hands of middle- and lower-status persons "on the make." High-status persons have therefore turned to ancient or foreign sports-cars.

It seems possible, however, that the trickle effect has other and perhaps more far-reaching functions for the society as a whole. Western (and particularly U.S.) society, with its stress upon the value of success through individual achievement, poses a major motivational problem: The occupational system is primarily organized about the norm of technical efficiency. Technical efficiency is promoted by recruiting and rewarding persons on the basis of their objective competence and performance in occupational roles. The field of opportunity for advancement, however, is pyramidal in shape; the number of available positions

decreases as differential rewards increase. But for the few most competent to be chosen, the many must be "called," that is, motivated to strive for competence and hence success. This, of course, involves relative failure by the many, hence the problem: How is the widespread motivation to strive maintained in the face of the patent likelihood of failure for all but the few? In a widely quoted paper, Merton has recognized that this situation is a serious focus of strain in the social system and has pointed to some structured types of deviant reaction to it.[4] I should like to suggest the hypothesis that *the trickle effect is a mechanism for maintaining the motivation to strive for success, and hence for maintaining efficiency of performance in occupational roles, in a system in which differential success is possible for only a few*. Status-symbolic consumption goods trickle down, thus giving the "illusion" of success to those who fail to achieve *differential* success in the opportunity and status pyramid. From this point of view, the trickle effect becomes a "treadmill."

There are, of course, other hypotheses to account for the maintenance of motivation to strive against very unfavorable odds. Perhaps the most common is the notion that the "myth of success," perhaps maintained by the mass-communications media under the control of the "vested interests," deceives people into believing that their chances for success are greater than is in fact the case. Merton seems to accept this explanation in part while denying that the ruse is entirely effective.[5] Somewhat similar is another common explanation, put forward, for example, by Schumpeter, that though the chances for success are not great, the rewards are so glittering as to make the struggle attractive.[6] Undoubtedly both the "success myth" theory and the "gambling" theory contain elements of truth. Individual achievement certainly *is* a major value in the society and dominates much of its ideology, while risk-taking is clearly institutionalized, at any rate in the business segment of the occupational system. Taken by themselves, however, these explanations do not seem sufficient to account for the situation. At any rate, if it is possible to show that the system *does* "pay off" for the many in the form of "trickle-down" status-symbolic consumption goods, one need not lean so heavily upon such arguments.

It seems a sound principle of sociological analysis to assume "irrationality" as a motivation for human action only where exhaustive analysis fails to reveal a "realistic" pay-off for the observed behavior. To be sure, the explanation put forward here also assumes "irrationality," but in a different sense. The individual who is rewarded for his striving by the trickling-down of status-symbolic consumption goods has the *illusion*, and not the *fact*, of status mobility among his fellows. But in terms of his life history, he nevertheless *has* been rewarded with things which are valued and to this degree his striving is quite "realistic."[7] Though his status position *vis-a-vis* his fellows has not changed, he can look back upon his own life history and say to himself (perhaps not explicitly since the whole status-mobility motivational complex is probably quite often wholly or in part unconscious): "I (or my family) have succeeded. I now have things which five (or ten or twenty) years ago I could not have had, things which were then possessed only by persons of higher status." To the degree that status is *defined* in terms of consumption of goods and services one should perhaps say, not that such an individual has only the *illusion* of mobility, but rather that the entire population has been upwardly mobile. From this point of view, status-symbolic goods and services do not "trickle-down" but rather remain in fixed positions; the population moves up through the hierarchy of status-symbolic consumption patterns.

The accompanying diagram illustrates the various possibilities in terms of the life-histories of individuals. The two half-pyramids represent the status hierarchy at two points in time (X and Y). A, B, C and D are individuals occupying different levels in the status hierarchy. Roman numerals I through V represent the hierarchy of status-symbolic consumption patterns. Between time periods X and Y, a new high-status consumption pattern has developed and has been taken over by the elite. All status levels have "moved up" to "higher" consumption patterns. During the elapsed time, individual C has "succeeded" in the sense of having become able to consume goods and services which were unavailable to him before, though he

has remained in the same relative status level. Individual B has been downwardly mobile in the status hierarchy, but this blow has been softened for him because the level into which he has dropped has in the meantime taken over consumption patterns previously available only to persons in the higher level in which B began. Individual D has been sufficiently downwardly mobile so that he has also lost ground in the hierarchy of consumption patterns. Finally, individual A, who has been a spectacular success, has risen to the very top of the status hierarchy where he is able to consume goods and services which were unavailable even to the elite at an earlier time period. Needless to say, this diagram is not meant to represent the actual status levels, the proportions of persons in each level, or the frequencies of upward and downward mobility in the U.S. social system. It is simply meant to illustrate diagrammatically the tendency of the system, in terms of status-symbolic consumption goods, to reward even those who are not status mobile and to provide a "cushion" for those who are slightly downward mobile.

Undoubtedly this view of the system misrepresents "the facts" in one way as much as the notion of status-symbolic goods and services "trickling down" through a stable status hierarchy does in another. Consumption patterns do not retain the same status-symbolic value as they become available to more people. Certainly to some degree the "currency becomes inflated." A more adequate diagram would show both consumption patterns trickling down and the status hierarchy moving up. Nonetheless, I would suggest that to *some degree* particular consumption goods have "absolute" value in terms of the individual's life history and his motivation to succeed. To the degree that this is so, the system pays off even for the person who is not status-mobile.

This pay-off, of course, is entirely dependent upon constant innovation and expansion in the industrial system. New goods and services must be developed and existing ones must become more widely available through mass-production. Average "real income" must constantly rise. If status-symbolic consumption patterns remained stationary both in

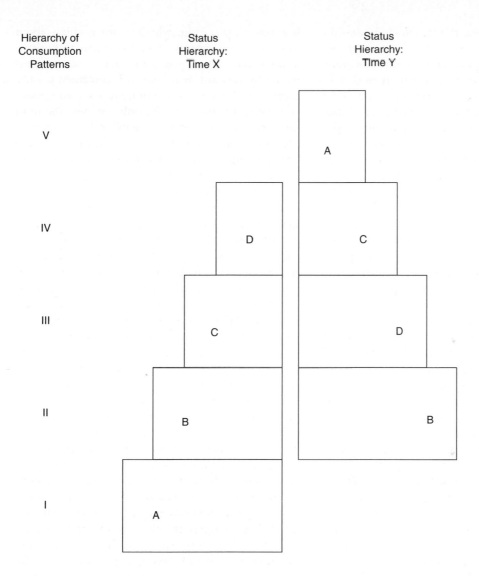

SOURCE: "A Note on the 'Trickle Effect,'" by Lloyd A. Fallers, pp. 314–322, 1954, *Public Opinion Quarterly*.

kind and in degree of availability, the system would pay off only for the status-mobile and the achievement motive would indeed be unrealistic for most individuals. Were the productive system to shrink, the pay-off would become negative for most and the unrealism of the motivation to achieve would be compounded. Under such circumstances, the motivational complex of striving–achievement–occupational efficiency would be placed under great strain. Indeed, Merton seems to have had such circumstances in mind when he described "innovation," "ritualism," "rebellion," and "passive withdrawal" as common patterned deviations from the norm.[8]

This suggests a "vicious circle" relationship between achievement motivation and industrial productivity. It seems reasonable to suppose that a high level of achievement motivation is both a cause and a result of efficiency in occupational role performance. Such an assumption underlies much of our thinking about the modern Western occupational system and indeed is perhaps little more than common sense. One British sociologist, commenting upon the reports of the British "Productivity Teams" which have recently been visiting American factories, is impressed by American workers' desire for status-symbolic consumption, partly the result of pressure upon husbands by their wives, as a factor in the greater "per man hour" productivity of American industry.[9] Greater productivity, of course, means more and cheaper consumption goods and hence a greater pay-off for the worker. Conversely, low achievement motivation and inefficiency in occupational role performance would seem to stimulate one another. The worker has less to work for, works less efficiently, and in turn receives still less reward. Presumably these relationships would tend to hold, though in some cases less directly, throughout the occupational system and not only in the sphere of the industrial worker.

To the degree that the relationships suggested here between motivation to status-symbolic consumption, occupational role performance, and expanding productivity actually exist, they should be matters of some importance to the theory of business cycles. Although they say nothing about the genesis of upturns and down-turns in business activity, they do suggest some social structural reasons why upward or downward movements, once started, might tend to continue. It is not suggested, of course, that these are the only, or even the most important, reasons. More generally, they exemplify the striking degree to which the stability of modern industrial society often depends upon the maintenance of delicate equilibria.

The hypotheses suggested here are, it seems to me, amenable to research by a number of techniques. It would be most useful to discover more precisely just which types of status-symbolic consumption goods follow the classical trickle-down pattern and which do not. Television sets, introduced in a period of relative prosperity, seem to have followed a different pattern, spreading laterally across the middle-income groups rather than trickling down from above. This example suggests another. Some upper-income groups appear to have shunned television on the grounds of its "vulgarity"—a valuation shared by many academics. To what degree are preferences for other goods and services introduced, not at the upper-income levels, but by the "intelligentsia," who appear at times to have greater pattern-setting potential than their relatively low economic position might lead one to believe? Finally, which consumption items spread rapidly and which more slowly? Such questions might be answered by the standard techniques of polling and market analysis.

More difficult to research are hy-significance of consumption goods. Hypotheses concerning the motivational have suggested that the significance for the individual of the trickling down of consumption patterns must be seen in terms of his life-history and not merely in terms of short-term situations. It seems likely that two general patterns may be distinguished. On the one hand, individuals for whom success means primarily rising above their fellows may be more sensitive to those types of goods and services which must be chosen and consumed according to relatively subtle and rapidly changing standards of taste current at any one time at higher levels. Such persons must deal successfully with the more rapid devaluations of status- symbolic currency which go on among those actively battling for dominance. Such persons it may be who are responsible for the more short-term fluctuations in consumption patterns. On the other hand, if my hypothesis is correct, the great mass of the labor force may be oriented more to long-term success in terms of their own life-histories—success in the sense of achieving a "better standard of living" without particular regard to *differential* status. Interviews centered upon the role of consumption patterns in individuals' life aspirations should reveal such differences if they exist, while differences in perception of symbols of taste might be tested by psychological techniques.

Most difficult of all would be the testing of the circular relationship between motivation and productivity. Major fluctuations in the economy are relatively long term and could be studied only through research planned on an equally long-term basis. Relatively short-term and localized fluctuations, however, do occur at more frequent intervals and would provide possibilities for research. One would require an index of occupational performance which could be related to real income and the relationship between these elements should ideally be traced through periods of both rising and falling real income.

NOTES

1.   Barber, Bernard and Lyle S. Lobel, "'Fashion' in Women's Clothes and the American Social System," *Social Forces*, Vol. 31, pp. 124–131. Reprinted in Bendix, Reinhard, and S. M. Lipset, *Class, Status and Power: A Reader in Social Stratification*, Free Press, 1953, pp. 323–332.

2.   It is not suggested that women are *solely* in charge of status-symbolic expenditure, merely that they play perhaps the major role in this respect. See also: Parsons, Talcott, *Essays in Sociological Theory*, Free Press, 1949, p. 225.

3.   Our thinking concerning the status-symbolic role of consumption patterns owes a great debt, of course, to Veblen's notion of "conspicuous consumption" and more recently to the work of W. L. Warner and his colleagues.

4.   Merton, R. K. "Social Structure and Anomie," reprinted as Chapter IV, *Social Theory and Social Structure*, Free Press, 1949.

5.   *Ibid.*

6.   Schumpeter, J. A., *Capitalism, Socialism and Democracy*, Harpers, 1947, pp. 73–74.

7.   By "irrationality" is meant here irrationality *within the framework of a given value system*. Values themselves, of course, are neither "rational" nor "irrational" but "non-rational." The value of individual achievement is non-rational. Action directed toward achievement may be termed rational to the degree that, in terms of the information available to the actor, it is likely to result in achievement; it is irrational to the degree that this is not so.

8.   Merton, R. K., *op. cit.*

9.   Balfour, W. C., "Productivity and the Worker," *British Journal of Sociology*, Vol. IV, No. 3, 1953, pp. 257–265.

40

No, Poverty Has Not Disappeared

HERBERT J. GANS

As noted in other readings in this book, we tend to blame the cause of poverty mostly on personal attributes rather than the restrictions of reality. Gans indicates another reason for the persistence of poverty: Poor people are fundamentally needed.

As you read this selection, ask yourself the following questions:

1. *The author believes that poverty will continue until functional alternatives are developed. What are these alternatives?*
2. *Do you believe that the alternatives suggested or your own will eliminate poverty? Why?*
3. *Regardless of any efforts, poverty will always be with us. Why?*

GLOSSARY **Functional analysis** A means of analysis that examines the objective consequences of an action, a law, or the like. **Latent function** A consequence that is not readily apparent. **Dysfunction** An objective consequence that hinders the fulfillment of a goal. **Negative income tax** Receipt of tax dollars when income falls below a set figure. **Family assistance plan** Various programs for aiding needy families. **Vicarious** Participating in another person's experience through the imagination.

Some twenty years ago Robert K. Merton applied the notion of functional analysis[1] to explain the continuing though maligned existence of the urban political machine: if it continued to exist, perhaps it fulfilled latent—unintended or unrecognized—positive functions. Clearly it did. Merton pointed out how the political machine provided central authority to get things done when a decentralized local government could not act, humanized the services of the impersonal bureaucracy for fearful citizens, offered concrete help (rather than abstract law or justice) to the poor, and

otherwise performed services needed or demanded by many people but considered unconventional or even illegal by formal public agencies.

Today, poverty is more maligned than the political machine ever was; yet it, too, is a persistent social phenomenon. Consequently, there may be some merit in applying functional analysis to poverty, in asking whether it also has positive functions that explain its persistence.

Merton defined functions as "those observed consequences [of a phenomenon] which make for the adaptation or adjustment of a given [social]

Source: Reprinted from *Social Policy*, July–August 1971, pp. 20–24, published by Social Policy Corporation, New York, NY 10036. Copyright 1971 by Social Policy Corporation.

system." I shall use a slightly different definition; instead of identifying functions for an entire social system, I shall identify them for the interest groups, socioeconomic classes, and other population aggregates with shared values that "inhabit" a social system. I suspect that in a modern heterogeneous society, few phenomena are functional or dysfunctional for the society as a whole, and that most result in benefits to some groups and costs to others. Nor are any phenomena indispensable; in most instances, one can suggest what Merton calls "functional alternatives" or equivalents for them, in other words, other social patterns or policies that achieve the same positive functions but avoid the dysfunctions.[2]

Associating poverty with positive functions seems at first glance to be unimaginable. Of course, the slumlord and the loan shark are commonly known to profit from the existence of poverty, but they are viewed as evil men, so their activities are classified among the dysfunctions of poverty. However, what is less often recognized, at least by the conventional wisdom, is that poverty also makes possible the existence or expansion of respectable professions and occupations, for example, penology, criminology, social work, and public health. More recently, the poor have provided jobs for professional and paraprofessional "poverty warriors," and for journalists and social scientists, this author included, who have supplied the information demanded by the revival of public interest in poverty.

Clearly, then, poverty and the poor may well satisfy a number of positive functions for many nonpoor groups in American society. I shall describe thirteen such functions—economic, social, and political—that seem to me most significant.

THE FUNCTIONS OF POVERTY

First, the existence of poverty ensures that society's "dirty work" will be done. Every society has such work: physically dirty or dangerous, temporary, dead-end and underpaid, undignified and menial jobs. Society can fill these jobs by paying higher wages than for "clean" work, or it can force people who have no other choice to do the dirty work—and

at low wages. In America, poverty functions to provide a low-wage labor pool that is willing—or, rather, unable to be *un*willing—to perform dirty work at low cost. Indeed, this function of the poor is so important that in some Southern states, welfare payments have been cut off during the summer months when the poor are needed to work in the fields. Moreover, much of the debate about the Negative Income Tax and the Family Assistance Plan has concerned their impact on the work incentive, by which is actually meant the incentive of the poor to do the needed dirty work if the wages therefrom are no larger than the income grant. Many economic activities that involve dirty work depend on the poor for their existence: restaurants, hospitals, parts of the garment industry, and "truck farming," among others, could not persist in their present form without the poor.

Second, because the poor are required to work at low wages, they subsidize a variety of economic activities that benefit the affluent. For example, domestics subsidize the upper middle and upper classes, making life easier for their employers and freeing affluent women for a variety of professional, cultural, civic, and partying activities. Similarly, because the poor pay a higher proportion of their income in property and sales taxes, among others, they subsidize many state and local governmental services that benefit more affluent groups. In addition, the poor support innovation in medical practice as patients in teaching and research hospitals and as guinea pigs in medical experiments.

Third, poverty creates jobs for a number of occupations and professions that serve or "service" the poor, or protect the rest of society from them. As already noted, penology would be minuscule without the poor, as would the police. Other activities and groups that flourish because of the existence of poverty are the numbers game, the sale of heroin and cheap wines and liquors, pentecostal ministers, faith healers, prostitutes, pawn shops, and the peacetime army, which recruits its enlisted men mainly from among the poor.

Fourth, the poor buy goods others do not want and thus prolong the economic usefulness of such goods—day-old bread, fruit and vegetables that would otherwise have to be thrown out, second-hand clothes, and deteriorating automobiles and buildings. They also provide incomes for doctors, lawyers, teachers, and others who are too old, poorly trained, or incompetent to attract more affluent clients.

In addition to economic functions, the poor perform a number of social functions.

Fifth, the poor can be identified and punished as alleged or real deviants in order to uphold the legitimacy of conventional norms. To justify the desirability of hard work, thrift, honesty, and monogamy, for example, the defenders of these norms must be able to find people who can be accused of being lazy, spendthrift, dishonest, and promiscuous. Although there is some evidence that the poor are about as moral and law-abiding as anyone else, they are more likely than middle-class transgressors to be caught and punished when they participate in deviant acts. Moreover, they lack the political and cultural power to correct the stereotypes that other people hold of them and thus continue to be thought of as lazy, spendthrift, and so on, by those who need living proof that moral deviance does not pay.

Sixth, and conversely, the poor offer vicarious participation to the rest of the population in the uninhibited sexual, alcoholic, and narcotic behavior in which they are alleged to participate and which, being freed from the constraints of affluence, they are often thought to enjoy more than the middle classes. Thus many people, some social scientists included, believe that the poor not only are more given to uninhibited behavior (which may be true, although it is often motivated by despair more than lack of inhibition) but derive more pleasure from it than affluent people (which research by Lee Rainwater, Walter Miller, and others shows to be patently untrue). However, whether the poor actually have more sex and enjoy it more is irrelevant; so long as middle-class people believe this to be true, they can participate in it vicariously when instances are reported in factual or fictional form.

Seventh, the poor also serve a direct cultural function when culture created by or for them is adopted by the more affluent. The rich often collect artifacts from extinct folk cultures of poor people; and almost all Americans listen to the blues, Negro spirituals, and country music, which originated among the Southern poor. Recently they have enjoyed the rock styles that were born, like the Beatles, in the slums; and in the last year, poetry written by ghetto children has become popular in literary circles. The poor also serve as culture heroes, particularly, of course, to the left, but the hobo, the cowboy, the hipster, and the mythical prostitute with a heart of gold perform this function for a variety of groups.

Eighth, poverty helps to guarantee the status of those who are not poor. In every hierarchical society someone has to be at the bottom; but in American society, in which social mobility is an important goal for many and people need to know where they stand, the poor function as a reliable and relatively permanent measuring rod for status comparisons. This is particularly true for the working class, whose politics is influenced by the need to maintain status distinctions between themselves and the poor, much as the aristocracy must find ways of distinguishing itself from the *nouveaux riches*.

Ninth, the poor also aid the upward mobility of groups just above them in the class hierarchy. Thus a goodly number of Americans have entered the middle class through the profits earned from the provision of goods and services in the slums, including illegal or nonrespectable ones that upper-class and upper-middle-class businessmen shun because of their low prestige. As a result, members of almost every immigrant group have financed their upward mobility by providing slum housing, entertainment, gambling, narcotics, etc., to later arrivals—most recently to blacks and Puerto Ricans.

Tenth, the poor help to keep the aristocracy busy, thus justifying its continued existence. "Society" uses the poor as clients of settlement houses and beneficiaries of charity affairs; indeed, the aristocracy must have the poor to demonstrate its superiority over other elites who devote themselves to earning money.

Eleventh, the poor, being powerless, can be made to absorb the costs of change and growth in American society. During the nineteenth century, they did the backbreaking work that built the cities; today, they are pushed out of their neighborhoods to make room for "progress." Urban renewal projects to hold middle-class taxpayers in the city and expressways to enable suburbanites to commute downtown have typically been located in poor neighborhoods, since no other group will allow itself to be displaced. For the same reason, universities, hospitals, and civic centers also expand into land occupied by the poor. The major costs of the industrialization of agriculture have been borne by the poor, who are pushed off the land without recompense; and they have paid a large share of the human cost of the growth of American power overseas, for they have provided many of the foot soldiers for Vietnam and other wars.

Twelfth, the poor facilitate and stabilize the American political process. Because they vote and participate in politics less than other groups, the political system is often free to ignore them. Moreover, since they can rarely support Republicans, they often provide Democrats with a captive constituency that has no other place to go. As a result, the Democrats can count on their votes, and be more responsive to voters—for example, the white working class—who might otherwise switch to the Republicans.

Thirteenth, the role of the poor in upholding conventional norms (see the fifth point, above) also has a significant political function. An economy based on the ideology of laissez-faire requires a deprived population that is allegedly unwilling to work or that can be considered inferior because it must accept charity or welfare in order to survive. Not only does the alleged moral deviancy of the poor reduce the moral pressure on the present political economy to eliminate poverty but socialist alternatives can be made to look quite unattractive if those who will benefit most from them can be described as lazy, spendthrift, dishonest, and promiscuous.

THE ALTERNATIVES

I have described thirteen of the more important functions poverty and the poor satisfy in American society, enough to support the functionalist thesis that poverty, like any other social phenomenon, survives in part because it is useful to society or some of its parts. This analysis is not intended to suggest that because it is often functional, poverty *should* exist, or that it *must* exist. For one thing, poverty has many more dysfunctions than functions; for another, it is possible to suggest functional alternatives.

For example, society's dirty work could be done without poverty, either by automation or by paying "dirty workers" decent wages. Nor is it necessary for the poor to subsidize the many activities they support through their low-wage jobs. This would, however, drive up the costs of these activities, which would result in higher prices to their customers and clients. Similarly, many of the professionals who flourish because of the poor could be given other roles. Social workers could provide counseling to the affluent, as they prefer to do anyway; and the police could devote themselves to traffic and organized crime. Other roles would have to be found for badly trained or incompetent professionals now relegated to serving the poor, and someone else would have to pay their salaries. Fewer penologists would be employable, however. And pentecostal religion could probably not survive without the poor—nor would parts of the second- and third-hand-goods market. And in many cities, "used" housing that no one else wants would then have to be torn down at public expense.

Alternatives for the cultural functions of the poor could be found more easily and cheaply. Indeed, entertainers, hippies, and adolescents are already serving as the deviants needed to uphold traditional morality and as devotees of orgies to "staff" the fantasies of vicarious participation.

The status functions of the poor are another matter. In a hierarchical society, some people must be defined as inferior to everyone else with respect to a variety of attributes, but they need not be poor in the absolute sense. One could conceive of a society in which the "lower class," though last in the pecking order, received 75 percent of the median income, rather than 15–40 percent, as is now the case. Needless to say, this would require considerable income redistribution.

The contribution the poor make to the upward mobility of the groups that provide them with goods and services could also be maintained without the poor's having such low incomes. However, it is true that if the poor were more affluent, they would have access to enough capital to take over the provider role, thus competing with, and perhaps rejecting, the "outsiders." (Indeed, owing in part to anti-poverty programs, this is already happening in a number of ghettos, where white storeowners are being replaced by blacks.) Similarly, if the poor were more affluent, they would make less willing clients for upper-class philanthropy, although some would still use settlement houses to achieve upward mobility, as they do now. Thus "society" could continue to run its philanthropic activities.

The political functions of the poor would be more difficult to replace. With increased affluence the poor would probably obtain more political power and be more active politically. With higher incomes and more political power, the poor would be likely to resist paying the costs of growth and change. Of course, it is possible to imagine urban renewal and highway projects that properly reimbursed the displaced people, but such projects would then become considerably more expensive, and many might never be built. This, in turn, would reduce the comfort and convenience of those who now benefit from urban renewal and expressways. Finally, hippies could serve also as more deviants to justify the existing political economy—as they already do. Presumably, however, if poverty were eliminated, there would be fewer attacks on that economy.

In sum, then, many of the functions served by the poor could be replaced if poverty were eliminated, but almost always at higher costs to others, particularly more affluent others. Consequently, a functional analysis must conclude that poverty persists not only because it fulfills a number of positive functions but also because many of the functional alternatives to poverty would be quite dysfunctional for the affluent members of society. A functional analysis thus ultimately arrives at much the same conclusion as radical sociology, except that radical thinkers treat as manifest what I describe as latent; that social phenomena that are functional for affluent or powerful groups and dysfunctional for poor or powerless ones persist; that when the elimination of such phenomena through functional alternatives would generate dysfunctions for the affluent or powerful, they will continue to persist; and that phenomena like poverty can be eliminated only when they become dysfunctional for the affluent or powerful, or when the powerless can obtain enough power to change society.

NOTES

1. "Manifest and Latent Functions," in *Social Theory and Social Structure* (Glencoe, Ill.: Free Press, 1949), p. 71.

2. I shall henceforth abbreviate positive functions as functions and negative functions as dysfunctions.

I shall also describe functions and dysfunctions, in the planner's terminology, as benefits and costs.

Chapter 10

Race and Ethnicity

The Problem of Inequality

Jeanne Ballantine

The process of stratification results in inequalities between individuals and groups. This chapter explores major sources of inequality and results of inequality.

How easy it is to judge by stereotypes rather than fact, and to maintain our ethnocentric attitude favoring the "we," the in-group, *our* group. If categorization serves as the primary means of ordering our lives, is that bad? Potentially, yes. In the process of categorizing, we accept some people and ideas, and rule out others—often by using arbitrary stereotypes that discriminate against individuals who belong to certain groups.

Inequality is a worldwide problem. Being in a low caste, or a low class, or a minority can be a matter of life or death. Minority groups can be distinguished from dominant groups by factors that make them different; race, ethnic group, religion, tribal affiliation, and political beliefs are a few factors that cause minorities to be judged less favorably than the dominant group. Minorities are excluded or denied full participation in society, which can mean fewer opportunities in education, politics, jobs, health, or recreational facilities. "Minority groups are stereotyped, ridiculed, condemned, or otherwise defamed, allowing dominant group members to justify and not feel guilty about unequal and poor treatment based on political or religious ideologies and ethnocentric beliefs" (Ballantine and Roberts 2007:224). Because of exclusion or hostile treatment, minority group members often insulate themselves from the hostility by creating ethnic or racial enclaves or by otherwise segregating themselves from the daily hostilities.

Consider the national and international news stories you hear: conflicts in Darfur between tribal groups; religious factions fighting in Iraq; conflicts in the United States between immigrants and other groups. Looking around the world, it becomes obvious that many conflicts are rooted in class, racial, or ethnic tensions. The question is, How can we understand these conflicts, their historical origins and the form they take today, and how can we tackle them? The articles in this chapter present some possible answers to these questions.

In the first reading, Michael Omi and Howard Winant discuss the concept of *social race*, a core concept in the sociological understanding of how some groups came to be treated differently. This concept is based in part on stereotypes and the way they affect the dominant group's reactions toward those who differ.

Where there is money to be made, someone will take advantage of the opportunity. Unfortunately, that principle underlies many cases of international human trafficking. Most of those who find themselves on the wrong side of the deal are poor minorities, often women and children, from poor families and countries. They are bought, kidnapped, and sold into slavery, often into the sex trade. In the second article, Kevin Bales discusses the plight of young girls and women, mostly from minority groups in countries around the world, who are bought and sold like cattle at market.

The United States—the so-called land of opportunity—has proven to be anything but that for certain segments of the population. The country was founded on principles of freedom and justice, but some groups, including Africans, Asians, and Hispanics, were left out of the formula. Africans came as slaves and have a long road to travel before the remnants of their suppression—prejudice, discrimination, and racism—are overcome. What has been referred to as the "black underclass" (Wilson 1984) is due mainly to historical discrimination and economic trends. These problems have causes that lie in the economic and social structure of society; they are difficult to change and, in fact, are increasing in many urban areas. Many U.S. citizens are excited about the latest step forward, an African American elected to the top office in the land; others resist this change that legitimizes minority claims to the presidency.

Asians often supplied the needed labor for building railroads, but when economic times were hard they were no longer welcomed by the dominant group in the United States. Asian Americans have been called the "model minority" because of the educational and business success achieved by certain groups, especially Japanese and Chinese immigrants; most of the success stories, however, involve families that were middle class or higher to begin with. Many Asian Americans are not so fortunate—they entered the United States as refugees with few skills to cope with U.S. society. A recent report by New York University and the College Board (Teranishi 2008) points out that the terms *Asian American* and *Pacific Islander* are very broad and embrace a number of different ethnic groups, each with its own language and culture. Not all of these groups qualify for the label "model minority." The problems faced by many Asian groups whose members were displaced by war or poverty, and who have tried to make a new home in a foreign land, are enormous.

Latinos, the fastest-growing minority group in the United States, experience prejudice and discrimination because of differences in culture and language, and because of the economic threat some of these immigrants pose to workers in low-paying jobs. In the third article, Joan Moore and Raquel Pinderhughes trace the history of Latino groups and reveal the roots of their current isolation and stigmatization in society, factors that will not make change easy. These authors consider the many frustrated and disaffected youth who see no opportunities and face daily hostile environments; they may find belonging, acceptance, and protection in gang membership.

If inequality stemming from racism, classism, prejudice, stereotyping, and discrimination is to be reduced, then individuals and groups must recognize the results of prejudice and discrimination embedded in the societal structure and culture and understand how they affect their individual beliefs

and actions. Only in this manner can individuals overcome their own stereotypes and ethnocentrism and view others from a cultural relativist's perspective, understanding and respecting other ways of life. In the final article, James Crone asks, "What might racial/ethnic equality mean?" and "What more can we do?" He discusses inequality in the United States and what can be done, both directly and indirectly, to solve the problems presented by inequality.

As you read the articles in this chapter, think of ways the United States, which has been populated by immigrants over time, can improve minority relations and provide equal opportunity for all. Consider why some minorities have had more success than others. Also note other examples you have observed of groups segregating themselves for self-protection, of false stereotyping, and of economic conflicts between groups that lead to discrimination.

REFERENCES

Ballantine, Jeanne H., and Keith A. Roberts. 2007. *Our Social World: Introduction to Sociology*. Thousand Oaks, CA: Pine Forge Press, p. 224.

Teranishi, Robert T. June 18, 2008. "Facts, Not Fiction: Setting the Record Straight." New York University and The College Board. In: Tamar Lewin, "Report Takes Aim at 'Model Minority' Stereotype of Asian-American Students." *The New York Times*.

Wilson, William Julius. Spring 1984. "The Black Underclass." *Wilson Quarterly*.

41

Racial Formations

MICHAEL OMI AND HOWARD WINANT

Race—a much discussed and debated topic! Sociologists focus on the effects of social race, or using race to categorize people into groups against which prejudice and discrimination may be directed. Omi and Winant open this section with a discussion of this social concept and of racial ideology and identity that results from how people view race.

As you read, think about the following:

1. *The idea of race is relatively new in human history. How did it develop?*
2. *What are examples of social race? Of hypo-descent?*
3. *How does the idea of race affect* all *groups in the United States?*

GLOSSARY **Social race** Categories defining social relations between ethnic groups based on historical contexts and events. **Hypo-descent** Defining anyone with one drop of blood (African, Asian, Native American) as a marker of the ethnic group. **Passing** Attempting to be seen or recognized as being a member of a different group, usually white.

WHAT IS RACE?

Race consciousness, and its articulation in theories of race, is largely a modern phenomenon. When European explorers in the New World "discovered" people who looked different than themselves, these "natives" challenged then existing conceptions of the origins of the human species, and raised disturbing questions as to whether *all* could be considered in the same "family of man."[1] Religious debates flared over the attempt to reconcile the Bible with the existence of "racially distinct" people. Arguments took place over creation itself, as theories of polygenesis questioned whether God had made only one species of humanity ("monogenesis").

Europeans wondered if the natives of the New World were indeed human beings with redeemable souls. At stake were not only the prospects for conversion, but the types of treatment to be accorded them. The expropriation of property, the denial of political rights, the introduction of slavery and other forms of coercive labor, as well as outright extermination, all presupposed a worldview which distinguished Europeans—children of God, human beings, etc.—from "others." Such a worldview was needed to explain why some should be "free" and others enslaved, why some had rights to land and property while others did not. Race, and the interpretation of racial differences, was a central factor in that worldview.

In the colonial epoch science was no less a field of controversy than religion in attempts to comprehend the concept of race and its meaning. Spurred on by the classificatory scheme of living organisms devised by Linnaeus in *Systema Naturae,* many scholars in the eighteenth and nineteenth centuries dedicated themselves to the identification and ranking of variations in humankind. Race was thought of as a *biological* concept, yet its precise definition was the subject of debates which, as we have noted, continue to rage today. Despite efforts ranging from Dr. Samuel Morton's studies of cranial capacity[2] to contemporary attempts to base racial classification on shared gene pools,[3] the concept of race has defied biological definition....

Attempts to discern the *scientific meaning* of race continue to the present day. Although most physical anthropologists and biologists have abandoned the quest for a scientific basis to determine racial categories, controversies have recently flared in the area of genetics and educational psychology. For instance, an essay by Arthur Jensen arguing that hereditary factors shape intelligence not only revived the "nature or nurture" controversy, but raised highly volatile questions about racial equality itself.[4] Clearly the attempt to establish a *biological* basis of race has not been swept into the dustbin of history, but is being resurrected in various scientific arenas. All such attempts seek to remove the concept of race from fundamental social, political, or economic determination. They suggest instead that the truth of race lies in the terrain of innate characteristics, of which skin color and other physical attributes provide only the most obvious, and in some respects most superficial, indicators.

RACE AS A SOCIAL CONCEPT

The social sciences have come to reject biologistic notions of race in favor of an approach which regards race as a *social* concept. Beginning in the eighteenth century, this trend has been slow and uneven, but its direction clear. In the nineteenth century Max Weber discounted biological explanations for racial conflict and instead highlighted the social and political factors which engendered such conflict.[5] The work of pioneering cultural anthropologist Franz Boas was crucial in refuting the scientific racism of the early twentieth century by rejecting the connection between race and culture, and the assumption of a continuum of "higher" and "lower" cultural groups. Within the contemporary social science literature, race is assumed to be a variable which is shaped by broader societal forces.

Race is indeed a pre-eminently *socio-historical* concept. Racial categories and the meaning of race are given concrete expression by the specific social relations and historical context in which they are embedded. Racial meanings have varied tremendously over time and between different societies.

In the United States, the black/white color line has historically been rigidly defined and enforced. White is seen as a "pure" category. Any racial intermixture makes one "nonwhite." In the movie *Raintree County*, Elizabeth Taylor describes the worst of fates to befall whites as "havin' a little Negra blood in ya'—just one little teeny drop and a person's all Negra."[6] This thinking flows from what Marvin Harris has characterized as the principle of *hypo-descent*:

> By what ingenious computation is the genetic tracery of a million years of evolution unraveled and each man [sic] assigned his proper social box? In the United States, the mechanism employed is the rule of hypo-descent. This descent rule requires Americans to believe that anyone who is known to have had a Negro ancestor is a Negro. We admit nothing in between.... "Hypo-descent" means affiliation with the subordinate rather than the super-ordinate group in order to avoid the ambiguity of intermediate identity.... The rule of hypo-descent is, therefore, an invention, which we in the United States have made in order to keep biological facts from intruding into our collective racist fantasies.[7]

By contrast, a striking feature of race relations in the lowland areas of Latin America since the abolition of slavery has been the relative absence of sharply defined racial groupings. No such rigid descent rule characterizes racial identity in many Latin American societies. Brazil, for example, has historically had less rigid conceptions of race, and thus a variety of "intermediate" racial categories exist. Indeed, as Harris notes, "One of the most striking consequences of the Brazilian system of racial identification is that parents and children and even brothers and sisters are frequently accepted as representatives of quite opposite racial types."[8] Such a possibility is incomprehensible within the logic of racial categories in the United States.

To suggest another example: the notion of "passing" takes on new meaning if we compare various American cultures' means of assigning racial identity. In the United States, individuals who are actually "black" by the logic of hypo-descent have attempted to skirt the discriminatory barriers imposed by law and custom by attempting to "pass" for white.[9] Ironically, these same individuals would not be able to pass for "black" in many Latin American societies.

Consideration of the term "black" illustrates the diversity of racial meanings which can be found among different societies and historically within a given society. In contemporary British politics the term "black" is used to refer to all nonwhites. Interestingly this designation has not arisen through the racist discourse of groups such as the National Front. Rather, in political and cultural movements, Asian as well as Afro-Caribbean youth are adopting the term as an expression of self-identity.[10] The wide-ranging meanings of "black" illustrate the manner in which racial categories are shaped politically.[11]

The meaning of race is defined and contested throughout society, in both collective action and personal practice. In the process, racial categories themselves are formed, transformed, destroyed and reformed. We use the term *racial formation* to refer to the process by which social, economic and political forces determine the content and importance of racial categories, and by which they are in turn shaped by racial meanings. Crucial to this formulation is the treatment of race as a *central axis* of social relations which cannot be subsumed under or reduced to some broader category or conception.

RACIAL IDEOLOGY AND RACIAL IDENTITY

The seemingly obvious, "natural" and "common sense" qualities which the existing racial order exhibits themselves testify to the effectiveness of the racial formation process in constructing racial meanings and racial identities.

One of the first things we notice about people when we meet them (along with their sex) is their race. We utilize race to provide clues abut *who* a person is. This fact is made painfully obvious when we encounter someone whom we cannot conveniently racially categorize—someone who is, for example, racially "mixed" or of an ethnic/racial group with which we are not familiar. Such an encounter becomes a source of discomfort and momentarily a crisis of racial meaning. Without a racial identity, one is in danger of having no identity.

Our compass for navigating race relations depends on preconceived notions of what each specific racial group looks like. Comments such as, "Funny, you don't look black," betray an underlying image of what black should be. We also become disoriented when people do not act "black," "Latino," or indeed "white." The content of such stereotypes reveals a series of unsubstantiated beliefs about who these groups are and what "they" are like.[12]

In U.S. society, then, a kind of "racial etiquette" exists, a set of interpretative codes and racial meanings which operate in the interactions of daily life. Rules shaped by our perception of race in a comprehensively racial society determine the "presentation of self,"[13] distinctions of status, and appropriate modes of conduct. "Etiquette" is not mere universal adherence to the dominant group's rules, but a more dynamic combination of these rules with the values and beliefs of subordinated groupings. This racial "subjection" is quintessentially ideological. Everybody learns some combination, some version, of the rules of racial classification, and of their own racial identity, often without obvious teaching or conscious inculcation. Race becomes "common sense"—a way of comprehending, explaining and acting in the world.

Racial beliefs operate as an "amateur biology," a way of explaining the variations in "human nature."[14] Differences in skin color and other obvious physical characteristics supposedly provide visible clues to differences lurking underneath. Temperament, sexuality, intelligence, athletic ability, aesthetic preferences and so on are presumed to be fixed and discernible from the palpable mark of race. Such diverse questions as our confidence and trust in others (for example, clerks or salespeople, media figures, neighbors), our sexual preferences and romantic images, our tastes in music, films, dance, or sports, and our very ways of talking, walking, eating and dreaming are ineluctably shaped by notions of race. Skin color "differences" are thought to explain perceived differences in intellectual, physical and artistic temperaments, and to justify distinct treatment of racially identified individuals and groups.

The continuing persistence of racial ideology suggests that these racial myths and stereotypes cannot be exposed as such in the popular imagination. They are, we think, too essential, too integral, to the maintenance of the U.S. social order. Of course, particular meanings, stereotypes and myths can change, but the presence of a *system* of racial meanings and stereotypes, of racial ideology, seems to be a permanent feature of U.S. culture.

Film and television, for example, have been notorious in disseminating images of racial minorities which establish for audiences what people from these groups look like, how they behave, and "who they are."[15] The power of the media lies not only in their ability to reflect the dominant racial ideology, but in their capacity to shape that ideology in the first place. D. W. Griffith's epic *Birth of a Nation*, a sympathetic treatment of the rise of the Ku Klux Klan during Reconstruction, helped to generate, consolidate and "nationalize" images of blacks which had been more disparate (more regionally specific, for example) prior to the film's appearance.[16] In U.S. television, the necessity to define characters in the briefest and most condensed manner has led to the perpetuation of racial caricatures, as racial stereotypes serve as shorthand for scriptwriters, directors and actors, in commercials, etc. Television's tendency to address the "lowest common denominator" in order to render programs "familiar" to an enormous and diverse audience leads it regularly to assign and reassign racial characteristics to particular groups, both minority and majority.

These and innumerable other examples show that we tend to view race as something fixed and immutable—something rooted in "nature." Thus we mask the historical construction of racial categories, the shifting meaning of race, and the crucial role of politics and ideology in shaping race relations. Races do not emerge full-blown. They are the results of diverse historical practices and are continually subject to challenge over their definition and meaning.

NOTES

1. Thomas F. Gossett notes:

   Race theory... had up until fairly modern times no firm hold on European thought. On the other hand, race theory and race prejudice were by no means unknown at the time when the English colonists came to North America. Undoubtedly, the age of exploration led many to speculate on race differences at a period when neither Europeans nor Englishmen were prepared to make allowances for vast cultural diversities. Even though race theories had not then secured wide acceptance or even sophisticated formulation, the first contacts of the Spanish with the Indians in the Americas can now be recognized as the beginning of a struggle between conceptions of the nature of primitive peoples which has not yet been wholly settled. (Thomas F. Gossett, *Race: The History of an Idea in America* (New York: Schocken Books, 1965), p. 16.)

   Winthrop Jordan provides a detailed account of early European colonialists' attitudes about color and race in *White over Black: American Attitudes Toward the Negro, 1550–1812* (New York: Norton, 1977 [1968]), pp. 3–43.

2. Pro-slavery physician Samuel George Morton (1799–1851) compiled a collection of 800 crania from all parts of the world which formed the sample for his studies of race. Assuming that the larger the size of the cranium translated into greater intelligence, Morton established a relationship between race and skull capacity. Gossett reports that:

> In 1849, one of his studies included the following results: The English skulls in his collection proved to be the largest, with an average cranial capacity of 96 cubic inches. The Americans and Germans were rather poor seconds, both with cranial capacities of 90 cubic inches. At the bottom of the list were the Negroes with 83 cubic inches, the Chinese with 82, and the Indians with 79. (Ibid., p. 74.)

On Morton's methods, see Stephen J. Gould, "The Finagle Factor," *Human Nature* (July 1978).

3. Definitions of race founded upon a common pool of genes have not held up when confronted by scientific research which suggests that the differences *within* a given human population are greater than those *between* populations. See L. L. Cavalli-Sforza, "The Genetics of Human Populations," *Scientific American*, September 1974, pp. 81–89.

4. Arthur Jensen, "How Much Can We Boost IQ and Scholastic Achievement?", *Harvard Educational Review 39* (1969):1–123.

5. Ernst Moritz Manasse, "Max Weber on Race," *Social Research 14* (1947):191–221.

6. Quoted in Edward D. C. Campbell, Jr., *The Celluloid South: Hollywood and the Southern Myth* (Knoxville: University of Tennessee Press, 1981), pp. 168–70.

7. Marvin Harris, *Patterns of Race in the Americas* (New York: Norton, 1964), p. 56.

8. Ibid., p. 57.

9. After James Meredith had been admitted as the first black student at the University of Mississippi, Harry S. Murphy announced that he, and not Meredith, was the first black student to attend "Ole Miss." Murphy described himself as black but was able to pass for white and spent nine months at the institution without attracting any notice (ibid., p. 56).

10. A. Sivanandan, "From Resistance to Rebellion: Asian and Afro-Caribbean Struggles in Britain," *Race and Class 23*(2–3) (Autumn–Winter 1981).

11. Consider the contradictions in racial status which abound in the country with the most rigidly defined racial categories—South Africa. There a race classification agency is employed to adjudicate claims for upgrading of official racial identity. This is particularly necessary for the "coloured" category. The apartheid system considers Chinese as "Asians" while the Japanese are accorded the status of "honorary whites." This logic nearly detaches race from any grounding in skin color and other physical attributes and nakedly exposes race as a juridical category subject to economic, social and political influences. (We are indebted to Steve Talbot for clarification of some of these points.)

12. Gordon W. Allport, *The Nature of Prejudice* (Garden City, NY: Doubleday, 1958), pp. 184–200.

13. We wish to use this phrase loosely, without committing ourselves to a particular position on such social psychological approaches as symbolic interactionism, which are outside the scope of this study. An interesting study on this subject is S. M. Lyman and W. A. Douglass, "Ethnicity: Strategies of Individual and Collective Impression Management," *Social Research 40*(2) (1973).

14. Michael Billig, "Patterns of Racism: Interviews with National Front Members," *Race and Class 20*(2) (Autumn 1978):161–79.

15. "Miss San Antonio USA Lisa Fernandez and other Hispanics auditioning for a role in a television soap-opera did not fit the Hollywood image of real Mexicans and had to darken their faces before filming." Model Aurora Garza said that their faces were bronzed with powder because they looked too white. "I'm a real Mexican [Garza said] and very dark anyway. I'm even darker right now because I have a tan. But they kept wanting me to make my face darker and darker'" (*San Francisco Chronicle*, 21 September 1984). A similar dilemma faces Asian American actors who feel that Asian character lead roles inevitably go to white actors who make themselves up to be Asian. Scores of Charlie Chan films, for example, have been made with white leads (the last one was the 1981 *Charlie Chan and the Curse of the Dragon Queen*). Roland Winters, who played in six Chan features, was asked by playwright Frank Chin to explain the logic of casting a white man in the role of Charlie Chan: "The only thing I can think of is, if you want to cast a homosexual in a

show, and get a homosexual, it'll be awful. It won't be funny… and maybe there's something there…" [Frank Chin, "Confessions of the Chinatown Cowboy," *Bulletin of Concerned Asian Scholars* 4(3) (Fall 1972)].

16. Melanie Martindale-Sikes, "Nationalizing 'Nigger' Imagery Through 'Birth of a Nation'," paper prepared for the 73rd Annual Meeting of the American Sociological Association, 4–8 September 1978, San Francisco.

42

Because She Looks like a Child

KEVIN BALES

Kevin Bales discusses a topic that crosses race, class, and gender lines—sex trafficking. Who are the targets? Poor, young minority women who are helpless to resist enslavement. They are bought by traders and sold by poor families, or sometimes just kidnapped. The experiences of these young women are horrific and illustrate the deep divisions in humanity.

Think about the following ideas as you read this selection:

1. *Who are the girls engaged in the sex trade who serve Thai and foreign men? What are their backgrounds?*
2. *How can families and governments allow sex trafficking to happen? Who is benefitting and how?*
3. *What should the world community do about cross-border trafficking, if anything?*
4. *Are you aware of any trafficking in your area?*

GLOSSARY **Brothel** Place where men pay for sex with prostitutes (in this case, often sex slaves).

When Siri wakes it is about noon.[1] In the instant of waking she knows exactly who and what she has become. As she explained to me, the soreness in her genitals reminds her of the fifteen men she had sex with the night before. Siri is fifteen years old. Sold by her parents a year ago, she finds that her resistance and her desire to escape the brothel are breaking down and acceptance and resignation are taking their place.

In the provincial city of Ubon Ratchathani, in northeastern Thailand, Siri works and lives in a brothel. About ten brothels and bars, dilapidated and dusty buildings, line the side street just around the corner from a new Western-style shopping mall. Food and noodle vendors are scattered between the brothels. The woman behind the noodle stall outside the brothel where Siri works is also a spy, warder, watchdog, procurer, and dinner lady to Siri and the other twenty-four girls and women in the brothel.

The brothel is surrounded by a wall, with iron gates that meet the street. Within the wall is a dusty yard, a concrete picnic table, and the ubiquitous spirit house, a small shrine that stands outside all Thai buildings. A low door leads into a windowless concrete room that is thick with the smell of

Source: Bales, Kevin. "Because she Looks like a Child." In Barbara Ehrenreich and Arlie Russell Hochschild, eds. *Global Woman: Nannies, Maids, and Sex Workers in the New Economy.* New York: Henry Holt and Company. 2002.

cigarettes, stale beer, vomit, and sweat. This is the "selection" room (*hong du*). On one side of the room are stained and collapsing tables and booths; on the other side is a narrow elevated platform with a bench that runs the length of the room. Spotlights pick out this bench, and at night the girls and women sit here under the glare while the men at the tables drink and choose the one they want.

Passing through another door, at the far end of the bench, the man follows the girl past a window, where a bookkeeper takes his money and records which girl he has selected. From there he is led to the girl's room. Behind its concrete front room, the brothel degenerates even further, into a haphazard shanty warren of tiny cubicles where the girls live and work. A makeshift ladder leads up to what may have once been a barn. The upper level is now lined with doors about five feet apart, which open into rooms of about five by seven feet that hold a bed and little else.

Scraps of wood and cardboard separate one room from the next, and Siri has plastered her walls with pictures of teenage pop stars cut from magazines. Over her bed, as in most rooms, there also hangs a framed portrait of the king of Thailand; a single bare lightbulb dangles from the ceiling. Next to the bed a large tin can holds water; there is a hook nearby for rags and towels. At the foot of the bed, next to the door, some clothes are folded on a ledge. The walls are very thin, and everything can be heard from the surrounding rooms; a shout from the bookkeeper echoes through all of them, whether their doors are open or closed.

After rising at midday, Siri washes herself in cold water from the single concrete trough that serves the brothel's twenty-five women. Then, dressed in a T-shirt and skirt, she goes to the noodle stand for the hot soup that is a Thai breakfast. Through the afternoon, if she does not have any clients, she chats with the other girls and women as they drink beer and play cards or make decorative handicrafts together. If the pimp is away the girls will joke around, but if not they must be constantly deferential and aware of his presence, for he can harm them or use them as he pleases. Few men visit in the afternoon, but those who do tend to

have more money and can buy a girl for several hours if they like. Some will even make appointments a few days in advance.

At about five, Siri and the other girls are told to dress, put on their makeup, and prepare for the night's work. By seven the men will be coming in, purchasing drinks, and choosing girls; Siri will be chosen by the first of the ten to eighteen men who will buy her that night. Many men choose Siri because she looks much younger than her fifteen years. Slight and round faced, dressed to accentuate her youth, she could pass for eleven or twelve. Because she looks like a child, she can be sold as a "new" girl at a higher price, about $15, which is more than twice that charged for the other girls.

Siri is very frightened that she will get AIDS. Long before she understood prostitution she knew about HIV, as many girls from her village returned home to die from AIDS after being sold into the brothels. Every day she prays to Buddha, trying to earn the merit that will preserve her from the disease. She also tries to insist that her clients use condoms, and in most cases she is successful, because the pimp backs her up. But when policemen use her, or the pimp himself, they will do as they please; if she tries to insist, she will be beaten and raped. She also fears pregnancy, but like the other girls she receives injections of the contraceptive drug Depo-Provera. Once a month she has an HIV test. So far it has been negative. She knows that if she tests positive she will be thrown out to starve.

Though she is only fifteen, Siri is now resigned to being a prostitute. The work is not what she had thought it would be. Her first client hurt her; and at the first opportunity she ran away. She was quickly caught, dragged back, beaten, and raped. That night she was forced to take on a chain of clients until the early morning. The beatings and the work continued night after night, until her will was broken. Now she is sure that she is a very bad person to have deserved what has happened to her. When I comment on how pretty she looks in a photograph, how like a pop star, she replies, "I'm no star; I'm just a whore, that's all." She copes as best she can. She takes a dark pride in her higher price and the large number of men who choose her. It is the adjustment of the concentration camp, an effort to make sense of horror.

In Thailand prostitution is illegal, yet girls like Siri are sold into sex slavery by the thousands. The brothels that hold these girls are but a small part of a much wider sex industry. How can this wholesale trade in girls continue? What keeps it working? The answer is more complicated than we might think. Thailand's economic boom and its social acceptance of prostitution contribute to the pressures that enslave girls like Siri.

RICE IN THE FIELD. FISH IN THE RIVER. DAUGHTERS IN THE BROTHEL.

Thailand is blessed with natural resources and sufficient food. The climate is mild to hot, there is dependable rain, and most of the country is a great plain, well watered and fertile. The reliable production of rice has for centuries made Thailand a large exporter of grains, as it is today. Starvation is exceedingly rare in its history, and social stability very much the norm. An old and often-repeated saying in Thai is "There is always rice in the fields and fish in the river." And anyone who has tried the imaginative Thai cuisine knows the remarkable things that can be done with those two ingredients and the local chili peppers.

One part of Thailand that is not so rich in necessities of life is the mountainous north. In fact, that area is not Thailand proper; originally the kingdom of Lanna, it was integrated into Thailand only in the late nineteenth century. The influence of Burma here is very strong—as are the cultures of the seven main hill tribes, which are distinctly foreign to the dominant Thai society. Only about a tenth of the land of the north can be used for agriculture, though what can be used is the most fertile in the country. The result is that those who control good land are well-off; those who live in the higher elevations, in the forest, are not. In another part of the world this last group might be called hillbillies,

and they share the hardscrabble life of mountain dwellers everywhere.

The harshness of this life stands in sharp contrast to that on the great plain of rice and fish. Customs and culture differ markedly as well, and one of those differences is a key to the sexual slavery practiced throughout Thailand today. For hundreds of years many people in the north, struggling for life, have been forced to view their own children as commodities. A failed harvest, the death of a key breadwinner, or any serious debt incurred by a family might lead to the sale of a daughter (never a son) as a slave or servant. In the culture of the north it was a life choice not preferred but acceptable and one that was used regularly. In the past these sales fed a small, steady flow of servants, workers, and prostitutes south into Thai society.

ONE GIRL EQUALS ONE TELEVISION

The small number of children sold into slavery in the past has become a flood today. This increase reflects the enormous changes in Thailand over the past fifty years as the country has gone through the great transformation of industrialization—the same process that tore Europe apart over a century ago. If we are to understand slavery in Thailand, we must understand these changes as well, for like so many other parts of the world, Thailand has always had slavery, but never before on this scale.

The economic boom of 1977 to 1997 had a dramatic impact on the northern villages. While the center of the country, around Bangkok, rapidly industrialized, the north was left behind. Prices of food, land, and tools all increased as the economy grew, but the returns for rice and other agriculture were stagnant, held down by government policies guaranteeing cheap food for factory workers in Bangkok. Yet visible everywhere in the north is a flood of consumer goods—refrigerators, televisions, cars and trucks, rice cookers, air conditioners—all of which are extremely tempting. Demand for these goods is high as families try to join the ranks of the prosperous. As it happens, the cost of participating in this consumer boom can be met from an old source that has become much more profitable: the sale of children.

In the past, daughters were sold in response to serious family financial crises. Under threat of losing its mortgaged rice fields and facing destitution, a family might sell a daughter to redeem its debt, but for the most part daughters were worth about as much at home as workers as they would realize when sold. Modernization and economic growth have changed all that. Now parents feel a great pressure to buy consumer goods that were unknown even twenty years ago; the sale of a daughter might easily finance a new television set. A recent survey in the northern provinces found that of the families who sold their daughters, two-thirds could afford not to do so but instead preferred to buy color televisions and video equipment.[2] And from the perspective of parents who are willing to sell their children, there has never been a better market.

The brothels' demand for prostitutes is rapidly increasing. The same economic boom that feeds consumer demand in the northern villages lines the pockets of laborers and workers in the central plain. Poor economic migrants from the rice fields now work on building sites or in new factories earning many times what they did on the land. Possibly for the first time in their lives, these laborers can do what more well-off Thai men have always done: go to a brothel. The purchasing power of this increasing number of brothel users strengthens the call for northern girls and supports a growing business in their procurement and trafficking.

Siri's story was typical. A broker, a woman herself from a northern village, approached the families in Siri's village with assurances of well-paid work for their daughters. Siri's parents probably understood that the work would be as a prostitute, since they knew that other girls from their village had gone south to brothels. After some negotiation they were paid 50,000 baht (US$2,000) for Siri, a very significant sum for this family of rice farmers.[3] This exchange began the process of debt bondage

that is used to enslave the girls. The contractual arrangement between the broker and the parents requires that this money be paid by the daughter's labor before she is free to leave or is allowed to send money home. Sometimes the money is treated as a loan to the parents, the girls being both the collateral and the means of repayment. In such cases the exorbitant interest charged on the loan means there is little chance that a girl's sexual slavery will ever repay the debt.

Siri's debt of 50,000 baht rapidly escalated. Taken south by the broker, Siri was sold for 100,000 baht to the brothel where she now works. After her rape and beating Siri was informed that the debt she must repay to the brothel equaled 200,000 baht. In addition, Siri learned of the other payments she would be required to make, including rent for her room, at 30,000 baht per month, as well as charges for food and drink, fees for medicine, and fines if she did not work hard enough or displeased a customer.

The total debt is virtually impossible to repay, even at Siri's higher rate of 400 baht. About 100 baht from each client is supposed to be credited to Siri to reduce her debt and pay her rent and other expenses; 200 goes to the pimp and the remaining 100 to the brothel. By this reckoning, Siri must have sex with three hundred men a month just to pay her rent, and what is left over after other expenses barely reduces her original debt. For girls who can charge only 100 to 200 baht per client, the debt grows even faster. This debt bondage keeps the girls under complete control as long as the brothel owner and the pimp believe they are worth having. Violence reinforces the control and any resistance earns a beating as well as an increase in the debt. Over time if the girl becomes a good and cooperative prostitute, the pimp may tell her she has paid off the debt and allow her to send small sums home. This "paying off" of the debt usually has nothing to do with an actual accounting of earnings but is declared at the discretion of the pimp, as a means to extend the brothel's profits by making the girl more pliable. Together with rare visits home, money sent back to the family operates to keep her at her job.

Most girls are purchased from their parents, as Siri was, but for others the enslavement is much more direct. Throughout Thailand agents travel to villages, offering work in factories or as domestics. Sometimes they bribe local officials to vouch for them, or they befriend the monks at the local temple to gain introductions. Lured by the promise of good jobs and the money that the daughters will send back to the village, the deceived families dispatch their girls with the agent, often paying for the privilege. Once they arrive in a city, the girls are sold to a brothel, where they are raped, beaten, and locked in. Still other girls are simply kidnapped. This is especially true of women and children who have come to visit relatives in Thailand from Burma or Laos. At bus and train stations, gangs watch for women and children who can be snatched or drugged for shipment to brothels.

Direct enslavement by trickery or kidnapping is not really in the economic interest of the brothel owners. The steadily growing market for prostitutes, the loss of girls to HIV infection, and the especially strong demand for younger and younger girls make it necessary for brokers and brothel owners to cultivate village families so that they can buy more daughters as they come of age. In Siri's case this means letting her maintain ties with her family and ensuring that after a year or so she send a monthly postal order for 10,000 baht to her parents. The monthly payment is a good investment, since it encourages Siri's parents to place their other daughters in the brothel as well. Moreover, the young girls themselves become willing to go when their older sisters and relatives returning for holidays bring stories of the rich life to be lived in the cities of the central plain. Village girls lead a sheltered life, and the appearance of women only a little older than themselves with money and nice clothes is tremendously appealing. They admire the results of this thing called prostitution with only the vaguest notion of what it is. Recent research found that young girls knew that their sisters and neighbors had become prostitutes, but when asked what it means to be a prostitute their most common answer was "wearing Western clothes in a restaurant."[4] Drawn by this glamorous life, they put up little

opposition to being sent away with the brokers to swell an already booming sex industry.

By my own conservative estimate there are perhaps thirty-five thousand girls like Siri enslaved in Thailand. Remarkably, this is only a small proportion of the country's prostitutes. In the mid-1990s the government stated that there were 81,384 prostitutes in Thailand—but that official number is calculated from the number of registered (though still illegal) brothels, massage parlors, and sex establishments. One Thai researcher estimated the total number of prostitutes in 1997 to be around 200,000.[5] Every brothel, bar, and massage parlor we visited in Thailand was unregistered, and no one working with prostitutes believes the government figures. At the other end of the spectrum are the estimates put forward by activist organizations such as the Center for the Protection of Children's Rights. These groups assert that there are more than 2 million prostitutes. I suspect that this number is too high in a national population of 60 million. My own reckoning, based on information gathered by AIDS workers in different cities, is that there are between half à million and 1 million prostitutes.

Of this number, only about one in twenty is enslaved. Most become prostitutes voluntarily, through some start out in debt bondage. Sex is sold everywhere in Thailand: barbershops, massage parlors, coffee shops and cafes, bars and restaurants, nightclubs and karaoke bars, brothels, hotels, and even temples traffic in sex. Prostitutes range from the high-earning "professional" women who work with some autonomy, through the women working by choice as call girls or in massage parlors, to the enslaved rural girls like Siri. Many women work semi-independently in bars, restaurants, and night clubs—paying a fee to the owner, working when they choose, and having the power to decide whom to take as a customer. Most bars and clubs cannot use an enslaved prostitute like Siri, as the women are often sent out on call and their clients expect a certain amount of cooperation and friendliness. Enslaved girls serve the lowest end of the market: the laborers, students, and workers who can afford only the 100 baht per half hour. It is low-cost sex in volume, and the demand is always there. For a Thai man, buying a woman is much like buying a round of drinks. But the reasons why such large numbers of Thai men use prostitutes are much more complicated and grow out of their culture, their history, and a rapidly changing economy.

"I DON'T WANT TO WASTE IT, SO I TAKE HER"

Until it was officially disbanded in 1910, the king of Thailand maintained a harem of hundreds of concubines, a few of whom might be elevated to the rank of "royal mother" or "minor wife." This form of polygamy was closely imitated by status-hungry nobles and emerging rich merchants of the nineteenth century. Virtually all men of any substance kept at least a mistress or a minor wife. For those with fewer resources, prostitution was a perfectly acceptable option, as renting took the place of out-and-out ownership.

Even today everyone in Thailand knows his or her place within a very elaborate and precise status system. Mistresses and minor wives continue to enhance any man's social standing, but the consumption of commercial sex has increased dramatically.[6] If an economic boom is a tide that raises all boats, then vast numbers of Thai men have now been raised to a financial position from which they can regularly buy sex. Nothing like the economic growth in Thailand was ever experienced in the West, but a few facts show its scale: in a country the size of Britain, one-tenth of the workforce moved from the land to industry in just the three years from 1993 to 1995; the number of factory workers doubled from less than 2 million to more than 4 million in the eight years from 1988 to 1995; and urban wages doubled from 1986 to 1996. Thailand is now the world's largest importer of motorcycles and the second-largest importer of pickup trucks, after the United States. Until the economic downturn of late 1997, money flooded Thailand, transforming poor rice farmers into wage laborers and fueling consumer demand.

With this newfound wealth, Thai men go to brothels in increasing numbers. Several recent studies show that between 80 and 87 percent of Thai men have had sex with a prostitute. Most report that their first sexual experience was with a prostitute. Somewhere between 10 and 40 percent of married men have paid for commercial sex within the past twelve months, as have up to 50 percent of single men. Though it is difficult to measure, these reports suggest something like 3 to 5 million regular customers for commercial sex. But it would be wrong to imagine millions of Thai men sneaking furtively on their own along dark streets lined with brothels; commercial sex is a social event, part of a good night out with friends. Ninety-five percent of men going to a brothel do so with their friends, usually at the end of a night spent drinking. Groups go out for recreation and entertainment, and especially to get drunk together. That is a strictly male pursuit, as Thai women usually abstain from alcohol. All-male groups out for a night on the town are considered normal in any Thai city, and whole neighborhoods are devoted to serving them. One man interviewed in a recent study explained, "When we arrive at the brothel; my friends take one and pay for me to take another. It costs them money; I don't want to waste it so I take her."[7] Having one's prostitute paid for also brings an informal obligation to repay in kind at a later date. Most Thais, men and women, feel that commercial sex is an acceptable part of an ordinary outing for single men, and about two-thirds of men and one-third of women feel the same about married men.[8]

For most married women, having their husbands go to prostitutes is preferable to other forms of extramarital sex. Most wives accept that men naturally want multiple partners, and prostitutes are seen as less threatening to the stability of the family.[9] Prostitutes require no long-term commitment or emotional involvement. When a husband uses a prostitute he is thought to be fulfilling a male role, but when he takes a minor wife or mistress, his wife is thought to have failed. Minor wives are usually bigamous second wives, often married by law in a district different than that of the men's first marriage (easily done, since no national records are kept). As wives, they require upkeep, housing, and regular support, and their offspring have a claim on inheritance; so they present a significant danger to the well-being of the major wife and her children. The potential disaster for the first wife is a minor wife who convinces the man to leave his first family, and this happens often enough to keep first wives worried and watchful.

For many Thai men, commercial sex is a legitimate form of entertainment and sexual release. It is not just acceptable: it is a clear statement of status and economic power. Such attitudes reinforce the treatment of women as mere markers in a male game of status and prestige. Combined with the new economy's relentless drive for profits, the result for women can be horrific. Thousands more must be found to feed men's status needs, thousands more must be locked into sexual slavery to feed the profits of investors. And what are the police, government, and local authorities doing about slavery? Every case of sex slavery involves many crimes—fraud, kidnap, assault, rape, sometimes murder. These crimes are not rare or random; they are systematic and repeated in brothels thousands of times each month. Yet those with the power to stop this terror instead help it continue to grow and to line the pockets of the slaveholders.

MILLIONAIRE TIGER AND BILLIONAIRE GEESE

Who are these modern slaveholders? The answer is anyone and everyone—anyone, that is, with a little capital to invest. The people who *appear* to own the enslaved prostitutes—the pimps, madams, and brothel keepers—are usually just employees. As hired muscle, pimps and their helpers provide the brutality that controls women and makes possible their commercial exploitation. Although they are just employees, the pimps do rather well for themselves. Often living in the brothel, they receive a salary and add to that income by a number of scams; for example, food and drinks are sold to customers at inflated prices, and the pimps pocket the difference.

Much more lucrative is their control of the price of sex. While each woman has a basic price, the pimps size up each customer and pitch the fee accordingly. In this way a client may pay two or three times more than the normal rate, and all of the surplus goes to the pimp. In league with the bookkeeper, the pimp systematically cheats the prostitutes of the little that is supposed to be credited against their debt. If they manage the sex slaves well and play all of the angles, pimps can easily make ten times their basic wage—a great income for an ex-peasant whose main skills are violence and intimidation, but nothing compared to the riches to be made by the brokers and the real slaveholders.

The brokers and agents who buy girls in the villages and sell them to brothels are only short-term slaveholders. Their business is part recruiting agency, part shipping company, part public relations, and part kidnapping gang. They aim to buy low and sell high while maintaining a good flow of girls from the villages. Brokers are equally likely to be men or women, and they usually come from the regions in which they recruit. Some are local people dealing in girls in addition to their jobs as police officers, government bureaucrats, or even schoolteachers. Positions of public trust are excellent starting points for buying young girls. In spite of the character of their work, they are well respected. Seen as job providers and sources of large cash payments to parents, they are well known in their communities. Many of the women brokers were once sold themselves; some spent years as prostitutes and now, in their middle age, make their living by supplying girls to the brothels. These women are walking advertisements for sexual slavery. Their lifestyle and income, their Western clothes and glamorous, sophisticated ways promise a rosy economic future for the girls they buy. That they have physically survived their years in the brothel may be the exception—many more young women come back to the village to die of AIDS—but the parents tend to be optimistic.

Whether these dealers are local people or traveling agents, they combine the business of procuring with other economic pursuits. A returned prostitute may live with her family, look after her parents, own a rice field or two, and buy and sell girls on the side.

Like the pimps, they are in a good business, doubling their money on each girl within two or three weeks; but also like the pimps, their profits are small compared to those of the long-term slaveholders.

The real slaveholders tend to be middle-aged businessmen. They fit seamlessly into the community, and they suffer no social discrimination for what they do. If anything, they are admired as successful, diversified capitalists. Brothel ownership is normally only one of many business interests for the slaveholder. To be sure, a brothel owner may have some ties to organized crime, but in Thailand organized crime includes the police and much of the government. Indeed, the work of the modern slaveholder is best seen not as aberrant criminality but as a perfect example of disinterested capitalism. Owning the brothel that holds young girls in bondage is simply a business matter. The investors would say that they are creating jobs and wealth. There is no hypocrisy in their actions, for they obey an important social norm: earning a lot of money is good enough reason for anything.

The slaveholder may in fact be a partnership, company, or corporation. In the 1980s, Japanese investment poured into Thailand, in an enormous migration of capital that was called "Flying Geese."[10] The strong yen led to buying and building across the country, and while electronics firms built television factories, other investors found that there was much, much more to be made in the sex industry. Following the Japanese came investment from the so-called Four Tigers (South Korea, Hong Kong, Taiwan, and Singapore), which also found marvelous opportunities in commercial sex. (All five of these countries further proved to be strong import markets for enslaved Thai girls, as discussed below.) The Geese and the Tigers had the resources to buy the local criminals, police, administrators, and property needed to set up commercial sex businesses. Indigenous Thais also invested in brothels as the sex industry boomed; with less capital, they were more likely to open poorer, working-class outlets.

Whether they are individual Thais, partnerships, or foreign investors, the slaveholders share many characteristics. There is little or no racial or ethnic difference between them and the slaves they own

(with the exception of the Japanese investors). They feel no need to rationalize their slaveholding on racial grounds. Nor are they linked in any sort of hereditary ownership of slaves or of the children of their slaves. They are not really interested in their slaves at all, just in the bottom line on their investment.

To understand the business of slavery today we have to know something about the economy in which it operates. Thailand's economic boom included a sharp increase in sex tourism tacitly backed by the government. International tourist arrivals jumped from 2 million in 1981 to 4 million in 1988 to over 7 million in 1996.[11] Two-thirds of tourists were unaccompanied men; in other words, nearly 5 million unaccompanied men visited Thailand in 1996. A significant proportion of these were sex tourists.

The recent downturn in both tourism and the economy may have slowed, but not dramatically altered, sex tourism. In 1997 the annual illegal income generated by sex workers in Thailand was roughly $10 billion, which is more than drug trafficking is estimated to generate.[12] According to ECPAT, an organization working against child prostitution, the economic crisis in Southeast Asia may have increased the exploitation of young people in sex tourism:

> According to Professor Lae Dilokvidhayarat from Chulalongkorn University, there has been a 10 percent decrease in the school enrollment at primary school level in Thailand since 1996. Due to increased unemployment, children cannot find work in the formal sector, but instead are forced to "disappear" into the informal sector. This makes them especially vulnerable to sexual exploitation. Also, a great number of children are known to travel to tourist areas and to big cities hoping to find work.
> We cannot overlook the impact of the economic crisis on sex tourism, either. Even though travelling costs to Asian countries are approximately the same as before mid 1997, when the crisis began,

the rates for sexual services in many places are lower due to increased competition in the business. Furthermore, since there are more children trying to earn money, there may also be more so-called situational child sex tourists, i.e., those who do not necessarily prefer children as sexual partners, but who may well choose a child if the situation occurs and the price is low."[13]

In spite of the economic boom, the average Thai's income is very low by Western standards. Within an industrializing country, millions still live in rural poverty. If a rural family owns its house and has a rice field, it might survive on as little as 500 baht ($20) per month. Such absolute poverty means a diet of rice supplemented with insects (crickets, grubs, and maggots are widely eaten), wild plants, and what fish the family can catch. If a family's standard of living drops below this level, which can be sustained only in the countryside, it faces hunger and the loss of its house or land. For most Thais, an income of 2,500 to 4,000 baht per month ($100 to $180) is normal. Government figures from December 1996 put two-thirds of the population at this level. There is no system of welfare or health care, and pinched budgets allow no space for saving. In these families, the 20,000 to 50,000 baht ($800 to $2,000) brought by selling a daughter provides a year's income. Such a vast sum is a powerful inducement that often blinds parents to the realities of sexual slavery.

DISPOSABLE BODIES

Girls are so cheap that there is little reason to take care of them over the long term. Expenditure on medical care or prevention is rare in the brothels, since the working life of girls in debt bondage is fairly short—two to five years. After that, most of the profit has been drained from the girl and it is more cost-effective to discard her and replace her with someone fresh. No brothel wants to take on the responsibility of a sick or dying girl.

Enslaved prostitutes in brothels face two major threats to their physical health and to their lives: violence and disease. Violence—their enslavement enforced through rape, beatings, or threats—is always present. It is a girl's typical introduction to her new status as a sex slave. Virtually every girl interviewed repeated the same story: after she was taken to the brothel or to her first client as a virgin, any resistance or refusal was met with beatings and rape. A few girls reported being drugged and then attacked; others reported being forced to submit at gunpoint. The immediate and forceful application of terror is the first step in successful enslavement. Within hours of being brought to the brothel, the girls are in pain and shock. Like other victims of torture they often go numb, paralyzed in their minds if not in their bodies. For the youngest girls, who understand little of what is happening to them, the trauma is overwhelming. Shattered and betrayed, they often have few clear memories of what occurred.

After the first attack, the girl has little resistance left, but the violence never ends. In the brothel, violence and terror are the final arbiters of all questions. There is no argument; there is no appeal. An unhappy customer brings a beating, a sadistic client brings more pain; in order to intimidate and cheat them more easily, the pimp rains down terror randomly on the prostitutes. The girls must do anything the pimp wants if they are to avoid being beaten. Escape is impossible. One girl reported that when she was caught trying to escape, the pimp beat her and then took her into the viewing room; with two helpers he then beat her again in front of all the girls in the brothel. Afterward she was locked into a room for three days and nights with no food or water. When she was released she was immediately put to work. Two other girls who attempted escape told of being stripped naked and whipped with steel coat hangers by pimps. The police serve as slave catchers whenever a girl escapes; once captured, girls are often beaten or abused at the police station before being sent back to the brothel. For most girls it soon becomes clear that they can never escape, that their only hope for release is to please the pimp and to somehow pay off their debt.

In time, confusion and disbelief fade, leaving dread, resignation, and a break in the conscious link between mind and body. Now the girl does whatever it takes to reduce the pain, to adjust mentally to a life that means being used by fifteen men a day. The reaction to this abuse takes many forms: lethargy, aggression, self-loathing, suicide attempts, confusion, self-abuse, depression, full-blown psychoses, and hallucinations. Girls who have been freed and taken into shelters exhibit all of these disorders. Rehabilitation workers report that the girls suffer emotional instability; they are unable to trust or to form relationships, to readjust to the world outside the brothel, or to learn and develop normally. Unfortunately, psychological counseling is virtually unknown in Thailand, as there is a strong cultural pressure to keep mental problems hidden. As a result, little therapeutic work is done with girls freed from brothels. The long-term impact of their experience is unknown.

The prostitute faces physical dangers as well as emotional ones. There are many sexually transmitted diseases, and prostitutes contract most of them. Multiple infections weaken the immune system and make it easier for other infections to take hold. If the illness affects a girl's ability to have sex, it may be dealt with, but serious chronic illnesses are often left untreated. Contraception often harms the girls as well. Some slaveholders administer contraceptive pills themselves, continuing them without any break and withholding the monthly placebo pills so that the girls can work more nights of the month. These girls stop menstruating altogether.

Not surprisingly, HIV/AIDS is epidemic in enslaved prostitutes. Thailand now has one of the highest rates of HIV infection in the world. Officially, the government admits to 800,000 cases, but health workers insist there are at least twice that many. Mechai Veravaidya, a birth-control campaigner and expert who has been so successful that *mechai* is now the Thai word for condom, predicts there will be 4.3 million people infected with HIV by 2001.[14] In some rural villages from which girls are regularly trafficked, the infection rate is over 60 percent. Recent research suggests that the younger the

girl, the more susceptible she is to HIV, because her protective vaginal mucous membrane has not fully developed. Although the government distributes condoms, some brothels do not require their use.

BURMESE PROSTITUTES

The same economic boom that has increased the demand for prostitutes may, in time, bring an end to Thai sex slavery. Industrial growth has also led to an increase in jobs for women. Education and training are expanding rapidly across Thailand, and women and girls are very much taking part. The ignorance and deprivation on which the enslavement of girls depends are on the wane, and better-educated girls are much less likely to fall for the promises made by brokers. The traditional duties to family, including the debt of obligation to parents, are also becoming less compelling. As the front line of industrialization sweeps over northern Thailand, it is bringing fundamental changes. Programs on the television bought with the money from selling one daughter may carry warning messages to her younger sisters. As they learn more about new jobs, about HIV/AIDS, and about the fate of those sent to the brothels, northern Thai girls refuse to follow their sisters south. Slavery functions best when alternatives are few, and education and the media are opening the eyes of Thai girls to a world of choice.

For the slaveholders this presents a serious problem. They are faced with an increase in demand for prostitutes and a diminishing supply. Already the price of young Thai girls is spiraling upward. The slaveholders' only recourse is to look elsewhere, to areas where poverty and ignorance still hold sway. Nothing, in fact, could be easier: there remain large, oppressed, and isolated populations desperate enough to believe the promises of the brokers. From Burma to the west and Laos to the east come thousands of economic and political refugees searching for work; they are defenseless in a country where they are illegal aliens. The techniques that worked so well in bringing Thai girls to brothels are again deployed, but now across borders. Investigators from Human Rights

Watch, which made a special study of this trafficking in 1993, explain:

> The trafficking of Burmese women and girls into Thailand is appalling in its efficiency and ruthlessness. Driven by the desire to maximize profit and the fear of HIV/AIDS, agents acting on behalf of brothel owners infiltrate ever more remote areas of Burma seeking unsuspecting recruits. Virgin girls are particularly sought after because they bring a higher price and pose less threat of exposure to sexually transmitted disease. The agents promise the women and girls jobs as waitresses or dishwashers, with good pay and new clothes. Family members or friends typically accompany the women and girls to the Thai border, where they receive a payment ranging from 10,000 to 20,000 baht from someone associated with the brothel.
> This payment becomes the debt, usually doubled with interest, that the women and girls must work to pay off, not by waitressing or dishwashing, but through sexual servitude.[15]

Once in the brothels they are in an even worse situation than the enslaved Thai girls: because they do not speak Thai their isolation is increased, and as illegal aliens they are open to even more abuse. The pimps tell them repeatedly that if they set foot outside the brothel, they will be arrested. And when they are arrested, Burmese and Lao girls and women are afforded no legal rights. They are often held for long periods at the mercy of the police, without charge or trial. A strong traditional antipathy between Thais and Burmese increases the chances that Burmese sex slaves will face discrimination and arbitrary treatment. Explaining why so many Burmese women were kept in brothels in Ranong, in southern Thailand, the regional police commander told a reporter for the *Nation*: "In my opinion it is disgraceful to let Burmese men [working in the local fishing industry] frequent Thai prostitutes. Therefore I have been flexible in allowing Burmese prostitutes to work here."[16]

A special horror awaits Burmese and Lao women once they reach the revolving door at the border. If they escape or are dumped by the brothel owners, they come quickly to the attention of the police, since they have no money for transport and cannot speak Thai. Once they are picked up, they are placed in detention, where they meet women who have been arrested in the periodic raids on brothels and taken into custody with only the clothes they are wearing. In local jails, the foreign women might be held without charge for as long as eight months while they suffer sexual and other abuse by the police. In time, they might be sent to the Immigrant Detention Center in Bangkok or to prison. In both places, abuse and extortion by the staff continue, and some girls are sold back to the brothels from there. No trial is necessary for deportation, but many women are tried and convicted of prostitution or illegal entry. The trials take place in Thai without interpreters, and fines are charged against those convicted. If they have no money to pay the fines, and most do not, they are sent to a factory-prison to earn it. There they make lightbulbs or plastic flowers for up to twelve hours a day; the prison officials decide when they have earned enough to pay their fine. After the factory-prison the women are sent back to police cells or the Immigrant Detention Center. Most are held until they can cover the cost of transportation (illegal aliens are required by law to pay for their own deportation); others are summarily deported.

The border between Thailand and Burma is especially chaotic and dangerous. Only part of it is controlled by the Burmese military dictatorship; other areas are in the hands of tribal militias or warlords. After arriving at the border, the deportees are held in cells by immigration police for another three to seven days. Over this time, the police extort money and physically and sexually abuse the inmates. The police also use this time to make arrangements with brothel owners and brokers, notifying them of the dates and places of deportation. On the day of deportation, the prisoners are driven in cattle trucks into the countryside along the border, far from any village, and then pushed out. Abandoned in the jungle, miles from any major road, they are given no food or water and have no idea where they are or how to proceed into Burma. As the immigration police drive away, the deportees are approached by agents and brokers who followed the trucks from town by arrangement with the police. The brokers offer work and transportation back into Thailand. Abandoned in the jungle, many women see the offer as their only choice. Some who don't are attacked and abducted. In either case, the cycle of debt bondage and prostitution begins again.

If they do make it into Burma, the women face imprisonment or worse. If apprehended by Burmese border patrols they are charged with "illegal departure" from Burma. If they cannot pay the fine, and most cannot, they serve six months' hard labor. Imprisonment applies to all those convicted—men, women, and children. If a girl or woman is suspected of having been a prostitute, she can face additional charges and long sentences. Women found to be HIV-positive have been imprisoned and executed. According to Human Rights Watch, there are consistent reports of "deportees being routinely arrested, detained, subjected to abuse and forced to porter for the military. Torture, rape and execution have been well documented by the United Nations bodies, international human rights organizations and governments."[17]

The situation on Thailand's eastern border with Laos is much more difficult to assess. The border is more open, and there is a great deal of movement back and forth. Lao police, government officials, and community leaders are involved in the trafficking, working as agents and making payments to local parents. They act with impunity, as it is very difficult for Lao girls to escape back to their villages; those who do find it dangerous to speak against police or officials. One informant told me that if a returning girl did talk, no one would believe her *and* she would be branded as a prostitute and shunned. There would be no way to expose the broker and no retribution; she would just have to resign herself to her fate. It is difficult to know how many Lao women and girls are brought into Thailand. In the northeast many Thais speak Lao, which makes it difficult to tell whether a prostitute is a local Thai or has actually come from Laos. Since they are illegal aliens, Lao girls will always claim to be local Thais and will often have false identity cards to prove it. In the brothels their lives are indistinguishable from those of Thai women.

TO JAPAN, SWITZERLAND, GERMANY, THE UNITED STATES

Women and girls flow in both directions over Thailand's borders.[18] The export of enslaved prostitutes is a robust business, supplying brothels in Japan, Europe, and America. Thailand's Ministry of Foreign Affairs estimated in 1994 that as many as 50,000 Thai women were living illegally in Japan and working in prostitution. Their situation in these countries parallels that of Burmese women held in Thailand. The enticement of Thai women follows a familiar pattern. Promised work as cleaners, domestics, dishwashers, or cooks, Thai girls and women pay large fees to employment agents to secure jobs in rich, developed countries. When they arrive, they are brutalized and enslaved. Their debt bonds are significantly larger than those of enslaved prostitutes in Thailand, since they include airfares, bribes to immigration officials, the costs of false passports, and sometimes the fees paid to foreign men to marry them and ease their entry.

Variations on sex slavery occur in different countries. In Switzerland girls are brought in on "artist" visas as exotic dancers. There, in addition to being prostitutes, they must work as striptease dancers in order to meet the carefully checked terms of their employment. The brochures of the European companies that have leaped into the sex-tourism business leave the customer no doubt about what is being sold:

> Slim, sunburnt, and sweet, they love the white man in an erotic and devoted way. They are masters of the art of making love by nature, an art that we Europeans do not know. (Life Travel, Switzerland) [M]any girls from the sex world come from the poor

north-eastern region of the country and from the slums of Bangkok. It has become a custom that one of the nice looking daughters goes into the business in order to earn money for the poor family... [Y]ou can get the feeling that taking a girl here is as easy as buying a package of cigarettes... little slaves who give real Thai warmth. (Kanita Kamha Travel, the Netherlands)[19]

In Germany they are usually bar girls, and they are sold to men by the bartender or bouncer. Some are simply placed in brothels or apartments controlled by pimps. After Japanese sex tours to Thailand began in the 1980s, Japan rapidly became the largest importer of Thai women. The fear of HIV in Japan has also increased the demand for virgins. Because of their large disposable incomes, Japanese men are able to pay considerable sums for young rural girls from Thailand. Japanese organized crime is involved throughout the importation process, sometimes shipping women via Malaysia or the Philippines. In the cities, the Japanese mob maintains bars and brothels that trade in Thai women. Bought and sold between brothels, these women are controlled with extreme violence. Resistance can bring murder. Because the girls are illegal aliens and often enter the country under false passports, Japanese gangs rarely hesitate to kill them if they have ceased to be profitable or if they have angered their slaveholders. Thai women deported from Japan also report that the gangs will addict girls to drugs in order to manage them more easily.

Criminal gangs, usually Chinese or Vietnamese, also control brothels in the United States that enslave Thai women. Police raids in New York, Seattle, San Diego, and Los Angeles have freed more than a hundred girls and women.[20] In New York, thirty Thai women were locked into the upper floors of a building used as a brothel. Iron bars sealed the windows and a series of buzzer-operated armored gates blocked exit to the street. During police raids, the women were herded into a secret basement room. At her trial, the brothel owner testified that she'd bought the women outright, paying between $6,000 and $15,000 for each. The women were charged $300 per week for room and board; they worked from 11:00 A.M until 4:00 A.M. and were sold by the hour to clients. Chinese and Vietnamese gangsters were also involved in the brothel, collecting protection money and hunting down escaped prostitutes. The gangs owned chains of brothels and massage parlors through which they rotated the Thai women in order to defeat law enforcement efforts. After being freed from the New York brothel, some of the women disappeared—only to turn up weeks later in similar circumstances three thousand miles away, in Seattle. One of the rescued Thai women, who had been promised restaurant work and then enslaved, testified that the brothel owners "bought something and wanted to use it to the full extent and they didn't think those people were human beings."[21]

OFFICIAL INDIFFERENCE AND A GROWTH ECONOMY

In many ways, Thailand closely resembles another country, one that was going through rapid industrialization and economic boom one hundred years ago. Rapidly shifting its labor force off the farm, experiencing unprecedented economic growth, flooded with economic migrants, and run by corrupt politicians and a greedy and criminal police force, the United States then faced many of the problems confronting Thailand today. In the 1890s, political machines that brought together organized crime with politicians and police ran the prostitution and protection rackets, drug sales, and extortion in American cities. Opposing them were a weak and disorganized reform movement and a muckraking press. I make this comparison because it is important to explore why Thailand's government is so ineffective when faced with the enslavement of its own citizens, and also to remember that conditions *can* change over time. Discussions with Thais about the horrific nature of sex slavery often end with their assertion that "nothing will ever change this... the problem is just too big... and those with power will never allow change." Yet the social and economic underpinnings of slavery in Thailand are always changing, sometimes for the worse and sometimes for the better. No society can remain static, particularly one undergoing such upheavals as Thailand.

As the country takes on a new Western-style materialist morality, the ubiquitous sale of sex sends a clear message: women can be enslaved and exploited for profit. Sex tourism helped set the stage for the expansion of sexual slavery.

Sex tourism also generates some of the income that Thai men use to fund their own visits to brothels. No one knows how much money it pours into the Thai economy, but if we assume that just one-quarter of sex workers serve sex tourists and that their customers pay about the same as they would pay to use Siri, then 656 billion baht ($26.2 billion) a year would be about right. This is thirteen times more than the amount Thailand earns by building and exporting computers, one of the country's major industries, and it is money that floods into the country without any concomitant need to build factories or improve infrastructure. It is part of the boom raising the standard of living generally and allowing an even greater number of working-class men to purchase commercial sex.

Joining the world economy has done wonders for Thailand's income and terrible things to its society. According to Pasuk Phongpaichit and Chris Baker, economists who have analyzed Thailand's economic boom,

> Government has let the businessmen ransack the nation's human and natural resources to achieve growth. It has not forced them to put much back. In many respects, the last generation of economic growth has been a disaster. The forests have been obliterated. The urban environment has deteriorated. Little has been done to combat the growth in industrial pollution and hazardous wastes. For many people whose labour has created the boom, the conditions of work, health, and safety are grim.
>
> Neither law nor conscience has been very effective in limiting the social costs of growth. Business has reveled in the atmosphere of free-for-all. The machinery for social protection has proved very pliable. The legal framework is defective. The judiciary is suspect. The police are unreliable. The authorities have consistently tried to block popular organizations to defend popular rights.[22]

The situation in Thailand today is similar to that of the United States in the 1850s; with a significant part of the economy dependent on slavery, religious and cultural leaders are ready to explain why this is all for the best. But there is also an important difference: this is the new slavery, and the impermanence of modern slavery and the dedication of human-rights workers offer some hope.

NOTES

1. Siri is, of course, a pseudonym; the names of all respondents have been changed for their protection. I spoke with them in December 1996.
2. "Caught in Modern Slavery: Tourism and Child Prostitution in Thailand," Country Report Summary prepared by Sudarat Sereewat-Srisang for the Ecumenical Consultation held in Chiang Mai in May 1990.
3. Foreign exchange rates are in constant flux. Unless otherwise noted, dollar equivalences for all currencies reflect the rate at the time of the research.
4. From interviews done by Human Rights Watch with freed child prostitutes in shelters in Thailand, reported in Jasmine Caye, *Preliminary Survey on Regional Child Trafficking for Prostitution in Thailand* (Bangkok: Center for the Protection of Children's Rights, 1996), p. 25.

5. Kulachada Chaipipat, "New Law Targets Human Trafficking," *Bangkok Nation*, November 30, 1997.

6. Thais told me that it would be very surprising if a well-off man or a politician did not have at least one mistress. When I was last in Thailand there was much public mirth over the clash of wife and mistress outside the hospital room of a high government official who had suffered a heart attack, as each in turn barricaded the door.

7. Quoted in Mark Van Landingham, Chanpen Saengtienchai, John Knodel, and Anthony Pramualratana, *Friends, Wives, and Extramarital Sex in Thailand* (Bangkok: Institute of Population Studies, Chulalongkorn University, 1995), p. 18.

8. Van Landingham et al., 1995, pp. 9–25.

9. Van Landingham et al., 1995, p. 53.

10. Pasuk Phongpaichit and Chris Baker, *Thailand's Boom* (Chiang Mai: Silkworm Books, 1996), pp. 51–54.

11. Center for the Protection of Children's Rights, *Case Study Report on Commercial Sexual Exploitation of Children in Thailand* (Bangkok, October 1996), p. 37.

12. David Kyle and John Dale, "Smuggling the State Back In: Agents of Human Smuggling Reconsidered," in *Global Human Smuggling: Comparative Perspectives*, ed. David Kyle and Rey Koslowski (Baltimore: Johns Hopkins University Press, 2001).

13. "Impact of the Asian Economic Crisis on Child Prostitution," *ECPAT International Newsletter 27* (May 1, 1999), found at http://www.ecpat.net/eng/Ecpat_inter/IRC/articles.asp?articleID=143&NewsID=21.

14. Mechai Veravaidya, address to the International Conference on HIV/AIDS, Chiang Mai, September 1995. See also Gordon Fairclough, "Gathering Storm," *Far Eastern Review*, September 21, 1995, pp. 26–30.

15. Human Rights Watch, *A Modern Form of Slavery*, p. 3.

16. "Ranong Brothel Raids Net 148 Burmese Girls," *Nation* (July 16, 1993), p. 12.

17. Dorothy O. Thomas, ed., *A Modern Form of Slavery: Trafficking of Burmese Women and Girls into Brothels in Thailand* (New York: Human Rights Watch, 1993), p. 112.

18. *International Report on Trafficking in Women (Asia-Pacific Region)* (Bangkok: Global Alliance Against Traffic in Women, 1996); Sudarat Sereewat, *Prostitution: Thai–European Connection* (Geneva: Commission on the Churches' Participation in Development, World Council of Churches, n.d.). Women's rights and antitrafficking organizations in Thailand have also published a number of personal accounts of women enslaved as prostitutes and sold overseas. These pamphlets are disseminated widely in the hope of making young women more aware of the threat of enslavement. Good examples are Siriporn Skrobanek, *The Diary of Prang* (Bangkok: Foundation for Women, 1994); and White Ink (pseud.), *Our Lives, Our Stories* (Bangkok: Foundation for Women, 1995). They follow the lives of women "exported," the first to Germany and the second to Japan.

19. The brochures are quoted in Truong, *Sex, Money, and Morality: Prostitution and Tourism in Southeast Asia* (London: Zed Books, 1990), p. 178.

20. Carey Goldberg, "Sex Slavery, Thailand to New York," *New York Times* (September 11, 1995), p. 81.

21. Quoted in Goldberg.

22. Phongpaichit and Baker, 1996, p. 237.

43

In the Barrios

Latinos and the Underclass Debate

JOAN MOORE AND RAQUEL PINDERHUGHES

In an important work by sociologist William Julius Wilson called The Truly Disadvantaged, *the term* underclass *was introduced to mean persistent poverty due largely to economic restructuring. The use of this term has been debated by scholars. Some see it as blaming the victim, the poor, for their condition because of their values and behaviors. Others see it as a debate over who is responsible for the poor—the individuals themselves or society? Behavioral pathology or economic structure?*

In this discussion by Moore and Pinderhughes, the concept of underclass is considered as it applies to Latinos.

As you read this selection, consider the following questions:

1. *Does the term* underclass *apply to the Latino population?*
2. *To what does "Latino population" refer?*
3. *What makes the Latino population unique as a minority group in the United States?*
4. *What might be done to alleviate problems for Latinos?*

GLOSSARY **Underclass** Meaning is debated, but it often refers to the poorest of the poor in the United States. **Polarization of the labor market** High- and low-level jobs but few in the middle. **Rustbelt** Area of the country (Midwest) where jobs are being lost. **Sunbelt** Area of the country (mostly south) where jobs are increasing. **Informal economic activities** Outside government control, small-scale.

In the publication *The Truly Disadvantaged*, William Julius Wilson's seminal work on persistent, concentrated poverty in Chicago's black neighborhoods, Wilson used the term "underclass" to refer to the new face of poverty, and traced its origins to economic restructuring. He emphasized the impact of persistent, concentrated poverty not only on individuals but on communities.

Source: Joan Moore and Raquel Pinderhughes, eds. *In the Barrios: Latinos and the Underclass Debate.* New York: Russell Sage Foundation, 1993. Excerpts from pp. xi to xxxix.

... The term "Hispanic" is used particularly by state bureaucracies to refer to individuals who reside in the United States who were born in, or trace their ancestry back to, one of twenty-three Spanish-speaking nations. Many of these individuals prefer to use the term "Latino." ...

No matter what the details, when one examines the history of the term underclass among sociologists, it is clear that Wilson's 1987 work seriously jolted the somewhat chaotic and unfocused study of poverty in the United States. He described sharply increasing rates of what he called "pathology" in Chicago's black ghettos. By this, Wilson referred specifically to female headship, declining marriage rates, illegitimate births, welfare dependency, school dropouts, and youth crime. The changes in the communities he examined were so dramatic that he considered them something quite new.

Two of the causes of this new poverty were particularly important, and his work shifted the terms of the debate in two respects. First, Wilson argued effectively that dramatic increases in joblessness and long-term poverty in the inner city were a result of major economic shifts—economic restructuring. "Restructuring" referred to changes in the global economy that led to deindustrialization, loss and relocation of jobs, and a decline in the number of middle-level jobs—a polarization of the labor market. Second, he further fueled the debate about the causes and consequences of persistent poverty by introducing two neighborhood-level factors into the discussion. He argued that the outmigration of middle- and working-class people from the urban ghetto contributed to the concentration of poverty. These "concentration effects" meant that ghetto neighborhoods showed sharply increased proportions of very poor people. This, in turn, meant that residents in neighborhoods of concentrated poverty were isolated from "mainstream" institutions and role models. As a result, Wilson postulates, the likelihood of their engaging in "underclass behavior" was increased. Thus the social life of poor communities deteriorated because poverty intensified. ...

THE LATINO POPULATION—SOME BACKGROUND

American minorities have been incorporated into the general social fabric in a variety of ways. Just as Chicago's black ghettos reflect a history of slavery, Jim Crow legislation, and struggles for civil and economic rights, so the nation's Latino barrios reflect a history of conquest, immigration, and a struggle to maintain cultural identity.

In 1990 there were some 22 million Latinos residing in the United States, approximately 9 percent of the total population. Of these, 61 percent were Mexican in origin, 12 percent Puerto Rican, and 5 percent Cuban. These three groups were the largest, yet 13 percent of Latinos were of Central and South American origin and another 9 percent were classified as "other Hispanics." Latinos were among the fastest-growing segments of the American population, increasing by 7.6 million, or 53 percent, between 1980 and 1990. There are predictions that Latinos will outnumber blacks by the twenty-first century. If Latino immigration and fertility continue at their current rate, there will be over 54 million Latinos in the United States by the year 2020.

This is an old population: as early as the sixteenth century, Spanish explorers settled what is now the American Southwest. In 1848, Spanish and Mexican settlers who lived in that region became United States citizens as a result of the Mexican–American War. Although the aftermath of conquest left a small elite population, the precarious position of the masses combined with the peculiarities of southwestern economic development to lay the foundation for poverty in the current period (see Barrera 1979; Moore and Pachon 1985).

In addition to those Mexicans who were incorporated into the United States after the Treaty of Guadalupe Hidalgo, Mexicans have continually crossed the border into the United States, where they have been used as a source of cheap labor by U.S. employers. The volume of immigration from Mexico has been highly dependent on fluctuations in certain segments of the U.S. economy.

This dependence became glaringly obvious earlier in this century. During the Great Depression of the 1930s state and local governments "repatriated" hundreds of thousands of unemployed Mexicans, and just a few years later World War II labor shortages reversed the process as Mexican contract-laborers (*braceros*) were eagerly sought. A little later, in the 1950s, massive deportations recurred when "operation Wetback" repatriated hundreds of thousands of Mexicans. Once again, in the 1980s, hundreds of thousands crossed the border to work in the United States, despite increasingly restrictive legislation.

High levels of immigration and high fertility mean that the Mexican-origin population is quite young—on the average, 9.5 years younger than the non-Latino population—and the typical household is large, with 3.8 persons, as compared with 2.6 persons in non-Latino households (U.S. Bureau of the Census 1991b). Heavy immigration, problems in schooling, and industrial changes in the Southwest combine to constrain advancement. The occupational structure remains relatively steady, and though there is a growing middle class, there is also a growing number of very poor people....

Over the past three decades the economic status of Puerto Ricans dropped precipitously. By 1990, 38 percent of all Puerto Rican families were below the poverty line. A growing proportion of these families were concentrated in poor urban neighborhoods located in declining industrial centers in the Northeast and Midwest, which experienced massive economic restructuring and diminished employment opportunities for those with less education and weaker skills. The rising poverty rate has also been linked to a dramatic increase in female-headed households. Recent studies show that the majority of recent migrants were not previously employed on the island. Many were single women who migrated with their young children (Falcon and Gurak 1991). Currently, Puerto Ricans are the most economically disadvantaged group of all Latinos. As a group they are poorer than African Americans.

Unlike other Latino migrants, who entered the United States as subordinate workers and were viewed as sources of cheap labor, the first large waves of Cuban refugees were educated middle- and upper-class professionals. Arriving in large numbers after Castro's 1959 revolution, Cubans were welcomed by the federal government as bona fide political refugees fleeing communism and were assisted in ways that significantly contributed to their economic well-being. Cubans had access to job-training programs and placement services, housing subsidies, English-language programs, and small-business loans. Federal and state assistance contributed to the growth of a vigorous enclave economy (with Cubans owning many of the businesses and hiring fellow Cubans) and also to the emergence of Miami as a center for Latin American trade. Cubans have the highest family income of all Latino groups. Nevertheless, in 1990, 16.9 percent of the Cuban population lived below the poverty line.

In recent years large numbers of Salvadorans and Guatemalans have come to the United States in search of refuge from political repression. But unlike Cubans, few have been recognized by the U.S. government as bona fide refugees. Their settlement and position in the labor market have been influenced by their undocumented (illegal) status. Dominicans have also come in large numbers to East Coast cities, many also arriving as undocumented workers. Working for the lowest wages and minimum job security, undocumented workers are among the poorest in the nation.

Despite their long history and large numbers, Latinos have been an "invisible minority" in the United States. Until recently, few social scientists and policy analysts concerned with understanding stratification and social problems in the United States have noticed them. Because they were almost exclusively concerned with relations between blacks and whites, social scientists were primarily concerned with generating demographic information on the nation's black and white populations, providing almost no information on other groups. Consequently, it has been difficult, sometimes impossible, to obtain accurate data about Latinos.

Latinos began to be considered an important minority group when census figures showed a huge increase in the population. By 1980 there were significant Latino communities in almost every metropolitan area in the nation. As a group, Latinos have low education, low family incomes, and are more clustered in low-paid, less-skilled occupations. Most Latinos live in cities, and poverty has become an increasing problem. On the whole, Latinos are more likely to live in poverty than the general U.S. population: poverty is widespread for all Latino subgroups except Cubans. They were affected by structural factors that influenced the socioeconomic status of all U.S. workers. In 1990, 28 percent were poor as compared with 13 percent of all Americans and 32 percent of African Americans (U.S. Bureau of the Census 1991b). Puerto Ricans were particularly likely to be poor....

THE IMPORTANCE OF ECONOMIC RESTRUCTURING

The meaning of economic restructuring has shaped the debate about the urban underclass....

First, there is the "Rustbelt in the Sunbelt" phenomenon. Some researchers have argued that deindustrialization has been limited to the Rustbelt, and that the causal chain adduced by Wilson therefore does not apply outside that region. But the fact is that many Sunbelt cities developed manufacturing industries, particularly during and after World War II. Thus Rustbelt-style economic restructuring—deindustrialization, in particular—has also affected them deeply. In the late 1970s and early 1980s cities like Los Angeles experienced a major wave of plant closings that put a fair number of Latinos out of work (Morales 1985).

Second, there has been significant reindustrialization and many new jobs in many of these cities, a trend that is easily overlooked. Most of the expanding low-wage service and manufacturing industries, like electronics and garment manufacturing, employ Latinos (McCarthy and Valdez 1986; Muller and Espenshade 1986), and some depend almost completely on immigrant labor working at minimum wage (Fernandez-Kelly and Sassen 1991). In short, neither the Rustbelt nor the Sunbelt has seen uniform economic restructuring.

Third, Latinos are affected by the "global cities" phenomenon, particularly evident in New York and Chicago. This term refers to a particular mix of new jobs and populations and an expansion of both high- and low-paid service jobs (see Sassen-Koob 1984). When large multinational corporations centralize their service functions, upper-level service jobs expand. The growing corporate elite want more restaurants, more entertainment, more clothing, and more care for their homes and children, but these new consumer services usually pay low wages and offer only temporary and part-time work. The new service workers in turn generate their own demand for low-cost goods and services. Many of them are Latino immigrants and they create what Sassen calls a "Third World city... located in dense groupings spread all over the city": this new "city" also provides new jobs (1989, p. 70).

Los Angeles...has experienced many of these patterns. The loss of manufacturing jobs has been far less visible than in New York or Chicago, for although traditional manufacturing declined, until the 1990s high-tech manufacturing did not. Moreover, Los Angeles' international financial and trade functions flourished (Soja 1987). The real difference between Los Angeles on the one hand and New York and Chicago on the other was that more poor people in Los Angeles seemed to be working. In all three cities internationalization had similar consequences for the *structure* of jobs for the poor. More of the immigrants pouring into Los Angeles were finding jobs, while the poor residents of New York and Chicago were not.

Fourth, even though the deindustrialization framework remains of overarching importance in understanding variations in the urban context of Latino poverty, we must also understand that economic restructuring shows many different faces. It is different in economically specialized cities. Houston, for example, has been called "the oil capital of the world," and most of the devastating economic shifts in that city were due to "crisis and reorganization in the world oil-gas industry" (Hill and Feagin 1987, p. 174). Miami is another special case. The economic changes that have swept Miami have little to do with deindustrialization, or with

Europe or the Pacific Rim, and much to do with the overpowering influence of its Cuban population, its important "enclave economy," and its "Latino Rim" functions (see Portes and Stepick 1993).

Finally, economic change has a different effect in peripheral areas. Both Albuquerque and Tucson are regional centers in an economically peripheral area. Historically, these two cities served the ranches, farms, and mines of their desert hinterlands. Since World War II, both became military centers, with substantial high-tech defense industrialization. Both cities are accustomed to having a large, poor Latino population, whose poverty is rarely viewed as a crisis. In Tucson, for example, unemployment for Mexican Americans has been low, and there is stable year-round income. But both cities remain marginal to the national economy, and this means that the fate of their poor depends more on local factors.

Laredo has many features in common with other cities along the Texas border, with its substantial military installations, and agricultural and tourist functions. All of these cities have been affected by general swings in the American and Texan economy. These border communities have long been the poorest in the nation, and their largely Mexican American populations have suffered even more from recent economic downturns. They are peripheral to the U.S. economy, but the important point is that their economic well-being is intimately tied to the Mexican economy. They were devastated by the collapse of the peso in the 1980s. They are also more involved than most American cities in international trade in illicit goods, and poverty in Laredo has been deeply affected by smuggling. Though Texas has a long history of discrimination against Mexican Americans, race is not an issue within Laredo itself, where most of the population—elite as well as poor—is of Mexican descent....

THE INFORMAL AND ILLICIT ECONOMIES

The growth of an informal economy is part and parcel of late twentieth-century economic restructuring.

Particularly in global cities, a variety of "informal" economic activities proliferates—activities that are small-scale, informally organized, and largely outside government regulations (cf. Portes, Castells, and Benton 1989). Some low-wage reindustrialization, for example, makes use of new arrangements in well-established industries (like home work in the garment industry, as seamstresses take their work home with them). Small-scale individual activities such as street vending and "handyman" house repairs and alterations affect communities in peripheral as well as global cities.... These money-generating activities are easily ignored by researchers who rely exclusively on aggregate data sources: they never make their way into the statistics on labor-market participation, because they are "off the books." But they play a significant role in the everyday life of many African American neighborhoods as well as in the barrios.

And, finally, there are illicit activities—most notoriously, a burgeoning drug market. There is not much doubt that the new poverty in the United States has often been accompanied by a resurgence of illicit economic activities. It is important to note that most of the Latino communities... have been able to contain or encapsulate such activities so that they do not dominate neighborhood life. But in most of them there is also little doubt that illicit economic activities form an "expanded industry." They rarely provide more than a pittance for the average worker: but for a very small fraction of barrio households they are part of the battery of survival strategies.

Researchers often neglect this aspect of the underclass debate because it is regarded as stigmatizing. However, some...make it clear that the neglect of significant income-generating activities curtails our understanding of the full range of survival strategies in poor communities. At the worst (as in Laredo) it means that we ignore a significant aspect of community life, including its ramifications in producing yet more overpolicing of the barrios. Even more important, many of these communities have been able to encapsulate illicit economic activities so that they are less disruptive. This capacity warrants further analysis.

IMMIGRATION

Immigration—both international and from Puerto Rico—is of major significance for poor Latino communities in almost every city in every region of the country. Further, there is every reason to believe that immigration will continue to be important.

First, it has important economic consequences. Immigration is a central feature of the economic life of global cities: for example, Los Angeles has been called the "capital of the Third World" because of its huge Latino and Asian immigration (Rieff 1991). In our sample, those cities most bound to world trends (New York, Los Angeles, Chicago, Houston, and Miami) experienced massive Latino immigration in the 1980s. In the Los Angeles, Houston, and Miami communities…immigration is a major factor in the labor market, and the residents of the "second settlement" Puerto Rican communities described in New York and Chicago operate within a context of both racial and ethnic change and of increased Latino immigration. The restructured economy provides marginal jobs for immigrant workers, and wage scales seem to drop for native-born Latinos in areas where immigration is high. This is a more complicated scenario than the simple loss of jobs accompanying Rustbelt deindustrialization. Immigrants are ineligible for most government benefits, are usually highly motivated, and are driven to take even the poorest-paying jobs. They are also more vulnerable to labor-market swings.

These may be construed as rather negative consequences, but in addition, immigrants have been a constructive force in many cities. For example, these authors point to the economic vitality of immigrant-serving businesses. Socially and culturally, there are references…to the revival of language and of traditional social controls, the strengthening of networks, and the emergence of new community institutions. Recent research in Chicago focuses on the "hard work" ethos of many Mexican immigrants and the extensive resource base provided by kinship networks, a pattern that is echoed and amplified…. Most of Tucson's Chicano poor—not just immigrants—are involved in such helping networks.

Though immigrants have been less important in the peripheral cities of Albuquerque, Laredo, and Tucson, each of these cities is special in some way. Albuquerque has attracted few Mexican immigrants, but it draws on a historical Latino labor pool—English-speaking rural *Manitos*—who are as economically exploitable as are Spanish-speaking immigrants from Mexico. Until recently Tucson was also largely bypassed by most Mexican immigrants. Instead, there is an old, relatively self-contained set of cross-border networks, with well-established pathways of family movement and mutual aid. Similar networks also exist in Laredo. Laredo's location on the border means that many of its workers are commuters—people who work in Laredo but live in Mexico.

In recent years, immigration has not been very significant in most African American communities, and as a consequence it is underemphasized in the underclass debate. It is also often interpreted as wholly negative. This is partly because the positive effects can be understood only by researchers who study immigrant communities themselves, partly because in some places large numbers of immigrants have strained public resources, and partly because immigrants have occasionally become a source of tension among poor minority populations. Though the specific contouring of immigration effects varies from place to place, in each city…immigration is a highly significant dimension of Latino poverty, both at the citywide level and also in the neighborhoods. It is an issue of overriding importance for the understanding of Latino poverty, and thus for the understanding of American urban poverty in general….

The concentration of poverty comes about not only because of market forces or the departure of the middle classes for better housing; in Houston, Rodriguez shows that restructuring in real estate had the effect of concentrating poverty. Concentrated poverty can also result from government planning. Chicago's decision decades ago to build a concentration of high-rise housing projects right next to one another is a clear case in point. Another is in New York's largely Latino South Bronx, where the city's ten-year-plan created neighborhoods in which the least enterprising of the poor are concentrated, and in which a set of undesirable "Not-In-My-Back-Yard" institutions, such as drug-treatment clinics and permanent shelters for the homeless, were located. These neighborhoods are likely to remain as pockets of unrelieved poverty for many generations to come (Vergara 1991). It was not industrial decline and the exodus of stable working people that created these pockets: the cities of Chicago and New York chose to segregate their problem populations in permanent buildings in those neighborhoods....

In addition, studies demonstrate that it is not just poverty that gets concentrated. Most immigrants are poor, and most settle in poor communities, thus further concentrating poverty. But, as Rodriguez shows, immigrant communities may be economically, culturally, and socially vital. Social isolation early in the immigration process, he argues, can strengthen group cohesion and lead to community development, rather than to deterioration. Los Angeles portrays institution-building among immigrants in poor communities, and institutional "resilience" characterizes many of the communities... especially New York and Chicago. Analysis of poverty in Tucson points to the overwhelming importance of "funds of knowledge" shared in interdependent household clusters. Although a priori it makes sociological sense that concentrated poverty should destroy communities, these studies offer evidence that a different pattern emerges under certain circumstances. To use Grenier and Stepick's term, "social capital" also becomes concentrated.

In short, the concentration of poverty need not plunge a neighborhood into disarray.... This line of reasoning raises other issues. If it isn't just demographic shifts that weaken neighborhoods, then what is it? These questions strike at the heart of the underclass debate. The old, rancorous controversy about the usefulness of the "culture of poverty" concept questioned whether the poor adhered to a special set of self-defeating values, and if so, whether those values were powerful enough to make poverty self-perpetuating. That argument faded as research focused more effectively on the situational and structural sources of poverty. We do not intend to revive this controversy. It is all too easy to attribute the differences between Latino and black poverty to

"the culture." This line can be invidious, pitting one poor population against another in its insinuation that Latino poverty is somehow "better" than black poverty. (Ironically, this would reverse another outdated contention—i.e., that Latinos are poor *because* of their culture.)...

OTHER ASPECTS OF URBAN SPACE...

Where a poor neighborhood is located makes a difference.

First, some are targets for "gentrification." This is traditionally viewed as a market process by which old neighborhoods are revitalized and unfortunate poor people displaced. But there is a different perspective. Sassen (1989) argues that gentrification is best understood in the context of restructuring, globalization, and politics. It doesn't happen everywhere...gentrification, along with downtown revitalization and expansion, affects Latino neighborhoods in Chicago, Albuquerque, New York, and west side Los Angeles. In Houston, a variant of "gentrification" is documented. Apartment owners who were eager to rent to Latino immigrants when a recession raised their vacancy rates were equally eager to "upgrade" their tenants when the economy recovered and the demand for housing rose once again. Latinos were "gentrified" out of the buildings.

Second, Latinos are an expanding population in many cities, and they rub up against other populations. Most of the allusions to living space center on ethnic frictions accompanying the expansion of Latino areas of residence. Ethnic succession is explicit in Albuquerque and in Chicago.... It is implicit in East Los Angeles, with the Mexicanization of Chicano communities, and in Houston, with the immigration of Central Americans to Mexican American neighborhoods and the manipulated succession of Anglos and Latinos. In Albuquerque and East Los Angeles, Latinos are "filling-in" areas of the city, in a late phase of ethnic succession. Ethnic succession is *not* an issue in Laredo because the city's population is primarily of Mexican origin. It is crucial in Miami, where new groups of immigrants are establishing themselves within the Latino community: newer immigrants tend to move into areas vacated by earlier Cuban arrivals, who leave for the suburbs. In Brooklyn, a different kind of urban ecological function is filled by the Puerto Rican barrio—that of an ethnic buffer between African American and Anglo communities. Los Angeles' Westlake area is most strongly affected by its location near downtown: it is intensely involved in both gentrification and problems of ethnic succession. Here the Central Americans displaced a prior population, and, in turn, their nascent communities are pressured by an expanding Koreatown to the west and by gentrification from the north and from downtown.

These details are important in themselves, but they also have implications for existing theories of how cities grow and how ethnic groups become segregated (and segregation is closely allied to poverty). Most such theories take the late nineteenth-century industrial city as a point of departure—a city with a strong central business district and clearly demarcated suburbs. In these models, immigrants initially settle in deteriorating neighborhoods near downtown. Meanwhile, earlier generations of immigrants, their predecessors in those neighborhoods, leapfrog out to "areas of second settlement," often on the edge of the city....

Thus it is no surprise that the "traditional" Rustbelt pattern of ethnic location and ethnic succession fails to appear in most cities discussed in this volume. New Latino immigrants are as likely to settle initially in communities on the edge of town (near the new jobs) as they are to move near downtown; or their initial settlement may be steered by housing entrepreneurs, as in Houston. The new ecology of jobs,

housing, and shopping malls has made even the old Rustbelt cities like Chicago less clearly focused on a central downtown business district.

Housing for the Latino poor is equally distinctive. Poor communities in which one-third to one-half of the homes are owner-occupied would seem on the face of it to provide a different ambience from public housing—like the infamous phalanx of projects on Chicago's South Side that form part of Wilson's focus....

Finally, space is especially important when we consider Mexican American communities on the border. Mexican Americans in most border communities have important relationships with kin living across the border in Mexico, and this is certainly the case in Tucson and Laredo. But space is also important in economic matters. Shopping, working, and recreation are conditioned by the proximity of alternative opportunities on both sides of the border. And in Laredo the opportunities for illicit economic transactions also depend on location. The Laredo barrios in which illicit activities are most concentrated are located right on the Rio Grande River, where cross-border transactions are easier.

In sum, when we consider poor minority neighborhoods, we are drawn into a variety of issues that go well beyond the question of how poverty gets concentrated because middle-class families move out. We must look at the role of urban policy in addition to the role of the market. We must look at the factors that promote and sustain segregation. We must look at how housing is allocated, and where neighborhoods are located within cities. And, finally, we must look at how the location of a neighborhood facilitates its residents' activity in licit and illicit market activities.

REFERENCES

Editors' Note: For documentation, see the source note on page 348.

AFL-CIO Industrial Union Department, 1986. *The Polarization of America.* Washington, DC: AFL-CIO Industrial Union Department.

Auletta, Ken, 1982. *The Underclass.* New York: Random House.

Barrera, Mario, 1979. *Race and Class in the Southwest.* Notre Dame, IN: University of Notre Dame Press.

Bluestone, Barry, and Bennett Harrison, 1982. *The Deindustrialization of America*. New York: Basic Books.

Chenault, Lawrence Royce, 1938. *The Puerto Rican Migrant in New York*. New York: Columbia University Press.

Clark, Margaret, 1959. *Health in the Mexican American Culture*. Berkeley: University of California Press.

Crawford, Fred, 1961. *The Forgotten Egg*. San Antonio, TX: Good Samaritan Center.

Edmundson, Munro S., 1957. *Los Manitos: A Study of Institutional Values*. New Orleans: Tulane University, Middle American Research Institute.

Ellwood, David T., 1988. *Poor Support: Poverty in the American Family*. New York: Basic Books.

Falcon, Luis, and Douglas Gurak, 1991. "Features of the Hispanic Underclass: Puerto Ricans and Dominicans in New York." Unpublished manuscript.

Fernandez-Kelly, Patricia, and Saskia Sassen, 1991. "A Collaborative Study of Hispanic Women in the Garment and Electronics Industries: Executive Summary." New York: New York University, Center for Latin American and Caribbean Studies.

Galarza, Ernesto, 1965. *Merchants of Labor*. San Jose, CA: The Rosicrucian Press Ltd.

Goldschmidt, Walter, 1947. *As You Sow*. New York: Harcourt, Brace.

Gosnell, Patricia Aran, 1949. *Puerto Ricans in New York City*. New York: New York University Press.

Handlin, Oscar, 1959. *The Newcomers: Negroes and Puerto Ricans*. Cambridge, MA: Harvard University Press.

Hill, Richard Child, and Joe R. Feagin, 1987. "Detroit and Houston: Two Cities in Global Perspective." In Michael Peter Smith and Joe R. Feagin, eds. In *The Capitalist City*, pp. 155–177. New York: Basil Blackwell.

Kluckhohn, Florence, and Fred Strodtbeck, 1961. *Variations in Value Orientations*. Evanston, IL: Row, Peterson.

Leonard, Olen, and Charles Loomis, 1938. *Culture of a Contemporary Rural Community: El Cerito, NM*. Washington, DC: U.S. Department of Agriculture.

Levy, Frank, 1977. "*How Big Is the Underclass?*" Working Paper 0090-1. Washington, DC: Urban Institute.

Maldonado-Denis, Manuel, 1972. *Puerto Rico: A Sociohistoric Interpretation*. New York: Random House.

Massey, Douglas, and Mitchell Eggers, 1990. "The Ecology of Inequality: Minorities and the Concentration of Poverty." *American Journal of Sociology* 95:1153–1188.

Matza, David, 1966. "The Disreputable Poor" In Reinhardt Bendix and Seymour Martin Lipset, eds. *Class, Status and Power*, pp. 289–302. New York: Free Press.

McCarthy, Kevin, and R. B. Valdez, 1986. *Current and Future Effects of Mexican Immigration in California*. Santa Monica, CA: Rand Corporation.

McWilliams, Carey, 1949. *North from Mexico*. New York: J. B. Lippincott.

Menefee, Seldon, and Orin Cassmore, 1940. *The Pecan Shellers of San Antonio*. Washington: WPA, Division of Research.

Mills, C. Wright, Clarence Senior, and Rose K. Goldsen, 1950. *The Puerto Rican Journey*. New York: Harper.

Montiel, Miguel, 1970. "The Social Science Myth of the Mexican American Family" *El Grito* 3:56–63.

Moore, Joan, 1989. "Is There a Hispanic Underclass?" *Social Science Quarterly* 70:265–283.

Moore, Joan, and Harry Pachon, 1985. *Hispanics in the United States*. Englewood Cliffs, NJ: Prentice Hall.

Morales, Julio, 1986. *Puerto Rican Poverty and Migration: We Just Had to Try Elsewhere*. New York: Praeger.

Morales, Rebecca, 1985. "Transitional Labor: Undocumented Workers in the Los Angeles Automobile Industry." *International Migration Review* 17:570–96.

Morris, Michael, 1989. "From the Culture of Poverty to the Underclass: An Analysis of a Shift in Public Language." *The American Sociologist* 20:123–133.

Muller, Thomas, and Thomas J. Espenshade, 1986. *The Fourth Wave*. Washington, DC: Urban Institute Press.

Murray, Charles, 1984. *Losing Ground*. New York: Basic Books.

Padilla, Elena, 1958. *Up from Puerto Rico*. New York: Columbia University Press.

Perry, David, and Alfred Watkins, 1977. *The Rise of the Sunbelt Cities*. Beverly Hills, CA: Sage.

Portes, Alejandro, Manuel Castells, and Lauren A. Benton, 1989. *The Informal Economy*. Baltimore: Johns Hopkins University Press.

Portes, Alejandro, and Alex Stepick, 1993. *City on the Edge: The Transformation of Miami*. Berkeley: University of California Press.

Rand, Christopher, 1958. *The Puerto Ricans*. New York: Oxford University Press.

Ricketts, Erol, and Isabel V. Sawhill, 1988. "Defining and Measuring the Underclass." *Journal of Policy Analysis and Management* 7:316–325.

Rieff, David, 1991. *Los Angeles: Capital of the Third World*. New York: Simon and Schuster.

Rodriguez, Clara, 1989. *Puerto Ricans: Born in the U.S.A.* Boston: Unwin Hyman.

Romano-V, Octavio I, 1968. "The Anthropology and Sociology of the Mexican Americans." *El Grito* 2:13–26.

Russell, George, 1977. "The American Underclass." *Time Magazine* 110 (August 28):14–27.

Sanchez, George, 1940. *Forgotten People: A Study of New Mexicans*. Albuquerque: University of New Mexico Press.

Sassen, Saskia, 1989. "New Trends in the Sociospatial Organization of the New York City Economy." In Robert Beauregard, ed. *Economic Restructuring and Political Response*. Newberry Park, CA.

Sassen-Koob, Saskia, 1984. "The New Labor Demand in Global Cities." In Michael Smith, ed. *Cities in Transformation*. Beverly Hills, CA: Sage.

Saunders, Lyle, 1954. *Cultural Differences and Medical Care*. New York: Russell Sage Foundation.

Senior, Clarence Ollson, 1965. *Our Citizens from the Caribbean*. New York: McGraw Hill.

Soja, Edward, 1987. "Economic Restructuring and the Internationalization of the Los Angeles Region." In Michael Peter Smith and Joe R. Feagin, eds.

The Capitalist City, pp. 178–198. New York: Basil Blackwell.

Stevens Arroyo, and M. Antonio, 1974. *The Political Philosophy of Pedro Abizu Campos: Its Theory and Practice*. Ibero American Language and Area Center. New York: New York University Press.

Sullivan, Mercer L., 1989. *Getting Paid: Youth Crime and Work in the Inner City*. Ithaca: Cornell University Press.

Taylor, Paul, 1928. *Mexican Labor in the U.S.: Imperial Valley*. Berkeley: University of California Publications in Economics.

———, 1930. *Mexican Labor in the U.S.: Dimit County, Winter Garden District, South Texas*. Berkeley: University of California Publications in Economics.

———, 1934. *An American–Mexican Frontier*. Chapel Hill, NC: University of North Carolina Press.

U.S. Bureau of the Census, 1991. *The Hispanic Population in the United States: March 1991*. Current Population Reports, Series P-20, No. 455. Washington, DC: U.S. Government Printing Office.

Vaca, Nick, 1970. "The Mexican American in the Social Sciences." *El Grito* 3:17–52.

Vergara, Camilo Jose, 1991. "Lessons Learned: Lessons Forgotten: Rebuilding New York City's Poor Communities." *The Livable City* 15:3–9.

Wagenheim, Kal, 1975. *A Survey of Puerto Ricans on the U.S. Mainland*. New York: Praeger.

Wakefield, Dan, 1959. *Island in the City*. New York: Corinth Books.

Wilson, William Julius, 1987. *The Truly Disadvantaged: The Inner City, the Underclass, and Public Policy*. Chicago: The University of Chicago Press.

———, 1990. "Social Theory and Public Agenda Research: The Challenge of Studying Inner-City Social Dislocations." Paper presented at Annual Meeting of the American Sociological Association.

44

The Problem of Racial/Ethnic Inequality

JAMES CRONE

The media bombard us with stories of the plight of minority groups around the world. Whether due to ethnic, racial, religious, political, or other differences, the results of stereotypes, prejudice, and discrimination are all too often conflicts, wars, ethnic cleansing, refugees, poverty, hunger, disease, and other human suffering. Crone asks why we can't all get along, and how we might begin to breech the chasm that divides so many people, using both individual and institutional means.

As you read this article, think about the following questions:

1. *What is meant by racial/ethnic equality?*
2. *What can individuals and society do to reduce discrimination and bring about greater equality between groups?*
3. *What are direct and indirect means of bringing about equality?*
4. *How does Crone argue we can bring about a change in the treatment of minorities?*

GLOSSARY **Direct measures** Measures that have the main goal of decreasing inequality. **Indirect measures** Measures whose main goal is other than decreasing inequality but that would have the side effect of decreasing inequality.

WHAT MIGHT RACIAL/ ETHNIC EQUALITY MEAN?

What might we mean by achieving racial/ethnic equality? Such equality could include some of the following changes. There is no longer racial/ethnic prejudice, meaning a negative attitude by one group of people that prejudges another group of people. There is no longer discrimination, meaning unequal treatment by the majority group of people toward a minority group of people. There is no longer institutional discrimination where the existing societal way of doing things works to the disadvantage of minorities. For example, we, as Americans, pay for our public schools through local property taxes, and this puts minority people at a disadvantage because they are more likely to live in poverty-laden areas of the country that cannot afford to collect as much tax revenue from property taxes as do whites and therefore cannot afford to have as good-quality public schools as do whites; therefore, racial/ethnic equality would mean that there is equal opportunity

Source: Crone, James. 2007. "What Might Racial/Ethnic Equality Mean?" in *How Can We Solve Our Social Problems?* Thousand Oaks: Pine Forge Press, pp. 84–93.

for minorities to attend good-quality public schools that prepare them just as much as white Americans to go to college or trade schools so that they are qualified to get decent-paying jobs. Attaining such racial/ethnic equality would also mean that there are comparable proportions of minorities in each social class as compared with the white population. Right now, there is a smaller proportion of African Americans, Hispanic Americans, and Native Americans in the middle, upper middle, and upper social classes and there is a larger proportion of minorities in the lower social classes. Attaining racial/ethnic equality would mean that minorities have similar proportions of their populations in various social classes as does the white majority.

Achieving all of these kinds of equalities together will indicate that we have reached racial/ethnic equality. Realistically, it may take a while to achieve all of these equalities, but these are goals toward which we can certainly work.

WHAT MORE CAN WE DO?

I want to address two general areas that I believe we could address as a way to solve, or at least greatly decrease, racial/ethnic inequality. The first area deals with what we can do to decrease racial/ethnic inequality directly, whereas the second area deals with what we can do to decrease this type of inequality indirectly.

Direct Measures to Solve or Greatly Decrease Racial/Ethnic Inequality

Teach More Tolerance and Acceptance

Schools, Teachers, and Teaching.   To address racial/ethnic prejudice and discrimination directly, we can increase the effort to teach tolerance and acceptance of other races and ethnic groups in schools. Teachers can emphasize the need for all Americans to go by some of the core values and beliefs of our society such as equal opportunity, fairness, justice, and freedom. Teachers can devise various teaching techniques to teach children that all Americans deserve

to have these values and beliefs fully applied to them—not just to whites, middle-class people, or males. The more young students continually hear this message from kindergarten on into elementary, middle, and high school, the more such a consistent type of socialization is imprinted on the minds of young people, the more likely this type of socialization will be internalized and accepted, and the more likely any socialization of intolerance from the home environment or elsewhere can be negated.

Government.   One way to address racial prejudice and discrimination is for the government to sponsor public service messages on television and radio and to place ads in newspapers, in magazines, and on billboards along highways. By taking such steps, the government can communicate clearly and visibly to the American people that such a stance represents the will of the American people. In other words, to hearken back to our theory of conflict and social change, the government establishes a new legitimacy; that is, it is seen as right to act in a tolerant and accepting way toward fellow Americans of another racial/ethnic group.

Private Organizations.   Private organizations can sponsor various activities with the intent of teaching tolerance and acceptance. We already have numerous organizations that help various groups of people such as Big Brothers and Big Sisters, Habitat for Humanity, the Salvation Army, and Boys' and Girls' Clubs. Many of these organizations, in their own way, teach tolerance and acceptance now. These organizations, in cooperation with each other, can brainstorm to see what different ways they can work together to present a common, visible, and intentional theme of tolerance and acceptance. So, we have existing resources and organizations that could see what more they could do collectively in a local community to promote more racial/ethnic tolerance and acceptance.

We as Individuals.   We, as individuals, in our daily lives can be constantly alert to opportunities that we personally have to say or do something that promotes racial/ethnic tolerance and acceptance. An easy step to take is the basic treating of every person we meet with dignity and respect. This is something that each person can do in his or her daily life. In other words, we can serve as daily role models of tolerance and acceptance.

Redress Grievances.   We can continue to enforce the existing laws against racial/ethnic discrimination so that minority Americans know that they can go to court to have a racial/ethnic transgression redressed. This option can be made more well known by teaching about it in schools and by the government's incorporating such information into its public service messages.

Accept More Interracial and Interethnic Dating and Marriage.   Up to and including the 1950s, there was a strong norm or taboo against whites and blacks dating and marrying. A few people did this, but most parents, friends, and others strongly advised against dating and marriage. Given such an atmosphere of informal norms and strong social pressure, most blacks and whites (and other combinations of minorities such as whites/Mexican Americans and whites/Native Americans) did not venture into dating. They might have seen and been around someone of another racial/ethnic group to whom they were attracted, but they knew their parents "would have a fit" if they even brought up the subject. So, for most people, the social structure of informal norms, customs, and family and community social pressure were, as Durkheim would put it, external to and yet coercive on anyone who might even consider asking out someone of another racial/ethnic group or even religion (during the 1950s, there was still a strong norm among many families that Catholics did not date or marry Protestants and vice versa, and "heaven forbid" dating Jews, atheists, or anyone else who did not have the "right" faith).

During the 21st century, we see more and more interracial/interethnic dating and marriage. In shopping malls, restaurants, movie theaters, and other public places, we see mixed couples. In schools, we see students who are of different shades and colors that represent mothers and fathers from different racial backgrounds. So, we are in the process of seeing and accepting more individuals and couples of mixed racial/ethnic and religious backgrounds.

As our society in general and as we, as individuals, get used to and accept more interracial/interethnic dating and marriage, this process will, over time, work to promote more racial/ethnic equality. We will get used to seeing and interacting with people of different combinations of races and ethnicities and will come to see this as less and less of a "big deal." Such interracial/interethnic dating and marriage and procreation of offspring will, I hypothesize, continue to break down barriers between people, and there will continue to be more tolerance and acceptance. To put it in Merton's (1967) terms, this process will act as a latent function for racial/ethnic minorities in that as there is more interracial/interethnic dating and marriage, minorities will be more accepted, will be more upwardly mobile, and will attain more money, power, and prestige. The consequence will be that they will increase their survival.

Promote a New Kind of Affirmative Action.   The U.S. Supreme Court affirmed affirmative action in college admissions, to a degree, during the summer of 2003 (Greenhouse, 2003). In a 5 to 4 decision, the justices stated that race can be considered as one factor in admission to law school. The intent of the Supreme Court is that although admissions offices cannot use quotas with regard to race, they can use race as a factor in achieving diversity of a student body "because such policies promote cross-racial understanding and break down racial stereotypes" (p. A4) (stereotypes are generalizations about a group of people that are unfavorable and oversimplified).

Although affirmative action helps minority students to gain admission to undergraduate and graduate schools, helps to give them more opportunity to get ahead, and creates a more diverse student body that allows for more interracial/interethnic interaction and understanding, there are major criticisms of this method of attempting to solve racial/ethnic inequality. First, it can leave out well-qualified white students who earn higher academic grades and SAT scores and have more extracurricular experiences than do minority applicants, thereby polarizing society with accusations of reverse discrimination against whites. Second, it can help middle- and upper middle-class minorities who might not need the help. Rather, lower-income minorities and whites are the ones who need help in having a chance to get ahead in our society.

Many people in our country have thought that we needed to go through a time of having affirmative action in college and graduate school admissions and in hiring for jobs because there was so much prejudice and discrimination within individuals and organizations (e.g., businesses, factories, unions, schools, churches). In other words, we needed to take strong measures to break these barriers so that minorities could get a chance to go to college and get decent-paying jobs.

It seems that a number of African Americans have benefited from affirmative action and have joined the middle class, as indicated by their incomes and middle-class occupations. The children of these families will, like white middle- and upper middle-class families, be able to have sufficient resources during their childhoods so that they will be able to attend good schools, go to college and graduate school, and get good jobs without further assistance from affirmative action programs.

However, minorities who are still left out of the chance for upward mobility in our country, especially poor African Americans, Hispanic Americans, and Native Americans, are those who come from poor families and neighborhoods, regions, or reservations that have high poverty, high unemployment, and poor-quality public schools. So, problems continue to persist among poor minority Americans.

We can continue affirmative action, especially for poor minorities and whites. This type of affirmative action would be more acceptable to most Americans because they would be more sympathetic toward poor people in general (e.g., black, white, brown, red, yellow) having the chance to get ahead—regardless of race/ethnicity.

Get More Minorities into the Middle Class or Higher Classes.   The more we can get minorities into the middle class or higher classes, the less racial/ethnic prejudice and discrimination there will be because there will be more interaction and understanding between minorities and whites as well as acceptance of whites with minorities and of minorities with whites due to their living in the same neighborhoods, parents having similar kinds of occupations, parents joining similar voluntary associations, children going to the same schools, and children being involved together in sports, choir, band, and other extracurricular activities. All of these commonalities together will cause minorities and whites to have more things in common, and this should result in more acceptance of each other.

However, to get more minorities into middle or higher social classes, we can also carry out a number of indirect measures such as developing good-quality public schools for all Americans, creating more decent-paying jobs, building a tax system that takes less from poor and near-poor minorities and whites, and providing more social services such as public transportation and child care subsidies. All of these measures will create the social structural conditions for minorities to move into higher social classes, thereby creating more racial/ethnic equality.

Indirect Measures to Solve or Greatly Decrease Racial/Ethnic Inequality

Develop Good-Quality Public Schools.   One indirect way we can create more racial/ethnic equality is to create good-quality public schools for all children in the United States…. Currently, we have unequal public education in our country, and this in turn perpetuates inequality.

The main reason for educational inequality is how schools are funded. Public schools in our country are funded mainly through people paying taxes on the property they own. These local property taxes pay for the building and maintenance of school buildings, teacher salaries, and books, computers, and other materials. The problem is that a disproportionate percentage of African American, Hispanic American, and Native American school-children live in poverty areas where not as much property tax can be collected for each child. As a result, these children, along with poor white children, frequently do not have the quality of schools that middle- and upper middle-class children have. These young students are many times not academically prepared for trade school or college and therefore are not as able to get good-paying jobs and are more likely to end up in low-paying, minimum-wage kinds of jobs that typically have few, if any, benefits such as health and retirement benefits.

In other words, to have the chance to be upwardly mobile in our society by having a good job, young people today, more than ever before, need to get a good education. Otherwise, they are more likely to remain in poverty. As you know, we have lost many good-paying unskilled factory jobs to other countries (called *deindustrialization*) because corporations can pay factory workers in these other countries lower wages with no health or retirement benefits and hence can make a larger profit. From the mid-1800s to the mid-1900s, our country had many decent-paying unskilled factory jobs that provided sufficient incomes for many American families. However, these days there are fewer of these kinds of jobs and more of the lower paying kinds of service jobs.

So, if we want to help poor African Americans, Hispanic Americans, and Native Americans—as well as poor whites—to have a chance at getting ahead and have a chance at getting a good job with a good salary, one major step our country can take is to create good-quality public schools so that poor minorities and poor whites have a chance to get ahead. By the way, it is also in our vested interests as a country to create such schools for all American children so that young people will be prepared to take the skilled kinds of jobs that our economy is creating such as those in the computer field, engineering, and medical technology.

Providing for such schools means that we, as a country, will need to invest more money in our public school systems from kindergarten through the 12th grade. With more money invested, we can hire more teachers and more qualified teachers,

meaning teachers who are certified in the areas they teach (e.g., a biology teacher has a major in biology in college and has ample educational background to know how to teach biology). Many poor children of today, and poor minority children in particular, do not have certified teachers teaching them. Also, there might not be enough certified teachers for all public schools in our country because many people who go to college do not want to become teachers because they believe that teachers do not make much money. Even if there are certified teachers available, these teachers might not want to teach in schools that have overcrowded classrooms, do not have enough equipment, and/or are located in dangerous neighborhoods.

With more money, we can motivate more people to go into teaching and provide for enough teachers and enough certified teachers for all children. With enough money, we can create well-maintained school buildings for all children. With enough money, we can provide enough books, computers, and other materials for teachers to be able to do a good job and for students to be able to learn well and prepare themselves for college, trade schools, and graduate schools so that they can get better paying jobs with health and retirement benefits. Consequently, a major key to more racial/ethnic equality is to create excellent public schools for all children in the United States, and these in turn will provide minority children with a greater chance to be upwardly mobile and to be prepared to work at better-paying jobs and attain a middle-class lifestyle.

This raises the following question: Where will the money come from? Currently, I see two possibilities that will help us to improve our public schools. One possibility is that we can make the federal income tax more progressive, especially on higher-income people. Also, we could tax the wealth of rich people more. Either a more progressive income tax or taxing wealth more, or a combination of these ways, could supply the money needed for good-quality public schools.

The problem, as you might suspect, is that the wealthy will not want either their incomes or their wealth taxed more. Yet of all the places to get additional tax revenue to pay for schools, these two places would be the least hurtful to people.

Create More Decent-Paying Jobs.   We need to find some way to have a strong and vibrant economy that provides enough good-paying jobs for all of the Americans who want them. If we do not create enough jobs or if we do not create enough good-paying jobs, Americans, even with a good education and well-developed skills, will be unable to get jobs appropriate for their educational and skill levels. So, this leads to the following crucial question: How do we create enough good-paying jobs in a capitalistic economy that does not necessarily provide enough jobs and enough decent-paying jobs?

One key method is to invest more money into what is called *research and development* with the intent of finding more ways to create good-paying jobs. That is, governments, corporations, colleges and universities, private foundations, and "think tanks" can invest money hiring people to do the following that can increase the number of good-paying jobs in our country: (a) invent new products to market and thus give new or existing businesses the reason to expand their plants and hire more people to produce these products, (b) invent new services that people would want to receive and thus create a new demand for jobs, (c) invent new ways to make a profit by recycling existing products, (d) fund research to see how our American companies can sell more goods and services to people in other countries, and (e) expand current services to meet the needs of people and, by so doing, create more jobs in the service sector; for example, expand the child care subsidy to a waiting list of 2,700 applicants who are working in low-income jobs in the state of Kentucky and, at the same time, add new jobs to administer this expanded service (Yetter, 2003).

A second way to create more decent-paying jobs is to have less tax on corporate earnings and more tax on personal income and wealth. The reason for doing this is that corporations will have more incentive to remain in the United States, and hence our country will retain more good-paying jobs.

Build a Tax System.   Another area that our country could look into to see how we could create more racial/ethnic equality is how we tax people…. African Americans, Hispanic Americans, and Native Americans are disproportionately poor. During the last 30 years of the 20th century, these three groups typically had poverty rates of approximately 30% of their respective populations, whereas whites had poverty rates of approximately 10%. Given that racial/ethnic minorities are disproportionately poor, we could decrease various taxes on poor people, the near poor, and the working poor, and this would help them to keep more of their take-home pay and hence help them to get out of their poverty. This action would be a step forward for many African Americans, Native Americans, Hispanic Americans, and other racial/ethnic minorities who are disproportionately poor.

Another way we could change the tax system so that poor minorities and poor whites could become more equal to nonpoor people in our country would be to change the way we tax people on their social security…. [P]eople who have jobs pay 6.2% of their wages in social security tax if they earn $94,200 or less per year. However, people who earn more than $94,200 do not pay social security tax on the money they make above this amount. We could create more racial/ethnic equality by having a progressive social security tax where poor and near-poor Americans pay a lower percentage in social security tax. Because a higher proportion of minorities have poverty and near-poverty incomes, they will have more take-home pay and hence not have need to live so close to the margin of subsistence. As a result, we, as a country, will have more racial/ethnic equality.

Another way to create more racial/ethnic equality is to provide a larger earned income tax credit, where the federal government gives additional money to people whose incomes are below the poverty line. We could continue to do this and increase this amount so that poor people who are racial/ethnic minorities, as well as poor whites, receive enough money to moves them above the poverty line. Gans (1995) predicted that our country might need to provide more income for people in this way if our economy does not provide enough income from jobs or does not provide enough jobs for everyone. Because African Americans, Hispanic Americans, and Native Americans are disproportionately poor, increasing the earned income tax credit would increase their incomes and help our country to achieve more racial/ethnic equality.

Provide More Social Services. If our country could provide more funding in areas such as (a) Section 8 subsidized housing for the poor or near poor..., (b) more public transportation to allow many of the poor who cannot afford personal transportation to get to jobs..., and (c) more subsidized health care..., these services, along with others, would disproportionately help minority groups and poor whites to move us in the direction of more racial/ethnic equality.

In other words, any way we can help poor Americans to increase their standard of living, and hence create less inequality in our country overall, will also help minority Americans to have a higher standard of living and thus create less racial/ethnic inequality. We all know that increasing funding for these social services will mean that more tax revenue will need to come in to pay for these services.

And this means that some people will need to be taxed more to pay for these programs. As I stated previously, the least sacrifice would occur if we increased taxes on those with the highest incomes and the most wealth. Those Americans with the highest incomes and the most wealth would still retain much of their incomes and wealth and hence would still enjoy a very high standard of living relative to the vast majority of other Americans.

... Our American history has been filled with much prejudice and discrimination; there is still leftover prejudice and discrimination within some Americans even today; institutional discrimination, such as the policy of local property taxes funding public schools and the policy of "last hired, first fired," continues to work to the disadvantage of minorities; and the historical effects of prejudice and discrimination have resulted in minorities living disproportionately in geographic areas of high unemployment, leading to higher rates of poverty, homelessness, and stress in the family, and living in areas of higher crime rates. Given these social and historical factors, our country could implement a number of the preceding policies as a way to achieve more racial/ethnic equality.

REFERENCES

Gans, H. J. (1995). Joblessness and antipoverty policy in the twenty-first century. In *The war against the poor: The underclass and antipoverty policy* (pp. 133–147). New York: Basic Books.

Greenhouse, L. (2003, June 24). Affirmative action upheld, with limits. *Louisville Courier–Journal,* pp. A1, A4.

Merton, R. K. (1967). Manifest and latent functions. In *On theoretical sociology: Five essays, old and new* (pp. 73–138). New York: Free Press.

Yetter, D. (2003, June 23). Many needy parents denied child-care help: State's long waiting list may push some on welfare. *Louisville Courier–Journal,* pp. A1, A4.

Some Processes of Social Life

An early pioneer in sociology, Auguste Comte, divided sociology into two major divisions: statics and dynamics—that is, order and social processes. Although the terms used by Comte have fallen out of favor over time, the basic division of sociology remains in use today as the study of social structure/institutions and social change.

Earlier parts of this text dealt mostly with structure and functions. We noted that individuals were transformed into social beings through the process of socialization via group interactions with the five major social institutions in a specific cultural context. In this manner, socialization transforms people into social beings with a gender identity and makes them viable members of society by imbuing them with the culture of the society.

Society, however, does not consist just of structure and institutions. It also includes processes that may bring about change that can aid or hinder individuals or society in meeting their needs. This final part of the text deals with some of those processes or dynamics in society.

Perhaps, surprisingly, Part V starts out with a chapter on deviancy (Chapter 11). Part IV, on inequality, noted that many of the processes in these areas have been and are unequal and, in some cases, are becoming more so. To deal with the negatives of this inequality, some people may resort to deviant acts via social movements or commit crimes to supposedly even out the score. In both situations, a process is in effect that will attempt to bring about change.

The process of aging has a mixed reception among various groups. In some societies, the process of aging is accompanied by enhanced status due to that person's accumulated knowledge. In other societies, such as the United States, the elderly are not valued as contributors to society but rather are seen as a handicap to be resented and taken care of. It is youth that is praised for their "new" knowledge in U.S. society. The need to take care of the aged has arisen mainly because we are now living much longer. The result is increasing health care needs and costs that will lead to changes in health care policies and options in the dying and death process.

Two processes have led to a vastly increased population in this country: the baby boom of the 1950s and the improvements in health care that have led to longer life spans. These population processes have also led to another one—urbanization, in which large numbers of people move to cities and their surrounding areas. This, in turn, leads to urban governments being stretched to provide the infrastructure and services needed to meet the needs of the burgeoning population. Population growth can also mean increased consumption of energy and food resources, which may bring about migration to other areas or other countries to find these resources. Of course, people may migrate to other areas because of the belief that better opportunities exist in those areas.

The last chapter in this text is, in a sense, a summary of the process that is happening in all the structures and institutions examined—change. All societies undergo change. When it is a slow process, it can be adjusted to with time. However, change is now occurring rapidly because new accomplishments can be built upon a past base. Earlier in this book, we gave the example of working for a penny a day and having over 10 million pennies at the end of 30 days. Similarly, it took only 50 years after the Wright Brothers' adventure to fly supersonic jets. Unfortunately, very rapid social change can lead to a breakdown in norms of society and a resulting lack of new ones to guide social behavior. The early scientific sociologist, Emile Durkheim, coined the term "anomie" to describe this situation. Yet, change is seen as something good when it is viewed in the positive light of bringing about improvement. Thus many times dissatisfied people will join together in a social movement geared toward accomplishing desired changes.

This part of the text highlights some of the forecasts made by social scientists about deviance, aging, population, and social movements, among other issues. Some people fear change and the disruptions it may bring about in their lives. Others believe that change can be positive rather than dysfunctional. By understanding the process of change and the future it promises, we are better prepared to lead productive and rewarding lives.

Chapter 11

Deviance

Violating the Norms of Society

Dr. Alan W. McEvoy

Deviance is one of the most myth-laden of social issues. Most often, myths are simplistic explanations based on an individual's own set of biases. The difficulty in dealing with myths is that they often contain a grain of truth, which makes it difficult to convince people that their belief is really a myth.

One popular myth-belief is that crime is genetic in origin. This belief states that there is a relationship between one's genes and one's behavior when it comes to deviance. For some, this belief explains why some types of people seemingly commit crimes at higher rates—both certain groups and some physical types. However, research shows that crime is not limited to any ethnic, racial, or physical group. This belief is related to the idea that because crime is deviant, it must be abnormal behavior. But, as Randall Collins explains in the first article of this chapter, crime is so prevalent it should be considered a normative behavior of society—a behavior, like all human behavior, that is learned.

Actually, this belief reflects the fact that the public likes to put people into categories and label them: autistic, demented, aggressive, shy, and so forth. The problem is that once the label is applied, it seemingly cannot be removed, even if the person is found to not suffer from the disease or is found innocent of the deviant behavior. For example, a number of child-care workers were accused of being child molesters. Despite the fact that many were found innocent, the so-called perpetrators were never able to shake the label and so were never able to obtain jobs or continue working in the child-care field. It appears that once labeled in this manner, a person will never be considered to be otherwise, despite the absence of the labeled behavior. In short, the label defines the person in society's eyes rather than the actions of that person. In the second article of this chapter, D. L. Rosenhan describes the situation of labeling in a mental institution.

A major belief is that we are in the midst of a massive crime wave. In fact, there is little scientific support for this view. Barry Glassner claims that we actually fear the wrong things. What is true is that there is now greater confidence in the police and so less reluctance to call them when a crime is suspected, there are now quicker and easier ways of reporting crime, there is the necessity to report crime to collect on theft insurance, and there is improved record keeping for all types of crime. In addition, some police departments tend to artificially increase the number of crimes reported to justify requests for more money and personnel. It is these features of

the current crime scene that leads to more crime being reported and, in turn, gives the impression that we are in the midst of a crime wave. Another factor that may drive the impression that we are in the midst of a crime wave is the fact that the media tends to mostly report only certain types of crime. For example, the frequent reporting of drug busts leads to the impression that drug abuse is pandemic. In reality, drug abuse and other crime rates have been on a downward spiral for many years. However, the United States may also have more crime because studies have shown that the higher the level of inequality in a society, the higher the rate of personal deviance—and the United States has a larger gap between the rich and the poor than any other Western nation.

Related to the myth that crime rates are climbing is the myth-belief that the only way to reduce crime numbers is through incarceration of the perpetrators. Adding to this belief in punishment for the "evil doers" is the desire for revenge—revenge against those who would dare violate the norms of society. The desire for stronger punishments is also propelled by the public's belief that they are safer from criminals while the latter are in prison and by politicians' recognition of the public's desire for revenge and their subsequent promotion of tougher sentencing programs. Thus the increase in the number of people imprisoned is more a reflection of tougher sentencing laws and not, as many believe, increasing crime rates. In fact, when one considers the high rates of repeated criminal behavior despite increasing severe punishment, it can be concluded that there is little or no relationship between incarceration and crime reduction. As one prisoner stated: How can I learn to be a good citizen when I am locked up with a bunch of unfit people?

It appears that our legal system has not learned the lesson that incarceration does not reduce crime, because new prisons are constantly being built to accommodate the burgeoning number of prisoners due to longer mandatory prison terms. In attempting to deal with this problem, it has been suggested that society decriminalize certain behaviors. A crime is defined as a violation of criminal law. If it is not listed in the criminal code, then it is not a crime. The idea behind the criminalization proposal is to remove behaviors that have no victims. Such crimes as marijuana smoking and prostitution affect only those doing the behavior, so "revenge" by society for causing harm is unnecessary, according to its proponents. What do you think of the idea of decriminalizing crimes that have no victims?

Strangely, the United States does have a large number of crimes that have countless victims but for which most people are seemingly unaware of their extent and impact. White-collar crime is crime committed in the course of one's business dealings. It is labeled as criminal behavior because it fits the legal definition of a socially harmful act and for which the law provides a penalty. For example, in the 1990s there were 20 antitrust crimes, 7 campaign finance crimes, 38 environmental crimes, 6 food and drug crimes, and 13 fraud crimes among the top 100 corporate crimes discovered. Interestingly enough, 6 of these top corporate criminals were recidivists. Yet most of the perpetrators of this kind of crime will receive minimal sentences and their fines will be written off as a tax-deductable business expense.

Why? An interesting question to ask at this point is why so little is known about this widely occurring, widely repeated criminal behavior. Probably the best-known reasons for the public's lack of knowledge about white-collar crime are that it is committed by widely respected persons, the companies are prominent corporate advertisers, the criminal cases tend to last for years, and the news of such criminal behavior and trials appears on the business pages and almost nowhere else—despite the fact that this type of criminal behavior affects and even kills hundreds of people each time that it happens.

The extensiveness of both recognized and unrecognized criminal behavior causes people to wonder about how to deal with this normative behavior. As mentioned earlier, incarceration does not appear to work, nor do the fines typically levied as punishments for widespread and even dangerous white-collar crimes. James Crone, in the final article of this chapter, suggests some ways to "solve" the problem of crime. Do you believe that these ideas will "solve" crime or even reduce it? What means would you use to reduce both types of criminal crime?

45

The Normalcy of Crime

RANDALL COLLINS

The author discusses three all-encompassing explanations for crime: conservative, liberal, and radical. These explanations ignore the possibility that crime is normal and even functional. As you read this selection, consider the following questions as guides:

1. *Do you believe that the broad reasons given for committing crime are too simplified? Why, or why not?*
2. *Defend or dispute the author's claim that crime is normal and functional.*
3. *If, indeed, crime is normal and functional, what should be done about it?*
4. *If crime is normal and functional, can anything be done about it?*

GLOSSARY **Altruistic** Unselfishly concerned.

There have been several widely accepted views about crime. The more obvious explanations begin at the level of common sense. The trouble with common sense, though, is that there are usually opposite opinions on any subject, both of which are equally commonsensical to those who believe in them. These views have generally corresponded to popular political beliefs. Roughly speaking, we may refer to them as the conservative and liberal views on crime....

The most sophisticated and least obvious theory of crime, I will suggest, goes back to Durkheim. The problem has not turned out to be just what we once thought it was. We may have to face a paradox: crime exists because it is built into the structure of society itself. This does not mean that nothing can be done about it, but the social costs of controlling crime may involve more difficult change than we have been aware of.

CONSERVATIVE EXPLANATIONS OF CRIME

One view of crime is that criminals are simply bad people; the only way to deal with them is to punish them. The more crime there is, the harder we should crack down on it. This position has been held for many centuries, and it keeps on being restated today. The trouble is that it has never really worked. In Europe during the 1600s and 1700s, punishments were as severe as one could imagine. People were hung for stealing a loaf of bread; others were branded or had their ears cut off. Some offenders, especially people accused of religious or political crimes, were tortured to death. All these punishments were public spectacles. A crowd would gather around to watch a good execution, while vendors sold refreshments and people made bets on

Source: From *Sociological Insight: An Introduction to Nonobvious Sociology*, 2nd ed., by Randall Collins. © 1992 by Oxford University Press, Inc. Used by permission of the publisher.

how long the criminal would yell while he or she was burning at the stake. People today who advocate severe punishments as a deterrent for crime would have been delighted by the situation.

But the brutal punishments did not work. Crime kept right on occurring at a tremendous rate for hundreds of years, despite the hangings and the mutilations....

The same kind of situation still can be found in some parts of the world today. In Saudi Arabia and some other Muslim countries, theft is punished by cutting off a hand, and many other offenses by death. Executions are carried out in public, with the whole community required to attend. But the results are the same as they were in medieval Europe....

We can begin to see, therefore, that the philosophy of punishing criminals as violently as possible is not really a policy that people advocate because it has proven effective. It is a political position, or what comes to the same thing, a moral philosophy, which declares it is good to be tough and even brutal or malicious to offenders. Just why people hold this position is itself a question for sociology to explain, since they must hold it for some other reason than its practical effects. The holders of this position doubtless consider it rational, but here again we see that their rationality has a nonrational foundation. They do not bother to look at the evidence for whether severe deterrents work but already "know" their policy is right. This sense of rightness is the mark of a partisan position, in this case political conservatism.

A somewhat more scientific version of this political position has tried to tie crime to biology. Today some assert that criminals have bad genes; their propensities to crime are inborn and, hence, nothing can be done about them. Society could only pick them out at an early age by appropriate testing and then presumably get rid of them in some way. Just how this is to be done is not yet worked out: whether the police would hold a complete dossier on all people with bad genes, or whether such people would be locked up for life, or be sterilized, or even exterminated. The issue hasn't really gotten to this point because the position so far is completely theoretical. No one knows how to make a test for

bad genes, and there is no real comparative evidence that such genes are causes of crime. The modern genetic theory of crime is another version of conservative political ideology. This is easy to see, since the same arguments about criminals are also applied to welfare recipients and other social types who are anathema in conservative thinking....

LIBERAL EXPLANATIONS

If there is a conservative version of common sense about crime, there is a liberal common sense as well. The liberal position makes an effort to understand what it is like being in the criminal's shoes. Why would someone enter a life of crime, and what can be done to help them out of it? There have been several answers to these questions.

One is that criminals are people who have gotten in with the wrong crowd. Youths hang around with a delinquent gang and start to pick up delinquent values themselves. Soon they are committing petty thefts, small acts of vandalism, and the like. This moves them more and more into the delinquent culture, and eventually they move on to serious crimes and become full-fledged criminals.

A similar type of explanation is that criminals come from broken homes and run-down neighborhoods. These childhood stresses and strains make people hostile and insecure, and lead them to a life of crime. Growing up in an area of poverty and disillusionment, these youths have no reason to be attached to normal society. They feel that society has no use for them, and they have every reason to take revenge in any way they can.

Sometimes this argument is taken one step further to propose that it isn't just their background that makes some people become criminals, but also the lack of opportunities to change their social condition. If children from poor families or racial minorities had a chance to rise in the world, they would become normal, productive members of society. It is because they are trapped by the lack of opportunities to get ahead that they turn to crime. It is proposed, moreover, that the social atmosphere of the United States makes this feeling particularly strong. For the United States is an achievement-oriented culture, where people are expected to make a success on their own....

Some of these arguments, we can see, get to be rather complicated. Nevertheless, they all share the notion that crime is not really the fault of the criminal. He (or she—although in fact the great majority of criminals are male) would rather not be a criminal if he could help it. It is only the adverse social conditions that force them into a criminal career.

This type of explanation certainly has the appeal of sounding altruistic, and it has given rise to a great many efforts at reform and rehabilitation to set criminals back on the path to normal social participation....

In this way, all of the various social causes that are believed to account for crime are to be counteracted by an appropriate social reform. If it is a delinquent milieu that starts youth on their evil ways, we provide youth services and group workers to try to lure the gangs off the streets and onto supervised playgrounds. For broken homes and run-down neighborhoods, there are social workers and urban renewal projects. For blocked mobility opportunities, there are various efforts to improve the life chances of the disadvantaged, to keep them in school longer, to provide remedial services, and the like.

As I said, all of this is very altruistic, but it has one big drawback. It simply has not worked very well....

These sorts of facts are a fairly serious indictment of the liberal theories of crime and its prevention, but this hasn't entirely convinced the proponents of these theories that they are wrong. They can continue to argue, for example, that the proper counteractive measures have not been applied vigorously enough. We need more youth group workers, they may reply, or a more extensive attack on the existence of poverty and racial discrimination, or a more serious effort to create career mobility opportunities for deprived youth and ex-convicts alike. This has some plausibility since it is certainly true that much more could be done in this altruistic direction. But the suspicion has been growing that the underlying theories just may not be accurate.

Take the broken-family-and-blighted-neighborhood hypothesis about crime. This explanation seems to fit our commonsense view of the world: stress and deprivation lead to crime. But the evidence does not exactly bear this out. Not everyone from a divorced family becomes a criminal; in fact, most such children do not. This is especially apparent today, when divorce has become a normal and accepted part of otherwise quite average families. Nor is it fair to say that everyone who lives in a poor neighborhood is a criminal: again, it is only a minority within this area who are. Hence it cannot be poverty per se that causes crime but some other factor. This becomes even clearer when we realize that by no means are all criminals poor or from racial minorities. Delinquent youths are found in middle-class areas as well as poor ones. Rich boys at fraternity parties commit acts of vandalism, too, as well as violence, rape, theft, and all the rest of it, although they are not always charged with these crimes. The same thing is true among adults. It is not just the poorer social classes that commit crimes. So-called white-collar crime is also a major problem, ranging from passing bad checks to embezzling business funds or conspiring to bribe government officials or to evade legal regulations.

The altruistic, liberal theories of crime are just not adequate to deal with these phenomena. What looked at first glance like a realistic sociological explanation of crime turns out on closer examination not to fit the facts very well at all. There is less crime in the deprived areas of society than the theory would predict and more crime elsewhere in society where these conditions do not hold. It is no wonder, one might conclude, that the liberal methods for preventing crime and rehabilitating criminals have not had much success.

RADICAL EXPLANATIONS
OF CRIME

In recent sociology there has been an upsurge of theorizing that rejects the more traditional kinds of theories in favor of a radically new look at the crime issue. Here the theories enter the realm of the nonobvious and even the paradoxical.

The basic turn in the argument has been to shift attention away from the criminal side and to look critically instead at the agents of law enforcement. For example, it is sometimes argued that increases in the crime rate have nothing to do with how many crimes are actually committed. All that has changed, it is suggested, is that more crimes are being reported. Sometimes a newspaper will create a crime wave by running crime stories more prominently on the front pages—perhaps for political purposes, to attack a city administration or make an issue of the crime problem for an upcoming election. The police, too, it is charged, inflate the crime rate by improving their record-keeping capabilities. Unsolved crimes that formerly were left unreported are now included. This makes a good argument for police appeals that they need an increased budget.

It does appear to be true that some alleged shifts in crime rates are produced in this way. Newspapers in particular are not a very reliable source of information on social trends, and official police statistics are also subject to biases due to shifts in reporting methods. Whenever one sees a rapid jump in crime rates over a space of a year, it is often due to a purely administrative change in the statistical accounting system. At the same time, it has to be said that not all of the shifts in crime rates can be attributed to causes of this sort.

But there is a much more radical sense in which it is proposed that crimes are created by the law-enforcement side. This is referred to as the labeling theory. The argument goes like this. All sorts of youths violate the laws. They engage in petty thefts and acts of vandalism. They get into fights, drink illegally, have illicit sex, smoke dope or use drugs, and so on. This is widespread and almost normal behavior at a certain age. What is crucial, though, is that some of these young people get caught. They are apprehended by the authorities for one thing or another. Now even at this point there is a possibility of heading off the negative social consequences. Some of these youths get off with a warning, because their school principal likes them, say, or because their parents intervene, or because the police are sympathetic to them. If so, then they have escaped going down into a long funnel at the end of which lies a full-fledged criminal identity.

If a young offender is actually arrested, charged with a crime, convicted, and all the rest, this has a crucial effect upon the rest of his or her career. This happens in several ways. One effect is psychological: those who had previously regarded themselves as more or less like anyone else, just goofing off perhaps, now are someone special. They are now labeled an offender, a juvenile delinquent, a criminal; they are caught up in a network of criminal-processing organizations. Every step along the way reinforces the sense that they have become someone different from the normal. They acquire a criminal identity....

In this way, a self-perpetuating chain of criminal activities builds up. The key point in the whole sequence is right at the beginning, where the labeling process begins. It is the first, dramatic confrontation with the law that makes all the difference, deciding which way the individuals will go. Either they will get by with a bit of normal goofing off, or they are embarked on a career of crime in which everything that is done to prevent it actually makes it all the more inevitable.

This is a rather psychological way of describing the dynamics of the labeling process. I could fill in the process from a different angle, one that does not stress so much the shift that takes place within the novice "criminal's" mind but within the organization of the law-enforcement world itself. Sociologists who have studied the police point out that the police constitute an organization, with administrative problems just like any other organization. A business organization needs to keep up its sales; a police organization needs to keep apprehending criminals and solving crimes. This is by no means an easy thing to do. Some crimes are relatively easy to solve, such as murder. But these make up only a small percentage of total crimes. The most common crimes, and those which most widely affect the public, are burglary, auto theft, and other types of larceny. These are hard to solve precisely because there are so many of them. There usually is little evidence left at the scene of the crime, and there are rarely any witnesses.... How, then, do the police try to control this large category of crime?

The best strategy they can follow is to try to get confessions from the criminals that they do apprehend. So whenever someone is arrested with goods from a burglary, say, a great deal of pressure is put on them to confess to other burglaries.... The most effective sort of pressure, though, is usually in the form of a bargain. The accused criminals are encouraged to confess to a list of unsolved thefts; in return for this, they are allowed to plead guilty to some restricted charge, e.g., one or two counts of burglary, or even some lesser offense. This is a typical plea bargain....

All this has a powerful effect in reinforcing the "labeling" process that keeps people going in criminal careers. The way police can make their system work is to keep tabs on people whom it is easiest to arrest.... The easiest people to arrest are people who have been arrested before. So one way police can "solve" a round of burglaries is to pay a surprise call on formerly convicted criminals in the area who are out on parole. One of the conditions of parole often is that the ex-convict should be subject to search. So the police arrive, look for stolen property, illegal drugs, or other violations. Often it is not hard to find these, especially since drugs of one kind or another are generally a part of the criminal culture. (Which is not to say that these same drugs may not also be part of the life-style of people who are not in the criminal world.)

So the police then are able to set the bargaining process in motion....

Thus, the chain of events that starts when someone is labeled a criminal for some initial offense, can end up as a kind of invisible prison in its own right. Once someone becomes known to the police, they are subject to organizational pressures that will send them through the system over and over again. Whether they come to strongly identify themselves personally with a criminal identity or not, the police will tend to do so, and that makes it all the harder to get out. Ex-convicts are trapped in a machine that constantly reprocesses them because they are its easiest materials to reach.

The labeling theory declares that crime is actually created by the process of getting caught. Unlike the previous types of theories that we looked at, the personal characteristics of the individuals, or their social class or ethnic or neighborhood background, is not a crucial point. It is assumed that all sorts of people violate the law. But only some of them get caught, are prosecuted, labeled and all the rest, thereby becoming full-fledged criminals. If criminals who go through the courts and the prisons are so often likely to be disproportionately poor, black, or otherwise fit someone's idea of "social undesirables" or the "socially deprived," it is because these are the types of people who are most likely to be apprehended and prosecuted. The fraternity boys stealing a college monument or raping the sorority girls at a party are let off with a reprimand because these are labeled "college pranks." The poor black youth who does the same sort of thing gets sent to juvenile court and a start on a career of serious crime.

There is an even stronger version of the radical approach to crime. This argues that it is not simply the police who create the criminals but the law itself. To cite an obvious example: possession of drugs such as narcotics was not a crime until laws were passed making private possession of them a felony. In the 1800s, the use of opium and opiate-based preparations such as laudanum was not illegal, and it was fairly widespread. The drugs could be bought over the counter at a pharmacy. Many people used them in patent medicines. Others used them for pain-killers, escape, or because they liked the sensations they produced. The same was true of hashish and marijuana, or of coca and cocaine, which were used in greater or lesser quantities by various kinds of people. In the early 1900s, the public use of opium and its derivatives was outlawed in the United States, and under a series of international agreements, by most of the modern states around the world. Other laws followed, outlawing cocaine and cannabinols.

These laws suddenly created a new category of crime. People who had previously been engaging in a purely private act were now breaking a fairly serious law. This had a great many social ramifications. For one thing, the labeling processes outlined above, both psychological and organizational, were set in motion....

The illegalizing of drugs, moreover, had an important economic effect. When drugs were sold on the open market, their cost was relatively low because they are relatively inexpensive to produce and transport. But when drugs became illegal, the whole business was greatly restricted. As one can see from a simple application of the economics of supply and demand, restricting the supply raised the price. Whereas a modest supply of opium in early nineteenth-century England cost a shilling (the equivalent of perhaps $25 today), heroin (a twentieth-century derivative of opium) now costs some $2000 an ounce. Drug dealers and smugglers incur much greater expenses, keeping their activities hidden as much as possible, paying out bribes, and also paying for the inevitable legal fees when they are caught. So the illegalizing of drugs, by raising the prices, ramified into many other crimes that had formerly been unconnected with the drug market. Smuggling and bribery expanded, of course, but so did burglary and robbery. Most drug addicts, unable to pay for the expense of supporting a costly opiate habit, turned to theft as a main way of keeping the money coming in. From the initial decision to outlaw drugs, then, many other crimes followed.

The same kind of analysis has been applied to many other sorts of crimes. The national prohibition of alcohol in the United States, which held sway between 1919 and 1933, for example, created a whole illegal culture of speak-easies, stills, alcohol smugglers, and an organized crime network to "protect" these operations....

The radical approach to the analysis of crime turns up a great many ironic interconnections between crime and the social structure. Actions taken by citizens in the name of morality and law-abidingness add up to vastly increasing the amount of criminality. Some sociologists have argued that an explanation of crime really boils down to an explanation of how certain things came to be defined as crimes. It has been suggested that crimes are manufactured by "moral entrepreneurs," people who try to create a morality and enforce it upon others. Other sociologists have gone farther, to look for the economic and organizational interests or the social movements that create crimes in this way. It may be suggested, for example, that the outlawing

of drugs in the early twentieth century was part of the efforts of the medical profession to monopolize control over all drugs for itself. The prohibitionist movement has been explained as a last-ditch effort of rural Anglo-American Protestants to try to head off what they saw as the degenerate alcoholic culture of the immigrants in the big cities. An analysis along these lines could be applied to current movements that are attempting to create new definitions of crime, such as the antiabortion movement.

At this point, one might step back and ask a question. The examples given have all been of the type of activities that offend some people's moral sense as to what is proper. Drug taking, drinking, gambling—one could add prostitution, pornography, homosexuality and other sexual practices—all involve people who willingly consent to these actions. These actions offend only outsiders. They are what are called "victimless crimes." Here the idea that society creates these crimes in a fairly arbitrary sense, just by passing a law against them, has a good deal of plausibility. But what about "real" crimes, such as robbery, murder, assault, rape, and all other actions that hurt someone's life, body, or property? One could well maintain that these actions would not be considered licit by most people, even if there were no laws prohibiting them. These seem to be "natural," rather than "artificial" categories of crime, and people would want to stop them without the necessity of some kind of moral crusade trying to have laws passed to outlaw them.

However, the most radical position in sociological theory attempts to show that these crimes, too, are socially created. For example, the crime of robbery is only a crime because of the system of property....

This is certainly a theory worth thinking about. It has the merit of seeing that "real" crimes are a matter of conflict between people in a stratified society, and especially that economic crimes are part of the system of economic stratification in general. Since economic crimes like robbery and auto theft make up the largest proportion of all crimes, this kind of theory can potentially explain a great deal.

Nevertheless, we cannot immediately jump to the conclusion that crime is class struggle of exactly the same sort as usually featured in the Marxian model. For one thing, when we look at who are the victims of crimes, we find a rather surprising pattern. The poorer classes are much more likely to be robbed or burglarized than the wealthier classes. And this is true, in the United States, for both whites and blacks. In fact, blacks with the lowest incomes are the most likely of all to be victims of crimes of virtually all sorts, including murder and rape as well as property crime.

Clearly, then, there is a stratified pattern of crime, but it is not primarily the poor robbing (and murdering and raping) the rich. Criminals are not Robin Hoods. What appears to be going on, rather, is that crime is mainly *local*. People rob, burglarize, murder, and rape in their own neighborhoods above all. The reason is fairly simple: these are the easiest opportunities, especially for teenagers, who commit the majority of all crimes.

The end result is that there is a social-class pattern in crime after all, but it comes out in the fact that neighborhoods tend to be segregated by social class, as well as by race and ethnicity. Hence it is the least privileged people who commit the largest number of crimes, but their victims are primarily people like themselves. It is mainly the poor robbing the poor....

There has been a lot of controversy over the death penalty in recent decades. If we leave aside the moral questions involved in this, and concentrate only on the research that has been done, we can see some interesting patterns. Some states in the United States have the death penalty, while others have abolished it. If we compare states that are similar in their social characteristics, it turns out that they have about the same murder rates, whether they have the death penalty or not. That implies that people do not decide to commit murder or not according to whether they expect to risk a severe penalty for it. Murders do not seem to be related to any social calculus. By the same token, none of the sociological theories given above seem to explain murder very well.

I mentioned earlier that murders are relatively easy for the police to solve. Why is this? It is because the large majority of murders are committed by people who know their victims personally. For that matter, the largest single category of murders happens within the family, especially one spouse murdering the other. Hence to solve a murder is not particularly difficult. The police need only look for someone who knew the victim and who had some motive to be especially angry with them. So if you are thinking about killing your husband or your wife, forget it; you will automatically be the number one suspect.

All this adds up to a picture in which crimes divide into quite different sorts. There are victimless crimes, very much created by social movements that define them as criminal; people who become labeled as criminal because of these sorts of offenses usually become involved in networks of other sorts of criminality as a result of the law-enforcement process. There are also property crimes, which have some relevance to the way in which individuals make their careers as criminals, but which would by no means disappear if laws stopped being enforced. And there are crimes of passion, which seem to be of a much more personal nature, and which do not seem to be related to any of the factors we have considered here.

Is there any perspective that encompasses all of this? Yes, I believe there is. But it is the most nonobvious of all, and one which does not resonate any too well in the hearts of either conservatives, liberals, or radicals. It is a perspective that declares that crime is a normal, and even necessary, feature of all societies.

THE SOCIAL NECESSITY OF CRIME

This perspective, like so many of the nonobvious ideas in sociology, traces back to Émile Durkheim. In this view, crime and its punishment are a basic part of the rituals that uphold any social structure. Suppose it is true that the process of punishing or reforming criminals is not very effective. The courts, the police, the parole system—none of these very effectively deter criminals from going on to a further life of crime. This would not surprise

Durkheim very much. It can be argued that the social purpose of these punishments is not to have a real effect upon the criminal, but to enact a ritual for the benefit of society.

Recall that a ritual is a standardized, ceremonial behavior, carried out by a group of people. It involves a common emotion, and it creates a symbolic belief that binds people closer to the group. Carrying out rituals over and over again is what serves to keep the group tied together. Now in the case of punishing criminals, the group that is held together is not the criminals' group. It is the rest of society, the people who punish the criminals. The criminal is neither the beneficiary of the ritual nor a member of the group that enacts the ritual, but only the raw material out of which the ritual is made....

The main object of a crime-punishment ritual, then, is not the criminal but the society at large. The trial reaffirms belief in the laws, and it creates the emotional bonds that tie the members of society together again. From this point of view, exactly how the criminal reacts to all this is irrelevant. The criminal is an outsider, an object of the ritual, not a member of it. He or she is the necessary material for this solidarity-producing machine, not the recipient of its benefits. It is the dramatics of the trial that counts, the moments when it is before the public eye. Afterwards, it may all come unraveled. The conviction may be reversed on appeal for some technical error. Criminals may go to an overcrowded prison where they make new criminal contacts and acquire a deeper commitment to the criminal role. Sooner than expected, the parole board may decide to relieve crowding in the prison by releasing them, and they are back out on parole and into the routine of police checks and parole officers and all the rest of an ongoing criminal career. If we look at the criminal justice system from the point of view of somehow doing something to deter the criminal, it appears ineffective, even absurd. It makes more sense once we realize that all the social pressure falls upon dramatizing the initiation of punishment, and that this is done to convince society at large of the validity of the rules, not necessarily to convince the criminal.

An even more paradoxical conclusion follows from this. Society needs crime, says Durkheim, if it is to survive; without crimes, there would be no punishment rituals. The rules could not be ceremonially acted out and would decay in the public consciousness. The moral sentiments that are aroused when the members of society feel a common outrage against some heinous violation would no longer be felt. If a society went too long without crimes and punishments, its own bonds would fade away and the group would fall apart.

For this reason, Durkheim explained, society is in the business of manufacturing crimes, if they do not already exist in sufficient abundance. Just what would count as a crime may vary a great deal, relative to what type of society it is. Even a society of saints would find things to make crimes out of: any little matters of falling off into less saintliness than the others would do. To put it another way, the saints, too, would have their central, especially sacred rules, and those who did not respect them as intensely as the others would be singled out for punishment rituals that served to dramatize and elevate the rules all the more....

Punishment rituals hold society together in a certain sense: they hold together the structure of domination. They do this partly by mobilizing emotional support for politicians and the police. Above all, they increase the feelings of solidarity within the privileged classes and enable them to feel superior to those who do not follow their own ideals. Outrage about crime legitimates the social hierarchy. The society that is held together by the ritual punishment of crime is the stratified society.

In this sense, crime is built into the social structure. Whatever resources the dominant group uses for control will have corresponding crimes attached to them. Since there is an ongoing struggle among groups over domination, some groups will violate other groups' standards. And those individuals who are least integrated into any groups will pursue their own individual aims without regard for the morality held by others. Therefore, there is usually no shortage of actions that are offensive to many groups in a society. And these violations are to a certain extent welcome by the dominant groups. Crime gives them an occasion for putting on ceremonies of punishment that dramatize the moral feelings of the community, which bolsters their group domination.

This means that every type of society will have its own special crimes. What is constant in all societies is that somehow the laws will be set in such a way that crimes and punishments do occur. A tribal society has its taboos, the violation of which calls down ferocious punishment. The Puritans of the New England colonies, with all their intense moral pressures, believed in the crime of witchcraft. Capitalist societies have endless definitions of criminality relating to property. Socialist societies have their crimes as well, especially political crimes of disloyalty to the state, as well as the individualistic crimes of failing to participate whole-heartedly in the collective. The ritual perspective finds that all societies manufacture their own types of crime. It may be possible to shift from one type of crime to another, but not to do away with crime altogether.

Crime is not simply a matter of poverty and social disorganization, nor of particularly evil or biologically defective individuals. The labeling theory is closer to the truth, but the processes are much wider than merely social-psychological occurrences within the minds of offenders. Criminals are only part of a larger system, which encompasses the whole society.

THE LIMITS OF CRIME

If the whole social structure is producing crime, we might wonder if there is any limit to how much crime it produces. If crime helps hold society together, doesn't it follow, paradoxically, that the more crime there is; the better integrated the society will be? Obviously, there must come a point at which the amount of crime is too great. There would be no one left to enforce the laws, and society would fall apart.

Nevertheless, this does not usually happen. If we look further into the matter, the reasons turn out to be not so much that the law-enforcement side effectively controls crime, but that crime tends to limit itself. Look at what happens when crime

becomes more and more successful. Individual criminals can do only so much. They are much more effective at stealing, embezzling, or whatever if they are organized. Individual thieves give way to gangs, and gangs to organized crime syndicates. But notice; organized crime now becomes a little society of its own. It creates its own hierarchy, its own rules, and it attempts to enforce these rules upon its own members. Organized crime tends toward regularity and normalcy. It begins to deplore unnecessary violence and strife. The more successful it is, the more it approximates an ordinary business. The very success of crime, then, tends to make it more law-abiding and less criminal. The same thing can be seen historically.

At some points in history, political power consisted of little more than marauding gangs of warriors or robber barons that plundered whoever came their way. The very success of some of these well-armed criminals, if we may call them that, meant that they had to take more responsibility for maintaining social order around them. At a minimum, the violent gang of warriors had to maintain discipline among themselves if it was to operate effectively in plundering others. The more successful a robber-baron became, the more he turned into an enforcer of laws. The state arose from a type of criminality but was forced to create a morality just to survive.

If social life creates crime, then crime also tends to create its own antithesis. Crime tends to drive out crime. It is not so easy, after all, to be a successful criminal. If you start out today to be a thief, let us say, how do you go about it? In many ways it is like learning any other occupation. You need to learn the tricks of the trade: how to break into a house, how to open a locked car. You need to know where to acquire the proper tools: where to get guns, if you want to be an armed robber. And you need to learn how to dispose of the loot once you have stolen it; it doesn't do you much good to steal a lot of television sets and stereos if you have no way of selling them for cash. And the more expensive the stolen goods, the more difficult it is to dispose of them profitably. To realize very much when stealing jewelry or artwork, for example, one needs both special training in how to recognize objects of value and special connections for getting rid of them. Stolen cars, too, because of the elaborate regulations of licensing and serial numbers, can only be profitably gotten rid of by tying in with a smoothly functioning criminal organization.

Any new criminal starting out on a life of crime has a lot to learn and many connections to make. Most novice criminals cannot make it very far in the crime world for exactly the same sort of reasons that most people in legitimate business never make it to the level of corporation executive. The average robbery nets less than $100, which is not exactly a fast way to get rich. Crime is a competitive world, too, as soon as one goes into it seriously in order to make a good living from it. Part of this is a kind of market effect, a process of sheer supply and demand. The more stolen goods show up at the fence, the less will be paid for them. The more criminals involved in any particular racket, the less take there will be for any one of them. Established criminals have no reason to want to help just anyone who wants to learn the trade and acquire the necessary connections. Hence, many novice criminals are simply "flunked out"; there isn't enough room for them in the world of crime.

Perhaps it is for this reason that crime rates peak for the youth population between ages fifteen and eighteen, and drop off rapidly thereafter. Youths at this age are not seriously committed to crime; they do not know much about the criminal world. They don't have much money of their own, or very much sense of what one can do with money. Small robberies may seem like an easy way to get a few luxuries. Auto theft, for example, is especially high at this age. But teenagers have little sense of how to market a stolen car; they are more likely to joy-ride around in it for a while and then abandon it. Obviously one can't make much of a living out of this sort of thing. If the crime rate starts dropping off in the late teens, and reaches a fairly low level by the age of thirty, it is not so much because of the effectiveness of the law enforcement system but simply because most youthful criminals wash out of a career in crime. (Again, as I mentioned, most crimes are committed by males, and that is the occupational pattern to pay attention to here.) Crime simply doesn't bring in enough income for them and they are forced to turn to something else to make their way in the adult world.

In the final analysis, the problem of crime, and its solution as well, is built much more deeply into the social structure than common sense would lead us to believe. Crime is so difficult to control because it is produced by large-scale social processes. The police, the courts, the prisons, the parole system are not very effective in counteracting criminality, and their very ineffectiveness seems foreordained by their largely ritualistic nature. Yet on the other side, crime has its own limitations. It works best the more it is organized, but the more organized it becomes, the more it becomes law-abiding and self-disciplining in its own ways. Individual criminals get squeezed out by the competitiveness of the world of crime itself, forced back into the world of ordinary society and its laws, whether they like it or not. Crime and society sway back and forth on this dialectic of opposing ironies.

46

On Being Sane in Insane Places

D. L. ROSENHAN

Rosenhan reveals a push toward deviance not noted by most theorists—the labeling of behavior as not normal. The question being raised in this reading is, "What is normality as opposed to abnormality?"

As you read this selection, ask yourself the following questions:

1. *What were the main factors affecting the treatment of the patients?*
2. *What other institutions are affected by labeling? Why do you say this?*
3. *Which activities would you remove from those considered abnormal?*

GLOSSARY   **Pseudopatient** A pretend patient.   **Type 1 error** A false-negative error; for example, diagnosing a sick person as healthy.   **Type 2 error** A false-positive error; for example, diagnosing a healthy person as sick. **Depersonalization** Removing a sense of the individual from treatment. **Prima facie evidence** Presumption of fact.   **Veridical** Truthful.

If sanity and insanity exist, how shall we know them?

The question is neither capricious nor itself insane. However much we may be personally convinced that we can tell the normal from the abnormal, the evidence is simply not compelling. It is commonplace, for example, to read about murder trials wherein eminent psychiatrists for the defense are contradicted by equally eminent psychiatrists for the prosecution on the matter of the defendant's sanity. More generally, there are a great deal of conflicting data on the reliability, utility, and meaning of such terms as "sanity," "insanity," "mental illness," and "schizophrenia."[1] Finally, as early as 1934, Benedict suggested that normality and abnormality are not universal.[2] What is viewed as normal in one culture may be seen as quite aberrant in another. Thus, notions of normality and abnormality may not be quite as accurate as people believe they are.

To raise questions regarding normality and abnormality is in no way to question the fact that some behaviors are deviant or odd. Murder is deviant. So, too, are hallucinations. Nor does raising such questions deny the existence of the personal anguish that is often associated with "mental illness." Anxiety and depression exist. Psychological suffering exists. But normality and abnormality, sanity and insanity, and the diagnoses that flow from them may be less substantive than many believe them to be.

At its heart, the question of whether the sane can be distinguished from the insane (and whether degrees of insanity can be distinguished from each

Source: Reprinted by permission from *Science*, Vol. 179, January 1973, pp. 250–258. Copyright © 1973 by the American Association for the Advancement of Science.

other) is a simple matter: do the salient characteristics that lead to diagnoses reside in the patients themselves or in the environments and contexts in which observers find them? From Bleuler, through Kretchmer, through the formulators of the recently revised *Diagnostic and Statistical Manual* of the American Psychiatric Association, the belief has been strong that patients present symptoms, that those symptoms can be categorized, and, implicitly, that the sane are distinguishable from the insane. More recently, however, this belief has been questioned. Based in part on theoretical and anthropological considerations, but also on philosophical, legal, and therapeutic ones, the view has grown that psychological categorization of mental illness is useless at best and downright harmful, misleading, and pejorative at worst. Psychiatric diagnoses, in this view, are in the minds of the observers and are not valid summaries of characteristics displayed by the observed.

Gains can be made in deciding which of these is more nearly accurate by getting normal people (that is, people who do not have, and have never suffered, symptoms of serious psychiatric disorders) admitted to psychiatric hospitals and then determining whether they were discovered to be sane and, if so, how. If the sanity of such pseudopatients were always detected, there would be prima facie evidence that a sane individual can be distinguished from the insane context in which he is found. Normality (and presumably abnormality) is distinct enough that it can be recognized wherever it occurs, for it is carried within the person. If on the other hand, the sanity of the pseudopatients were never discovered, serious difficulties would arise for those who support traditional modes of psychiatric diagnosis. Given that the hospital staff was not incompetent, that the pseudopatient had been behaving as sanely as he had been outside of the hospital, and that it had never been previously suggested that he belonged in a psychiatric hospital, such an unlikely outcome would support the view that psychiatric diagnosis betrays little about the patient but much about the environment in which an observer finds him.

This reading describes such an experiment. Eight sane people gained secret admission to 12 different hospitals. Their diagnostic experiences constitute the data of the first part of this article; the remainder is devoted to a description of their experiences in psychiatric institutions. Too few psychiatrists and psychologists, even those who have worked in such hospitals, know what the experience is like. They rarely talk about it with former patients, perhaps because they distrust information coming from the previously insane. Those who have worked in psychiatric hospitals are likely to have adapted so thoroughly to the settings that they are insensitive to the impact of that experience. And while there have been occasional reports of researchers who submitted themselves to psychiatric hospitalization,[3] these researchers have commonly remained in the hospitals for short periods of time, often with the knowledge of the hospital staff. It is difficult to know the extent to which they were treated like patients or like research colleagues. Nevertheless, their reports about the inside of the psychiatric hospital have been valuable. This reading extends those efforts.

PSEUDOPATIENTS AND THEIR SETTINGS

The eight pseudopatients were a varied group. One was a psychology graduate student in his twenties. The remaining seven were older and "established." Among them were three psychologists, a pediatrician, a psychiatrist, a painter, and a housewife. Three pseudopatients were women, five were men. All of them employed pseudonyms, lest their alleged diagnoses embarrass them later. Those who were in mental health professions alleged another occupation in order to avoid the special attentions that might be accorded by staff, as a matter of courtesy or caution, to ailing colleagues.[4] With the exception of myself (I was the first pseudopatient and my presence was known to the hospital administrator and chief psychologist and, so far as I can tell, to them alone), the presence of pseudopatients and the nature of the research program was not known to the hospital staff.[5]

The settings were similarly varied. In order to generalize the findings, admission into a variety of hospitals was sought. The 12 hospitals in the sample were located in five different states on the East and West coasts. Some were old and shabby, some were quite new. Some were research-oriented, others not. Some had good staff–patient ratios, others were quite understaffed. Only one was a strictly private hospital. All of the others were supported by state or federal funds or, in one instance, by university funds.

After calling the hospital for an appointment, the pseudopatient arrived at the admissions office complaining that he had been hearing voices. Asked what the voices said, he replied that they were often unclear, but as far as he could tell they said "empty," "hollow," and "thud." The voices were unfamiliar and were of the same sex as the pseudopatient. The choice of these symptoms was occasioned by their apparent similarity to existential symptoms. Such symptoms are alleged to arise from painful concerns about the perceived meaninglessness of one's life. It is as if the hallucinating person were saying, "My life is empty and hollow." The choice of these symptoms was also determined by the *absence* of a single report of existential psychoses in the literature.

Beyond alleging the symptoms and falsifying name, vocation, and employment, no further alterations of person, history, or circumstances were made. The significant events of the pseudopatient's life history were presented as they had actually occurred. Relationships with parents and siblings, with spouse and children, with people at work and in school, consistent with the aforementioned exceptions, were described as they were or had been. Frustrations and upsets were described along with joys and satisfactions. These facts are important to remember. If anything, they strongly biased the subsequent results in favor of detecting sanity, since none of their histories or current behaviors were seriously pathological in any way.

Immediately upon admission to the psychiatric ward, the pseudopatient ceased simulating *any* symptoms of abnormality. In some cases, there was a brief period of mild nervousness and anxiety, since none of the pseudopatients really believed that they would be admitted so easily. Indeed, their shared fear was that they would be immediately exposed as frauds and greatly embarrassed. Moreover, many of them had never visited a psychiatric ward; even those who had, nevertheless, had some genuine fears about what might happen to them. Their nervousness, then, was quite appropriate to the novelty of the hospital setting, and it abated rapidly.

Apart from that short-lived nervousness, the pseudopatient behaved on the ward as he "normally" behaved. The pseudopatient spoke to patients and staff as he might ordinarily. Because there is uncommonly little to do on a psychiatric ward, he attempted to engage others in conversation. When asked by staff how he was feeling, he indicated that he was fine, that he no longer experienced symptoms. He responded to instructions from attendants, to calls for medication (which was not swallowed), and to dining-hall instructions. Beyond such activities as were available to him on the admissions ward, he spent his time writing down his observations about the ward, its patients, and the staff. Initially these notes were written "secretly," but as it soon became clear that no one much cared, they were subsequently written on standard tablets of paper in such public places as the day-room. No secret was made of these activities.

The pseudopatient, very much as a true psychiatric patient, entered a hospital with no foreknowledge of when he would be discharged. Each was told that he would have to get out by his own devices, essentially by convincing the staff that he was sane. The psychological stresses associated with hospitalization were considerable, and all but one of the pseudopatients desired to be discharged immediately after being admitted. They were, therefore, motivated not only to behave sanely, but to be paragons of cooperation. That their behavior was in no way disruptive is confirmed by nursing reports, which have been obtained on most of the patients. These reports uniformly indicate that the patients were "friendly," "cooperative," and "exhibited no abnormal indications."

THE NORMAL ARE NOT DETECTABLY SANE

Despite their public "show" of sanity, the pseudo-patients were never detected. Admitted, except in one case, with a diagnosis of schizophrenia,[6] each was discharged with a diagnosis of schizophrenia "in remission." The label "in remission" should in no way be dismissed as a formality, for at no time during any hospitalization had any question been raised about any pseudopatient's simulation. Nor are there any indications in the hospital records that the pseudopatient's status was suspect. Rather, the evidence is strong that, once labeled schizophrenic, the pseudopatient was stuck with that label. If the pseudopatient was to be discharged, he must naturally be "in remission"; but he was not sane, nor, in the institution's view, had he ever been sane.

The uniform failure to recognize sanity cannot be attributed to the quality of the hospitals, for, although there were considerable variations among them, several are considered excellent. Nor can it be alleged that there was simply not enough time to observe the pseudopatients. Length of hospitalization ranged from 7 to 52 days, with an average of 19 days. The pseudopatients were not, in fact, carefully observed, but this failure clearly speaks more to traditions within psychiatric hospitals than to lack of opportunity.

Finally, it cannot be said that the failure to recognize the pseudopatients' sanity was due to the fact that they were not behaving sanely. While there was clearly some tension present in all of them, their daily visitors could detect no serious behavioral consequences—nor, indeed, could other patients. It was quite common for the patients to "detect" the pseudopatients' sanity. During the first three hospitalizations, when accurate counts were kept, 35 of a total of 118 patients on the admissions ward voiced their suspicions, some vigorously. "You're not crazy. You're a journalist, or a professor [referring to the continual notetaking]. You're checking up on the hospital." While most of the patients were reassured by the pseudopatient's insistence that he had been sick before he came in but was fine now, some

continued to believe that the pseudopatient was sane throughout his hospitalization.[7] The fact that the patients often recognized normality when staff did not raises important questions.

Failure to detect sanity during the course of hospitalization may be due to the fact that physicians operate with a strong bias toward what statisticians call the type 2 error. This is to say that physicians are more inclined to call a healthy person sick (a false positive, type 2) than a sick person healthy (a false negative, type 1). The reasons for this are not hard to find: it is clearly more dangerous to misdiagnose illness than health. Better to err on the side of caution, to suspect illness even among the healthy.

But what holds for medicine does not hold equally well for psychiatry. Medical illnesses, while unfortunate, are not commonly pejorative. Psychiatric diagnoses, on the contrary, carry with them personal, legal, and social stigmas.[8] It was therefore important to see whether the tendency toward diagnosing the sane insane could be reversed. The following experiment was arranged at a research and teaching hospital whose staff had heard these findings but doubted that such an error could occur in their hospital. The staff was informed that at some time during the following 3 months, one or more pseudopatients would attempt to be admitted into the psychiatric hospital. Each staff member was asked to rate each patient who presented himself at admission or on the ward according to the likelihood that the patient was a pseudopatient. A 10-point scale was used, with a 1 and 2 reflecting high confidence that the patient was a pseudopatient.

Judgments were obtained on 193 patients who were admitted for psychiatric treatment. All staff who had had sustained contact with or primary responsibility for the patient—attendants, nurses, psychiatrists, physicians, and psychologists—were asked to make judgments. Forty-one patients were alleged, with high confidence, to be pseudopatients by at least one member of the staff. Twenty-three were considered suspect by at least one psychiatrist. Nineteen were suspected by one psychiatrist *and* one other staff member. Actually, no genuine pseudopatient (at least from my group) presented himself during this period.

The experiment is instructive. It indicates that the tendency to designate sane people as insane can be reversed when the stakes (in this case, prestige and diagnostic acumen) are high. But what can be said of the 19 people who were suspected of being "sane" by one psychiatrist and another staff member? Were these people truly "sane," or was it rather the case that in the course of avoiding the type 2 error the staff tended to make more errors of the first sort—calling the crazy "sane"? There is no way of knowing. But one thing is certain: any diagnostic process that lends itself so readily to massive errors of this sort cannot be a very reliable one.

THE STICKINESS OF PSYCHODIAGNOSTIC LABELS

Beyond the tendency to call the healthy sick—a tendency that accounts better for diagnostic behavior on admission than it does for such behavior after a lengthy period of exposure—the data speak to the massive role of labeling in psychiatric assessment. Having once been labeled schizophrenic, there is nothing the pseudopatient can do to overcome the tag. The tag profoundly colors others' perceptions of him and his behavior.

From one viewpoint, these data are hardly surprising, for it has long been known that elements are given meaning by the context in which they occur. Gestalt psychology made this point vigorously, and Asch[9] demonstrated that there are "central" personality traits (such as "warm" versus "cold") which are so powerful that they markedly color the meaning of other information in forming an impression of a given personality.[10] "Insane," "schizophrenic," "manic-depressive," and "crazy" are probably among the most powerful of such central traits. Once a person is designated abnormal, all of his other behaviors and characteristics are colored by that label. Indeed, that label is so powerful that many of the pseudopatients' normal behaviors were overlooked entirely or profoundly misinterpreted. Some examples may clarify this issue.

Earlier I indicated that there were no changes in the pseudopatient's personal history and current status beyond those of name, employment, and, where necessary, vocation. Otherwise, a veridical description of personal history and circumstances was offered. Those circumstances were not psychotic. How were they made consonant with the diagnosis of psychosis? Or were those diagnoses modified in such a way as to bring them into accord with the circumstances of the pseudopatient's life, as described by him?

As far as I can determine, diagnoses were in no way affected by the relative health of the circumstances of a pseudopatient's life. Rather, the reverse occurred: the perception of his circumstances was shaped entirely by the diagnosis. A clear example of such translation is found in the case of a pseudopatient who had had a close relationship with his mother but was rather remote from his father during his early childhood. During adolescence and beyond, however, his father became a close friend, while his relationship with his mother cooled. His present relationship with his wife was characteristically close and warm. Apart from occasional angry exchanges, friction was minimal. The children had rarely been spanked. Surely there is nothing especially pathological about such a history. Indeed, many readers may see a similar pattern in their own experiences, with no markedly deleterious consequences. Observe, however, how such a history was translated in the psychopathological context, this from the case summary prepared after the patient was discharged.

> This white 39-year-old male ... manifests a long history of considerable ambivalence in close relationships, which begins in early childhood. A warm relationship with his mother cools during his adolescence. A distant relationship to his father is described as becoming very intense. Affective stability is absent. His attempts to control emotionality with his wife and children are punctuated by angry outbursts and, in the case of the children, spanking. And while he says that he has several good friends, one senses considerable ambivalence embedded in those relationships also ...

The facts of the case were unintentionally distorted by the staff to achieve consistency with a popular theory of the dynamics of a schizophrenic reaction.[11] Nothing of an ambivalent nature had been described in relations with parents, spouse, or friends. To the extent that ambivalence could be inferred, it was probably not greater than is found in all human relationships. It is true the pseudopatient's relationships with his parents changed over time, but in the ordinary context that would hardly be remarkable—indeed, it might very well be expected. Clearly, the meaning ascribed to his verbalizations (that is, ambivalence, affective instability) was determined by the diagnosis: schizophrenia. An entirely different meaning would have been ascribed if it were known that the man was "normal."

All pseudopatients took extensive notes publicly. Under ordinary circumstances, such behavior would have raised questions in the minds of observers, as, in fact, it did among patients. Indeed, it seemed so certain that the notes would elicit suspicion that elaborate precautions were taken to remove them from the ward each day. But the precautions proved needless. The closest any staff member came to questioning these notes occurred when one pseudopatient asked his physician what kind of medication he was receiving and began to write down the response. "You needn't write it," he was told gently. "If you have trouble remembering, just ask me again."

If no questions were asked of the pseudopatients, how was their writing interpreted? Nursing records for three patients indicate that the writing was seen as an aspect of their pathological behavior. "Patient engages in writing behavior" was the daily nursing comment on one of the pseudopatients who was never questioned about his writing. Given that the patient is in the hospital, he must be psychologically disturbed. And given that he is disturbed, continuous writing must be a behavioral manifestation of that disturbance, perhaps a subset of the compulsive behaviors that are sometimes correlated with schizophrenia.

One tacit characteristic of psychiatric diagnosis is that it locates the sources of aberration within the individual and only rarely within the complex of stimuli that surrounds him. Consequently, behaviors that are stimulated by the environment are commonly misattributed to the patient's disorder. For example, one kindly nurse found a pseudopatient pacing the long hospital corridors. "Nervous, Mr. X?" she asked. "No, bored," he said.

The notes kept by pseudopatients are full of patient behaviors that were misinterpreted by well-intentioned staff. Often enough, a patient would go "berserk" because he had, wittingly or unwittingly, been mistreated by, say, an attendant. A nurse coming upon the scene would rarely inquire even cursorily into the environmental stimuli of the patient's behavior. Rather, she assumed that his upset derived from his pathology, not from his present interactions with other staff members. Occasionally, the staff might assume that the patient's family (especially when they had recently visited) or other patients had stimulated the outburst. But never were the staff found to assume that one of themselves or the structure of the hospital had anything to do with a patient's behavior. One psychiatrist pointed to a group of patients who were sitting outside the cafeteria entrance half an hour before lunchtime. To a group of young residents he indicated that such behavior was characteristic of the oral-acquisitive nature of the syndrome. It seemed not to occur to him that there were very few things to anticipate in a psychiatric hospital besides eating.

A psychiatric label has a life and an influence of its own. Once the impression has been formed that the patient is schizophrenic, the expectation is that he will continue to be schizophrenic. When a sufficient amount of time has passed, during which the patient has done nothing bizarre, he is considered to be in remission and available for discharge. But the label endures beyond discharge, with the unconfirmed expectation that he will behave as a schizophrenic again. Such labels, conferred by mental health professionals, are as influential on the patient as they are on his relatives and friends, and it should not surprise anyone that the diagnosis acts on all of them as a self-fulfilling prophecy. Eventually, the patient himself accepts the diagnosis, with all of its surplus meanings and expectations, and behaves accordingly.

The inferences to be made from these matters are quite simple. Much as Zigler and Phillips have demonstrated that there is enormous overlap in the symptoms presented by patients who have been variously diagnosed,[12] so there is enormous overlap in the behaviors of the sane and the insane. The sane are not "sane" all of the time. We lose our tempers "for no good reason." We are occasionally depressed or anxious, again for no good reason. And we may find it difficult to get along with one or another person—again for no reason that we can specify. Similarly, the insane are not always insane. Indeed, it was the impression of the pseudopatients while living with them that they were sane for long periods of time—that the bizarre behaviors upon which their diagnoses were allegedly predicated constituted only a small fraction of their total behavior. If it makes no sense to label ourselves permanently depressed on the basis of an occasional depression, then it takes better evidence than is presently available to label all patients insane or schizophrenic on the basis of bizarre behaviors or cognitions. It seems more useful, as Mischel[13] has pointed out, to limit our discussion to *behaviors*, the stimuli that provoke them, and their correlates.

It is not known why powerful impressions of personality traits, such as "crazy" or "insane," arise. Conceivably, when the origins of and stimuli that give rise to a behavior are remote or unknown, or when the behavior strikes us as immutable, trait labels regarding the *behavior* arise. When, on the other hand, the origins and stimuli are known and available, discourse is limited to the behavior itself. Thus, I may hallucinate because I am sleeping, or I may hallucinate because I have ingested a peculiar drug. These are termed sleep-induced hallucinations, or dreams, and drug-induced hallucinations, respectively. But when the stimuli to my hallucinations are unknown, that is called craziness, or schizophrenia—as if that inference were somehow as illuminating as the others.

THE EXPERIENCE OF PSYCHIATRIC HOSPITALIZATION

The term "mental illness" is of recent origin. It was coined by people who were humane in their inclinations and who wanted very much to raise the station of (and the public's sympathies toward) the psychologically disturbed from that of witches and "crazies" to one that was akin to the physically ill. And they were at least partially successful, for the treatment of the mentally ill *has* improved considerably over the years. But while treatment has improved, it is doubtful that people really regard the mentally ill in the same way that they view the physically ill. A broken leg is something one recovers from, but mental illness allegedly endures forever.[14] A broken leg does not threaten the observer, but a crazy schizophrenic? There is by now a host of evidence that attitudes toward the mentally ill are characterized by fear, hostility, aloofness, suspicion, and dread.[15] The mentally ill are society's lepers.

That such attitudes infect the general population is perhaps not surprising, only upsetting. But that they affect the professionals—attendants, nurses, physicians, psychologists, and social workers—who treat and deal with the mentally ill is more disconcerting, both because such attitudes are self-evidently pernicious and because they are unwitting. Most mental health professionals would insist that they are sympathetic toward the mentally ill, that they are neither avoidant nor hostile. But it is more likely that an exquisite ambivalence characterizes their relations with psychiatric patients, such that their avowed impulses are only part of their entire attitude. Negative attitudes are there too and can easily be detected. Such attitudes should not surprise us. They are the natural offspring of the labels patients wear and the places in which they are found.

Consider the structure of the typical psychiatric hospital. Staff and patients are strictly segregated. Staff have their own living space, including their dining facilities, bathrooms, and assembly places.

The glassed quarters that contain the professional staff, which the pseudopatients came to call "the cage," sit out on every dayroom. The staff emerge primarily for care-taking purposes—to give medication, to conduct a therapy or group meeting, to instruct or reprimand a patient. Otherwise, staff keep to themselves, almost as if the disorder that afflicts their charges is somehow catching.

So much is patient–staff segregation the rule that, for four public hospitals in which an attempt was made to measure the degree to which staff and patients mingle, it was necessary to use "time out of the staff cage" as the operational measure. While it was not the case that all time spent out of the cage was spent mingling with patients (attendants, for example, would occasionally emerge to watch television in the dayroom), it was the only way in which one could gather reliable data on time for measuring.

The average amount of time spent by attendants outside of the cage was 11.3 percent (range, 3 to 52 percent). This figure does not represent only time spent mingling with patients, but also includes time spent on such chores as folding laundry, supervising patients while they shave, directing ward cleanup, and sending patients to off-ward activities. It was the relatively rare attendant who spent time talking with patients or playing games with them. It proved impossible to obtain a "percent mingling time" for nurses, since the amount of time they spent out of the cage was too brief. Rather, we counted instances of emergence from the cage. On the average, daytime nurses emerged from the cage 11.5 times per shift, including instances when they left the ward entirely (range, 4 to 39 times). Late afternoon and night nurses were even less available, emerging on the average 9.4 times per shift (range, 4 to 41 times). Data on early morning nurses, who arrived usually after midnight and departed at 8 A.M., are not available because patients were asleep during most of this period.

Physicians, especially psychiatrists, were even less available. They were rarely seen on the wards. Quite commonly, they would be seen only when they arrived and departed, with the remaining time being spent in their offices or in the cage. On the average, the physicians emerged on the ward 6.7 times per day (range, 1 to 17 times). It proved difficult to make an accurate estimate in this regard, since physicians often maintained hours that allowed them to come and go at different times.

The hierarchical organization of the psychiatric hospital has been commented on before,[16] but the latent meaning of that kind of organization is worth noting again. Those with the most power have least to do with patients, and those with the least power are most involved with them. Recall, however, that the acquisition of role-appropriate behaviors occurs mainly through the observation of others, with the most powerful having the most influence. Consequently, it is understandable that attendants not only spend more time with patients than do any other members of the staff—that is required by their station in the hierarchy—but also, insofar as they learn from their superiors' behavior, spend as little time with patients as they can. Attendants are seen mainly in the cage, which is where the models, the action, and the power are.

I turn now to a different set of studies, these dealing with staff response to patient-initiated contact. It has long been known that the amount of time a person spends with you can be an index of your significance to him. If he initiates and maintains eye contact, there is reason to believe that he is considering your requests and needs. If he pauses to chat or actually stops and talks, there is added reason to infer that he is individuating you. In four hospitals, the pseudopatient approached the staff member with a request which took the following form: "Pardon me, Mr. [or Dr. or Mrs.] X, could you tell me when I will be eligible for grounds privileges?" (or "… when I will be presented at the staff meeting?" or "… when I am likely to be discharged?"). While the content of the question varied according to the appropriateness of the target and the pseudopatient's (apparent) current needs the form was always a courteous and relevant request for information. Care was taken never to approach a particular member of the staff more than once a day, lest the staff member become suspicious or irritated.

In examining these data, remember that the behavior of the pseudopatients was neither bizarre nor disruptive. One could indeed engage in good conversation with them....

... Minor differences between these four institutions were overwhelmed by the degree to which staff avoided continuing contacts that patients had initiated. By far, their most common response consisted of either a brief response to the question, offered while they were "on the move" and with head averted, or no response at all.

The encounter frequently took the following bizarre form: (pseudopatient) "Pardon me, Dr. X. Could you tell me when I am eligible for grounds privileges?" (physician) "Good morning, Dave. How are you today?" (Moves off without waiting for a response.)

It is instructive to compare these data with data recently obtained at Stanford University. It has been alleged that large and eminent universities are characterized by faculty who are so busy that they have no time for students. For this comparison, a young lady approached individual faculty members who seemed to be walking purposely to some meeting or teaching engagement and asked them the following six questions.

1. Pardon me, could you direct me to Encina Hall?" (at the medical school: "... to the Clinical Research Center?")

2. "Do you know where Fish Annex is?" (there is no Fish Annex at Stanford).

3. "Do you teach here?"

4. "How does one apply for admission to the college?" (at the medical school: "... to the medical school?")

5. "Is it difficult to get in?"

6. "Is there financial aid?"

Without exception ... all of the questions were answered. No matter how rushed they were, all respondents not only maintained eye contact, but stopped to talk. Indeed, many of the respondents went out of their way to direct or take the questioner to the office she was seeking, to try to locate "Fish Annex," or to discuss with her the possibilities of being admitted to the university.

Similar data ... were obtained in the hospital. Here too, the young lady came prepared with six questions. After the first question, however, she remarked to 18 of her respondents... "I'm looking for a psychiatrist," and to 15 others..., "I'm looking for an internist." Ten other respondents received no inserted comment.... The general degree of cooperative responses is considerably higher for these university groups than it was for pseudopatients in psychiatric hospitals. Even so, differences are apparent within the medical school setting. Once having indicated that she was looking for a psychiatrist, the degree of cooperation elicited was less than when she sought an internist.

POWERLESSNESS AND DEPERSONALIZATION

Eye contact and verbal contact reflect concern and individuation: their absence, avoidance and depersonalization. The data I have presented do not do justice to the rich daily encounters that grow up around matters of depersonalization and avoidance. I have records of patients who were beaten by staff for the sin of having initiated verbal contact. During my own experience, for example, one patient was beaten in the presence of other patients for having approached an attendant and told him, "I like you." Occasionally, punishment meted out to patients for misdemeanors seemed so excessive that it could not be justified by the most radical interpretations of psychiatric canon. Nevertheless, they appeared to go unquestioned. Tempers were often short. A patient who had not heard a call for medication would be roundly excoriated, and the morning attendants would often wake patients with, "Come on, you m—f—s, out of bed!"

Neither anecdotal nor "hard" data can convey the overwhelming sense of powerlessness which invades the individual as he is continually exposed to the depersonalization of the psychiatric hospital. It hardly matters *which* psychiatric hospital—the excellent public ones and the very plush private hospital were better than the rural and shabby ones in this regard, but, again, the features that psychiatric hospitals had in common overwhelmed by far their apparent differences.

Powerlessness was evident everywhere. The patient is deprived of many of his legal rights by dint of his psychiatric commitment.[17] He is shorn of credibility by virtue of his psychiatric label. His freedom of movement is restricted. He cannot initiate contact with the staff, but may only respond to such overtures as they make. Personal privacy is minimal. Patient quarters and possessions can be entered and examined by any staff member, for whatever reason. His personal history and anguish is available to any staff member (often including the "gray lady" and "candy striper" volunteer) who chooses to read his folder, regardless of their therapeutic relationship to him. His personal hygiene and waste evacuation are often monitored. The water closets may have no doors.

At times, depersonalization reached such proportions that pseudopatients had the sense that they were invisible, or at least unworthy of account. Upon being admitted, I and other pseudopatients took the initial physical examinations in a semipublic room, where staff members went about their own business as if we were not there.

On the ward, attendants delivered verbal and occasionally serious physical abuse to patients in the presence of other observing patients, some of whom (the pseudopatients) were writing it all down. Abusive behavior, on the other hand, terminated quite abruptly when other staff members were known to be coming. Staff are credible witnesses. Patients are not.

A nurse unbuttoned her uniform to adjust her brassiere in the presence of an entire ward of viewing men. One did not have the sense that she was being seductive. Rather, she didn't notice us. A group of staff persons might point to a patient in the dayroom and discuss him animatedly, as if he were not there.

One illuminating instance of depersonalization and invisibility occurred with regard to medications. All told, the pseudopatients were administered nearly 2100 pills, including Elavil, Stelazine, Compazine, and Thorazine, to name but a few. (That such a variety of medications should have been administered to patients presenting identical symptoms is itself worthy of note.) Only two were swallowed. The rest were either pocketed or deposited in the toilet. The pseudopatients were not alone in this. Although I have no precise records on how many patients rejected their medications, the pseudopatients frequently found the medications of other patients in the toilet before they deposited their own. As long as they were cooperative, their behavior and the pseudopatients' own in this matter, as in other important matters, went unnoticed throughout.

Reactions to such depersonalization among pseudopatients were intense. Although they had come to the hospital as participant observers and were fully aware that they did not "belong," they nevertheless found themselves caught up in and fighting the process of depersonalization. Some examples: a graduate student in psychology asked his wife to bring his textbooks to the hospital so he could "catch up on his homework"—this despite the elaborate precautions taken to conceal his professional association. The same student, who had trained for quite some time to get into the hospital, and who had looked forward to the experience, "remembered" some drag races that he had wanted to see on the weekend and insisted that he be discharged by that time. Another pseudopatient attempted a romance with a nurse. Subsequently he informed the staff that he was applying for admission to graduate school in psychology and was very likely to be admitted, since a graduate professor was one of his regular hospital visitors. The same person began to engage in psychotherapy with other patients—all of this as a way of becoming a person in an impersonal environment.

THE SOURCES OF DEPERSONALIZATION

What are the origins of depersonalization? I have already mentioned two. First are attitudes held by all of us toward the mentally ill—including those who treat them—attitudes characterized by fear, distrust, and horrible expectations on the one hand, and benevolent intentions on the other. Our ambivalence leads, in this instance as in others, to avoidance.

Second, and not entirely separate, the hierarchical structure of the psychiatric hospital facilitates depersonalization. Those who are at the top have least to do with patients, and their behavior inspires the rest of the staff. Average daily contact with psychiatrists, psychologists, residents, and physicians combined ranged from 3.9 to 25.1 minutes, with an overall mean of 6.8 (six pseudopatients over a total of 129 days of hospitalization). Included in this average are time spent in the admissions interview, ward meetings in the presence of a senior staff member, group and individual psychotherapy contacts, case presentation conferences, and discharge meetings. Clearly, patients do not spend much time in interpersonal contact with doctoral staff. And doctoral staff serve as models for nurses and attendants.

There are probably other sources. Psychiatric installations are presently in serious financial straits. Staff shortages are pervasive, staff time at a premium. Something has to give, and that something is patient contact. Yet while financial stresses are realities, too much can be made of them. I have the impression that the psychological forces that result in depersonalization are much stronger than the fiscal ones and that the addition of more staff would not correspondingly improve patient care in this regard. The incidence of staff meetings and the enormous amount of record-keeping on patients, for example, have not been as substantially reduced as has patient contact. Priorities exist, even during hard times. Patient contact is not a significant priority in the traditional psychiatric hospital, and fiscal pressures do not account for this. Avoidance and depersonalization may.

Heavy reliance upon psychotropic medication tacitly contributes to depersonalization by convincing staff that treatment is indeed being conducted and that further patient contact may not be necessary. Even here, however, caution needs to be exercised in understanding the role of psychotropic drugs. If patients were powerful rather than powerless, if they were viewed as interesting individuals rather than diagnostic entities, if they were socially significant rather than social lepers, if their anguish truly and wholly compelled our sympathies and concerns, would we not *seek* contact with them, despite the availability of medications? Perhaps for the pleasure of it all?

THE CONSEQUENCES OF LABELING AND DEPERSONALIZATION

Whenever the ratio of what is known to what needs to be known approaches zero, we tend to invent "knowledge" and assume that we understand more than we actually do. We seem unable to acknowledge that we simply don't know. The needs for diagnosis and remediation of behavioral and emotional problems are enormous. But rather than acknowledge that we are just embarking on understanding, we continue to label patients "schizophrenic," "manic-depressive," and "insane," as if in those words we had captured the essence of understanding. The facts of the matter are that we have known for a long time that diagnoses are often not useful or reliable, but we have nevertheless continued to use them. We now know that we cannot distinguish insanity from sanity. It is depressing to consider how that information will be used.

Not merely depressing, but frightening. How many people, one wonders, are sane but not recognized as such in our psychiatric institutions? How many have been needlessly stripped of their privileges of citizenship, from the right to vote and drive to that of handling their own accounts? How many have feigned insanity in order to avoid the criminal consequences of their behavior, and, conversely, how many would rather stand trial than live interminably in a psychiatric hospital—but are wrongly thought to be mentally ill? How many have been stigmatized by well-intentioned, but nevertheless erroneous, diagnoses? On the last point, recall again that a "type 2 error" in psychiatric diagnosis does not have the same consequences it does in medical diagnosis. A diagnosis of cancer that has been found to be in error is cause for celebration. But psychiatric diagnoses are rarely found to be in error. The label sticks, a mark of inadequacy forever.

Finally, how many patients might be "sane" outside the psychiatric hospital but seem insane in it—not because craziness resides in them, as it were, but because they are responding to a bizarre setting, one that may be unique to institutions which harbor neither people? Goffman calls the process of socialization to such institutions "mortification"—an apt metaphor that includes the processes of depersonalization that have been described here. And while it is impossible to know whether the pseudopatients' responses to these processes are characteristic of all inmates—they were, after all, not real patients—it is difficult to believe that these processes of socialization to a psychiatric hospital provide useful attitudes or habits of response for living in the "real world."

SUMMARY AND CONCLUSIONS

It is clear that we cannot distinguish the sane from the insane in psychiatric hospitals. The hospital itself imposes a special environment in which the meanings of behavior can easily be misunderstood. The consequences to patients hospitalized in such an environment—the powerlessness, depersonalization, segregation, mortification, and self-labeling—seem undoubtedly countertherapeutic.

I do not, even now, understand this problem well enough to perceive solutions. But two matters seem to have some promise. The first concerns the proliferation of community mental health facilities, of crisis intervention centers, of human potential movement, and of behavior therapies that, for all of their own problems, tend to avoid psychiatric labels, to focus on specific problems and behaviors, and to retain the individual in a relatively nonpejorative environment. Clearly, to the extent that we refrain from sending the distressed to insane places,

our impressions of them are less likely to be distorted. (The risk of distorted perceptions, it seems to me, is always present, since we are much more sensitive to an individual's behaviors and verbalizations than we are to the subtle contextual stimuli that often promote them. At issue here is a matter of magnitude. And, as I have shown, the magnitude of distortion is exceedingly high in the extreme context that is a psychiatric hospital.)

The second matter that might prove promising speaks to the need to increase the sensitivity of mental health workers and researchers to the *Catch 22* position of psychiatric patients. Simply reading materials in this area will be of help to some such workers and researchers. For others, directly experiencing the impact of psychiatric hospitalization will be of enormous use. Clearly, further research into the social psychology of such total institutions will both facilitate treatment and deepen understanding.

I and other pseudopatients in the psychiatric setting had distinctly negative reactions. We do not pretend to describe the subjective experiences of true patients. Theirs may be different from ours, particularly with the passage of time and the necessary process of adaptation to one's environment. But we can and do speak to the relatively more objective indices of treatment within the hospital. It could be a mistake, and a very unfortunate one, to consider that what happened to us derived from malice or stupidity on the part of the staff. Quite the contrary, our overwhelming impression of them was of people who really cared, who were committed, and who were uncommonly intelligent. Where they failed, as they sometimes did painfully, it would be more accurate to attribute those failures to the environment in which they, too, found themselves than to personal callousness. Their perceptions and behavior were controlled by the situation, rather than being motivated by a malicious disposition. In a more benign environment, one that was less attached to global diagnosis, their behaviors and judgments might have been more benign and effective.

NOTES

1. ... For an analysis of these artifacts and summaries of the disputes, see J. Zubin, *Annu. Rev. Psychol. 18*, 373 (1967); L. Phillips and J. G. Draguns, *ibid. 22*, 447 (1971).

2. R. Benedict, *J. Gen. Psychol. 10*, 59 (1934).

3. A. Barry, *Bellevue Is a State of Mind* (Harcourt Brace Jovanovich, New York, 1971); I. Belknap, *Human Problems of a State Mental Hospital* (McGraw-Hill, New York, 1956); W. Caudill, F. C. Redlich, H. R. Gilmore, E. B. Brody, *Amer. J. Orthopsychiat. 22*, 314 (1952); A. R. Goldman, R. H. Bohr, T. A. Steinberg, *Prof. Psychol. 1*, 427 (1970); unauthored, *Roche Report 1* (No. 13), 8 (1971).

4. Beyond the personal difficulties that the pseudopatient is likely to experience in the hospital, there are legal and social ones that, combined, require considerable attention before entry....

5. However distasteful such concealment is, it was a necessary first step to examining these questions. Without concealment, there would have been no way to know how valid these experiences were; nor was there any way of knowing whether whatever detections occurred were a tribute to the diagnostic acumen of the staff or to the hospital's rumor network....

6. Interestingly, of the 12 admissions, 11 were diagnosed as schizophrenic and one, with the identical symptomatology, as manic-depressive psychosis. This diagnosis has a more favorable prognosis, and it was given by the only private hospital in our sample. On the relations between social class and psychiatric diagnosis, see A. deB. Hollingshead and F. C. Redlich, *Social Class and Mental Illness: A Community Study* (Wiley, New York, 1958).

7. It is possible, of course, that patients have quite broad latitudes in diagnosis and therefore are inclined to call many people sane, even those whose behavior is patently aberrant. However, although we have no hard data on this matter, it was our distinct impression that this was not the case. In many instances, patients not only singled us out for attention, but came to imitate our behaviors and styles.

8. J. Cumming and E. Cumming, *Community Ment. Health 1*, 135 (1965); A. Farina and K. Ring, *J. Abnorm. Psychol. 70*, 47 (1965); H. E. Freeman and O. G. Simmons, *The Mental Patient Comes Home* (Wiley, New York, 1963); W. J. Johannsen, *Ment. Hygiene 53*, 218 (1969); A. S. Linsky, *Soc. Psychiat. 5*, 1966 (1970).

9. S. E. Asch, *J. Abnorm. Soc. Psychol. 41*, 258 (1946); *Social Psychology* (Prentice-Hall, New York, 1952).

10. See also I. N. Mensh and J. Wishner, *J. Personality 16*, 188 (1947); J. Wishner, *Psychol. Rev. 67*, 96 (1960); J. S. Bruner and R. Tagiuri, in *Handbook of Social Psychology*, G. Lindzey, Ed. (Addison-Wesley, Cambridge Mass., 1954), vol. 2, pp. 634–654; J. S. Bruner, D. Shapiro, R. Tagiuri, in *Person Perception and Interpersonal Behavior*, R. Tagiuri and L. Petrullo, Eds. (Stanford Univ. Press, Stanford, Calif., 1958), pp. 277–288.

11. For an example of a similar self-fulfilling prophecy, in this instance dealing with the "central" trait of intelligence, see R. Rosenthal and L. Jacobson, *Pygmalion in the Classroom* (Holt, Rinehart & Winston, New York, 1968).

12. E. Zigler and L. Phillips, *J. Abnorm. Soc. Psychol. 63*, 69 (1961). See also R. K. Freudenberg and J. P. Robertson, *A.M.A. Arch. Neurol. Psychiatr. 76*, 14 (1956).

13. W. Mischel, *Personality and Assessment* (Wiley, New York, 1968).

14. The most recent and unfortunate instance of this tenet is that of Senator Thomas Eagleton.

15. T. R. Sarbin and J. C. Mancuso, *J. Clin. Consult. Psychol. 35*, 159 (1970); T. R. Sarbin, *ibid. 31*, 447 (1967); J. C. Nunnally, Jr., *Popular Conceptions of Mental Health* (Holt, Rinehart & Winston, New York, 1961).

16. A. H. Stanton and M. S. Schwartz, *The Mental Hospital: A Study of Institutional Participation in Psychiatric Illness and Treatment* (Basic, New York, 1954).

17. D. B. Wexler and S. E. Scoville, *Ariz. Law Rev. 13*, 1 (1971).

47

The Culture of Fear
Why Americans Fear the Wrong Things

BARRY GLASSNER

The author notes a number of factors that many Americans fear. He also explains why these fears are based on falsehoods and, therefore, are unnecessary.

As you read this selection, consider the following questions as guides:

1. *Consider each fear in turn and indicate the reasons that you agree or disagree with the author's contention that it is groundless.*
2. *Think of items that should be added to the list of real fears. List other false fears.*
3. *If a fear is false, what should or can be done about the problem to which it refers?*

Why are so many fears in the air, and so many of them unfounded? Why, as crime rates plunged throughout the 1990s, did two-thirds of Americans believe they were soaring? How did it come about that by mid-decade 62 percent of us described ourselves as "truly desperate" about crime—almost twice as many as in the late 1980s, when crime rates were higher? Why, on a survey in 1997, when the crime rate had already fallen for a half dozen consecutive years, did more than half of us disagree with the statement "This country is finally beginning to make some progress in solving the crime problem"?[1]

In the late 1990s the number of drug users had decreased by half compared to a decade earlier; almost two-thirds of high school seniors had never used any illegal drugs, even marijuana. So why did a majority of adults rank drug abuse as the greatest danger to America's youth? Why did nine out of ten believe the drug problem is out of control, and only one in six believe the country was making progress?[2]

Give us a happy ending and we write a new disaster story. In the late 1990s the unemployment rate was below 5 percent for the first time in a quarter century. People who had been pounding the pavement for years could finally get work. Yet pundits warned of imminent economic disaster. They predicted inflation would take off, just as they had a few years earlier—also erroneously—when the unemployment rate dipped below 6 percent.[3]

Source: From *The Culture of Fear: Why Americans Are Afraid of the Wrong Things* by Barry Glassner. New York: Basic Books, 1998, pp. x–xxv.

INTRODUCTION

We compound our worries beyond all reason. Life expectancy in the United States has doubled during the twentieth century. We are better able to cure and control diseases than any other civilization in history. Yet we hear that phenomenal numbers of us are dreadfully ill.... The scope of our health fears seems limitless. Besides worrying disproportionately about legitimate ailments and prematurely about would-be diseases, we continue to fret over already refuted dangers....

KILLER KIDS

When we are not worrying about deadly diseases we worry about homicidal strangers. Every few months for the past several years it seems we discover a new category of people to fear: government thugs in Waco, sadistic cops on Los Angeles freeways and in Brooklyn police stations, mass-murdering youths in small towns all over the country. A single anomalous event can provide us with multiple groups of people to fear. After the 1995 explosion at the federal building in Oklahoma City first we panicked about Arabs. "Knowing that the car bomb indicates Middle Eastern terrorists at work, it's safe to assume that their goal is to promote free-floating fear and a measure of anarchy, thereby disrupting American life," a *New York Post* editorial asserted. "Whatever we are doing to destroy Mideast terrorism, the chief terrorist threat against Americans, has not been working," wrote A. M. Rosenthal in the *New York Times*.[4]

When it turned out that the bombers were young white guys from middle America, two more groups instantly became spooky: right-wing radio talk show hosts who criticize the government—depicted by President Bill Clinton as "purveyors of hatred and division"—and members of militias. No group of disgruntled men was too ragtag not to warrant big, prophetic news stories.[5]

We have managed to convince ourselves that just about every young American male is a potential mass murderer—a remarkable achievement, considering the steep downward trend in youth crime throughout the 1990s. Faced year after year with comforting statistics, we either ignore them—adult Americans estimate that people under eighteen commit about half of all violent crimes when the actual number is 13 percent—or recast them as "The Lull Before the Storm"... (*Newsweek* headline).[6]

The more things improve, the more pessimistic we become. Violence-related deaths at the nation's schools dropped to a record low during the 1996–97 academic year (19 deaths out of 54 million children), and only one in ten public schools reported *any* serious crime. Yet *Time* and *U.S. News & World Report* both ran headlines in 1996 referring to "Teenage Time Bombs." In a nation of "Children Without Souls" (another *Time* headline that year), "America's beleaguered cities are about to be victimized by a paradigm shattering wave of ultraviolent, morally vacuous young people some call 'the super-predators,'" William Bennett, the former Secretary of Education, and John DiIulio, a criminologist, forecast in a book published in 1996.[7]

Instead of the arrival of superpredators, violence by urban youths continued to decline. So we went looking elsewhere for proof that heinous behavior by young people was "becoming increasingly more commonplace in America" (CNN). After a sixteen-year-old in Pearl, Mississippi, and a fourteen-year-old in West Paducah, Kentucky, went on shooting sprees in late 1997, killing five of their classmates and wounding twelve others, these isolated incidents were taken as evidence of "an epidemic of seemingly depraved adolescent murderers" (Geraldo Rivera). Three months later in March 1998 all sense of proportion vanished after two boys ages eleven and thirteen killed four students and a teacher in Jonesboro, Arkansas. No longer, we learned in *Time*, was it "unusual for kids to get back at the world with live ammunition." When a child psychologist on NBC's "Today" show advised parents to reassure their children that shootings at schools are rare, reporter Ann Curry corrected him. "But this is the fourth case since October," she said.[8]

Over the next couple of months young people failed to accommodate the trend hawkers. None committed mass murder.... Yet given what had happened in Mississippi, Kentucky, Arkansas, and Oregon, could anyone doubt that today's youths are "more likely to pull a gun than make a fist," as Katie Couric declared on the "Today" show?[9]

ROOSEVELT WAS WRONG

We had better learn to doubt our inflated fears before they destroy us. Valid fears have their place; they cue us to danger. False and over-drawn fears only cause hardship.

Even concerns about real dangers, when blown out of proportion, do demonstrable harm. Take the fear of cancer. Many Americans overestimate the prevalence of the disease, underestimate the odds of surviving it, and put themselves at greater risk as a result....[10]

Still more ironic, if harder to measure, are the adverse consequences of public panics. Exaggerated perceptions of the risks of cancer at least produce beneficial by-products, such as bountiful funding for research and treatment of this leading cause of death. When it comes to large-scale panics, however, it is difficult to see how potential victims benefit from the frenzy. Did panics a few years ago over sexual assaults on children by preschool teachers and priests leave children better off? Or did they prompt teachers and clergy to maintain excessive distance from children in their care, as social scientists and journalists who have studied the panics suggest? How well can care givers do their jobs when regulatory agencies, teachers' unions, and archdioceses explicitly prohibit them from any physical contact with children, even kindhearted hugs?[11] Was it a good thing for children and parents that male day care providers left the profession for fear of being falsely accused of sex crimes?...

We all pay one of the costs of panics: huge sums of money go to waste. Hysteria over the ritual abuse of children cost billions of dollars in police investigations, trials, and imprisonments. Men and women went to jail for years "on the basis of some of the most fantastic claims ever presented to an American jury," as Dorothy Rabinowitz of the *Wall Street Journal* demonstrated in a series of investigative articles for which she became a Pulitzer Prize finalist in 1996. Across the nation expensive surveillance programs were implemented to protect children from fiends who reside primarily in the imaginations of adults.[12]

The price tag for our panic about overall crime has grown so monumental that even law-and-order zealots find it hard to defend. The criminal justice system costs Americans close to $100 billion a year, most of which goes to police and prisons. In California we spend more on jails than on higher education. Yet increases in the number of police and prison cells do not correlate consistently with reductions in the number of serious crimes committed. Criminologists who study reductions in homicide rates, for instance, find little difference between cities that substantially expand their police forces and prison capacity and others that do not.[13]

The turnabout in domestic public spending over the past quarter century, from child welfare and antipoverty programs to incarceration, did not even produce reductions in *fear* of crime. Increasing the number of cops and jails arguably has the opposite effect: it suggests that the crime problem is all the more out of control.[14]

Panic-driven public spending generates over the long term a pathology akin to one found in drug addicts. The more money and attention we fritter away on our compulsions, the less we have available for our real needs, which consequently grow larger. While fortunes are being spent to protect children from dangers that few ever encounter, approximately 11 million children lack health insurance, 12 million are malnourished, and rates of illiteracy are increasing.[15]

I do not contend, as did President Roosevelt in 1933, that "the only thing we have to fear is fear itself." My point is that we often fear the wrong things....

One of the paradoxes of a culture of fear is that serious problems remain widely ignored even though they give rise to precisely the dangers that the populace most abhors....

TWO EASY EXPLANATIONS

In the following discussion I will try to answer two questions: Why are Americans so fearful lately, and why are our fears so often misplaced? To both questions the same two-word answer is commonly given by scholars and journalists: premillennial tensions. The final years of a millennium and the early years of a new millennium provoke mass anxiety and ill reasoning, the argument goes. So momentous does the calendric change seem, the populace cannot keep its wits about it.... In a classic study thirty years ago Alan Kerckhoff and Kurt Back pointed out that "the belief in a tangible threat makes it possible to explain and justify one's sense of discomfort."[16]...

Another popular explanation blames the news media. We have so many fears, many of them off-base, the argument goes, because the media bombard us with sensationalistic stories designed to increase ratings. This explanation, sometimes called the media-effects theory, is less simplistic than the millennium hypothesis and contains sizable kernels of truth. When researchers from Emory University computed the levels of coverage of various health dangers in popular magazines and newspapers they discovered an inverse relationship: much less space was devoted to several of the major causes of death than to some uncommon causes. The leading cause of death, heart disease, received approximately the same amount of coverage as the eleventh-ranked cause of death, homicide. They found a similar inverse relationship in coverage of risk factors associated with serious illness and death. The lowest-ranking risk factor, drug use, received nearly as much attention as the second-ranked risk factor, diet and exercise.[17]

Disproportionate coverage in the news media plainly has effects on readers and viewers.... Asked in a national poll why they believe the country has a serious crime problem, 76 percent of people cited stories they had seen in the media. Only 22 percent cited personal experience.[18]

When professors Robert Blendon and John Young of Harvard analyzed forty-seven surveys about drug abuse conducted between 1978 and 1997, they too discovered that the news media, rather than personal experience, provide Americans with their predominant fears.[19]...

Television news programs survive on scares. On local newscasts, where producers live by the dictum "if it bleeds, it leads," drug, crime, and disaster stories make up most of the news portion of the broadcasts. Evening newscasts on the major networks are somewhat less bloody, but between 1990 and 1998, when the nation's murder rate declined by 20 percent, the number of murder stories on network newscasts increased 600 percent (*not* counting stories about O. J. Simpson).[20]

After the dinnertime newscasts the networks broadcast newsmagazines, whose guiding principle seems to be that no danger is too small to magnify into a national nightmare.... A wide array of groups, including businesses, advocacy organizations, religious sects, and political parties, promote and profit from scares. News organizations are distinguished from other fear-mongering groups because they sometimes bite the scare that feeds them.

A group that raises money for research into a particular disease is not likely to negate concerns about that disease. A company that sells alarm systems is not about to call attention to the fact that crime is down. News organizations, on the other hand, periodically allay the very fears they arouse to lure audiences....

Several major newspapers parted from the pack in other ways. *USA Today* and the *Washington Post*, for instance, made sure their readers knew that what should worry them is the availability of guns. *USA Today* ran news stories explaining that easy access to guns in homes accounted for increases in the number of juvenile arrests for homicide in rural areas during the 1990s.... *USA Today* ran an op-ed piece proposing legal parameters for gun ownership akin to those for the use of alcohol and motor vehicles. And the paper published its own editorial in support of laws that require gun owners to lock their guns or keep them in locked containers. Adopted at that time by only fifteen states, the laws had reduced the number of deaths among children in those states by 23 percent.[21]

The *Washington Post*, meanwhile, published an excellent investigative piece by reporter Sharon

Walsh showing that guns increasingly were being marketed to teenagers and children.... "Seems like only yesterday that your father brought you here for the first time," reads the copy beside a photo of a child aiming a handgun, his father by his side. "Those sure were the good times—just you, dad and his Smith & Wesson."[22]

As a social scientist I am impressed and somewhat embarrassed to find that journalists, more often than media scholars, identify the jugglery involved in making small hazards appear huge and huge hazards disappear from sight. Take, for example, the scare several years ago over the Ebola virus.... A report by *Dateline NBC* on deaths in Zaire, for instance, interspersed clips from *Outbreak*, a movie whose plot involves a lethal virus that threatens to kill the entire U.S. population. Alternating between Dustin Hoffman's character exclaiming, "We can't stop it!" and real-life science writer Laurie Garrett, author of *The Coming Plague*, proclaiming that "HIV is not an aberration ... it's part of a trend," *Dateline*'s report gave the impression that swarms of epidemics were on their way....[23]

"It is one of the ironies of the analysis of alarmists such as Preston that they are all too willing to point out the limitations of human beings, but they neglect to point out the limitations of microscopic life forms," Gladwell notes....[24]

Among my personal favorites is an article published in 1996 titled "Fright by the Numbers," in which reporter Cynthia Crossen rebuts a cover story in *Time* magazine on prostate cancer. One in five men will get the disease, *Time* thundered. "That's scary. But it's also a lifetime risk—the accumulated risk over some 80 years of life," Crossen responds. A forty-year-old's chance of coming down with (not dying of) prostate cancer in the next ten years is 1 in 1,000, she goes on to report. His odds rise to 1 in 100 over twenty years. Even by the time he's seventy, he has only a 1 in 20 chance of *any* kind of cancer, including prostate.[25]

In the same article Crossen counters other alarmist claims as well, such as the much-repeated pronouncement that one in three Americans is obese. The number actually refers to how many are overweight, a less serious condition. Fewer are *obese*

(a term that is less than objective itself), variously defined as 20 to 40 percent above ideal body weight as determined by current standards....[26]

MORALITY AND MARKETING

From a psychological point of view extreme fear and outrage are often projections. Consider, for example, the panic over violence against children. By failing to provide adequate education, nutrition, housing, parenting, medical services, and child care over the past couple of decades we have done the nation's children immense harm. Yet we project our guilt onto a cavalcade of bogeypeople—pedophile preschool teachers, preteen mass murderers, and homicidal au pairs, to name only a few....[27]

Diverse groups used the ritual-abuse scares to diverse ends. Well-known feminists such as Gloria Steinem and Catharine MacKinnon took up the cause, depicting ritually abused children as living proof of the ravages of patriarchy and the need for fundamental social reform.[28] This was far from the only time feminist spokeswomen have mongered fears about sinister breeds of men who exist in nowhere near the high numbers they allege....

Within public discourse fears proliferate through a process of exchange. It is from crosscurrents of scares and counterscares that the culture of fear swells ever larger.... Samuel Taylor Coleridge was right when he claimed, "In politics, what begins in fear usually ends up in folly." Political activists are more inclined, though, to heed an observation from Richard Nixon: "People react to fear, not love. They don't teach that in Sunday school, but it's true." That principle, which guided the late president's political strategy throughout his career, is the sine qua non of contemporary political campaigning. Marketers of products and services ranging from car alarms to TV news programs have taken it to heart as well.[29]

The short answer to why Americans harbor so many misbegotten fears is that immense power and money await those who tap into our moral insecurities and supply us with symbolic substitutes....

NOTES

1. Crime data here and throughout are from reports of the Bureau of Justice Statistics unless otherwise noted. Fear of crime: Esther Madriz, *Nothing Bad Happens to Good Girls* (Berkeley: University of California Press, 1997), Ch. 1; and Richard Morin, "As Crime Rate Falls, Fears Persist," *Washington Post* National Edition, 16 June 1997, p. 35; David Whitman, "Believing the Good News," *U.S. News & World Report*, 5 January 1998, pp. 45–46.

2. Eva Bertram, Morris Blachman et al., *Drug War Politics* (Berkeley: University of California Press, 1996), p. 10; Mike Males, *Scapegoat Generation* (Monroe, ME: Common Courage Press, 1996), ch. 6; Karen Peterson, "Survey: Teen Drug Use Declines," *USA Today*, 19 June 1998, p. A6; Robert Blendon and John Young, "The Public and the War on Illicit Drugs," *Journal of the American Medical Association* 279 (18 March 1998): 827–32. In presenting these statistics and others I am aware of a seeming paradox: I criticize the abuse of statistics by fearmongering politicians, journalists, and others but hand down precise-sounding numbers myself. Yet to eschew all estimates because some are used inappropriately or do not withstand scrutiny would be as foolhardy as ignoring all medical advice because some doctors are quacks. Readers can be assured I have interrogated the statistics presented here as factual. As notes throughout the book make clear, I have tried to rely on research that appears in peer-reviewed scholarly journals. Where this was not possible or sufficient, I traced numbers back to their sources, investigated the research methodology utilized to produce them, or conducted searches of the popular and scientific literature for critical commentaries and conflicting findings.

3. Bob Herbert, "Bogeyman Economics," *New York Times*, 4 April 1997, p. A15; Doug Henwood, "Alarming Drop in Unemployment," *Extra*, September 1994, pp. 16–17; Christopher Shea, "Low Inflation and Low Unemployment Spur Economists to Debate 'Natural Rate' Theory," *Chronicle of Higher Education*, 24 October 1997, p. A13.

4. Jim Naureckas, "The Jihad That Wasn't," *Extra*, July 1995, pp. 6–10, 20 (contains quotes). See also Edward Said, "A Devil Theory of Islam," *Nation*, 12 August 1996, pp. 28–32.

5. Lewis Lapham, "Seen but Not Heard," *Harper's*, July 1995, pp. 29–36 (contains Clinton quote). See also Robin Wright and Ronald Ostrow, "Illusion of Immunity Is Shattered," *Los Angeles Times*, 20 April 1995, pp. A1, 18; Jack Germond and Jules Witcover, "Making the Angry White Males Angrier," column syndicated by Tribune Media Services, May 1995; and articles by James Bennet and Michael Janofsky in the *New York Times*, May 1995.

6. Tom Morganthau, "The Lull Before the Storm?" *Newsweek*, 4 December 1995, pp. 40–42; Mike Males, "Wild in Deceit," *Extra*, March 1996, pp. 7–9; *Progressive*, July 1997, p. 9 (contains Clinton quote); Robin Templeton, "First, We Kill All the 11-Year-Olds," *Salon*, 27 May 1998.

7. Statistics from "Violence and Discipline Problems in U.S. Public Schools: 1996–97," National Center on Education Statistics, U.S. Department of Education Washington, DC, March 1998; CNN, "Early Prime," 2 December 1997; and Tamar Lewin, "Despite Recent Carnage, School Violence Is Not on Rise," *New York Times*, 3 December 1997, p. A14. Headlines: *Time*, 15 January 1996; *U.S. News & World Report*, 25 March 1996; Margaret Carlson, "Children Without Souls," *Time*, 2 December 1996, p. 70; William J. Bennett, John J. DiIulio, and John Walters, *Body Count* (New York: Simon & Schuster, 1996).

8. CNN, "Talkback Live," 2 December 1997; CNN, "The Geraldo Rivera Show," 11 December 1997; Richard Lacayo, "Toward the Root of Evil," *Time*, 6 April 1998, pp. 38–39; NBC, "Today," 25 March 1998. See also Rick Bragg, "Forgiveness, After 3 Die in Shootings in Kentucky," *New York Times*, 3 December 1997, p. A14; Maureen Downey, "Kids and Violence," 28 March 1998, *Atlanta Journal and Constitution*, p. A12.

9. Jocelyn Stewart, "Schools Learn to Take Threats More Seriously," *Los Angeles Times*, 11 May 1998, pp. A1, 17; "Kindergarten Student Faces Gun Charges," *New York Times*, 11 May 1998, p. A11; Rick Bragg, "Jonesboro Dazed by Its Darkest Day" and "Past Victims Relive Pain as Tragedy Is Repeated," *New York Times*, 18 April 1998, p. A7, and idem, 25 May 1998, p. A8. Remaining quotes are from Tamar Lewin, "More Victims and Less Sense in

Shootings," *New York Times*, 22 May 1998, p. A20; NPR, "All Things Considered," 22 May 1998; NBC, "Today," 25 March 1998. See also Mike Males, "Who's Really Killing Our Schoolkids," *Los Angeles Times*, 31 May 1998, pp. M1, 3; Michael Sniffen, "Youth Crime Fell in 1997, Reno Says," Associated Press, 20 November 1998.

10. Overestimation of breast cancer: Willam C. Black et al., "Perceptions of Breast Cancer Risk and Screening Effectiveness in Women Younger Than 50," *Journal of the National Cancer Institute* 87 (1995): 720–31; B. Smith et al., "Perception of Breast Cancer Risk Among Women in Breast and Family History of Breast Cancer," *Surgery* 120 (1996): 297–303. Fear and avoidance: Steven Berman and Abraham Wandersman, "Fear of Cancer and Knowledge of Cancer," *Social Science and Medicine* 31 (1990): 81–90; S. Benedict et al., "Breast Cancer Detection by Daughters of Women with Breast Cancer," *Cancer Practice* 5 (1997): 213–19; M. Muir et al., "Health Promotion and Early Detection of Cancer in Older Adults," *Cancer Oncology Nursing Journal* 7 (1997): 82–89. For a conflicting finding see Kevin McCaul et al., "Breast Cancer Worry and Screening," *Health Psychology* 15 (1996): 430–33.

11. Philip Jenkins, *Pedophiles and Priests* (New York: Oxford University Press, 1996), see esp. Ch. 10; Debbie Nathan and Michael Snedeker, *Satan's Silence* (New York: Basic Books, 1995), see esp. Ch. 6; Jeffrey Victor, "The Danger of Moral Panics," *Skeptic* 3 (1995): 44–51. See also Noelle Oxenhandler, "The Eros of Parenthood," *Family Therapy Networker* (May 1996): 17–19.

12. Dorothy Rabinowitz, "A Darkness in Massachusetts," *Wall Street Journal*, 30 January 1995, p. A20 (contains quote); "Back in Wenatchee" (unsigned editorial), *Wall Street Journal*, 20 June 1996, p. A18; Dorothy Rabinowitz, "Justice in Massachusetts," *Wall Street Journal*, 13 May 1997, p. A19. See also Nathan and Snedeker, *Satan's Silence*; James Beaver, "The Myth of Repressed Memory," *Journal of Criminal Law and Criminology*, 86 (1996): 596–607; Kathryn Lyon, *Witch Hunt* (New York: Avon, 1998); Pam Belluck, "'Memory' Therapy Leads to a Lawsuit and Big Settlement," *New York Times*, 6 November 1997, pp. A1, 10.

13. Elliott Currie, *Crime and Punishment in America* (New York: Metropolitan, 1998); Tony Pate et al., *Reducing*

Fear of Crime in Houston and Newark (Washington, DC: Police Foundation, 1986); Steven Donziger, *The Real War on Crime* (New York: HarperCollins, 1996); Christina Johns, *Power, Ideology and the War on Drugs* (New York: Praeger, 1992); John Irwin et al., "Fanning the Flames of Fear," *Crime and Delinquency* 44 (1998): 32–48.

14. Steven Donziger, "Fear, Crime and Punishment in the U.S.," *Tikkun* 12 (1996): 24–27, 77.

15. Peter Budetti, "Health Insurance for Children," *New England Journal of Medicine* 338 (1998): 541–42; Eileen Smith, "Drugs Top Adult Fears for Kids' Well-being," *USA Today*, 9 December 1997, p. D1. Literacy statistic: Adult Literacy Service.

16. Alan Kerckhoff and Kurt Back, *The June Bug* (New York: Appleton-Century-Crofts, 1968), see esp. pp. 160–61.

17. Karen Frost, Erica Frank et al., "Relative Risk in the News Media," *American Journal of Public Health* 87 (1997): 842–45. Media-effects theory: Nancy Signorielli and Michael Morgan, eds., *Cultivation Analysis* (Newbury Park, CA: Sage, 1990); Jennings Bryant and Dolf Zillman, eds., *Media Effects* (Hillsdale, NJ: Erlbaum, 1994); Ronald Jacobs, "Producing the News, Producing the Crisis," *Media, Culture and Society* 18 (1996): 373–97.

18. Madriz, *Nothing Bad Happens to Good Girls*, see esp. pp. 111–14; David Whitman and Margaret Loftus, "Things Are Getting Better? Who Knew," *U.S. News & World Report*, 16 December 1996, pp. 30–32.

19. Blendon and Young, "War on Illicit Drugs." See also Ted Chiricos et al., "Crime, News and Fear of Crime," *Social Problems* 44 (1997): 342–57.

20. Steven Stark, "Local News: The Biggest Scandal on TV," *Washington Monthly* (June 1997): 38–41; Barbara Bliss Osborn, "If It Bleeds, It Leads," *Extra*, September–October 1994, p. 15; Jenkins, *Pedophiles and Priests*, pp. 68–71; "It's Murder," *USA Today*, 20 April 1998, p. D2; Lawrence Grossman, "Does Local TV News Need a National Nanny?" *Columbia Journalism Review* (May 1998): 33.

21. "Licensing Can Protect," *USA Today*, 7 April 1998, p. A11; Jonathan Kellerman, "Few Surprises When It Comes to Violence," *USA Today*, 27 March 1998, p. A13; Gary Fields, "Juvenile Homicide Arrest Rate on Rise in Rural USA," *USA Today*,

26 March 1998, p. A11; Karen Peterson and Glenn O'Neal, "Society More Violent, So Are Its Children," *USA Today*, 25 March 1998, p. A3; Scott Bowles, "Armed, Alienated and Adolescent," *USA Today*, 26 March 1998, p. A9. Similar suggestions about guns appear in Jonathan Alter, "Harnessing the Hysteria," *Newsweek*, 6 April 1998, p. 27.

22. Sharon Walsh, "Gun Sellers Look to Future—Children," *Washington Post*, 28 March 1998, pp. A1, 2.

23. John Schwartz, "An Outbreak of Medical Myths," *Washington Post* National Edition, 22 May 1995, p. 38.

24. Richard Preston, *The Hot Zone* (New York: Random House, 1994); Malcolm Gladwell, "The Plague Year," *New Republic*, 17 July 1995, p. 40.

25. Erik Larson, "A False Crisis: How Workplace Violence Became a Hot Issue," *Wall Street Journal*, 13 October 1994, pp. A1, 8; Cynthia Crossen, "Fright by the Numbers," *Wall Street Journal*, 11 April 1996, pp. B1, 8. See also G. Pascal

Zachary, "Junk History," *Wall Street Journal*, 19 September 1997, pp. A1, 6.

26. On variable definitions of obesity see also Werner Cahnman, "The Stigma of Obesity," *Sociological Quarterly* 9 (1968): 283–99; Susan Bordo, *Unbearable Weight* (Berkeley: University of California Press, 1993); Joan Chrisler, "Politics and Women's Weight," *Feminism and Psychology* 6 (1996): 181–84.

27. See Marina Warner, "Peroxide Mug-shot," *London Review of Books*, 1 January 1998, pp. 10–11.

28. Nathan and Snedeker, *Satan's Silence* (quote from p. 240). See also David Bromley, "Satanism: The New Cult Scare," in James Richardson et al., eds., *The Satanism Scare* (Hawthorne, NY: Aldine de Gruyter, 1991), pp. 49–71.

29. Henry Nelson Coleridge, ed., *Specimens of the Table Talk of the Late Samuel Taylor Coleridge* (London: J. Murray, 1935), entry for 5 October 1830. Nixon quote cited in William Safire, *Before the Fall* (New York: Doubleday, 1975), Prologue.

48

How Can We Solve the Problem of Crime?

JAMES CRONE

James Crone believes that the amount of crime can be lessened by improving social conditions for the underclass. These suggestions, however, do not deal with the large amount of crime that is committed in the course of one's business by people who have good jobs and wages—white-collar crime. What do you believe can be done to lessen this very pervasive crime which has many victims?

As you read this selection, consider the following questions as guides:

1. *Do you believe that the front-end tasks mentioned by the author as a means to lessen crime would work? Why or why not?*
2. *Do you believe that the back-end tasks mentioned by the author as a means to lessen crime would work? Why or why not?*
3. *What are some of the things you would do to lessen the crime rate? Any other suggestions for dealing with white-collar crime?*

... WHAT CAN WE DO?

There are a number of actions we can take to help us decrease crime because we will help people have more legitimate opportunities to get ahead....

Public Education and Opportunities

One step that we can take to give people more legal opportunities is to give them an excellent public education from Grade 1 through Grade 12.... I mentioned that we could have a much better public education system by hiring only certified teachers, hiring more teachers so that class sizes are smaller (especially for low-income children who need extra help), creating smaller schools so that a higher percentage of the student body can take part in the extracurricular activities that go on at school and hence feel like a part of the school community, making sure that we have well-maintained schools, and making sure that the teachers and students have the equipment they need to do a good job such as having enough chalk, pencils, paper, textbooks, microscopes, computers, and band instruments. As young Americans get to attend good-quality public schools and get a firm grounding during the first 12 years of their educational careers, they will be more prepared to pursue legal opportunities and less likely to resort to illegal opportunities, thereby reducing crime.

More Jobs and Opportunities

If people have more legal opportunities, they will be prepared to take on more legal jobs. It follows

Source: Crone, James. 2007. *"How Can We Solve Our Social Problems,"* Pine Forge Press, pp. 121–128.

that we will need to provide more legal jobs for people.... during the first few years of the 21st century, the United States has had an overall unemployment rate of between 5% and 6%. People living in low-income areas of our country (e.g., inner-city neighborhoods, rural areas, reservations) typically face even higher rates of unemployment. We need to find ways to create more jobs, especially more decent-paying jobs. This leads to the following crucial question: How do we do that?

There are a number of things we can do to connect people to legitimate jobs. One action we can take is to provide child care subsidies for poor people, especially single mothers, so that they can work outside the home. Without child care subsidies, many mothers either will not get these jobs or will not make enough money, especially at minimum-wage and lower-wage jobs, to be able to keep these jobs. The child care subsidy allows single mothers to get legitimate jobs rather than seek out illegitimate activities such as engaging in prostitution, selling drugs, and shoplifting. In other words, if we want to decrease crime, we need to give single mothers legal options that are better than illegal options.

Another step we can take via jobs and opportunities is to raise the minimum wage so that legal jobs have more of an enticement than do illegal activities. If people, after working 40 hours per week, do not have enough money to pay the rent, pay for food, and pay for other necessary expenses and they are still thousands of dollars below the poverty line, they will have greater incentives to consider illegal options as a way to survive and get ahead. In addition to increasing the minimum wage, we can increase the earned income tax credit to make up for the gap between the new minimum wage and the poverty line. Moreover, we can decrease income taxes and social security taxes on the poor. As we carry out these wage and tax policies, people who have more limited economic opportunities will have more money in their hands and will therefore have less incentive to commit crime.

Taxes, Corporations, Jobs, and Opportunities

Another action that our country can take is to decrease taxes on corporations in exchange for creating more jobs in the United States. A drop in corporate taxes will create an incentive for corporations to stay in the United States and provide more jobs for Americans. As this happens, there will be more legal opportunities for people and less likelihood of people choosing illegal opportunities, thus acting to decrease crime.

Unemployment Compensation, Job Training, Job Placement, and Crime

We can also increase unemployment compensation for people who have lost their jobs so that they are less likely to resort to crime to survive while out of work. We can also provide for better job training and retraining so that people who have lost their jobs can find new legal avenues of employment and feel hope in being able to go this route rather than the illegal crime route. Furthermore, we can get better at placing people in jobs that open up so that they believe they have a good chance to get a legal job, again increasing the incentive to do legal work while decreasing the incentive to resort to illegal activities. As we increase our unemployment compensation, our job training programs, and our job placement programs, we will develop a social structure that allows Americans to choose legal opportunities over illegal opportunities, thereby decreasing crime.

Remember, there is no guarantee that every adult desiring a job will be able to have a job in a capitalistic economy. Even during good years, there is usually some unemployment. And during other years, there can be substantial unemployment. For example, during the mid-to-late 1990s, we had fairly low unemployment—in the 3% to 4% range (University of Texas, n.d.). However, during the early years of the 21st century, we have had higher unemployment—in the 5% to 6% range. Consequently, our unemployment compensation system is extremely important for providing people with a temporary cushion to help them in their transition from one legitimate job to another legitimate job.

Skilled Trades and Opportunities

… Besides providing better quality public schools in general, which will help all students prepare for better jobs, providing better trade schools in particular, as one part of the improvement in the public school system, will give a number of high school students who do not want to go to college, or who are not capable of going to college, an avenue that will provide them with a skill that could give them good-paying legal jobs.

Decreasing Inequality and Increasing the Standard of Living

… So long as many Americans live considerably below the poverty line and many others live near the poverty line, there is an incentive for these Americans to find illegal means to survive. If they can have a decent standard of living via a legitimate job that pays above the poverty line, they will have less incentive to commit crime. So, it is in our vested interests as a country that we attempt to decrease inequality by redistributing resources; somewhat if we want to have less crime.…

It is also in the vested interests of nonpoor Americans to want less inequality in that these people will be less likely to be the victims of crimes such as burglary, robbery, and car theft. So, the society will be safer in addition to having less crime if we create a more equal society where the needs of poor and near-poor Americans are met.…

Health Care and Less Crime

Also, having a national health care system could provide health care for many poor and lower income people who do not currently have any health care.… During recent years, more than 40 million people in our country have had no health care coverage (Pear, 1999). Having a health care system would decrease inequality and would therefore provide a base for a decent standard of living, thereby decreasing the incentive for people to undertake illegal activities to survive.

The past few paragraphs have suggested that the way we tax ourselves and the kinds and amounts of social services we provide can lift up poor and near-poor people in our country and in turn create disincentives for them to commit crime, thereby decreasing crime. Although it might seem unrelated at first, on further reflection, we can begin to realize that the way we tax people and the amount and kind of social services we provide people are key factors that can decrease crime.

As you can see, I have focused mainly on what we, as a society, can do to help people get legal jobs and remain working at legal jobs. That is, we create a social structure that gives people more legal opportunities to have a decent standard of living. The more we can restructure society to give people more legal opportunities in the form of skills, education, and decent-paying jobs as well as more income via lower taxes and more and different kinds of services that provide a base for all Americans to live decently, the less incentive they will have to commit crime.

Registering of Guns and Crime

Another factor that we might consider is the registering and licensing of all guns in the United States. Although this action does not relate to the main thrust of the preceding paragraphs that emphasized giving people more legal opportunities, it may help to reduce crime in two ways. One way is by decreasing homicides in the United States. The total number of firearm homicides in 1995 was 15,835 in the United States versus 34 in Japan (Booth, 2004, p. 199). That is, although our U.S. population is only twice the size of Japan's population (Population Reference Bureau, 2002), the number of U.S. firearm homicides was 466 times that of Japan.

Besides decreasing the homicide rate in the United States, the registering and licensing of all guns in the country could help in not giving as easy access to firearms to people who intend to commit crime. This is not fool-proof; people can still get firearms in illegal ways. However, we can make it more difficult for these people to get fire-arms. Less access should help in decreasing crime.

"Front-End" Investment and Crime

I have focused mainly on what our society can do so that people will be less likely to commit crime in the first place. In other words, if we invest a lot into preventing crime, our investment should eventually pay off. If I am right, we would not need to invest so heavily in other things (e.g., police, courts, pris-ons) after people have already committed crime. For example, once people have committed crime, we need more police to catch suspects; more law-yers, judges, and courts to try and convict defen-dants; and more prisons and prison personnel to hold the convicted criminals. If we, as a society, could invest more in the "front end" of this process so that people have less incentive to commit crime in the first place, we could spend less on the "back end" of this process, that is, the police, courts, and prisons. No doubt, there will still be a need for police, courts, and prisons, but the need should be less as we invest in restructuring society to give people more legal opportunities.

Job Training in Prisons and Job Placement

Speaking of investment in the back end of this pro-cess, another action we could take is to provide more and better job training in prisons and more job placement for men and women who have fin-ished their prison terms. Many men and women who have completed their time in prison are let out of prison with nowhere to go and no job to help them stay legal. If our prisons could train peo-ple to become carpenters, electricians, plumbers, machinists, repair technicians (e.g., cars, appliances, televisions, computers), and other kinds of skilled and semiskilled workers, men and women leaving the prison system would have a better chance to find legal jobs and hence more incentive to "go straight," that is, to remain legal. Moreover, if the prison system had an effective job placement sys-tem, men and women leaving the prison system

would have a much better chance of getting legal jobs and "going straight" in their lives. With such a combination of job training and job placement pro-grams in our prison system, we could take another step toward decreasing crime.

The job training and job placement programs both are a part of the larger idea of rehabilitating people in prisons rather than punishing them only. In the past in the United States, our prisons have been mainly a place of punishment with little or no rehabilitation. If we want to decrease crime, espe-cially the recidivism among people who previously committed crimes and served time in prison, it makes sense to increase rehabilitation, especially in the form of providing job training and job place-ment. In other words, if we can rehabilitate people leaving prisons by giving them job skills and then placing them in jobs, former prisoners will have the opportunity to make a decent living in a legal man-ner, thereby decreasing crime.

Moreover, if people leaving prisons have gone through job training and job placement for months or even years, they have had a chance to think about the possibility and to anticipate and plan for the day when they will be free and have legal and decent-paying jobs. This set of social conditions will help them to go through what sociologists call *anticipatory socialization*, where they will have thought about what it will be like to be free and working in legal, decent-paying jobs. Once they are working at their new jobs, they will be members of new reference groups that should help them to get connected with other people who are working at legal jobs. It would seem, therefore, that providing job training, job placement, and new reference groups will work to prevent former prisoners from returning to crime, thereby decreasing crime.

Ideally, if we could get these people into an en-tirely new social environment, where the effects of differential association with former friends and ac-quaintances who had helped them to commit acts of crime in the past were no longer a problem, the chances for recidivism and hence crime would dimin-ish. Coming out of prison, having a new job skill, being placed in a decent-paying job, being a member of a new reference group at work, and being placed in a new social environment separated from their former reference group that promoted and reinforced acts of crime—altogether, these new social conditions should act, at the back end of the process to reduce recidivism, thereby decreasing crime.

To get former prisoners to move to another geographic location where they no longer interact with their former colleagues in crime, we might need to provide them with some kind of incentive such as a certain amount of money per month for a while. These people, with their new decent-paying jobs and new reference groups (and, it is hoped, new friends and legal role models), will have established a new legal way of living and can carve out a new legal way of life.

The policies I have suggested, both front end and back end, will require investment of money and expertise. To the degree that the actions we take in our society at the front end, where we provide better education, training, child care subsidies, more decent-paying jobs, a higher minimum wage, and lower taxes on poor and near-poor people, are combined with the actions we take at the back end, where we provide job training, job placement with new and legal reference groups and new and legal role models, and new geographic locations away from old illegal reference groups and role models, all of these front-end and back-end policies taken together should help us to substantially decrease crime in our country.

More Emphasis on the Front End

As you can see, my way of decreasing crime is a little different from others in that I put much more emphasis on changing the front end of the process by providing a lot more legal opportunities for people so that they have the incentive to do legal work and remain legal. The more we can invest in these front-end policies, the less we will need to invest in the back-end policies.

Other people will emphasize adding more police in total, more patrol cars, more police walking neighborhood beats, more surveillance cameras, more harsh punishment, and more prisons. We can go this way and put the money we invest into decreasing crime in doing these kinds of things. I am sure that these kinds of things will reduce crime somewhat because they will increase the certainty that someone committing a crime will get caught and be punished more severely. Research suggests that certainty of punishment adds to the deterrence of crimes (Paternoster, 1989). This is all well and good, and I think that these kinds of actions will be effective to a degree.

However, given the limited resources our society has to work with, investing in expanding the legal opportunities for Americans not only will address what causes crime but also will create the type of social structure that will reduce crime year after year and decade after decade....

Chapter 12

Health and Aging

The Problems of Growing Old

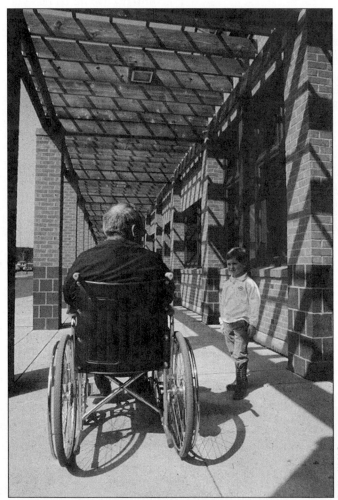

Dr. Alan W. McEvoy

Life and death parallel health, disease, and age. The readings in this chapter deal with problems in the health care system, disease epidemics, the status of the elderly, the aging process, and death and dying. Thanks to influences such as modernization, urbanization, and globalization, the aging process has been going through a period of rapid change. People are living longer, more productive lives in much of the world because of improved living standards and health care. Infant mortality rates are dropping and diseases, including epidemics, are being controlled.

Sociologists view health and medicine as one of the major institutions in society, along with family, education, religion, politics, and economics. Some form of each is found in every society. Medical sociology, a growing major subfield in sociology, studies the institution of health and medicine, and includes topics such as the roles and expectations of the sick person (that is, what happens when you miss social responsibilities such as work or school because you are sick); cultural differences in health care beliefs and practices; types and organization of health care systems; roles in health care systems; who has access to health care; and the cost and financing of health care.

One issue that has received attention from sociologists is the *medicalization* of certain conditions, including forms of deviance and normal human functioning. Conditions such as alcohol and drug abuse, smoking, pregnancy and childbirth, and even aging are being shifted from the family, legal, or religious arenas to the health care system. By treating them as medical problems, the responsibility is taken from the individual and control becomes a medical problem to be diagnosed and treated. As you read about the aging process, look for aspects that might indicate medicalization.

The first article, "The Crisis in American Health Care," discusses problems and dilemmas facing the U.S. health care system, and compares it to systems in other Western societies. It outlines the factors that are creating crises and suggests approaches to dealing with the crises.

In the second reading, Edward Kain considers a disease that does not discriminate on the basis of age. However, in the case of HIV/AIDS, younger sexually active individuals are at greater risk of contracting the disease. This disease is putting tremendous stress on health care systems, especially in poor countries that are already pressed to meet the basic health care needs of citizens. Mothers can pass the disease to their newborns. Those most affected by the disease vary by country and culture, and solutions also depend on cultural attitudes toward disease treatment and prevention.

The world's population *is* getting old. The numbers of the elderly are growing rapidly, and in most countries the percentage of the elderly is also increasing. As the population ages and reduces its productive capacity, this group becomes dependent on the younger population. Because of the growing number of older people in the world, the institution is taking on increasing significance as demand by the elderly for health services rises. In fact, a large and increasing number of sociologists are in the specialty of medical sociology. Among the many issues sociologists study in the field of health and medicine are doctors as professionals, other roles in health care, the patient (or sick) role, hospitals as organizations, national health care systems, and health care for elderly patients, a focus in this chapter.

As people live longer and modern health care becomes increasingly sophisticated—with technology to deal with many conditions that were death sentences in the past—questions of cost and delivery of health care increase. Medical technology has progressed to the point where we can diagnose illnesses and keep people alive for much longer periods than were possible in the past. This ability raises many practical and ethical questions. An ethical dilemma occurs, for example, when medical technology can keep individuals alive well beyond the point when they would have died naturally without intervention. The costs of keeping people alive may be depriving others of needed health care. What should be done in such situations?

Although workers have contributed to government health care systems in most modern societies, and these systems have often provided a safety net for older citizens, the safety net in many societies is under strain or failing because of the increasing numbers of elderly citizens. Add to these pressures the advances in medical care and technology that contribute significant costs to health care services, and many feel a crisis is pending.

Outlining some of the issues from an international perspective, Jeremy Seabrook discusses aging around the world. He reviews the changing status of the elderly, the growing number of elderly persons in societies, and the strain their presence creates on societies that need younger people to carry out the work that will support aging populations. Add to the problem the fact that birth rates have declined dramatically in many modern societies, and it becomes clear that there will not be enough workers to support the dependent young and old populations. Some have suggested that the political ramifications of the changing status of the elderly include a rise of religious fundamentalism and the influence of aging leaders in traditional societies who are attempting to maintain the status quo and influence of the elders in those societies. Seabrook

points out the changing status of the elderly around the world through examples of fulfillment in the senior years versus declining capacity and dependence. He contrasts the elderly who are honored for their wisdom, respected, and given obedience versus those who are discarded because their wisdom is no longer relevant in the modern world, losing power and influence as younger generations take control.

The costs of growing old continue their rapid rise: insurance, health care, the cost of dying. These costs create potential conflicts between generations as different age groups compete for limited governmental funds. Sam Ervin presents fourteen forecasts for an aging society, issues from work and independence to shortages of health and elder care, that reflect rising costs.

This chapter raises questions about health, disease, aging, and forcasts for aging societies. As you read, think about issues of access to health care and the "right to life," diseases such as HIV/AIDS, the increasing costs of health care, and the process of dying and death in society today.

49

The Crisis in American Health Care

JAMES WILLIAM COLEMAN AND HAROLD R. KERBO

Among developed nations, the United States is the only one without a broadly based system of government-supported health care that covers all citizens. We hear stories of citizens going without care until there is an emergency, and not being able to cover the costs of care or not being able to get insurance because of preexisting conditions. This article discusses several of the problems faced by the U.S. health care system: access denied to some who are ill; partial coverage systems; runaway costs; malpractice (legal) cases; and ethical dilemmas such as physician-assisted suicide, surrogate mothering, and cloning.

As you read this article, think about the following questions:

1. Do you know anyone who has faced problems with obtaining access to the U.S. medical system?
2. Why does the U.S. system have more problems than the health care systems of many other industrialized countries?
3. Why does health care in the United States cost more money than in other systems?
4. What might be done about the problems in the U.S. health care system?

GLOSSARY **Managed care** A form of health insurance that monitors and controls decisions of health care providers to keep costs down. **Health maintenance organizations** Organizations that employ their own medical personnel and offer services at a fixed rate. **Malpractice** Lawsuits against doctors for wrongdoing, resulting in high insurance costs for doctors that are passed on to patients.

THE CRISIS IN AMERICAN HEALTH CARE

The organization of America's health care system remains unique among industrialized nations. In response to ever-rising costs and the demand that competent health care be available to everyone, other industrialized countries have all adopted broadly based systems of government-supported health care. The United States, however, took a different course, creating a medical welfare system for the poor and the elderly but leaving the rest of the health care system in private hands. Direct payments by patients have been supplemented by both government programs and private insurance, which now pick up most of the

Source: Coleman, James William and Harold R. Kerbo. 2006. *Social Problems*, 9th ed. Upper Saddle River, NJ: Prentice Hall. 171–177.

tab, but these changes were not carried out in a systematic way.

How well does the American system work? Despite the efforts of some of the most dedicated and capable health care workers in the world, the problems of the American system remain serious and deep-seated, and many trends seem to be pointing in the wrong direction. Not only does the United States spend far more money per person than any other nation, but it also spends a much larger share of its total national income than any other industrial nation. (See Figure 1.) Yet despite spending more than $1.4 trillion on health care every year, the United States does rather poorly in a comparison of international health statistics.[1] Although the United States is the world leader in many branches of medical research, its infant mortality rate is the highest of any major industrialized nation, and so is its percentage of low-birthweight infants. Overall, the United States ranks only fifteenth in average life expectancy. It would be wrong to attribute all these differences to the health care system alone; variations in lifestyle, diet, and environment are also important. Nonetheless, it is clear that the American system often fails its neediest patients, while at the same time the overall cost of health care continues to escalate out of control. . . .

FAILING THE PATIENTS:
ACCESS DENIED

The American health care system offers some of the best care to be found anywhere on earth, but only for those who can pay for it. Those affluent enough

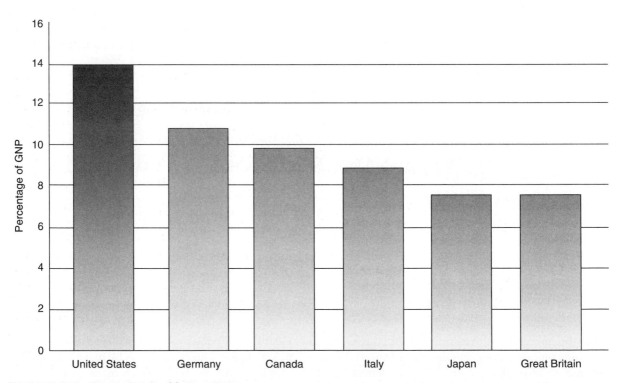

FIGURE 1 Comparing Health Care Costs

The United States has the most expensive system of health care in the world.

SOURCE: U.S. Bureau of the Census, *Statistical Abstract of the United States, 2003*. (Washington, DC: U.S. Government Printing Office, 2003), p. 848.

to have good insurance coverage pay little or nothing in out-of-pocket costs for even the best medical care. While most middle-class Americans have some kind of private insurance, there are usually significant gaps in coverage. Some policies have little or no coverage for office visits or preventive care, while others require patients to make large deductible payments before the insurance company contributes. According to the American Psychiatric Association, 98 percent of health insurance plans discriminate against those seeking treatment for mental problems by requiring higher out-of-pocket payments, allowing fewer doctor's visits and less time in the hospital, and even by denying coverage completely for some types of problems.[2] Most policies also limit the total amount the insurer will pay for any physical or mental illness. Thus, people with serious medical conditions may find that their coverage has run out or that their insurance company has canceled their policy and refuses to pay for any future treatment. Moreover, there have been numerous reports of insurance companies denying patients treatments they need when the costs are high. Some insurance companies, for example, classify established therapies as "experimental" in order to refuse coverage, or they demand that physicians use less costly treatments even when the more expensive ones are known to work better. Many policies lack adequate coverage for prescription drugs—an especially glaring problem because of the runaway inflation in the prices the public is charged for pharmaceutical products. Some insurance policies provide no drug benefits at all, and many others place strict restrictions on which prescription drugs they will pay for, thus preventing many patients from receiving the most effective medications for their problem.

Around three of every four Americans with health insurance are currently in some kind of managed care plan that monitors and controls the decisions health care providers make in order to keep costs down. Cost-conscious insurance companies provide financial incentives for general practitioners to limit the services they give to patients, and they require them to decide whether patients should be allowed to see specialists. Similar pressures are being felt by the physicians who work at health maintenance organizations (HMOs), which employ their own medical personnel and offer a complete range of health care services at a fixed rate. A 2000 Harris poll found that 59 percent of Americans felt that HMOs compromised their quality of health care—a jump from only four years before when that figure was 39 percent—and the public's attitudes would undoubtedly be much the same about other forms of managed care. Similar opinions are found among physicians as well. One survey found that a majority of American physicians felt that managed care hurt their patients because of limitations on such things as diagnostic tests, the length of hospital stays, and the choice of specialists.[3]

Government health care plans also have glaring deficiencies. Government payments come primarily from two programs: Medicare and Medicaid. Medicare buys medical services for people age 65 and older, while Medicaid is designed to help the poor, the blind, and the disabled. Medicare, a federal program, is relatively uniform throughout the nation. Medicaid, however, is administered by the states, and each state has its own standards of eligibility and levels of benefits. There are major gaps in the coverage of Medicaid, and they are growing wider year by year. When Medicaid was first established in the mid-1960s, it covered about 70 percent of those with incomes below the poverty line. Today, only 40 percent of the poor are covered.[4] In addition, state and federal governments have been placing tighter limits on the assistance the poor receive. Some states have been creating more restrictive lists of the kinds of treatment they are willing to pay for, bringing charges that they are rationing health care for the poor. For example, the state of Oregon decided that medical procedures such as organ and tissue transplants were too expensive and that poor people who needed them would either have to get them from charity or do without. But the most common approach has been to limit access to care informally by making it difficult or unattractive for physicians and hospitals to treat welfare patients. For one thing, the states pay far less for most medical procedures than physicians and hospitals usually charge. On top of that, states often impose a bewildering array of bureaucratic barriers that must be overcome before a physician can actually be paid. As a result, many physicians simply refuse to accept Medicaid patients, and that means long waits for and rushed service from those who do.

Medicare coverage for the elderly is far less restrictive than Medicaid—almost everyone age 65 and over qualifies. But Medicare still requires the elderly to make a substantial financial contribution of their own and excludes coverage for the costs of long-term care in a nursing home. In 2003 a bill to add voluntary prescription drug coverage to Medicare was passed. At the time of this writing, the plan had yet to go into full effect, but even when completely implemented it will provide only limited coverage for seniors. Another problem is that the government's efforts at cost control have made Medicare payments fall further and further behind those of private health insurance companies. As a result, physicians and hospitals are becoming increasingly reluctant to treat Medicare patients.

Yet despite all these problems, those with some kind of private or government-sponsored health care insurance are the lucky ones. Approximately 15 percent of the American population lack any health coverage at all, because they make too much money to qualify for welfare, don't have employer-provided insurance, and can't afford the cost of private insurance.[5]

As a result of all these problems, the poor and minorities often face severely restricted access to health services. Research shows that older whites are 3.5 times more likely to have heart bypass surgery than older African Americans and that African American kidney patients are only half as likely to receive a kidney transplant as white kidney patients. A study of patients on Medicaid found that they were less than half as likely as those with private insurance to receive several common surgical treatments, and the gap between the two groups has increased as the relative value of Medicaid payments has decreased over the years.[6] In many ways, the problems with our system of dental care are even worse. Over 100 million Americans have no dental insurance. A minority child is three times more likely than a white child to lose a secondary tooth by the age of 17, and tooth decay in over half of Latino and African American children goes untreated.[7]

One poll found that almost one-fourth of Americans had put off some medical treatment in the last year because they could not afford it.[8] This situation is not only unjust—it is also foolishly shortsighted. When people delay medical treatment until their problems are so severe that they have no choice but to seek help, the total cost is likely to be far greater than that of timely preventive care. The cost of prenatal care denied to many poor women, for example, is far less than the hundreds of thousands of dollars often necessary to help their gravely ill infants.

RUNAWAY COSTS

Not only is the American health care system the most expensive in the world, but its cost continues to go up year by year. In 1950, the United States spent 4.4 percent of its gross national product on health care; today that figure is around 13 percent.[9] The specter of runaway medical inflation led to a host of new measures by the government and private insurance companies in the last decade to institute stricter cost controls. The rate of medical inflation dipped from 6.3 percent in 1995 to around 3 percent for the next few years, but by 2000 it was back above 4 percent and is now almost 5 percent.[10] Many experts believe that much of the cost savings from this managed care have already been achieved, and that we can expect more sharp increases in the years ahead. (See Figure 2.)

Why is health care so much more expensive in the United States than in most other countries? The most fundamental cause is the way the health care industry is organized and financed. Unlike most other countries, medicine in the United States is largely a private business organized for individual profit. The inefficient system of competing private health insurance companies that duplicate each other's services wastes billions of dollars in unnecessary overhead costs. Although estimates vary, private health insurers spend somewhere between 26 to 33 cents of every dollar they take in on administration, marketing, and commissions. In sharp contrast, the Canadian system of national health care spends only 2 or 3 cents per

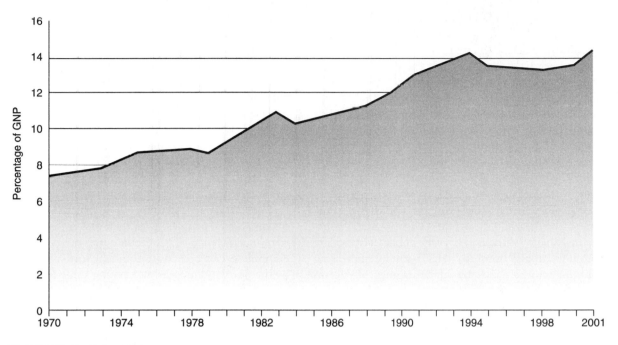

FIGURE 2   Rising Costs

Health care now takes twice as big a piece of the American economy as it did 25 years ago.

SOURCE: David R. Francis. "Rising Health Costs Limit Other Programs," *Christian Science Monitor*, December 12, 1993, p. 9; U.S. Department of the Census, *Statistical Abstract, 1991* (Washington. DC: U.S. Government Printing Office, 1991), p. 92; *Statistical Abstract, 1993*, p. 849; *Statistical Abstract, 2003*, p. 104.

dollar on overhead.[11] The lack of centralized buying power has also led to much higher prices for prescription drugs. All the countries in the European Union, for example, have some kind of national health care, and as a whole they spend *60 percent* less per person on prescription drugs.[12]

Another reason America's health care costs are so high is the staggering salaries paid to many of the people who work in the medical field. Although we would like to think that the primary motivation for someone to become a physician is to help patients, physicians have kept their salaries so high that it severely restricts the access many people have to medical care. Most physicians choose to work in the affluent areas that offer them the best pay, while the inner cities and poverty-stricken rural areas have a desperate shortage of all kinds of medical personnel. But the physicians look like paupers compared

to the executives who run the big medical corporations. One study that looked at the compensation of top executives in 17 health maintenance organizations in the United States found that they were paid an average of $42 million in cash and another $87 million in stock options in 1999![13]

A third factor is that patients have discovered that they can sue doctors for malpractice and win; as a result, the number of those suits has quadrupled since the late 1970s. All doctors—competent or not—must now pay high malpractice insurance premiums. The fear of being charged with malpractice has also forced many doctors to practice "defensive medicine," That is, they may order costly tests because they do not want to be accused in court of having forgotten something important.

The problem of runaway medical costs is not, however, in any way unique to the United States.

One common problem among all the industrialized nations is that their populations are aging and, of course, older people require more medical attention. A second factor has been the development of expensive new drugs and medical techniques. Such procedures as organ transplants are extremely costly and tend to drive up the overall price tag for health care. Some of the biggest increases have been in the costs of prescription drugs. A steady stream of powerful new drugs has come on the market in recent years, and the costs for their research and development can be staggering. Moreover, even though people can't buy their products without the approval of their physician, the pharmaceutical companies have discovered that public advertising provides a big boost to sales both by increasing public interest and by discouraging physicians from prescribing less expensive generic drugs. Not only have the big pharmaceutical corporations consistently been at or near the top of the list of the companies with the highest return on their investment, but their top executives earn even bigger incomes than those who run health maintenance organizations. In 1998, the heads of the ten largest drug companies earned an average compensation of $290 million each![14] Moreover, the fact that there are huge profits to be made in developing new drugs and medical procedures has diverted attention from less expensive, and often more effective, techniques of preventive medicine. As Jeffrey Klein and Michael Castleman put it, "The processes that drive medical research toward expensive treatments also turn it away from preventive measures that do not hold the promise of corporate profit."[15]

ETHICAL DILEMMAS

To most outsiders, the moral responsibility of the medical profession seems clear: to save lives and help patients be as healthy as possible. The increasing power of medical technology has created perplexing ethical dilemmas for which we have yet to find satisfactory answers, however. One such issue concerns the so-called heroic efforts physicians use to extend the lives of dying patients. Medical costs mount rapidly in the final weeks of a patient's life. One-third of Medicare's entire budget is spent on people in the last year of life. Most of us would say that cost should not be the standard to decide who lives or dies—but in a world in which millions of people starve to death every year, the money we use to keep one dying patient alive for another month could save a thousand hungry babies in the Third World. Is this fair? What about all the American babies who die because their mothers never received basic prenatal care?

Even if we ignore the costs, serious ethical issues remain. When people die, all their life-sustaining systems usually fail at about the same time, but medical equipment can take over the functions performed by the heart and lungs, thus keeping some gravely ill patients alive almost indefinitely. Although such patients are alive, the quality of their life is often pitifully low. They lie trapped in a hospital bed, connected by wires and tubes to a machine they totally depend on. If the patient wishes to continue under such conditions, there appear to be few ethical problems. In many cases, however, the patient is unconscious and unable to make any sort of decision. Some people now make out "living wills" that spell out how far they wish their physicians to go in using heroic means to extend their lives if they become gravely ill. Yet even the principle of self-determination implied in such wills is not universally accepted in our society. Moreover, many hospitals refuse to turn off patients' life-support machines even if they request it.

Oregon is the only U.S. state where physician-assisted suicide is legal. The law, first approved in 1994 and reaffirmed by Oregon voters in 1997, allows doctors to prescribe a lethal dose of drugs to patients with less than six months to live. In other states, a physician who helps a dying patient end a life of pain can be charged with murder. There is, however, growing nationwide support for legalizing "physician-assisted suicide," that is, allowing physicians to assist patients to end their lives if the patients are too ill to do it by themselves. A Harris poll found that 73 percent of the Americans polled agreed with the statement "The law should allow doctors to comply with the wishes of a dying patient in severe distress who asks to have his or her life ended."[16] However, although the majority of Americans support making physician-assisted suicide legal under certain conditions, the terminally ill tend to resist using this option and prefer instead to use their final time preparing for death and being with loved ones. In Oregon, just 16 terminally ill patients took advantage of the new law in 1998 and 27 in 1990.[17] It is evident, nonetheless, that knowing the option of euthanasia or physician-assisted suicide is available gives great psychological comfort to those facing the imminent end of their lives.

Just as many perplexing ethical questions surround the beginning of life as its end. One of the most controversial ethical issues of our time is abortion. This difficult matter revolves around two separate issues that are often confused. On a personal level, the question is under what circumstances a woman is morally justified in deciding to have an abortion. On a sociological level, the question is whether or not the government should attempt to prevent women from undergoing such procedures....

The abortion debate has been going on for decades, but advances in medical technology are also creating new dilemmas about human reproduction. The technique for artificial insemination has been used for decades to help women with infertile husbands. In contrast, the practice of surrogate mothering, in which a woman is hired to bear someone else's child, has raised a storm of protests. Although the surrogate mother signs a contract agreeing to give the child to its biological parents, bitter legal battles have arisen when surrogates have attempted to void those contracts and claim legal custody of the child. Even more troubling are the ethical issues posed by the possibility of human cloning—a scientific technique that allows reproduction with the DNA from a single cell. Scientists have already cloned several kinds of mammals and there is little doubt that similar techniques could be applied to human beings as well. We know that cloned animals have more health problems than others, but we have little idea of the social and physical consequences that might occur if such procedures were ever widely used. There is, however, an important difference between cloning individual cells for medical research and reproductive cloning intended to create another human. Obviously, there are no simple answers to the troubling ethical dilemmas posed by advancing medical technology, yet society must nonetheless formulate social policies to guide the medical and scientific communities in making such vital ethical decisions.

NOTES

1. U. S. Bureau of the Census, *Statistical Abstract of the United States*, 2003, p. 104; National Center for Health Statistics, *Health, United States, 2000,* Table 114.

2. American Psychiatric Association, "Mental Health Parity—Its Time Has Come," 1999, APA Online, www.psycho.org.

3. Physicians for a National Health Program, "PNHP Data Update, September 2000," www.pahp.org/data-update0900.htm; Debra S. Fledman, Dennis H. Novack, and Edward Gracely. "Effects of Managed Care on Physician Patient Relationships, Quality of Care, and the Ethical Practice of Medicine: A Physician Survey," *Archives of International Medicine* 158: 1626–1632.

4. U. S. Bureau of the Census, *Statistical Abstract of the United States*, 2003. p.111: Consumers Union, "The Crisis in Health Insurance," *Consumer Reports* 55 (September 1990): 608–617.

5. U. S. Bureau of the Census, *Statistical Abstract of the United States, 2003*, p. 114.

6. Sonia Nazario, "Treating Doctors for Prejudice," *Los Angeles Times.* December 29, 1993, pp. A1, A36, A37; Douglas P. Shuit."Black, Poor Medicare Patients Get Worse Care." *Los Angeles Times,* April 20, 1994, pp. B1, B4; Thomas H. Maugh II, "Surgery Study Finds Poor at a Disadvantage," *Los Angeles Times*, December 9, 1993, pp. A3, A36.

7. Physicians for a National Health Program. "PNHP Data Update, September 2000."

8. National Center for Health Statistics, *Health, United States, 2001.* Table 114. U.S. Bureau of the Census, *Statistical Abstract of the United States, 1996,* p. 834.

9. National Center for Health Statistics, *Health United States, 2001,* Table 115.

10. U.S. Bureau of the Census, *Statistical Abstract of the United States,* 2003, p. 108, Health Care Financing Administration. "The Clinton Administration's Comprehensive Strategy to Fight Health Care Fraud. Waste and Abuse." 1998, www.hcfa.gov/facts; Consumers Union, "The Crisis in Health Insurance," Associated Press, "Health Insurers' Efficiency Is Questioned," *Los Angeles Times.* October 19, 1990, p. D6.

11. Consumers Union, "Wasted Health Care Dollars."

12. "The Trouble with Cheap Drugs," *The Economist,* January 31, 2004, pp. 59–60.

13. Physicians for a National Health Program, "PNHP Data Update 2000."

14. Ibid.

15. Jeffrey Klein and Michael Castleman. "The Profit Motive in Breast Cancer," *Los Angeles Times,* April 4, 1994, p. B7.

16. Brad Knickerbocker, "Assisted-Suicide Issue More Active as Citizens Appear to Change Mood," *Christian Science Monitor,* May 2, 1994, p. 6.

17. *Denver Post,* February 24, 2000, p. 20A.

50

Some Sociological Aspects of HIV Disease

EDWARD L. KAIN

At first glance, the problem of AIDS might not appear to be an issue related to sociology. But if the sociologist is "interested in the action of human conduct," then AIDS—a disease spread through human behavior—is an object of sociological interest. Kain explains this relationship.

As you read, ask yourself the following questions:

1. *The social construction of AIDS can be seen from what two perspectives?*
2. *How does the notion of "stigma" relate to AIDS?*
3. *How would you deal with the problem of AIDS in the United States? In Africa?*
4. *What effect is HIV/AIDS having on health care systems around the world?*

GLOSSARY **Disease** biological aspects of sickness. **Illness** Social construct affecting society's reaction to disease. **Role of the other** Label and expectations attached by one individual to another individual or group. **Social construction** Process individuals use to create their reality through social interaction. **Stigma** Negative label that changes an individual's self-concept and social identity.

One of the most crucial social problems emerging in the last two decades of the twentieth century and following us into the twenty-first century is the epidemic of HIV disease. From an illness that at first was perceived as affecting only a small number of gay and bisexual men in a few U.S. cities, it has grown to a pandemic (a worldwide epidemic). Both the number of AIDS cases and the estimated number of people who are infected with HIV have grown at a staggering rate.[1]

In April 1991, scarcely a decade into the epidemic, the World Health Organization estimated that approximately 9 million people were infected worldwide (World Health Organization, 1991). By 2007, that estimate had reached just over 33 million, with over 22 million having died since the first cases were identified in the early 1980s (UNAIDS, 2007).

The epidemic has implications far beyond these numbers. Worldwide, there are over 6,800 additional infections per day and over 5,700 daily deaths from AIDS—resulting in 2.5 million new infections in 2007, with an estimated 2.1 million deaths from AIDS in the same year. Over three quarters of the deaths in 2007 occurred in sub-Saharan Africa (UNAIDS, 2007). When adults

Source: This reading was rewritten for the eleventh edition of *Sociological Footprints*, edited by Leonard Cargan and Jeanne H. Ballantine. Research for the paper was funded, in part, by a Fellowship from the Brown Foundation of Houston, Texas. The author would like to thank Dan Hilliard for helpful comments on an earlier version of the manuscript.

die from HIV disease, they often leave behind children who must be cared for by others. Worldwide, the pandemic has left over 13 million children with one or both parents dead; that number will increase to more than 25 million by the year 2010 (UNAIDS, 2002).

Although disease is a biological concept, illness is a social construct. As such, it has an impact upon how well a society is able to combat the disease as well as upon the personal experience of individuals who contract it. Discussion of HIV disease ultimately points to important sociological concepts from a number of areas within the discipline. The next few pages examine some of the ways in which a sociological perspective can help us understand the social impact of AIDS and HIV disease.

A full sociological analysis of the epidemic is beyond the scope of this reading. This discussion focuses upon three topics: (1) the social construction of illness, (2) HIV disease and stigma, and (3) shifting definitions of HIV disease.

THE SOCIAL CONSTRUCTION OF ILLNESS

The current social construction of HIV disease and AIDS must be understood in the context of historical changes in the broader social definition of disease. Rosenberg (1988) suggests that in late eighteenth- and early nineteenth-century America illness was conceived primarily in terms of the individual. "Even epidemic disease was understood to result from an unbalanced state in a particular individual…thus the conventional and persistent emphases on regimen and diet in the cause and cure of sickness" (Rosenberg, 1988:17). Using the example of cholera, Rosenberg points out that certain groups of people were seen as "predisposed" to falling ill— those who had poor nutrition, who were dirty, or who were gluttonous. Such an approach had the function of reducing the seeming randomness with which the epidemic struck.

This approach to disease in the West slowly shifted throughout the nineteenth century. The Paris clinical school argued that disease was something "lesion-based" which played itself out in each person who was afflicted. By the end of the century the germ theory of disease gave such a model an explanatory mechanism. At the same time, a broader range of behaviors, previously linked to concepts of sin and deviance, came under the purview of medicine. The medicalization of alcoholism, mental illness, and a variety of sexual behaviors expanded the authority of medicine to deal with behavior previously not thought of as illness.

One result of medicalization is to reduce the amount of individual blame. Defining alcoholism, for example, as a disease transforms its very character. The alcoholic is now someone who has an affliction rather than someone who is of weak moral character. The blame begins to shift from the individual to the disease.

Blame

Blame never shifts entirely away from the individual, however. In the West, although we may be able to trace an historical shift along a continuum from blaming the individual to blaming an outside causal agent of some type (whether it be a germ, a gene, or a virus), this shift has not occurred equally for various types of diseases. Illness or injury resulting from behaviors that are defined by the society as morally wrong receive much greater attribution of individual responsibility than those resulting from behaviors which do not have the same degree of moral censure.

Ultimately, many types of illness can be linked to individual behavior. Smoking and overeating both are behaviors that are "chosen" by individuals. Indeed it has been estimated that the "real" causes of one third of deaths in the United States are tobacco use, and diet and activity patterns (McGinnis and Foege, 1993). People who develop lung cancer, strokes, or heart attacks related to these behaviors are not, by and large, blamed for their illness. Sports like football (or more obviously hot-dog skiing and other extreme sports) lead to a large number of serious injuries and deaths each year. Yet these are viewed as "accidents." Typically, those who suffer the results are not defined as deserving their fate.

Sexually transmitted diseases are much more likely to elicit reactions of blame and deservedness because of their linkage to behavior that is defined as morally wrong by the community. Further, there is a long cultural history of dividing the innocent victims[2] (spouses and children who are unknowingly infected by a partner, usually a husband, who strayed) from guilty sufferers who deserve their illness as the wages of sin.

The stigmatization of those who are ill and a search for someone to blame are not contemporary phenomena. During the 1656 outbreak of bubonic plague in Rome, for example, foreigners, the poor, and Jews were blamed for the epidemic. Similarly, the poor and immigrants were blamed for the 1832 cholera outbreak in New York City. In this case, the moral failings that led to poverty were seen as the root cause of the illness. Indeed, in his examination of past responses to epidemics (1988:57), Guenter Risse concludes that "in the face of epidemic disease, mankind has never reacted kindly… the response to disease is a powerful tool to buttress social divisions and prejudices."

The Role of the Other

The cause of a disease is often understood so as to shift blame for an epidemic upon the other. Whether this other is an ethnic or racial group, a religious or social category, or a group stigmatized for behavior which is labeled as deviant, this conception of the other is powerful in shaping the social response to disease.

The most extreme social response is total isolation of those who are sick in an attempt to stop the spread of the illness from one segment of society to another. Historically, quarantine has been tried as a method for coping with a variety of illnesses. With tuberculosis, yellow fever, cholera, and leprosy, all efforts to quarantine large numbers of people have been failures. Rather than being effective public health measures, these mass quarantines have been expressions of fear of the other—attacks upon the

civil liberties of groups not accepted by the general public (Musto, 1988).

In the case of HIV disease, the complex interactions among blame, fear, discrimination, and stigma (a concept discussed in the next section) have led to a social construction of the disease in which "us" and "them" play a central role. The worldwide social impact of the epidemic, however, makes it clear that all of us are living with HIV disease (see Gilmore and Somerville, 1994, for an excellent discussion of this issue.)

The Role of "Deserving to Have the Disease"

The search for a group to blame for illness, combined with the tendency to blame the "other," often creates the conception that those who are sick deserve to have the disease. This is reflected in early cultural constructions of AIDS in the United States. Those who contracted the disease were often divided into "innocent victims" of AIDS and those who somehow deserved their illness. This ascription of personal responsibility for their illness affects not only people with AIDS but also their care givers (see Sosnowitz and Kovacs, 1992).

HIV DISEASE AND STIGMA

A key to understanding the social construction of HIV disease is the concept of stigma. Erving Goffman defines stigma as "an attribute that is deeply discrediting." He goes on to say that there are three types of stigma—what he calls "abominations of the body" (physical deformities); "blemishes of individual character" (some examples include a weak will, dishonesty, alcoholism and other addictions, homosexuality, radicalism); and "the tribal stigma of race, nation, and religion" (Goffman, 1963).

HIV disease has the potential of developing all three types of stigma described by Goffman. Some of the opportunistic infections associated with end-stage HIV disease (AIDS) can be disfiguring—the skin lesions of Kaposi's sarcoma being a prime example. The extreme weight loss associated with end-stage HIV disease also creates what Goffman called an "abomination of the body." The social and cultural construction of HIV disease involving blame, as noted, further links it to a number of "blemishes in individual character"—drug use, homosexuality, and inability to control one's own behavior in a safe manner. Finally, what Goffman called the "tribal stigma" of race, nation, and religion also applies to the cultural construction of AIDS and HIV disease. In the early years of the epidemic, one of the major "risk groups" was Haitians. A number of authors have suggested that the cultural constructions of AIDS are tinged with racism. In *AIDS and Accusation*, for example, Paul Farmer argues that ethnocentrism and racism in the United States were key factors in the theories about a Haitian origin for AIDS (Farmer, 1992).

Why Is HIV Disease Particularly Stigmatized?

Most of the literature on HIV disease and AIDS talks about the importance of stigma. From this literature we find a number of characteristics that predict whether or not a disease will be particularly stigmatized. One of the first analyses of the stigma associated with AIDS was by Peter Conrad (1986). After a discussion of the public hysteria surrounding this illness, Conrad identifies four social aspects of AIDS which lead to its peculiar status.

First, and foremost, throughout the early years of the epidemic AIDS was associated with "risk groups" which were both marginal and stigmatized. Because early cases of AIDS were found in homosexual men and intravenous drug users, a powerful cultural construction emerged that defined the illness as a gay disease.

Second, because a major mode of transmission involves sexual activity, the disease is thus further stigmatized. In his insightful book on the history of venereal disease in the United States, Allan Brandt suggests that human societies have never been very effective in dealing with sexually transmitted diseases. He begins his book by noting that "the

most remarkable change in patterns of health during the last century has been the largely successful conquest of infectious disease." The striking exception to this pattern has been an explosion in sexually transmitted diseases. He asks, "Why, if we have been successful in fighting infectious disease in this century, have we been unable to deal effectively with venereal disease?" His answer lies in an examination of the social and cultural responses to sexuality and sexually transmitted disease. He argues that venereal disease was a social symbol for "a society characterized by a corrupt sexuality" and was used as "a symbol of pollution and contamination." The power of this social construction has rendered efforts to control sexually transmitted diseases ineffective (Brandt, 1987).

Third, Conrad points out that contagion plays a major role in whether or not a disease stigmatizes the individual who is infected. If a disease is contagious, or if it is perceived as contagious, then the stigma of the illness increases.

Fourth, Conrad argues that the fact that AIDS is a deadly disease also adds to its stigma. Because of these social characteristics, Conrad concludes that AIDS is a disease with a triple stigma: "it is connected to stigmatized groups, it is sexually transmitted, and it is a terminal disease."

One of the best sociological discussions on the stigma associated with HIV disease is found in Rose Weitz's *Life with AIDS*. Weitz says that stigma is greatest when an illness evokes the strongest blame and dread. She delineates six conditions that increase blame and dread. Like Conrad, Weitz includes (1) linkage to stigmatized groups, (2) an association with sexuality, (3) if the illness is perceived to be contagious and if there is no vaccine available, and (4) if the illness is "'consequential,' producing death or extensive disability and appearing to threaten not just scattered individuals but society as a whole" (Weitz, 1991:45–48). To Conrad's list she adds two more—if the illness creates dehumanizing or disfiguring changes that "seem to transform the person into something beastly or alien" and "if mysteries remain regarding their natural history."

To this list I would add two more factors. First, industrial societies have not had to deal with fatal infectious diseases in several decades. We have come to expect that young people will not be struck down by disease in the prime of their life. Polio was the last major infectious disease to affect modern industrial societies, and much of the population has grown up in a world with no experience of consequential infectious diseases. This circumstance increases the fear and dread associated with HIV disease.

Finally, if a disease has an impact upon mental functioning, it increases stigma. HIV can directly infect brain cells. As prophylactic measures such as the use of inhalant pentamidine and anti-viral drugs such as AZT increase the time between exposure to HIV and first opportunistic infections, the number of people with HIV disease who have impaired mental functioning because of infection with the virus will increase … thus increasing the potential for stigma. Indeed, dementia is now the most common neuropsychiatric problem found among HIV patients (Buckingham, 1994).

The Future of Stigma as It Relates to HIV Disease

Just as the relative stigma of various illnesses can be predicted by examining the eight characteristics just delineated, the stigma of a particular illness varies by culture and will change over time within any particular culture as these variables change. In cultures where sexual behavior is more openly discussed and where attitudes are more tolerant of homosexuality, the stigma associated with HIV disease will be less. Similarly, in cultures where drug use is conceptualized differently (the Netherlands is an instructive example here), the stigma of HIV will be lessened.

There is substantial stigma of HIV in China, for example, where nearly half of nurses in one survey said they would rather not work in AIDS wards (UNAIDS, 2008a). In much of the world, assessment of the transmission of HIV is complicated by strong cultural taboos against homosexuality. One result is that same sex behavior may be underplayed as one mode of transmission. This leads to yet another pattern of importance; in many countries throughout the world, men who have sex with other men also

have sex with women, potentially spreading the virus to their female partners as well as any children who are born (UNAIDS, 2008b). Throughout many African countries, stigma is very high. Married women in Zambia, for example, typically do not tell their husbands that they are HIV positive, fearing they may be abandoned or blamed, despite the likelihood that they were infected by their husbands (UNAIDS, 2008b).

No matter what the cultural context, however, the point here is that no other disease has such a high potential for stigma on all eight characteristics, making HIV disease the prototypical example of stigmatized illness in modern times.

Shifting Definitions of HIV Disease

There is no single social definition of HIV disease and AIDS. Public perception of the disease varies among cultures and has changed over time. It also varies considerably from one segment of the population to another and from region to region in a population. Much of this variation is linked to epidemiology.

The social epidemiology of HIV disease in Africa, for example, is radically different from that in our country. Rather than being associated with certain high-risk groups, it is more equally distributed among males and females, and appears to be as common in the heterosexual population as it is in the homosexual and bisexual populations. The disease has continued to be "feminized" over time, so that by 2007, 61% of all adult cases in sub-Saharan Africa were among women (UNAIDS, 2007). These differences in epidemiology clearly have a major impact upon the social definition of the disease in different parts of the world as well as the treatment and ultimate social consequences of the disease.

It is interesting to note that coverage of the disease by the popular press shifted both in magnitude and in tone when it became clear that AIDS could be contracted by so-called "innocent" people—babies, hemophiliacs, and recipients of blood transfusions (Altman, 1986; Kain and Hart, 1987). Indeed, the social definition of AIDS shifted considerably over the first decade of its existence. Some of the early literature referred to the illness as GRID (gay-related immune deficiency), and there is evidence that when it was defined as a "gay plague" the scientific community joined the general public in its reluctance to take the disease as seriously as was warranted (Shilts, 1987).

Panic, fear, and rumor were common elements of popular press reports in the early 1980s. Indeed, many scientists and AIDS educators were frustrated by continued misconceptions about modes of HIV transmission. These early responses to AIDS reflect basic principles in research on collective behavior. Rumors are most likely to develop when there is ambiguity about something. They help to clarify the situation when data are unavailable (Shibutani, 1966; Macionis, 2008). The early years of the HIV epidemic fit this description perfectly. In addition, because the disease is fatal, people have a high degree of interest in the topic. Fear of contagion is a very predictable response.

From Acute to Chronic

Before the causal agent of AIDS was understood, and before effective treatments had been developed for some of the opportunistic infections, AIDS was defined as a short-term disease that led to death in a relatively short time. As the etiology of the disease has been clarified it is becoming defined as a long-term chronic illness (see Fee and Fox, 1992). This shift has implications for the cost of treatment, the stigma associated with the disease, and calculations of the social impact of the epidemic.

One change resulting from the shift to defining HIV disease as a chronic illness is the emergence of new issues for the delivery of health care. Chronic illness has typically been associated with the elderly. Long-term care facilities must rethink their methods of patient care when working with younger persons with HIV (Zablotsky and Ory, 1995). Further, policy makers and clinicians may view this shift to HIV as a chronic disease differently. Thus a simple redefinition of the disease as chronic may be inadequate for planning patient care in the case of HIV (Clarke, 1994).

Changing Epidemiology

As the epidemiology of the disease changes, so will social definitions. In the United States there has been a shift over time in which a larger proportion of new AIDS cases are linked to drug use rather than homosexual and bisexual activity. Further, African Americans and Hispanics are disproportionately affected by the disease. The general trend in this country has been that the epidemic increasingly affects women, the poor, people of color, and the heterosexual population (Karon et al., 2001). As in Africa, the proportion of new cases in the United States found among women has been increasing over time (UNAIDS, 2004). Because HIV has not been defined as a women's disease in the West, and because of women's lower social status, women with HIV have a shorter survival time than men. As the epidemiology changes, approaches to women's health issues must also change (Lea, 1994). The combination of race and gender are also important to examine. By the mid-1990s, women of color made up well over two-thirds of all cases of HIV-infected women in the United States (Land, 1994). By 2002, nearly half of all the HIV-positive people on the planet were women (UNAIDS, 2004), and by 2007 the estimate was that the numbers were equal for men and women (UNAIDS, 2007).

Higher rates of infection among African Americans and Hispanics in the United States have led researchers to explore ways in which HIV education programs may need to be targeted to specific populations. Different intervention techniques may be more effective with one population than another, and ethnographic analysis can help identify the best programs to use for a particular ethnic or racial group (Goicoechea-Balbona, 1994; Bletzer, 1995).

The history of the epidemic has also seen a shift in the worldwide distribution of reported cases of AIDS as well as projections concerning HIV infection. When the early reported cases were concentrated in North America, AIDS was defined as an American disease. In a number of communist countries it was defined as a disease of Western capitalism. As data on the epidemic improved, it became clear that large numbers of people were infected in poorer countries. Current estimates suggest that sub-Saharan Africa has nearly 68% of all worldwide HIV infections and accounted for over three quarters of the AIDS deaths in the world in 2007. It appears that the locus of the epidemic in the future may be Asia, where the population size holds the potential for large numbers of infections. The diversity of epidemiology in the region has led to a discussion of HIV epidemics rather than a single epidemic. Some areas (including Thailand and Cambodia) have been very successful in bringing the epidemic under control, whereas other areas (some parts of Indonesia and Vietnam) have increasing prevalence rates and thus far little success in limiting the spread of infection (UNAIDS, 2007).

As the epidemiology of HIV disease shifts, so will our cultural constructions of the disease. As women and heterosexuals become larger proportions of the HIV-positive population in Western industrialized countries, the stigma will decrease. In addition, infection rates are actually beginning to decline in some of the industrialized West as well as parts of Africa and Asia, and treatments using mixtures of "cocktail" drugs have been very promising in slowing (or even reversing) the replication of HIV within infected individuals. As the disease becomes more treatable, it will become less stigmatized. Worldwide, the epidemiology will continue to evolve. Although Asia currently has low infection rates, the rate of spread is rapid in some areas. In Latin America and the Caribbean the overwhelming proportion of infections are found in marginalized groups, which predicts high stigmatization in those regions. In Eastern Europe most cases are associated with drug use—again predicting stigmatization of the disease.

Data on patterns and trends in HIV infection change relatively rapidly. Luckily there are several excellent resources on the Internet that provide up-to-date information both for the United States and other countries. The Centers for Disease Control and Prevention maintains a number of websites related to HIV/AIDS. For general information, statistics, and other information, go to http://www.cdc.gov/. Once at the CDC website, you can use the search function, or click on HIV/AIDS in the "Diseases and Conditions" section.

The U.S. Census Bureau website maintains an international database that includes infection rates for various countries around the world. It also has links to maps of the worldwide pandemic. The general URL for the Census Bureau is http://www.census.gov/. Once you are at the website, search within "Subjects A–Z."

Perhaps the most comprehensive site for reports on the global pandemic is http://www.unaids.org/.

CONCLUSION

A full sociological analysis of the HIV epidemic would examine a wide variety of issues. The economic and demographic impact of the disease has already been devastating in a number of central and southern African nations. Gains in infant mortality that have taken four decades to achieve have been wiped out in less than a decade. Worldwide, women are particularly vulnerable to HIV disease both because of their lack of power in sexual relationships and because of the relative ease of viral transmission between sexual partners (Panos Institute, 1990). Indeed, issues of social stratification are central to understanding the epidemic. In most societies, race, class, and gender are critical variables in predicting who is more likely to become infected, and once infected, who will receive adequate treatment. Worldwide, the poorest countries are among the hardest hit in the epidemic. Unfortunately, much of the progress made in the developed countries relies upon very expensive drug treatments, which will be unavailable to the majority of those infected with HIV throughout the world.

Although this reading has not covered nearly all of the issues involved in such an analysis of the HIV pandemic, it has illustrated how a sociological perspective informs our understanding of one of the greatest social problems facing the world today.

NOTES

1. This reading makes a distinction between AIDS (acquired immune deficiency syndrome) and HIV disease. HIV disease begins when a person is infected with the human immunodeficiency virus. For most adults there is an extended period of as much as 10 to 15 years during which there are very few symptoms of infection. AIDS is the name associated with the end stages of HIV disease.

2. The use of the word *victim* is politically charged because of its implication of powerlessness. Much of the literature on HIV disease and AIDS suggests that *HIV-positive, seropositive*, and *persons with AIDS (PWA)* be used rather than *victim*. In this context I use the word *victim* because the popular consciousness clearly separates "innocent victims" from "guilty sufferers."

REFERENCES

Altman, Dennis. 1986. *AIDS in the Mind of America.* Garden City, NY: Anchor Press/Doubleday.

Bletzer, Keith V. 1995. Use of ethnography in the evaluation and targeting of HIV/AIDS education among Latino farm workers. *AIDS Education and Prevention* 7(2): 178–91.

Brandt, Allan M. 1987. *No Magic Bullet: A Social History of Venereal Disease in the United States since 1880*, expanded ed. New York: Oxford.

Buckingham, Stephan L. 1994. HIV-associated dementia: A clinician's guide to early detection, diagnosis, and intervention. *Families in Society* 75(6): 333–45.

Clarke, Aileen. 1994. What is a chronic disease? The effects of a re-definition in HIV and AIDS. *Social Science and Medicine* 39(4): 591–97.

Conrad, Peter. 1986. The social meaning of AIDS. *Social Policy* (Summer 1986): 51–56.

Farmer, Paul. 1992. *AIDS and Accusation: Haiti and the Geography of Blame.* Berkeley: University of California Press.

Fee, Elizabeth, and Daniel M. Fox. 1992. *AIDS: The Making of a Chronic Disease.* Berkeley: University of California Press.

Gilmore, Norbert, and Margaret A. Somerville. 1994. Stigmatization, scapegoating and discrimination in sexually transmitted diseases: Overcoming "them" and "us." *Social Science and Medicine* 39(9): 1339–58.

Goffman, Erving. 1963. *Stigma: Notes on the Management of Spoiled Identity.* Englewood Cliffs, NJ: Prentice-Hall.

Goicoechea-Balbona, Anamaria. 1994. Why we are losing the AIDS battle in rural migrant communities. *AIDS and Public Policy Journal* 9(1): 36–48.

Kain, Edward L., and Shannon Hart. 1987. *AIDS and the Family: A Content Analysis of Media Coverage.* Presented to the National Council on Family Relations, Atlanta.

Karon, John M., Patricia L. Fleming, Richard W. Steketee, and Kevin M. De Cock. 2001. HIV in the United States at the turn of the century: An epidemic in transition. *American Journal of Public Health* 91(7): 1060–68.

Land, Helen. 1994. AIDS and women of color. *Families and Society* 75(6): 355–61.

Lea, Amandah. 1994. Women with HIV and their burden of caring. *Health Care for Women International* 15(6): 489–501.

Macionis, John J. 2008. *Sociology*, 12th ed. Upper Saddle River, NJ: Pearson, Prentice Hall.

McGinnis, J. Michael, and William H. Foege. 1993. Actual causes of death in the United States. *Journal of the American Medical Association* 270: 2207–12.

Musto, David F. 1988. Quarantine and the problem of AIDS. Pp. 67–85 in Elizabeth Fee and Daniel M. Fox (eds.), *AIDS: The Burdens of History.* Berkeley: University of California Press.

Panos Institute. 1990. *Triple Jeopardy: Women and AIDS.* London: Author.

Risse, Guenter B. 1988. Epidemics and history: Ecological perspectives and social responses. Pp. 33–66 in Elizabeth Fee and Daniel M. Fox (eds.), *AIDS: The Burdens of History.* Berkeley: University of California Press.

Rosenberg, Charles E. 1988. Disease and social order in America: Perceptions and expectations. Pp. 12–32 in Elizabeth Fee and Daniel M. Fox (eds.), *AIDS: The Burdens of History.* Berkeley: University of California Press.

Shibutani, Tomotsu. 1966. *Improvised News: A Sociological Study of Rumor.* Indianapolis, IN: Bobbs-Merrill.

Shilts, Randy. 1987. *And the Band Played On: Politics, People, and the AIDS Epidemic.* New York: St. Martin's Press.

Sosnowitz, Barbara G., and David R. Kovacs. 1992. From burying to caring: Family AIDS support groups. Pp. 131–44 in Joan Huber and Beth E. Schneider (eds.), *The Social Context of AIDS.* Newbury Park, CA: Sage.

UNAIDS. 2002. *Children on the Brink 2002: A Joint Report on Orphan Estimates and Program Strategies.*

UNAIDS. 2004. *2004 Report on the Global HIV/AIDS Epidemic: 4th Global Report.*

UNAIDS. 2007. AIDS epidemic update. Accessed May 30, 2008, at www.unaids.org.

UNAIDS. 2008a. Asia AIDS epidemic update regional summary. Accessed May 29, 2008, at www.unaids.org.

UNAIDS. 2008b. Sub-Saharan Africa AIDS epidemic update regional summary. Accessed May 29, 2008, at www.unaids.org.

Weitz, Rose. 1991. *Life with AIDS.* New Brunswick, NJ: Rutgers University Press.

World Health Organization. 1991. *Current and Future Dimensions of the HIV/AIDS Pandemic: A Capsule Summary.*

Zablotsky, Diane L., and Marcia G. Ory. 1995. Fulfilling the potential: Modifying the current long-term care system to meet the needs of persons with AIDS. *Research in the Sociology of Health Care* 12: 313–28.

51

A World Growing Old

JEREMY SEABROOK

Once upon a time, the elderly held high status in societies around the world. Today the number of societies in which this is true has dwindled, leaving the elderly in many societies feeling left behind, ignored, and disenfranchised. Seabrook analyzes the reasons for this change, the current status of the elderly, and the impact globalization is having on the elderly. One problem he points to is the lack of younger people to offset the growing percentage of elderly persons in most Western societies and the problem this imbalance causes for economies, especially in trying to support their older populations.

As you read about populations growing older, think about the following questions:

1. *What is producing the increase in the percentage of the elderly in many societies?*
2. *What is the impact on societies of this increase?*
3. *What can or should societies do about the growing number of elderly persons and the issues of support for the elderly?*
4. *Have you noted an increase in the elderly in your community? If so, which changes is this bringing about?*

GLOSSARY **Globalization** Increased economic, political, and social interconnectedness and interdependence among societies in the world. **Ageism** Prejudice and discrimination against the elderly.

RESPONSES TO AGEING

The world has, over time, produced a vast range of responses towards old age. These often contradict one another, as well they might, given the ambiguities surrounding old age itself. Growing old may be regarded as a time of ripeness and fulfilment or a period of declining health and failing powers. The storehouse of human societies has amassed a great variety of ways and means of coming to terms with an experience which remains essentially *that of other people*, until, at last, it catches up with us too.

There are good reasons not to anticipate the decline that comes with ageing, not least the tendency to avoid meeting trouble halfway. Received ideas about ageing are often a means of evasion and denial. "I'll worry about that when the time comes." "I'm not going to live that long." "I believe in living in the present." Our own old age is almost inconceivable until it is upon us. That it is a time of serenity, or that it holds all the terrors associated with standing on the edge of eternity, are beliefs of convenience, a mechanism to distance our younger selves from our own fate.

Source: From *A World Growing Old*, by Jeremy Seabrook. London and Sterling, VA: Pluto Press, 2003, pp. 1–13.

Throughout most of recorded time old bones were rare, and the great majority of people would have died by the time they reached what we would now consider middle age. In Britain, in 1901 8 percent of the population were over 60. By 1941 this had risen to 14 percent. In 1991 it was 20 percent. Today, although the old are present in increasing numbers, they nevertheless suffer a different kind of invisibility. They have become part of the landscape, obstacles on the sidewalk, impediments to the accelerating tempo of life, delaying the swiftly moving crowds in their urgent forward movement. Although they constitute one-fifth of the population, as one elderly woman in North London said, "People look through you. If you are old and a woman, you are doubly invisible. We have become like ghosts before we die."

The present moment inflects the ancient puzzle of old age and its meaning in ways that are historically unprecedented. In Britain, in 2002 it was remarked that for the first time there are more people over 60 than under 16. This ought, in a democracy, to give greater power to the elderly.

Yet the testimony of the old suggests something different. Paradoxically, as they become more numerous, they observe a growing indifference towards them. It seems to them that the rich reservoir of their accumulated experience is a wasting—and often wasted—resource. They find themselves speaking an alien language to those who have little wish to understand. They no longer recognize the world they live in. "We have lived too long" is a recurring theme.

It is remarkable that, now that the elderly are so numerous in the world, they should lament their loss of influence and power. Although in the past there were cultures which exiled or even killed their old, for the most part, when they were comparatively few, they commanded both respect and obedience. It is, perhaps, easier to create myths of wisdom and discernment in hoary heads when these are uncommon; and the nodding of senescence might well frequently have passed for sagacity.

But when life expectancy rises well into the 70s—and in Japan now, for women it is over 80—the scarcity value of the old is undermined. The growing numbers of elderly in the world, far from representing a precious store of wisdom, are often perceived as a constraint upon the freedom and development of the young. It is not that large numbers of older people are abandoned or institutionalized. The myth of a more caring past persists, even though it has been rare for elderly parents to live with their families. In 1929–30, for instance, less than one-fifth of over-60s lived in extended families, and only 7 percent lived in three-generation households. It was more common for people to live closer to their elderly parents than is now the case: in the dense mesh of the streets of industrial Britain, relatives often lived a few doors, or a couple of streets, away. The distance between people, which some observe today, is only partly spatial. It is also psychological, since the destinies of individuals diverge more obviously than they did when most people worked in the staple industry of a single town and expected their children to do likewise.

Conflict between the generations is no new thing. All cultures tell of a new generation, eager to play its part in the life of society, excluded and often humiliated by those in positions of power and influence. And that means the old, seniors, chiefs and headmen. Youthful energy, repressed by elders, is a persistent theme.

In many societies, the authority and prestige of elders were often unlimited. In Thailand, traditional law stated that wives and children were liable for the commission of crimes by the (senior, male) head of the family. "The liability was not due to the fact that they were members of a family, but because their status in the family was property owned by the head of the family. Which was not so different from the manner by which slaves were owned."[1] In some cultures a child could be given as payment to a creditor. A girl might be given to cancel a debt, and she would become the mistress of the individual to whom she was given.

Feudalism in Europe was a hierarchical system which was believed to reflect on earth the hierarchy of heaven, with its archangels, angels and saints, and an omnipotent God at its apex. Social reconstructions of this belief in the arrangements of religious institutions, and the societies that evolved around them, have shown a remarkable persistence through time.

Veneration of the elderly, especially of men, had an even more direct significance in tribal societies where the hierarchy of the dead and living was blurred. The ancestors were closest to God and had to be propitiated in order to earn their goodwill towards the living. Among the living, the oldest members of the tribe, being close to death, had a privileged relationship to all those who had gone before. Ancestor worship was an extension into the supernatural of existing family structures, in which the older members enjoyed a high level of authority. The family comprised both the material world and the invisible, but no less real, world of the spirits. The family and the tribe transcended mortality, and the oldest were the bridge between the living and the dead.

Nor is this unintelligible to us. Even today, many people in the West think of the dead as "looking down," "watching over" the living, a mixture of guardian angel and moral police. The dead are granted the compensatory privilege of supervising our mortal lives. I was much struck, at the time of the death of Diana, Princess of Wales, by the number of cards and mementoes left by people outside Kensington Palace referring to her caring for people, and her ability to do so now from her place in heaven. Speaking ill of the dead remains a taboo, even if much weakened by a market avid for revelations and the true story of dead celebrities.

The idea of the patriarch, the paterfamilias, the head of family, has been remarkably tenacious in all castes and classes. Their power was not uncontested—the resentment it created in the young may be read in the almost universal severity of the laws against parricide. The next generation must have been often tempted to put an end to the tyranny of those who lived on, denying them their inheritance, land and the power that went with it. This temptation had to be limited by the threat of the most draconian punishments.

Nor was the power of the patriarch curbed by the coming of industrial society. Industrial discipline only strengthened the authority of senior males in all social classes, exemplified by the often tyrannical, though sometimes paternalistic, mill or factory owner. The industrial workers, who were at the mercy of the arbitrary power of employers, visited their own victimhood on those over whom they had control, their wives and children.

STATUS OF THE ELDERLY

Now, everywhere in the world, gerontocracy is dying, although faster in some cultures than others. In certain areas of the world, the weakening powers of the old have called forth a vigorous reaction and a sometimes violent reassertion of authority. This is one possible reading of the emergence of religious fundamentalism: the reclamation of traditional forms of social and spiritual control by priests, imams and all the other—usually aged—intermediaries between this world and the next. A reaffirmation of dominance expresses itself in a hardening of old faiths: fundamentalism, ostensibly "a return to tradition," is a very contemporary phenomenon, a response to a modernization which robs elders of power and undermines sources of authority.

In Africa, where rural, clan-based societies bestowed social and religious knowledge on elders, and where the main productive resource—land—was controlled by them, these patterns were first disrupted by colonialism. Later, Western-style education discredited ancient patterns of lordship by shamans, traditional healers and priests, and empowered those who had acquired the skills and knowledge appropriate to a new, urbanizing and industrial society.

In Asia, joint and extended families are rapidly decaying under the same influences. The knowledge of the old is perceived increasingly as of dwindling use to, and an encroachment upon, the lives of a generation formed for a quite different way of living from anything known to their forebears. That the young should see this as liberation, and the elderly as evidence of deterioration, is scarcely surprising. But contemporary shifts in sensibility go far beyond a familiar cross-generational friction. They are symptomatic of more profound social and economic movements in the world, which have caught up whole cultures and civilizations in the compulsions of globalization.

These have their origin in convulsive changes that have occurred in the West, where accelerating technological innovation, "de-industrialization" and economic restructuring have rapidly removed the skills and competences of an older generation in favor of the flexibility and adaptability of the young. The "virtues" of frugality, thrift and self-denial have been eclipsed, since these are an embarrassment to a consumer society where status reflects spending power, and extravagance is a sign of success. Youth has acquired a social supremacy it has hitherto rarely enjoyed. This has been at the expense of the old.

LIFE EXPECTANCY AND GLOBALIZATION

The dramatic rise in life expectancy is, to a considerable extent, a result of the application of medical technologies, which have prolonged life far beyond anything foreseen by the introduction of the welfare state in the mid-twentieth century. But in the rich countries, other factors have contributed to the rising proportion of elderly people, some of which are puzzling.

It was not anticipated that populations would fail to replenish themselves in the "developed" world. In Britain, in 2002 the birth rate fell to 1.6, which is just below the level at which the population will maintain itself. Wolfgang Lutz of Austria's International Institute for Applied Systems Analysis estimates that almost half the population of Western Europe and Japan will be over 60 by the end of the twenty-first century.[2] This forecast may, of course, prove false, as demographic extrapolations often have been in the past. (There was, for instance, a scare in Britain in the 1930s about the future depopulation of the country. It was forecast then that the total population of Britain by 2000 would be a mere 35 million. This prediction was swiftly overtaken after the Second

World War, when the birth rate rose again, affluence became widespread and, above all, young and healthy migrants from the Caribbean, India and Pakistan came to ease labor shortages, and in the process rejuvenated the population.) In spite of this, however, there is no doubt that a reduction in the proportion of people of working age in relation to the retired is imminent.

The social, economic and moral consequences of these developments are far-reaching, although there is by no means unanimity on their meaning. Some researchers find nothing disturbing in the projections....[3] Optimists argue that with a healthier older population and their desire to go on working longer, with continuing economic growth and improving productivity, there is no reason for excessive concern.

...Others argue that globalization endangers the collective social transfers that are essential to elders in later life, pointing out that work, the family and collective institutions are all jeopardized by the neo-liberal ideology that presently dominates the global economy: work is decreasingly available to older people in the West (despite the current talk of raising the retirement age), as well as in the South, as the informal economy is replacing a "liberalized" formal sector; family support is eroded by growing individualism, while resistance to public spending is part of the global ideological curb on state provision for old age.

REPLACING THE GENERATIONS

The United States is the only industrialized country which has a fertility rate above the replacement level of 2.1 children per woman. The United States has also maintained a fairly steady flow of immigrants from all over the world. About 30 million people in the United States were born outside the country, while there are an estimated 6 million undocumented migrants. These factors combine to protect the United States against the threat of drastic population decline or a very high proportion of elderly. In spite of this, however, it is estimated that by 2020 23 percent of the U.S. population will be over 60. After the trauma of September 11, it may be that migration into the country will become more tightly controlled; the effect of this on the population profile and, consequently, on the dynamism and energy of the United States is not yet clear.

In the United States, the proportion of the population over 65 is expected to double by 2030 to 70 million, while the number of people over 80 will rise from 9.3 million in 2000 to 19.5 million in 2030. This will lead to increased health-care costs. In 1997, the United States had the highest per capita health-care spending per person over 65 (U.S. $12,100), by far greater than that of Canada (U.S. $6,800) and the United Kingdom (U.S. $3,600). In the United States, nursing home and home health-care spending doubled between 1990 and 2001, when it reached U.S. $132 billion.

In North America, on average individuals between the ages of 65 and 69 have a further life expectancy of about 15 years. Between 75 and 79 it reaches ten years, and even at 80 it is six or seven years.... International agencies, governments, national charities and local organizations now routinely commit themselves to policies against ageism. These remain largely declaratory, although legislation against age discrimination in employment has been effective in the United States, where the over-60s make up a larger proportion of the workforce than in any other Western country.

However this may be, the *social* power of the elderly shows little sign of being enhanced by their numbers. Youth, as an increasingly scarce commodity, is likely to go on appreciating in the demographic marketplace. If it has traditionally been the destiny of the young to rail against the authoritarianism and tyranny of age, there is little evidence that when the young are in the ascendant they are likely to be more merciful to their elders than these were to those subordinated to *them* in the past.

Nevertheless, the capacity to prolong life yet further, into the tenth and eleventh decades, is constantly advertised by enthusiasts of technological progress. These promises of a provisional immortality are limited only by questions sometimes raised about the purpose and function of superfluous aged populations, their unproductiveness and their dead weight on the declining number of earners of the future. It seems we are likely to hear much more about the desirability—or otherwise—of shortening, rather than extending, the lifespan by a further 20 years....

What is certain is that, within little more than a generation, the population of much of the developed world will be ageing and falling. It will also be fat (more than 20 percent are expected to be obese). These mutations in European society are unparalleled in modern times, and it is scarcely surprising that the policies to deal with them are both improvised and inadequate.

What does it mean, if rich societies fail to replenish themselves? Have they become too—what?—selfish? frightened? liberated? Does it matter? Are declining populations a blessing to the crowded lands of Japan or the Netherlands? Has child-bearing become too burdensome? Should we celebrate the freedom of women from an ancient cycle of pregnancy and childbirth, subservience and enslavement to the will of men? What are the consequences of elective childlessness for the future structure and cohesion of society?

Or have children simply become too expensive? In the United States, where most aspects of human life have been meticulously costed, the Department of Agriculture estimates that it now costs between U.S. $121,000 and U.S. $241,000 to bring up a child. A baby born today will be even more costly. By the age of 17, these omnivorous infants will have devoured between U.S. $171,000 and U.S. $340,000. It seems that the privileged people of the world are coming to regard children as something of a luxury. The comfort of the present depends not only upon growing inequality in the distribution of the wealth of the world, but is also constructed on the absence of the unborn. How future—and possibly depleted—generations will regard the legacy bequeathed by their begetters scarcely troubles a world which feels the pressing problems of today weigh upon it quite heavily enough without having to think about those of a distant tomorrow.

The Western model of development has now usurped all others and is presented as the sole source of hope and renewal to the whole world. What are the implications of this, when it creates a Japan or an Italy peopled by shadows, whose lives have been prolonged by technology far beyond anything that can be understood as their "natural term"? What will these people do, sitting in the low-watt penumbra of old-age homes, their hearing ruined by decades of hyper-decibel music, their eyesight dimmed by long years of voyeuristic television, their memories all but erased by the media-crowded images of the day before yesterday? Even in the West such an achievement chills the spirit. Can it be, should it be, exported globally?

In its example to the world of abstention from increasing its population, we have a rare case of Europe and Japan practizing what they preach—a birth control so effective that we can see future generations dwindling before our eyes. In the meantime, it is clear that the people of the rich countries will be able to ransack the countries of the South for urgently needed personnel to service our dereliction. Having already extracted maximum profit from their crop lands, forests, seas and mineral riches, and having taken advantage of the cheapness of their labor in the slums of Mexico City, Jakarta, and Dhaka, we shall now pluck out the people they depend on most to help their own countries deal with the asperities of globalization—doctors and nurses, carers for the old and infirm. Of course, people-stealing is not new. It was once known as slavery, but in the transformed circumstances of globalization, this now appears as privilege.

NOTES

1. *The History of Thai Laws*, Chanvit Kasetsiri and Vikul Pongpanitanondha (eds), Bangkok, n.d.

2. *Guardian*, August 20, 2002.

3. *Guardian*, August 16, 2002.

52

Fourteen Forecasts for an Aging Society

SAM L. ERVIN

A population trend that accompanies modernization is lengthened life spans; in other words, people live longer. And the "oldest-old" groups are expanding most rapidly. This change creates new challenges for societies as they provide for the needs of aging citizens. Ervin addresses forecasts for societies in the future.

As you read this selection, think about the following questions:

1. *How does an aging population affect society as a whole?*
2. *How are societies likely to change due to an aging population?*
3. *What is the relationship between an aging population and a society's health care system?*
4. *Which of the 14 forecasts have you seen come true? Give examples.*

GLOSSARY   **Medicare** U.S. government program to fund health care for the elderly.   **HMO** Health maintenance organization; a form of group insurance. **Baby boomers** Members of the large post-World War II cohort, born between 1946 and 1964.

As the baby-boom generation ages and the pool of retirees increases exponentially, a period of great change in elder care looms.

The median age of the U.S. population has been steadily rising. In 1900, one American in 25 was 65 or over. By 2050, that figure will increase to one in five. The U.S. Census Bureau projects that the over-65 population will more than double between 2000 and 2050. The proportion of "oldest-old" Americans, those 85 and over, will grow even more rapidly—quadrupling over the same period. By the year 2020, the ratio of over-65 individuals to the working-age adult population will be about one to four.

As a result, we will see sweeping changes in health care, including Medicare, the government program that currently covers 39 million Americans. There will also be major growth in options for elder care, as well as a flood of new products and services aimed directly at this swelling segment of the population.

Source: Originally published in the November/December 2000 issue of *The Futurist*. Used with permission from the World Future Society, 7910 Woodmont Avenue, Suite 450, Bethesda, Maryland 20814. Telephone: 301/656-8274; Fax: 301/951-0394; http://www.wfs.org

Here are 14 forecasts based on recent surveys and studies, many of them conducted by the SCAN Health Plan, a not-for-profit plan serving about 39,000 seniors in Southern California.

1. THE RETIRED WILL WORK AGAIN

More seniors are likely to reenter the labor force, thanks to new legislation allowing those 65 to 69 to earn without penalizing Social Security benefits. Currently, 23% of the 9.2 million people in this age bracket are in the labor force. That is over 2 million senior workers.

Seniors' job-search success will no doubt be boosted by their increased comfort and proficiency with computers.

In addition to the financial benefits of earning a steady paycheck, seniors might get some health benefits, too. A SCAN Health Plan study among its members found that nonworking seniors are more likely to have significant health and daily living problems.

2. TECH-SAVVY SENIORS WILL MAINTAIN THEIR INDEPENDENCE

Elder-friendly technology will significantly improve access to resources and information to assist those who are frail and vulnerable. It will also reduce isolation among those living in rural or hard-to-reach areas of the country.

Products such as the multifunctional pager, which alerts seniors when it is time to take a particular medication, will be readily available. The increased use of technology will be a key factor in helping tech-savvy seniors to remain living independently, as it will enhance their ability to communicate and obtain valuable health-care information. Technology will

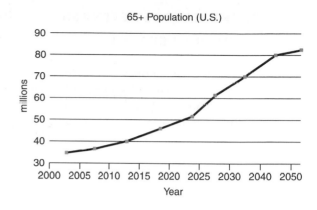

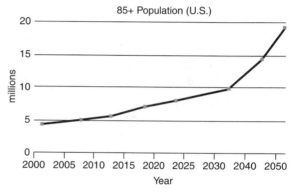

The population of senior 65 ad older will more than double by 2050. Seniors 85 and up, the most in need of elder care, will quadruple.

SOURCE: SCAN Health Plan. Data supplied by U.S. Census Bureau.

also allow health-care providers to better monitor their patients.

Currently, nearly a quarter of the 20 million seniors aged 60 to 69 own and use a computer, as do 14% of the 16 million seniors 70 to 79, according to AgeLight Institute. Computer ownership among members of SCAN Health Plan is even greater. A survey found that 36% of seniors 65 to 74 reported having a PC in the home. Of those 74 to 89, 34% owned a computer.

While technology will play a wider role for seniors in the twenty-first century, the potential for isolation will increase the need for service fostering human interaction.

3. THE HOTTEST FITNESS BUFFS? SENIORS!

Health plans may begin to offer health club memberships and personal trainers as part of their coverage for seniors because of the proven benefit regular exercise has on seniors' overall health. Health clubs report that seniors are the fastest-growing group of members.

A recent survey finds more than two-thirds of seniors engage in regular physical exercise, double the national average for younger adults.

A SCAN survey of 2,035 seniors aged 65 to 90 found that 68% maintain a regular regimen of exercise ranging from moderate activities like walking to more intense workouts including weight lifting, jogging, cycling, tennis, and even heart-pounding handball. Forty percent of the survey respondents said they spend more than four hours per week on these activities. Another 44% exercise one to four hours each week.

Among a group of seniors in a SCAN-sponsored mall-walking program, an insulin-dependent diabetic was able to reduce insulin intake by four units per day. Others reported improved heart conditions, lower blood pressure, less pain from arthritis, and the elimination of leg cramps.

4. SENIOR-FRIENDLY CARS WILL OFFER INDEPENDENCE

Automakers may one day market cars that are easier and safer for America's growing population of seniors to drive.

The "senior-mobiles" may include such features as higher seats, larger numbers on the speedometer, and slower acceleration.

Retaining the ability to drive is the chief concern among aging seniors, according to SCAN research. For many, losing the ability to drive means losing one's independence and being forced to rely on others for transport.

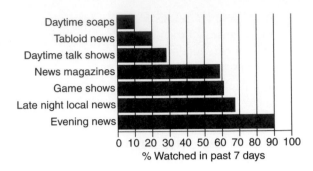

Seniors stay informed. In a SCAN Health Plan survey, seniors reported watching some type of news program more frequently than any other type of programming. Only 29% of respondents said they watched daytime talk shows, and soap operas drew just 10% of the seniors.
SOURCE: SCAN Health Plan.

5. SENIORS WILL BE IMPORTANT VOTERS

Seniors will wield the most power of any demographic group in the voting booth.

Seniors are big-time voters. In the 1996 federal elections, more than two-thirds of those 65 and older voted, according to the Census Bureau. That is 36% better than the 25 to 44 age group.

Seniors not only rank as the top voters, but they are likely to be the most-informed voters. Contrary to the popular belief that seniors are not interested in current events, a SCAN Health Plan survey reveals that they are major consumers of the daily news.

According to the survey, four of the top five television programs most frequently watched by seniors are some type of news program. Only 29% of respondents said they watched daytime talk shows. Soap operas drew just 10% of the seniors. By comparison, 90% said they had watched the evening news.

Seniors are also avid readers. Eighty-seven percent of seniors ranked reading the newspaper among their most favored regular activities. Magazine reading was favored by 75%.

6. MORE ALTERNATIVES TO NURSING HOMES WILL EMERGE

One of the government's biggest tasks in the new millennium will be an extensive education campaign to increase awareness of the wide range of alternatives to nursing homes.

Because of Medicaid's reliance on publicly funded nursing homes and hospital care, most seniors are not aware of alternatives, such as assisted living, independent living, life-care communities, and adult day care. A Harvard School of Public Health survey found that a majority of adults over 50 had never heard or read about six of 10 alternatives to nursing homes listed in the survey.

Nursing-home care cost Medicaid $40.6 billion in 1998 (24% of total outlays), compared to $14.7 billion in 1985. Clearly, more cost-effective options are needed, although these alternatives must provide a high standard of care.

7. BOOMERS COULD END UP IMPOVERISHED

Many aging baby boomers who thought they would be spending their golden years in relative financial comfort may actually find themselves impoverished because they did not prepare for the costs of long-term care.

A survey by the American Health Care Association found that 68% of baby boomers are not financially prepared for long-term care should they need it later in life. Half of the boomers polled had not even given any thought to how they will pay for long-term-care needs.

Part of the problem may stem from a lack of understanding about how long-term care is paid for.

A separate survey by the National Council on Aging, in conjunction with John Hancock Mutual Life Insurance, found that one-third of baby boomers incorrectly assume that Medicare is the primary source for long-term-care funding. The fact is, Medicare covers only about 53% of all health-care costs. Most people

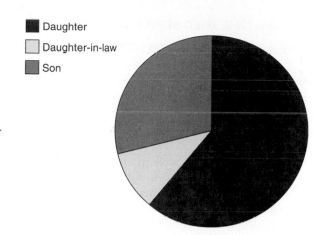

Legend:
■ Daughter
□ Daughter-in-law
▨ Son

Women's climb up the corporate ladder may be slowed as they remain the predominant caregivers for aging parents. A SCAN Health Plan survey found that, of seniors relying on a child for caregiving, 71% rely on a daughter or daughter-in-law.
SOURCE: SCAN Health Plan.

are surprised to find that it does not cover long-term care, prescription drugs, or vision or dental care.

8. ELDER CARE SHORTAGE IS COMING

As the population of those 85 and older doubles to 8.4 million by 2030, the demand of professional home-care aides will skyrocket. By 2008, personal care and home health aide jobs will rank second only to systems analysts in terms of sheer numbers (1.19 million versus 1.18 million), according to the Bureau of Labor Statistics.

Twenty-nine percent of seniors age 65 and up rely on a daughter or daughter-in-law for caregiving assistance compared to 12% who rely on a son, according to a SCAN survey. However, daughters will become increasingly unable to take on the caregiver's role; more and more women will take on time-consuming managerial and executive positions as they continue to outpace men in earning bachelor's degrees (637,000 versus 500,000 in 2000).

As the need for in-home elder-care-service providers soars, a major shortage of caregivers could result.

9. AGING BOOMERS WILL FORCE HEALTH-CARE-POLICY CHANGES

Baby boomers can be expected to use their influence to ensure that quality health-care services and programs are available for their parents now, not to mention for themselves in the not-too-distant future.

There is power in numbers among the baby boomers. Approximately 76 million boomers born between 1946 and 1964 will soon join the ranks of older Americans. Many of these boomers are beginning to provide care for their aging parents.

However, they also have a significant vested interest in ensuring that quality health care will be available for themselves. According to the Census Bureau, one in every nine boomers will live at least 90 years.

Perhaps the biggest concern among boomers is long-term care, which is currently not covered by Medicare. In fact, one survey found that long-term care has replaced child care as the number-one concern among baby boomers. Nearly 66% of boomers polled in the 1999 survey by the National Council on the Aging and John Hancock Life Insurance said they would expand Medicare to cover long-term care, even if it means paying higher taxes.

10. ELDER CARE WILL HURT WOMEN'S CAREERS

Women's progress toward finally shattering the glass ceiling will be slowed because of the increasing burden of elder care.

Despite women's advancement up the corporate ladder, a SCAN survey indicates that women are still disproportionately affected by elder care.

Strong evidence of elder care's adverse impact on career advancement comes from a national study conducted by the National Alliance for Caregiving and the American Association of Retired Persons. The study found that 31% of caregivers significantly alter their career paths; some leave the work force altogether.

According to the study, which polled 1,200 people who provided caregiving assistance to someone 50 or older, 11% of caregivers took a leave of absence, 7% opted to work fewer hours, 4% lost job benefits, 3% turned down a promotion, and 10% took early retirement or quit their jobs.

11. TELECOMMUTING WILL ASSIST FAMILY CAREGIVERS

Increased telecommuting will allow working adults to move closer to their aging parent and thus ease the burden of long-distance caregiving.

The National Council on the Aging (NCOA) estimates that nearly 7 million caregivers provide care for someone who lives at least one hour away. As more and more companies offer telecommuting, employees will be able to work from anywhere in the country and will have the flexibility to meet elder-caregiving responsibilities.

The percentage of companies with telecommuting rose steadily during the 1990s. In 1993, just 6% of the companies surveyed by William M. Mercer Inc. offered telecommuting. That figure grew to 14% by 1995 and 33% by 1998.

12. MORE EMPLOYERS WILL OFFER ELDER CARE

Elder-care benefits will become a major issue for workers and their employers as increased elder care-related absences and falling productivity begin to take a toll on the workplace.

The Census Bureau projects that over 40 million people in the United States will be older than 65 by the year 2010, an increase of 19% from 1995. As a result, about 12% of informal caregivers will quit their jobs to provide care full time, estimates the Family Caregiver Alliance. In a tight labor market, employers will offer more elder-care benefits to combat employee turnover.

In order to attract and retain quality employees, as well as strengthen productivity, companies will develop ways to help their employees deal with the burden of caring for aging parents. This trend will mirror the movement among corporations to provide child-care assistance in the 1980s and early 1990s.

In 1999, 47% of companies offered some type of elder care, up from 40% in 1998, according to an annual survey of over 1,000 employers by benefits consultant Hewitt Associates of Lincolnshire, Illinois. By comparison, 90% of companies offered some type of child-care assistance.

13. CAREGIVERS WILL NEED INTERVIEWING SKILLS

The sandwich generation, those caught between raising children and taking care of aging parents, will have to learn a lesson from human-resource executives.

As the senior population grows, there will be a constant stream of new business start-ups offering products and services to this group. Caregivers will be forced to make decisions on whom to hire and which organization or firm to use, much as human-resource professionals do on a daily basis.

Employers, working in concert with their human-resource departments, will set up training sessions for employees to allow them to make better decisions when it comes to hiring home aides or choosing elder-care-service providers. These sessions will stress that employees need to:

- Brush up on interviewing skills. Most of us have been through an interview process, but only as a candidate.

- Ask about availability of services. Twenty-four-hour, seven-day-a-week services are a must.

- Have potential caregivers submit résumés, complete with references.

- Ask, "Will the qualified care manager have a qualified back-up during vacations and time off?"

- View parents and relatives receiving care as "upper management," in that no decision on hiring should be made without upper management's input.

14. WORKING FAMILIES WILL GAIN STATE ALLIES

As the balance between work and family becomes an increasingly major issue, more state legislatures are likely to follow in the footsteps of the four that have already taken steps to ensure that companies accommodate employees' caregiving responsibilities.

While the legislation covers family caregiving, it is especially valuable to workers caring for an aging parent.

Terms of the legislation approved in California, Oregon, Washington, and Minnesota allow workers not covered by union contracts to use up to one-half of their paid sick leave to care for an ill child, spouse, or parent.

Elder care is likely to become an increasingly worrisome issue for employers as aging seniors eschew nursing homes in favor of independent living. Annual growth in nursing home residents has slowed to just 0.4%, down from 4.8% in the mid-1970s, according to the National Bureau of Economic Research.

However, many independently living seniors still need some type of caregiving assistance. SCAN surveyed 1,453 members, averaging age 82: None lived in a nursing home, despite meeting state qualifications. The survey found that 76% rely on a caregiver for assistance with daily activities. In 39.5% of these situations, the primary caregiver is a son or daughter.

THE FACTS OF LONG-TERM LIFE

Consider the following concluding statistics:

- In the twenty-first century, one in five Americans will be 65 or older. One in nine current baby boomers will live to at least age 90. The number of those 85 years old and over will quadruple by 2050.

- About 6.5 million older people need assistance with activities of daily living (e.g., bathing, cooking, cleaning, dressing). That number is expected to double by 2020.

- Women account for 72% (18 million) of the approximately 25 million family caregivers in the United States.

- The U.S. Census Bureau projects that the number of caregivers will drop from 11 for each person needing long-term care in 1990 to four in 2050.

- Collectively, family caregivers spend $2 billion of their own assets each month to assist relatives.

- By 2005, noninstitutionalized people over age 65 may spend an average of $14,000 annually on health care.

- Nearly 90% of baby boomers say taking care of their parents is among their top three life priorities.

- Ninety-four percent of seniors believe their health conditions do not affect their adult children's quality of life, but 80% of children say they do.

- Less than one-quarter of seniors expect to move in with their children; more than half of baby boomers anticipate having their parents move in at some point.

- Eighty-one percent of seniors do not believe their children will have to provide a great deal of financial support for their care; one-third of children believe they will.

The number of older Americans is increasing rapidly, while the human and financial resources to care for them are dwindling. Clearly then, elder care will become a major political issue in local, state, and national elections in the decades ahead.

Chapter 13

The Human Environment
Population and Urbanization

Dr. Alan W. McEvoy

Population expansion and related ecological problems are with us today and will remain with us for many years to come—if we survive. This is the verdict of many experts, called *demographers*, who study changes in human populations. Three variables are crucial to population change: fertility (the birth rate), mortality (the death rate), and migration (population movement). In studying populations, demographers focus on how these variables affect three different areas:

1. Population growth and decline: size of the population.
2. Population distribution: where the population is located.
3. Population composition and structure: characteristics of the population, such as age, sex, education, and so forth.

To understand the dynamics that influence everything from the size of populations in cities, regions, and countries, to the opportunities available to each generation—education, health care, jobs—the study of demography is crucial. Why do some people have large families even though they are very poor? Why do people in some countries live, on average, into their 80s while the average life expectancy in other countries is in the 40s? And why do people move from one part of the world, leaving their family and friends, to other parts of the world where they often receive a cold reception and experience discrimination?

Let us look at each of the major population variables. *Fertility rates* are influenced by whether a country is urban or rural. Typically, people in agricultural societies have large families to help with the farmwork. Values and religious beliefs may also stress large families, partly because in past times many children did not survive childhood. Today with modern medicines, many of these large families remain large due to better food, control of diseases, and health care.

Mortality is still very high in some countries of the world, especially in countries faced with drought, famine, lack of health care, poverty, and disease. Malnutrition weakens the most vulnerable, so that the death rate is very high among these groups. Developed countries face a different kind of problem, as many people are now living into their 80s, 90s, and even 100s. This longevity brings up other issues. The health care systems of these countries have developed ways to treat many problems of old age, but at great expense to the societies. Thus we can keep people alive—but at a cost. In addition, the birth rate in many developed countries is low, meaning that there are fewer workers being born to help support the dependent aging populations.

The third variable is *migration*. When people move, they are generally looking for better opportunities or trying to escape persecution or war. Theorists speak of these forces as *push-pull factors* that influence decisions to move. Think about why your family might move. For some people around the world, finding opportunities may mean life or death for themselves and their families.

Each of these areas must be examined to understand the population in the world today. The readings in this chapter deal with these factors and with the impact of population on the environment.

We must be concerned with the rapid growth rate that many countries are experiencing, both for humanitarian reasons and because of the increasing demand on scarce resources. The Population Reference Bureau estimates that in developing countries population doubles every 20 to 35 years. This means that natural resources, food-production capacity, and other essentials must double in the same period to maintain present lifestyles. Yet most countries are demanding more food, improved communications, better education, scientific advances, and higher standards of living at the same time that resources and capacities to produce are already severely strained or declining.

The themes running through these readings are (1) factors that change the population of nations, such as birth rates and death rates, (2) the stress of increased population on the environment and natural resources, (3) human migration patterns, (4) the results of migration to urban areas, and (5) major events that have affected the world's population and environment in recent years.

As you read these articles, consider the implications of population conditions and urbanization along with the benefits and value of protecting our environment. In the first article, David Bloom and David Canning discuss the "booms, busts, and echoes" that affect population dynamics in the global system. As countries move from one stage of development to the next, changes in population dynamics inevitably occur.

The "third technological revolution" continues to bring about changes in birth and death rates through technological advances in medicine, and population migrations to urban areas where technology is concentrated. In the second article, Daniel Bell discusses this revolution and its impact on social structures and populations.

Urbanization refers to the movement of people from rural areas to urban areas and between countries, resulting in most immigrants moving to urban areas. Cities developed centuries ago as agricultural surplus freed people from farming. Those not needed in the vital task of feeding the population were attracted to urban areas by work opportunities and the excitement of city living. This process of migration to cities continues today in many parts of the world. Louis Wirth wrote about urbanism in 1934 in an essay still considered a classic today for its insights on variables affecting urban life. However, the rapid expansion of cities has caused strains on communications, transportation, and distribution of needed goods and services. Rapidly growing urban areas struggle to meet the needs of swelling populations. At the same time, those who can afford to move to the suburbs cause increased problems for cities by reducing the urban tax base. Unable to cope with deteriorating infrastructures and the demands of newcomers, cities are decaying in both physical structure and social control.

Finally, we consider the impact of human populations on our planet. A major factor in the depletion of the earth's resources is overconsumption by developed countries. Exploitation of resources by the world's richest countries threatens the well-being of all. Lester Brown raises questions about the relationships between birth rates and population growth, use of the world's resources, and the human condition.

As you read these selections, consider how population dynamics relate to economic crises, environmental degradation, and city problems. How might we change our consumption patterns to influence and help solve world problems?

53

Boom, Busts, and Echoes

DAVID E. BLOOM AND DAVID CANNING

Understanding demographic (population) patterns—the births, deaths, and migrations of humans—helps us understand the history of the world. Bloom and Canning discuss the variables that affect our lives, especially population trends around the world and their effects on world economies.

As you read this article, think about the following questions:

1. What is the biggest demographic upheaval in history?
2. What caused the sharp rise in global population?
3. What is meant by "the missing link" and what evidence of it do you see?
4. What do the authors predict for the future of the world's population?

GLOSSARY **Missing link** A factor ignored by many analyses of social and economic impacts on countries—"demographic effects." **Total fertility rates** The number of children born per woman in the world. **Life expectancy** The average number of years people in the world or in specific countries live.

For much (and perhaps most) of human history, demographic patterns were fairly stable: the human population grew slowly, and age structures, birth rates, and death rates changed very little. The slow long-run growth in population was interrupted periodically by epidemics and pandemics that could sharply reduce population numbers, but these events had little bearing on long-term trends.

Over the past 140 years, however, this picture has given way to the biggest demographic upheaval in history, an upheaval that is still running its course. Since 1870 death rates and birth rates have been declining in developed countries. This long-term trend toward lower fertility was interrupted by a sharp, post–World War II rise in fertility, which was followed by an equally sharp fall (a "bust"), defining the "baby boom." The aging of this generation and continued declines in fertility are shifting the population balance in developed countries from young to old. In the developing world, reductions in mortality resulting from improved nutrition, public health infrastructure, and medical care were followed by reductions in birth rates. Once they began, these declines proceeded much more rapidly than they did in the developed countries. The fact that death rates decline before birth rates has led to a population explosion in developing countries over the past 50 years.

Even if the underlying causes of rapid population growth were to suddenly disappear, humanity

SOURCE: Bloom, David E., and David Canning. September 2006. "Booms, Busts, and Echoes." *Finance and Development*, pp. 8–13. From The International Monetary Fund and the World Bank, Copyright Clearance Center.

would continue to experience demographic change for some time to come. Rapid increases in the global population over the past few decades have resulted in large numbers of people of childbearing age (whose children form an "echo" generation). This creates "population momentum," where the populations of most countries, even those with falling birth rates, will grow for many years, particularly in developing countries.

These changes have huge implications for the pace of economic development. Economic analysis has tended to focus on the issue of population numbers and growth rates as factors that can put pressure on scarce resources, dilute the capital–labor ratio, or lead to economies of scale. However, demographic change has important additional dimensions. Increasing average life expectancy can change life-cycle behavior affecting education, retirement, and savings decisions—potentially boosting the financial capital on which investors draw and the human capital that strengthens economies. Demographic change also affects population age structure, altering the ratio of workers to dependents. This [article] looks at many facets of the impact of demographic change on the global economy and examines the policy adjustments needed in both the developed and the developing world.

SHARP RISE IN GLOBAL POPULATION

The global population, which stood at just over 2.5 billion in 1950, has risen to 6.5 billion today, with 76 million new inhabitants added each year (representing the difference, in 2005, for example, between 134 million births and 58 million deaths). Although this growth is slowing, middle-ground projections suggest the world will have 9.1 billion inhabitants by 2050.

These past and projected additions to world population have been, and will increasingly be, distributed unevenly across the world. Today, 95 percent of population growth occurs in developing countries. The populations of the world's 50 least developed countries are expected to more than double by the middle of this century, with several poor countries tripling their populations over the period. By contrast, the population of the developed world is expected to remain steady at about 1.2 billion, with declines in some wealthy countries.

The disparity in population growth between developed and developing countries reflects the considerable heterogeneity in birth, death, and migration processes, both over time and across national populations, races, and ethnic groups. The disparity has also coincided with changes in the age composition of populations. An overview of these factors illuminates the mechanisms of population growth and change around the world.

Total Fertility Rate

The total world fertility rate, that is, the number of children born per woman, fell from about 5 in 1950 to a little over 2.5 in 2006 (see Figure 13.1). This number is projected to fall to about 2 by 2050. This decrease is attributable largely to changes in fertility in the developing world and can be ascribed to a number of factors, including declines in infant mortality rates, greater levels of female education and increased labor market opportunities, and the provision of family-planning services.

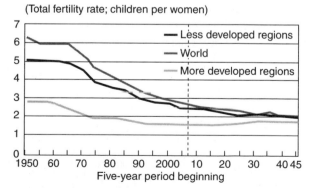

(Total fertility rate; children per women)

— Less developed regions
— World
— More developed regions

Five-year period beginning

F I G U R E 13.1 Smaller Families

Fertility rates are tending to converge at lower levels after earlier sharp declines.

SOURCE: United Nations, *World Population Prospects*, 2004.

Infant and Child Mortality Decline

The developing world has seen significant reductions in infant and child mortality over the past 50 years. These gains are primarily the result of improved nutrition, public health interventions related to water and sanitation, and medical advances, such as the use of vaccines and antibiotics. Infant mortality (death prior to age 1) in developing countries has dropped from 180 to about 57 deaths per 1,000 live births. It is projected to decline to fewer than 30 by 2050. By contrast, developed countries have seen infant mortality decline from 59 deaths per 1,000 live births to 7 since 1950, and this is projected to decline further still, to 4 by 2050. Child mortality (death prior to age 5) has also fallen in both developed and developing countries.

Life Expectancy and Longevity

For the world as a whole, life expectancy increased from 47 years in 1950–55 to 65 years in 2000–05. It is projected to rise to 75 years by the middle of this century, with considerable disparities between the wealthy industrial countries, at 82 years, and the less developed countries, at 74 years (see Figure 13.2). (Two major exceptions to the upward trend are sub-Saharan Africa, where the AIDS epidemic has drastically lowered life expectancy, and some of the countries of the former Soviet Union, where

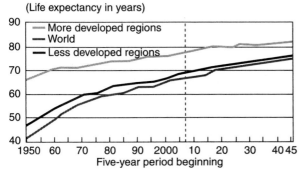

F I G U R E 13.2 Living Longer
Life expectancy is continuing to rise, but there are big differences between rates in well-off and poorer countries.
SOURCE: United Nations, *World Population Prospects*, 2004.

economic dislocations have led to significant health problems.) As a result of the global decline in fertility, and because people are living longer, the proportion of the elderly in the total population is rising sharply. The number of people over the age of 60, currently about half the number of those aged 15 to 24, is expected to reach 1 billion (overtaking the 15–24 age group) by 2020 and almost 2 billion by 2050. The proportion of individuals aged 80 or over is projected to rise from 1 percent to 4 percent of the global population by 2050.

Age Distribution: Working-Age Population

Baby booms have altered the demographic landscape in many countries. As the experiences of several regions during the past century show, an initial fall in mortality rates creates a boom generation in which high survival rates lead to more people at young ages than in earlier generations. Fertility rates fall over time, as parents realize they do not need to give birth to as many children to reach their desired family size, or as desired family size contracts for other reasons. When fertility falls and the baby boom stops, the age structure of the population then shows a "bulge" or baby-boom age cohort created by the nonsynchronous falls in mortality and fertility. As this generation moves through the population age structure, it constitutes a share of the population larger than the cohorts that precede or follow. This creates particular challenges and opportunities for countries, such as a large youth cohort to be educated, followed by an unusually large working-age (approximately ages 15–64) population, with the prospect of a "demographic dividend," and characterized eventually by a large elderly population, which may burden the health and pension systems (see Figure 13.3).

Migration

Migration also alters population patterns. Globally, 191 million people live in countries other than the one in which they were born. On average during the next 45 years, the United Nations projects that

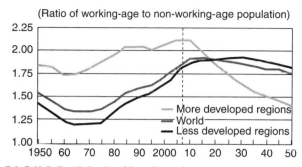

FIGURE 13.3   Tracking the Bulge

Developing countries are nearing the peak of their opportunity to benefit from a high ratio of workers to dependents.

SOURCE: United Nations, *World Population Prospects*, 2004.

over 2.2 million individuals will migrate annually from developing to developed countries. It also projects that the United States will receive by far the largest number of immigrants (1.1 million a year), and China, Mexico, India, the Philippines, and Indonesia will be the main sources of emigrants.

Urbanization

In both developed and developing countries, there has been huge movement from rural to urban areas since 1950. Less developed regions, in aggregate, have seen their population shift from 18 percent to 44 percent urban, while the corresponding figures for developed countries are 52 percent to 75 percent. A new UN report says that in 2007 the worldwide balance will tip and more than half of all people will be living in urban areas. This shift—and the concomitant urbanization of areas that were formerly peri-urban or rural—is consistent with the shift in most countries away from agriculturally based economies.

The existence and growth of megacities (that is, those with 10 million or more residents) is a late-20th-century phenomenon that has brought with it special problems. There were 20 such cities in 2003, 15 in developing countries. Tokyo is by far the largest, with 35 million people, followed by (in descending order) Mexico City, New York, São Paulo, and Mumbai (all with 17 to 19 million).

Cities in general allow for economies of scale—and, most often, for a salutary mix of activities, resources, and people—that make them centers of economic growth and activity and account, in some measure, for their attractiveness. As continued movement to urban areas leads to megacities, however, these economies of scale and of agglomeration seem to be countered, to some extent, by problems that arise in transportation, housing, air pollution, and waste management. In some instances, socioeconomic disparities are particularly exacerbated in megacities.

WHAT IS THE IMPACT ON ECONOMIES?

The economic consequences of population growth have long been the subject of debate. Early views on the topic, pioneered by Thomas Malthus, held that population growth would lead to the exhaustion of resources. In the 1960s, it was proposed that population growth aided economic development by spurring technological and institutional innovation and increasing the supply of human ingenuity. Toward the end of the 1960s, a neo-Malthusian view, focusing again on the dangers of population growth, became popular. Population control policies in China and India, while differing greatly from each other, can be seen in this light. Population neutralism, a middle-ground view, based on empirical analysis of the link between population growth and economic performance, has held sway for the past two decades. According to this view, the net impact of population growth on economic growth is negligible.

Population neutralism is only recently giving way to a more fine-grained view of the effects of population dynamics in which demographic change does contribute to or detract from economic development. To make their case, economists and demographers point to both the "arithmetic accounting" effects of age structure change and the effects of behavioral change caused by longer life spans …

Arithmetic Accounting Effects

These effects assume constant behavior within age and sex groups, but allow for changes in the relative size of those groups to influence overall outcomes. For example, holding age- and sex-specific labor force participation rates constant, a change in age structure affects total labor supply.

As a country's baby-boom generation gets older, for a time it constitutes a large cohort of working-age individuals and, later, a large cohort of elderly people. The span of years represented by the boom generation (which determines how quickly this cohort moves through the age structure) and the size of the population bulge vary greatly from one country to another. In all circumstances, there are reasons to think that this very dynamic age structure will have economic consequences. A historically high proportion of working-age individuals in a population means that, potentially, there are more workers per dependent than previously. Production can therefore increase relative to consumption, and GDP per capita can receive a boost.

Life-cycle patterns in savings also come into play as a population's age structure changes. People save more during their working-age years, and if the working-age cohort is much larger than other age groups, savings per capita will increase.

Behavioral Effects

Declining rates of adult mortality and the movement of large cohorts through the global population pyramid will lead to a massive expansion in the proportion of elderly in the world population (see the projections for 2050 in Figure 13.4). Some simple economic projections show catastrophic effects of this aging. But such projections tend to be based on an "accounting" approach, which assumes that age-specific behavior remains unchanged and ignores the potentially significant effects of behavior change.

The aging of the baby-boom generation potentially promotes labor shortages, creating upward pressure on wages and downward pressure on the real incomes of retirees. In response, people may adjust their behavior, resulting in increased labor force participation, the immigration of workers

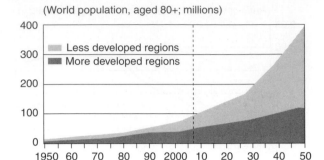

(World population, aged 80+; millions)

FIGURE 13.4 Retiree Boom

The number of people living past 80 is projected to rise sharply, but labor shortages could drive up living costs for retirees.

SOURCE: United Nations, *World Population Prospects*, 2004.

from developing countries, and longer working lives. Child mortality declines can also have behavioral effects, particularly for women, who tend to be the primary caregivers for children. When the reduced fertility effect of a decrease in child mortality is in place, more women participate in the workforce, further boosting the labor supply.

THE MISSING LINK

Demographic effects are a key missing link in many macroeconomic analyses that aim to explain cross-country differences in economic growth and poverty reduction. Several empirical studies show the importance of demographics in explaining economic development.

East Asia's Baby Boom

East Asia's remarkable economic growth in the past half century coincided closely with demographic change in the region. As infant mortality fell from 181 to 34 per 1,000 births between 1950 and 2000, fertility fell from six to two children per woman. The lag between falls in mortality and fertility created a baby-boom generation: between 1965 and 1990, the region's working-age population grew nearly four times faster than the dependent population.

Several studies have estimated that this demographic shift was responsible for one-third of East Asia's economic growth during the period (a welcome demographic dividend).

Labor Supply and the Celtic Tiger

From 1960 to 1990, the growth rate of income per capita in Ireland was approximately 3.5 percent per year. In the 1990s, it jumped to 5.8 percent, well in excess of any other European economy. Demographic change contributed to the country's economic surge. In the decade following the legalization of contraceptives in 1979, Ireland saw a sharp fall in the crude birth rate. This led to decreasing youth dependency and a rise in the working-age share of the total population. By the mid-1990s, the dependency burden in Ireland had dropped to a level below that in the United Kingdom.

Two additional demography-based factors also helped fuel economic growth by increasing labor supply per capita. First, while male labor force participation rates remained fairly static, the period 1980–2000 saw a substantial increase in female labor force participation rates, particularly among those aged between 25 and 40. Second, Ireland historically had high emigration levels among young adults (about 1 percent of the population per year) because its economy was unable to absorb the large number of young workers created by its high fertility rate. The loss of these young workers exacerbated the problem of the high youth dependency rate. The decline in youth cohort sizes and rapid economic growth of the 1990s led to a reversal of this flow, resulting in net in-migration of workers, made up partly of return migrants and also, for the first time, substantial numbers of foreign immigrants.

Continued High Fertility in Sub-Saharan Africa

Demographic change of a very different type can account for slow economic development. Much of sub-Saharan Africa remains stalled at the first stage of a demographic transition. Fertility rates actually increased a bit from the 1950s through the 1970s and only recently have begun a slow fall. As swollen youth cohorts have entered the labor force, an inadequate policy and economic environment in most countries has prevented many young people from being able to engage in productive employment. The existence of large dependent populations (in this case, of children) has kept the proportion of working-age people low, making it more difficult for these economies to rise out of poverty.

LOOKING TO THE FUTURE

Based on the indicators that are available, we can make a few important points:

- *All signs point to continued but slowing population growth.* This growth will result in the addition of roughly 2.5 billion people to the world population, before it stabilizes around 2050 at about 9 billion. Managing this increase will be an enormous challenge, and the economic consequences of failing to do so could be severe.

- *The world's population is aging rapidly.* The United Nations predicts that 31 percent of China's population in 2050—432 million people—will be age 60 or older. The corresponding figures for India are 21 percent and 330 million. No longer can aging be thought of as just a developed-world phenomenon.

- *International migration will continue, but the extent is unclear.* The pressures that encourage people to migrate—above all, the lure of greater economic well-being in the developed countries—will undoubtedly persist, but the strength of countervailing policy restrictions that could substantially stanch the flow of migrants is impossible to predict.

- *Urbanization will continue, but the pace is also hard to predict.* Greater economic opportunities in the cities will surely continue to attract migrants from rural areas, but environmental and social problems may stymie growth.

GETTING THE FOCUS RIGHT

Rapid and significant demographic change places new demands on national and international policy-making. Transitions from high mortality and fertility to low mortality and fertility can be beneficial to economies as large baby-boom cohorts enter the workforce and save for retirement. Rising longevity also tends to increase the incentives to save for old age.

The ability of countries to realize the potential benefits of the demographic transition and to mitigate the negative effects of aging depends crucially on the policy and institutional environment. Focusing on the following areas is likely to be key.

Health and Nutrition

Although it has long been known that increased income leads to improved health, recent evidence indicates that good health may also be an important factor in economic development. Good nutrition in children is essential for brain development and for allowing them to become productive members of society. Health improvements—especially among infants and children—often lead to declines in fertility, above and beyond the heightened quality of life they imply. Focusing on the diseases of childhood can therefore increase the likelihood of creating a boom generation and certain positive economic effects. Countries wishing to accelerate fertility declines may benefit from focusing on access to family-planning services and education about fertility decisions.

Education

Children are better able to contribute to economic growth as they enter the workforce if they have received an effective education. East Asia capitalized on its baby boom by giving its children a high-quality education, including both general schooling and technical skills, that equipped them to meet the demands of an ever-changing labor market. Ireland also profited from its baby boomers by introducing free secondary schooling and expanding tertiary education.

Labor Market Institutions

Restrictive labor laws can limit a country's ability to benefit from demographic change, particularly when they make it unduly difficult to hire and fire workers or to work part-time. International outsourcing, another controversial subject, may become an increasingly important means of meeting the demand for labor.

Trade

One way that East Asian countries provided their baby-boom cohorts with productive opportunities was by carefully opening up to international trade. By providing a new avenue for selling the region's output, this opening helped countries avoid the unemployment that could have arisen. We have found that open economies benefit much more from demographic change than the average, and that closed economies do not derive any statistically significant benefit from age structure changes.

Retirement

Population aging will require increased savings to finance longer retirements. This will likely affect financial markets, rates of return, and investment. In addition, as more people move into old age, health care costs will tend to increase, with the expansion of health care systems and growth in long-term care for the elderly. As nontradable, labor-intensive sectors with a low rate of technical progress, health care and elder care may slow economic growth. The ability of individuals to contribute to the financing of their retirement may be hampered by existing social security systems, many of which effectively penalize individuals who work beyond a fixed retirement age.

Although demographic changes are generally easier to predict than economic changes, the big picture outlook is nonetheless unclear. Indeed, many forces that affect the world's demographic profile are highly unpredictable. Will an outbreak of avian flu or another disease become pandemic, killing many millions and decimating economies? What happens if these diseases are, or become, drug-resistant? Conversely, scientific advances in areas such as genomics, contraceptive methods, or vaccines for diseases such as AIDS or malaria could save and improve millions of lives. Global warming and other environmental change could completely alter the context of demographic and economic predictions. Or—to take things to extremes—wars could result in massive premature mortality, thereby rendering irrelevant most predictions about demographic and related economic changes.

REFERENCES

Bloom, David E., and David Canning, 2004, "Global Demographic Change: Dimensions and Economic Significance," in *Global Demographic Change: Economic Impacts and Policy Challenges*, proceedings of a symposium, sponsored by the Federal Reserve Bank of Kansas City, Jackson Hole, Wyoming, August 26–28, pp. 9–56.

Lee, Ronald, 2003, "The Demographic Transition: Three Centuries of Fundamental Change," *Journal of Economic Perspectives*, Vol. 17 (Fall), pp. 167–90.

National Research Council, 1986, *Population Growth and Economic Development: Policy Questions* (Washington: National Academies Press).

54

The Third Technological Revolution and Its Possible Socioeconomic Consequences

DANIEL BELL

Change can be abrupt and unplanned, long term and planned, or combinations of these. Some change can only be seen over long periods; historians speak of the Dark and Middle Ages, each characterized by major changes in the social and technological structures. Societies have progressed through a number of stages with major inventions such as the plow, the printing press, steam engines, and now the computer. Bell discusses the latest technological revolution and how it is affecting societal structures.

As you read this article, consider the following questions:

1. *The author notes changes from mechanical to electrical systems. Name several other of these types of changes.*
2. *The author also speaks of changes in miniaturization and digitalization. Indicate other areas in which these processes are occurring.*
3. *Think about the rapidity of change and describe the world as you think it will be in 25 years; in 50 years.*

GLOSSARY **Postindustrial economy and society** Based on service work, communications industries, and high technology. **Interdependent international economy** Changes in the scale of human activities such that new technology allows all countries to communicate and become dependent on each other.

We are today on the rising slope of a third technological revolution. It is a rising slope, for we have passed from the plus–minus stage of invention and innovation into the crucial period of diffusion. The rates of diffusion will vary, depending upon the economic conditions and political stabilities of societies. Yet the phenomenon cannot be reversed, and its consequences may be even greater than the previous two technological revolutions that reshaped the West and now, with the spread of industrialization, other parts of the world as well....

What I hope to do in this reading is to identify the salient aspects of the "third technological revolution," sketch a number of social frameworks that may allow us to see how this technological

SOURCE: Reprinted from *DISSENT* Spring 1989, pp. 164–176.

revolution may proceed in the reorganization of basic structures, and describe the choices we may have....

... The third technological revolution. If we think of the changes that are beginning to occur, we think, inevitably of things and the ways we seek to use them: computers, telecommunications, and the like. But to think in these terms is to confuse applications or instruments with some underlying processes that are the crucial understandings for this revolution, and only by identifying the relevant underlying processes can we begin to "track" the vast number of changes in socioeconomic and political structures that may take place. Four technological innovations underlie this new technological revolution, and I shall describe each briefly:

(1) *The change of all mechanical and electric and electromechanical systems to electronics.* The machines of industrial society were mechanical instruments, powered first by steam and later by electricity. Increasingly, electronic systems have taken over and replaced mechanical parts. A telephone system was basically a set of mechanical parts (e.g., a dial system) in which signals were converted into electricity. Today, the telephone is entirely electronic. Printing was a system in which mechanical type was applied, with inked surfaces, to paper; today, printing is electronic. So is television, with solid-state circuits. The changes mean a reduction in the large number of parts, and an incredible increase in the speed of transmission. In modern computers, we have; speeds of nanoseconds, or one-billionth (10^{-9}) of a second (or thirty years of seconds, if one sought to add these up), and even picoseconds, or one-trillionth (10^{-12}) of a second, permitting "lightning" calculation of problems.

(2) *Miniaturization.* One of the most remarkable changes is the "shrinkage" of units that conduct electricity or switch electrical impulses. Our previous modes were vacuum tubes, each, as in the old-fashioned radios, about two or three inches high. The invention of the transistor is akin to the invention of steam power, for it represented a quantum change in the ability to manufacture microelectronic devices for the hundreds of different functions

of control, regulation, direction, and memory that microprocessors perform. We had 4k (k = a thousand) bits on a chip, the size of a thin fingernail, then 32k, 64k, and now we begin to construct megabits, or a million binary digits, or bits, on a chip.

In the past two decades we have seen an exponential growth in components per chip, by a factor of one hundred per decade. Today the limit is almost a million components; by 1990 it will be about five million; and by the year 2000 between ten and one hundred million.

Today a tiny chip of silicon contains an electronic circuit consisting of hundreds of thousands of transistors and all the necessary interconnecting conductors, and it costs only a few dollars. The circuitry on that chip, now made by printed boards, is equivalent to about ten years' work by a person soldering discrete components onto that printed wiring board. A single chip can itself be a microcomputer with input/output processing capability and random-access memory and be, like the AT&T WE 32100, smaller than an American dime.

(3) *Digitalization.* In the new technology information is represented by digits. Digits are numbers, discrete in their relation to one another, rather than continuous variables. A telephone, for example, was an analogue system, for sound is a wave. Through digital switching a telephone becomes converted to the use of binary systems. One sees this in sound recordings, as on musical discs. The third technological revolution involves the conversion of all previous systems into digital form.

(4) *Software.* Older computers had the instructions or operating systems wired into the machine, and one had to learn a programming language, such as Cobol or Fortran, or the more specialized languages such as Pascal or Lisp, to use the machine. Software, an independent program, frees the user to do various tasks quickly. In distributed processing the software directing the work of a particular computer terminal operates independently of software in other terminals or in die central processing unit. Micro or personal computers have specific software programs—for financial analysis or information database retrieval—that tailor the system to particular

user needs and become, in the argot of the computer, "user-friendly."

Software—the basis of customization—is still a developing art. It takes a programmer about a year to produce a few thousand lines of code. In telecommunications, large electronic switching machines (to route the hundreds of thousands of calls onto different lines) use more than two million lines. Breaking the "bottleneck" of software programming is the key to the rapid spread of the personal computer into the small business and the home.

(One can point to a significant development that promises the enlargement and enhancement of the new technology: photonics. Photonics is the key technology for transmitting large amounts of digital information through laser and ultra-pure glass or optical fibers. Combined, they provide a transmission capability that far exceeds the copper wire and radio. In laboratory experiments, the AT&T Bell laboratories set a "distance record" by transmitting 420 million bits per second over 125 miles without amplification, and two billion bits per second over eighty miles without amplification. The pulse rate can transmit the entire thirty-volume *Encyclopedia Britannica* in a few seconds. But these are still in the development stage, and we are concerned here with already proven technologies that are in the process of marketing and diffusion.)

The most crucial fact about the new technology is that it is not a separate domain (such as the label "high-tech" implies), but a set of changes that pervade all aspects of society and reorganize all older relationships. The industrial revolution produced an age of motors—something we take for granted. Motors are everywhere, from automobiles to boats to power tools and even household devices (such as electric toothbrushes and electric carving knives) that can run on fractional horsepower—motors of one-half and one-quarter horsepower. Similarly, in the coming decades, we shall be "pervaded" by computers—not just the large ones, but the "computer on a chip," the microcomputer, which will transform all our equipment and homes. For automobiles, appliances, tools, home computers and the like, microcomputers will operate with computing power of ten MIPS (millions of instructions per second) per computer.

We can already see the shape of the manifold changes. The old distinctions in communication between telephone (voice), television (image), computer (data), and text (facsimile) have been broken down, physically interconnected by digital switching, and made compatible as a single unified set of teletransmissions. This is what my colleague Anthony Oettinger calls "compunications" and what Simon Nora and Hilary Minc, in their report to the president of France several years ago, called "télématique." The introduction of computer-aided design and simulation has revolutionized engineering and architectural practices. Computer-aided manufacturing and robotics are beginning to transform the production floor. Computers are now indispensable in record-keeping, inventory, scheduling, and other aspects of management information systems in business, firms, hospitals, universities, and any organization. Data-base and information-retrieval systems reshape analysis for decisions and intellectual work. The household is being transformed as digital devices begin to program and control household appliances and, in the newer home designs, all aspects of the household environment. Computers, linked to television screens, begin to change the way we communicate, make transactions, receive and apply information.

The intellectual task is how to "order" these changes in comprehensible ways, rather than just describing the multitude of changes, and thus to provide some basis of analysis rooted in sociological theory. What I intend to do, in the following sections, is to present a number of "social frameworks," or matrices, which may allow us to see how existing social structures come under pressure for change, and the ways in which such changes may occur. I repeat one caveat stated earlier: Technology does not determine social change; technology provides instrumentalities and potentialities. The ways that these are used are social choices. The frameworks that I sketch below, therefore, indicate the "areas" within which relevant changes may occur.

THE POSTINDUSTRIAL SOCIETY

The postindustrial society is not a projection or extrapolation of existing trends in Western society; it is a new principle of social-technical organization and ways of life, just as the industrial system (e.g., factories) replaced an agrarian way of life. It is, first, a shift in the centrality of industrial production, as it was organized on the basis of standardization and mass production. This does not mean the disappearance of manufacturing or the production of goods; the production of food and products from the soil does not disappear from the Western world (in fact, more food is produced than ever before), but there is a significant change in the way food is produced, and, more significantly, in the number of persons engaged in agricultural production. But more than all these, the idea is a "logical construct," in order to see what is central to the new social forms, rather than an empirical description. Postindustrial developments do not replace previous social forms as "stages" of social development. They often coexist, as a palimpsest, on top of the others, thickening the complexity of society and the nature of social structure.

One can think of the world as divided into three kinds of social organization. One is preindustrial. These are primarily extractive industries: farming, mining, fishing, timber. This is still the lot of most of Africa, Latin America, and Southeast Asia, where 60 percent or more of the labor force is engaged in these activities. These are largely what I call "games against nature," subject to the vicissitudes of the weather, the exhaustion of the soils, the thinning out of forests, or the higher costs of the recovery of minerals and metals.

Similar sections of the world have been industrial, engaged in fabrication, the application of energy to machines for the mass production of goods. These have been the countries around the Atlantic littoral: those of Western Europe and the United States, and then the Soviet Union and Japan. Work, here, is a game against fabricated nature: the hitching of men to machines, the organized rhythmic pacing of work in a highly co-ordinated fashion.

The third type is postindustrial. These are activities that are primarily processing, control, and information. It is a social way of life that is, increasingly, a "game between persons." More important, there is a new principle of innovation, especially of knowledge and its relation to technology.

Let me describe some of the lineaments of the postindustrial society. It is, first, a society of services. In the United States today, more than 70 percent of the labor force is engaged in services. Yet "services" is inherently an ambiguous term and, in economic analysis, one without shape because it has been used primarily as a "residual" term.

In every society there is a high component of services. In preindustrial society, it is primarily domestic or personal service. In a country such as India, most persons with a middle-class income would have one or two servants, because many persons simply wish for a roof to sleep and a place to eat. (In England, until 1870, the largest single occupational class was that of domestic servants.)

In an industrial society services are those activities auxiliary to industry: utilities, transportation (including garages and repairs), finance, and real estate.

In a postindustrial society there is an expansion of new kinds of service. These are human services—education, health, social work, social services—and professional services—analysis and planning, design, programming, and the like. In the older conceptions of classical economics (including Marxism), services were thought of as inherently unproductive, since wealth was identified with goods, and lawyers and priests or barbers or waiters did not contribute to the national wealth. Yet surely education and health services contribute to the increased skills and strengths of a population, while professional services (such as linear programming in the organization of production, or new modes of layout of work and social interaction) contribute to the productivity of an enterprise and society. And the important fact is that the expansion of a postindustrial sector of a society requires the expansion of higher education and the education of many more in the population in abstract conceptual, technical, and alphanumeric skills.

In the United States today more than 30 percent of the labor force (of more than one hundred million persons) is professional, technical, and managerial, an amazing figure in social history. About 17 percent of die labor force does factory work (the industrial proletariat, in the older Marxian sense of the term), and it is likely that this will shrink to about 10 percent within a decade. If one thinks this is small, consider the fact that fewer than 4 percent of the labor force are farmers, producing a glut of food for the United States—as against 50 percent in 1900.

An equally important change is in the role of women. In 1950 the "typical" picture for 70 percent of the labor force was a husband at work and his wife and two children at home. Today that is true of only 15 percent of the labor force. Today more than 50 percent of all wives are working outside the home.

Any social change is an intersection of cultural attitudes with the ability to institutionalize those attitudes in market terms. The cultural attitudes regarding equal rights of women go back a hundred years. But the ability to institutionalize those sentiments in market terms goes back only to the past twenty-five or so years—with the expansion of postindustrial employments, particularly in the "quinary" sector of services (health, education, research) and then back into the "quaternary" areas (trade, finance, real estate). The reason is, broadly, that industrial work has been largely considered men's work (including the corporate sectors of management). Postindustrial employments are open, in skills and capacities, to women.

The decisive change—what I call the axial principle of organization—is a change in the character of knowledge. Now, every human society has always existed on the basis of knowledge. The sources go far back, lost in the vistas of time, when the human animal was able, because of the voice box in the larynx, to take the sounds of communication made by all birds and animals, and to codify these into distinct vocables that could be combined, differentiated, and organized into complex meanings, and, through voice, to make intelligible signals that could be transmitted through an oral tradition. With the creation of alphabets we could take a few ideographic scratches and combine these into thousands of words that could be written in stylized forms, to be learned and read by others.

But what is radically new today is the codification of theoretical knowledge and its centrality for innovation, both of new knowledge and for economic goods and services....

...One has to distinguish technological changes (even when they are now not only in machine technology but in intellectual technology) from the more valuable changes in social structure. Changes in technology, as I have insisted, do not determine social changes; they pose problems that the political controllers of society have to deal with. It would take a book to begin to explore the many problems suggested by the possible changes we have seen. Some of these are explored in the following two sections on changes in infrastructure, or the social geography of societies, and changes in the nature of production systems. Let me briefly, however, with the more delimited framework of a postindustrial hypothesis, pose a number of questions.

1. The shrinkage of the traditional manufacturing sectors—augmented, in these instances, by the rising competition from Asia and the ease whereby the routinized, low-value-added production can be taken up by some of the Third World societies—raises the question whether Western societies (all or some) can reorganize their production to move toward the new "high-tech,high-value-added" kinds of specialized production, or whether they will be "headquarter economies" providing investment and financial services to the rest of the world.

2. The costs of transition. Can these be managed? And if so, by the "market," or by some kind of "industrial policy"?

3. The reorganization of an educational system to provide a greater degree of "alphanumeric" fluency in larger portions of the population who would be employed in these postindustrial sectors.

4. The character of "work." If character is defined by work, then we shall see a society where "nature" is largely excluded and "things" are largely excluded within the experience of persons. If more and more individuals are in work situations that involve a "game between persons," clearly more and more questions of equity and "comparable worth" will arise. The nature of hierarchy in work may be increasingly questioned, and new modes of participation may be called for. All of these portend huge changes in the structures of organization from those we have seen in the older models of the army and the church, or the industrial-factory organization, which have been the structures of organization (if not domination) until now.

SOCIETAL GEOGRAPHY AND INFRASTRUCTURES

Historically every society has been tied together by three kinds of infrastructure: These have been the nodes and highways of trade and transactions, of the location of cities and the connections between peoples. The first has been transportation: rivers, roads, canals, and, in modern times, railroads, highways, and airplanes. The second is energy systems: hydro-power, electricity grids, oil pipelines, gas pipelines, and the like. And the third has been communications: postal systems (which moved along highways), then telegraph (the first break in that linkage), telephone, radio, and now the entire panoply of new technological means from microwave to satellites.

The oldest system has been transportation. The breakdown between isolated segments of a society comes when roads are built to connect these, so that trade can commence. The location of human habitats has come with the crossing of roads or the merging of rivers and arms of lakes: traders stop with their wares, farmers bring their food, artisans settle down to provide services, and towns and cities develop.

Within the system of transport, the most important has been water routes. They are the easiest means for carrying bulk items; waterways weave around natural obstacles; tides and currents provide means of additional motion. It is striking to realize that almost every major city in the world, in the last millennia (leaving aside the fortified hill towns that arose during the breakdown of commerce and provided a means of protection against marauders) is located on water: Rome on the Tiber, Paris on the Seine, London on the Thames, not to mention the great cities located on the oceans, seas, and great lakes.

If one looks at industrial societies, the location of cities and the hubs of production come from the interplay of water and resources. Consider a map of the United States and look at the north-central area of the country. In the Mesabi range of Minnesota there was iron ore; in the fields of southern Illinois and western Pennsylvania there was coal. And these were tied together by a Great Lakes and river-valley system that connected them with ports on the oceans: the lakes of Superior, Huron, Michigan, Ontario, and Erie, the St. Lawrence waterway through Canada reaching out to the Atlantic, the Erie Canal across New York reaching down the Hudson River, and the Ohio River wending its way down to the Mississippi and the Gulf of Mexico.

Given the iron ore and coal, one has a steel industry and from it an automobile industry, a machine-tool industry, a rubber industry, and the like. And given the water-transport system tying these together, we get the locational reasons for the great industrial heartland of the United States, the bands of cities along the lakes and rivers of Chicago, Detroit, Cleveland, Buffalo, and Pittsburgh. Thus the imprint of economic geography.

Now all this is changing, as industrial society begins to give way. Communication begins to replace transportation as the major node of connection between people and as the mode of transaction.

Water and natural resources become less important as locational factors for cities, particularly as, with the newer technology, the size of manufacturing plants begins to shrink. Proximity to universities and culture becomes more important as a locational factor. If we look at the major development of high-tech in the United States, we see that the four major concentrations respond to these elements: Silicon Valley, in relation to Stanford University and San Francisco; the circumferential Route 128 around Boston, in relation to MIT and Harvard; Route 1 in New Jersey, from New Brunswick to Trenton, with Princeton University at its hub; and Minneapolis–St. Paul in Minnesota, clustering around the large state university and the Twin City metropolis.

What we see, equally, with communication networks becoming so cheap, is a great pull toward decentralization. In the past, central business districts concentrated the headquarters of large enterprises because of the huge "external economies" available through the bunching of auxiliary services. One could "walk across the street" and have easily available legal services, financial services, advertising services, printing and publishing, and the like. Today, with the increasing cheapness of communication and the high cost of land, density and the external economies become less critical. So we find that dozens of the major U.S. corporations, in the last decade or so, have moved their basic headquarters from New York to the suburban areas where land is cheaper, and transport to and from work easier: northeast to Fairfield County in Connecticut; north to Westchester County in New York; and west and southwest to Mercer County in New Jersey.

In Japan we see a major effort now under way, the Technopolis project, to create large, far-flung regional centers for the new computer and telecommunications industries. For status reasons, many corporations maintain a display building in New York or Tokyo; but the major managerial activities are now decentralized.

As geography is no longer the controller of costs, distance becomes a function not of space but of time; and the costs of time and rapidity of communication become the decisive variables. And, with the spread of mini- and microcomputers,

the ability to "down-load" databases and memories, and to place these in the small computers (as well as give them access to the large mainframes) means there is less of a necessary relation to fixed sites in the location of work.

As with habitats, so with markets. What is a market? Again, it is a place where roads crossed and rivers merged and individuals settled down to buy and sell their wares. Markets were places. Perhaps no longer.

Take the Rotterdam spot market for oil. It was the place where tankers carrying surplus oil would come so that oil could be sold "on the spot." They came to Rotterdam because it was a large, protected port, close to the markets of Western Europe; it had large storage capacity; there was a concentration of brokers who would go around and make their deals. It is still called the Rotterdam spot market for oil, but it is no longer in Rotterdam. But if not in Rotterdam, where? Everywhere. It is a telex-and-radio system whereby brokers in different parts of the world can make their deals and redirect the ships on the high seas to different ports for the sales they have made. In effect, markets are no longer places but networks.

And this is true for most commodities, especially for capital and currency markets. Today one can get in "real time" quotations for dollars, euros, yen, ... sterling, ... in Tokyo, Singapore, Hong Kong, Milan, Frankfurt, Paris, London, New York, Chicago, San Francisco, and money moves swiftly across national lines. Capital flows in response to differential interest rates or in reaction to news of political disturbances.

What we have here, clearly, are the nerves nodes, and ganglia of a genuine international economy tied together in ways the world has never seen before. What this means—and I shall return to the question at the close of this reading—is a widening of the arenas, the multiplication of the numbers of actors, and an increase in the velocity and volatility of transactions and exchanges. The crucial question is whether the older institutional structures are able to deal with this extraordinary volume of interactions.

THE SOCIAL ORGANIZATION OF PRODUCTION

The modern corporation—I take the United States as the model—is less than a hundred years old. Business, the exchange of goods and services, is as old as human civilization itself. But the modern corporation, as a social form to coordinate men, materials and markets for the mass production and mass consumption of goods, is an institution that has taken shape only in the past century.

There are three kinds of innovators who conjoined to create the modern industrial system. The greatest attention has been paid to those who have been the organizers of the production system itself: Eli Whitney, who created standardized forms and interchangeable parts in production; Frederick Taylor, who designed the measurement of work; and Henry Ford, who created the assembly line and mass production. (There were of course other forebears, and there were European counterparts: Siemens, Bedeaux, Renault, etc.)

Those who achieved the greatest notoriety were the capitalists, the men who by ruthless means put together the great enterprises: the Carnegies, the Rockefellers, Harriman, the men who initiated the large quasi-monopoly organizations, and the financiers, such as J. P. Morgan, who assembled the monies for the formation of such great corporations as U.S. Steel, General Electric, and the like.

But there was also a different social role, often unnoticed even in the history of business, played by men who, curiously, were probably just as important, and perhaps more so: the organizers of the corporate form, those who rationalized the system and gave it an ongoing structural continuity. I will discuss three individuals who symbolize the three crucial structural changes: one was Walter Teagle, of Standard Oil of New Jersey, who created vertical integration; another was Theodore N. Vail, who fashioned the American Telephone and Telegraph Co. and imposed the idea of a single uniform system; and the third was Alfred P. Sloan, of General Motors, who created the system of financial controls and budgetary

accounting that still rules the corporate world today.

These three men created modern industrial capitalism. It is my thesis, implicit in this article and which can be stated only schematically here, that this system, marvelously adaptive to a mass-production society, is increasingly dysfunctional in today's postindustrial world.

Vertical integration, the control of all aspects of a product—in the case of Teagle, from oil in the ground, to shipping, refining, and distribution to industrial customers and retail outlets—was created for the clear reasons of economies of scale, reduction of transaction costs, the utilization of information within the entire process, and the control of prices, from raw materials to finished goods. What vertical integration did, as Alfred Chandler has pointed out in his book *The Visible Hand*, was to destroy "producer markets" within the chain of production and impose uniform controls. In the previous system, one of merchant capitalism, production was in the hands of independent artisans or small-business companies, and all of this was funneled through the matrix of the merchant capitalist, who ordered the goods he needed, or contracted production to the small workshop, and sold finished products to the customers. But the creation of large-scale, mass-produced, identical goods made vertical integration a functional necessity.

The idea of a single system arose when Vail, seeking to build a telephone utility, beheld the railroad system in the United States, where railroad systems grew "higgledy-piggledy," without plan, and often for financial reasons, to sell inflated stock. Franchises were obtained from corrupt legislatures or from congressional land grants, and the roads were built in sprawling ways. Before the advent of coast-to-coast air flight, if a traveler wanted to go from New York to the West Coast by train, he could not do so on a single system. He could take one of two competitive railroads from New York to Chicago, where he changed trains and then took one of three competitive systems to the Coast. (If one wished to ship a hog, or freight, it was not necessary to change trains. Animals or freight goods, unlike human beings, could not pick themselves up and move to another freight car; it was cheaper to shuttle the freight car onto different lines.) Even today there is no unified rational rail system in the United States.

Vail, in building a telephone network, decided that if there was to be efficient service between a person calling from any point in the United States to any other point, there would have to be a single set of "long lines" connecting all the local telephones to one another. Until the recent federal court decision which broke up the American Telephone and Telegraph Co., it was a unified, single system.

Alfred P. Sloan's innovations came about when he took over the sprawling General Motors from William C. Durant, a Wall Street speculator who had put together the different automobile companies (named for their early founders: Chevrolet, Olds, Cadillac, and the like) into a single firm, General Motors. But Durant had little talent for creating a rational structure. Alfred P. Sloan, the MIT-trained engineer who was installed as head of the company by the Du Pont interests (the largest block of stockholders until the courts forced them to divest their holdings about twenty-five years ago), installed unit cost accounting and financial controls with a single aim: to obtain a clear return on investment for the monies given to the different divisions. Durant never knew which of the companies was making money, and which not; he did not know whether it was cheaper to make his own steel or buy outside, make his own parts or buy outside. Sloan rationalized the company. His key innovation was a pricing system for the different lines of automobiles that would provide a 20 percent return on investment based on a stipulated capacity, a break-even point based on overhead and fixed costs, and a market share for the particular line of car.

Together, these innovations were the corporate principles of modern industrial capitalism. Why are they now dysfunctional?

In the case of production, the older standardized, routinized, low-value-added forms of production are being increasingly taken over by the newly industrializing societies, where cheap wages provide the crucial cost differential in competition. More than that, the newer technologies—particularly computer-aided design (CAD), numerical-control machine tools (NC), and computer-aided manufacturing (CAM)—now make possible *flexible*, shorter-run, batch productions that can be easily adapted to different kinds of markets, and which can be responsive to specialized products and customized demands.

One of the great success stories in this respect is Italy, in such an "old-fashioned" industry as textiles. The textile district of Prato—the group of towns in Central Italy in the provinces of Florence and Pistoia—was able to survive and flourish because it could adapt. As two MIT scholars have pointed out (relying, of course, on Italian studies), "Prato's success rests on two factors: a long-term shift from standard to fashionable fabrics and a corresponding reorganization of production from large integrated mills to technologically sophisticated shops specializing in various phases of production—a modern *systeme* Motte."

But what holds true for textiles is true for a wide variety of industries as well. In steel, integrated production is now cumbersome and costly, and it is the minimills, with their specialized, flexible production, and the specialty steels that have become the basis for survival in the Western world. It is not, thus, deindustrialization, but a new form of industrialization, which is taking place.

In the case of telecommunications—to be brief—the breakdown of the old distinctions between telephone, computer, television, and facsimile (Xerox) means that new, highly differentiated systems—private branch exchanges, local area networks, "internal" communication networks between firms, international satellite communication—all emphasize diversity rather than uniformity, with many specialized systems rather than a single product such as the telephone.

In the case of Sloan's system of a return on investment through budgetary controls, the assumptions he made were those of a quasi monopoly or oligopoly in a "steady-state" market, and that kind of financial planning can scarcely adapt to a changing world where old product lines are breaking down (one need simply consider the old distinctions between banks, insurance companies, brokerage houses, credit firms, real estate investment, all of which become to some extent interchangeable under the rubric of financial-asset management), where substitutions of products provide price challenges, where market share and cash flow may be more important momentarily, and a long-term commitment necessary technologically, than the simple unit-cost accounting that Alfred Sloan introduced.

In effect, the world of the postindustrial society requires new modes of social organization, and these are only now being fashioned by the new entrepreneurs of the new technology.

THE QUESTION OF SCALE

The crucial question, as I have indicated, is how new social structures will be created in response to the different values of societies, to the new technological instruments of a postindustrial world. Beyond the structural frameworks I have tried to identify, there is one crucial variable that must be taken into account—the change in scale.

It is a cliché of our time that ours is an era of acceleration in the pace of change. I must confess that I do not understand what this actually means. If we seek to use this concept analytically, we find a lack of boundary and meaning. To speak of "change" is in itself meaningless, for the question remains: change of what? To say that "everything" changes is hardly illuminating. And if one speaks of a pace, or of an acceleration in pace, the words imply a metric—a unit of measurement. But what is being measured?

However, one can gain a certain perspective about what is happening by thinking of the concept of scale. A change in the scale of an institution is a change of form. Metaphorically, this goes back to Galileo's square-cube law: If you double the size of an object, you triple its volume. There is consequently a question of shape and proportion. A university with fifty thousand students may still be called by the same name it had thirty years before, with five thousand students, but the increase in numbers calls for a change in the institutional structure. And this is true of all social organizations.

What the revolutions in communication are doing is changing the scale of human activities. Given the nature of "real time" communication, we are for the first time forging an interdependent international economy with more and more characteristics of an unstable system in which changes in the magnitudes of some variables, or shocks and disturbances in some of the units, have immediate repercussions in all the others.

The management of scale has been one of the oldest problems in social institutions, whether it be the church, the army, or economic enterprise, let alone the political order. Societies have tended to function reasonably well when there is a congruence of scale between economic activities, social units and organization, and political and administrative control. But increasingly what is happening is a mismatch of scale. As I stated in an essay several years ago, the national state has become too small for the big problems of life, and too big for the small problems. The national state, with its political policies, is increasingly ineffective in dealing with the tidal waves of the international economy (coordination through economic summitry is only a charade) and too big, when political decisions are concentrated in a bureaucratic center, for the diversity and initiative of the varied local and regional units under its control. To that extent, if there is a single overriding sociological problem in the postindustrial society—particularly in the management of transition—it is the management of scale.

55

Urbanism as a Way of Life

LOUIS WIRTH

As world population trends show people moving from rural to urban areas, newcomers to cities find different lifestyles. In a classic discussion of the urban way of life, Wirth points out three characteristics of urbanism—size of urban populations, density of the populations, and heterogeneity. These three variables explain many of the differences between urban and rural life and life in different-sized cities.

As you read this classic article about urban life, consider the following questions:

1. *Which key variables describe differences in urban and rural life?*
2. *Which elements of urban life make this a logical area of study for sociologists?*
3. *What are some examples of these differences in your community versus the surrounding urban or rural areas?*

GLOSSARY   **Population aggregate** Those people who mass in certain areas, in this case urban areas.   **Density** Concentration of population in a limited space.   **Heterogeneity** A variety of different types of peoples (class, ethnic background, etc.).

A SOCIOLOGICAL DEFINITION OF THE CITY

For sociological purposes a city may be defined as a relatively large, dense, and permanent settlement of socially heterogeneous individuals. On the basis of the postulates which this minimal definition suggests, a theory of urbanism may be formulated in the light of existing knowledge concerning social groups....

The central problem of the sociologist of the city is to discover the forms of social action and organization that typically emerge in relatively permanent, compact settlements of large numbers of heterogeneous individuals. We must also infer that urbanism will assume its most characteristic and extreme form in the measure in which the conditions with which it is congruent are present. Thus the larger, the more densely populated, and the more heterogeneous a community, the more accentuated the characteristics associated with urbanism will be....

To say that large numbers are necessary to constitute a city means, of course, large numbers in relation to a restricted area or high density of settlement. There are, nevertheless, good reasons for treating large numbers and density as separate factors, because each may be connected with significantly different social consequences. Similarly the need for adding heterogeneity to numbers of population as a necessary and distinct criterion of urbanism might be questioned, since we should expect the range of differences to increase with numbers. In defense, it may be said that the city shows a kind and degree of heterogeneity of population which cannot be wholly accounted for by the law of large numbers or adequately represented by means of a normal distribution curve. Because the population of the city does not reproduce itself, it must recruit its migrants from other cities, the countryside, and … from other countries. The city has thus historically been the melting-pot of races, peoples, and cultures, and a most favorable breeding-ground of new biological and cultural hybrids. It has not only tolerated but rewarded individual differences. It has brought together people from the ends of the earth *because* they are different and thus useful to one another, rather than because they are homogeneous and like-minded.

A number of sociological propositions concerning the relationship between (a) numbers of population, (b) density of settlement, (c) heterogeneity of inhabitants and group life can be formulated on the basis of observation and research.

Size of the Population Aggregate

Ever since Aristotle's *Politics*, it has been recognized that increasing the number of inhabitants in a settlement beyond a certain limit will affect the relationships between them and the character of the city. Large numbers involve, as has been pointed out, a greater range of individual variation. Furthermore, the greater the number of individuals participating in a process of interaction, the greater is the *potential* differentiation between them. The personal traits, the occupations, the cultural life, and the ideas of the members of an urban community may, therefore, be expected to range between more widely separated poles than those of rural inhabitants.

That such variations should give rise to the spatial segregation of individuals according to color, ethnic heritage, economic and social status, tastes and preferences, may readily be inferred. The bonds of kinship, of neighborliness, and the sentiments arising out of living together for generations under a common folk tradition are likely to be absent or, at best, relatively weak in an aggregate the members of which have such diverse origins and backgrounds. Under such circumstances competition and formal control mechanisms furnish the substitutes for the bonds of solidarity that are relied upon to hold a folk society together.

Increase in the number of inhabitants of a community beyond a few hundred is bound to limit the possibility of each member of the community knowing all the others personally. Max Weber, in recognizing the social significance of this fact, explained that from a sociological point of view large numbers of inhabitants and density of settlement mean a lack of that mutual acquaintanceship which ordinarily inheres between the inhabitants in a neighborhood.[1] The increase in numbers thus involves a changed character of the social relationships. As Georg Simmel points out: "[If] the unceasing external contact of numbers of persons in the city should be met by the same number of inner reactions as in the small town, in which one knows almost every person he meets and to each of whom he has a positive relationship, one would be completely atomized internally and would fall into an unthinkable mental condition."[2] The multiplication of persons in a state of interaction under conditions which make their contact as full personalities impossible produces that segmentalization of human relationships which has sometimes been seized upon by students of the mental life of the cities as an explanation for the "schizoid" character of urban personality. This is not to say that the urban inhabitants have fewer acquaintances than rural inhabitants, for the reverse may actually be true; it means rather that in relation to the number of people whom they see and with whom they rub elbows in the course of daily life, they know a smaller proportion, and of these they have less intensive knowledge.

Characteristically, urbanites meet one another in highly segmental roles. They are, to be sure, dependent upon more people for the satisfactions of their life-needs than are rural people and thus are associated with a greater number of organized groups, but they are less dependent upon particular persons, and their dependence upon others is confined to a highly fractionalized aspect of the other's round of activity. This is essentially what is meant by saying that the city is characterized by secondary rather than primary contacts. The contacts of the city may indeed be face to face, but they are nevertheless impersonal, superficial, transitory, and segmental. The reserve, the indifference, and the blasé outlook which urbanites manifest in their relationships may thus be regarded as devices for immunizing themselves against the personal claims and expectations of others.

The superficiality, the anonymity, and the transitory character of urban social relations make intelligible, also, the sophistication and the rationality generally ascribed to city-dwellers. Our acquaintances tend to stand in a relationship of utility to us in the sense that the role which each one plays in our life is overwhelmingly regarded as a means for the achievement of our own ends. Whereas the individual gains, on the one hand, a certain degree of emancipation or freedom from the personal and emotional controls of intimate groups, he loses, on the other hand, the spontaneous self-expression, the morale, and the sense of participation that comes with living in an integrated society. This constitutes essentially the state of *anomie*, or the social void, to which Durkheim alludes in attempting to account for the various forms of social disorganization in technological society.

The segmental character and utilitarian accent of interpersonal relations in the city find their institutional expression in the proliferation of specialized tasks which we see in their most developed form in the professions. The operations of the pecuniary nexus lead to predatory relationships, which tend to obstruct the efficient functioning of the social order unless checked by professional codes and occupational etiquette. The premium put upon utility and efficiency suggests the adaptability of the corporate device for the organization of enterprises in which individuals can engage only in groups. The advantage that the corporation has over the individual entrepreneur and the partnership in the urban-industrial world derives not only from the possibility it affords of centralizing the resources of thousands of individuals or from the legal privilege of limited liability and perpetual succession, but from the fact that the corporation has no soul.

The specialization of individuals, particularly in their occupations, can proceed only, as Adam Smith pointed out, upon the basis of an enlarged market, which in turn accentuates the division of labor. This enlarged market is only in part supplied by the city's hinterland; in large measure it is found among the large numbers that the city itself contains. The dominance of the city over the surrounding hinterland becomes explicable in terms of the division of labor which urban life occasions and promotes. The extreme degree of interdependence and the unstable equilibrium of urban life are closely associated with the division of labor and the specialization of occupations. This interdependence and this instability are increased by the tendency of each city to specialize in those functions in which it has the greatest advantage.

In a community composed of a larger number of individuals than can know one another intimately and can be assembled in one spot, it becomes necessary to communicate through indirect media and to articulate individual interests by a process of delegation. Typically in the city, interests are made effective through representation. The individual counts for little, but the voice of the representative is heard with a deference roughly proportional to the numbers for whom he speaks.

While this characterization of urbanism, in so far as it derives from large numbers, does not by any means exhaust the sociological inferences that might be drawn from our knowledge of the relationship of the size of a group to the characteristic behavior of the members, for the sake of brevity the assertions made may serve to exemplify the sort of propositions that might be developed.

Density

As in the case of numbers, so in the case of concentration in limited space certain consequences of relevance in sociological analysis of the city emerge. Of these only a few can be indicated.

As Darwin pointed out for flora and fauna and as Durkheim noted in the case of human societies,[3] an increase in numbers when area is held constant (i.e., an increase in density) tends to produce differentiation and specialization, since only in this way can the area support increased numbers. Density thus reinforces the effect of numbers in diversifying men and their activities and in increasing the complexity of the social structure.

On the subjective side, as Simmel has suggested, the close physical contact of numerous individuals necessarily produces a shift in the media through which we orient ourselves to the urban milieu, especially to our fellow-men. Typically, our physical contacts are close but our social contacts are distant. The urban world puts a premium on visual recognition. We see the uniform which denotes the role of the functionaries, and are oblivious to the personal eccentricities hidden behind the uniform. We tend to acquire and develop a sensitivity to a world of artifacts, and become progressively farther removed from the world of nature.

We are exposed to glaring contrasts between splendor and squalor, between riches and poverty, intelligence and ignorance, order and chaos. The competition for space is great, so that each area generally tends to be put to the use which yields the greatest economic return. Place of work tends to become dissociated from place of residence, for the proximity of industrial and commercial establishments makes an area both economically and socially undesirable for residential purposes.

Density, land values, rentals, accessibility, healthfulness, prestige, aesthetic consideration, absence of nuisances such as noise, smoke, and dirt determine the desirability of various areas of the city as places of settlement for different sections of the population. Place and nature of work income, racial and ethnic characteristics, social status, custom, habit, taste, preference, and prejudice are among the significant factors in accordance with which the urban population is selected and distributed into more or less distinct settlements. Diverse population elements inhabiting a compact settlement thus become segregated from one another in the degree in which their requirements and modes of life are incompatible and in the measure in which they are antagonistic. Similarly, persons of homogeneous status and needs unwittingly drift into, consciously select, or are forced by circumstances into the same area. The different parts of the city acquire specialized functions, and the city consequently comes to resemble a mosaic of social worlds in which the transition from one to the other is abrupt. The juxtaposition of divergent personalities and modes of life tends to produce a relativistic perspective and a sense of toleration of differences which may be regarded as prerequisites for rationality and which lead toward the secularization of life.

The close living together and working together of individuals who have no sentimental and emotional ties foster a spirit of competition, aggrandizement, and mutual exploitation. Formal controls are instituted to counteract irresponsibility and potential disorder. Without rigid adherence to predictable routines a large compact society would scarcely be able to maintain itself. The clock and the traffic signal are symbolic of the basis of our social order in the urban world. Frequent close physical contact, coupled with great social distance, accentuates the reserve of unattached individuals toward one another and, unless compensated by other opportunities for response, gives rise to loneliness. The necessary frequent movement of great numbers of individuals in a congested habitat causes friction and irritation. Nervous tensions which derive from such personal frustrations are increased by the rapid tempo and the complicated technology under which life in dense areas must be lived.

Heterogeneity

The social interaction among such a variety of personality types in the urban milieu tends to break down the rigidity of caste lines and to complicate the class structure; it thus induces a more ramified and differentiated framework of social stratification than is found in more integrated societies. The heightened mobility of the individual, which brings him within the range of stimulation by a great number of diverse individuals and subjects him to fluctuating status in the differentiated social groups that compose the social structure of the city, brings him toward the acceptance of instability and insecurity in the world at large as a norm. This fact helps to account, too, for the sophistication and cosmopolitanism of the urbanite. No single group has the undivided allegiance of the individual. The groups with which he is affiliated do not lend themselves readily to a simple hierarchical arrangement. By virtue of his different interests arising out of different aspects of social life, the individual acquires membership in widely divergent groups, each of which functions only with reference to a single segment of his personality. Nor do these groups easily permit a concentric arrangement so that the narrower ones fall within the circumference of the more inclusive ones, as is more likely to be the case in the rural community or in primitive societies. Rather the groups with which the person typically is affiliated are tangential to each other or intersect in highly variable fashion.

Partly as a result of the physical footlooseness of the population and partly as a result of their social mobility, the turnover in group membership generally is rapid. Place of residence, place and character of employment, income, and interests fluctuate, and the task of holding organizations together and maintaining and promoting intimate and lasting acquaintanceship between the members is difficult.

This applies strikingly to the local areas within the city into which persons become segregated more by virtue of differences in race, language, income, and social status than through choice or positive attraction to people like themselves. Overwhelmingly the city-dweller is not a home-owner, and since a transitory habitat does not generate binding traditions and sentiments, only rarely is he a true neighbor. There is little opportunity for the individual to obtain a conception of the city as a whole or to survey his place in the total scheme. Consequently he finds it difficult to determine what is to his own "best interests" and to decide between the issues and leaders presented to him by the agencies of mass suggestion. Individuals who are thus detached from the organized bodies which integrate society comprise the fluid masses that make collective behavior in the urban community so unpredictable and hence so problematical.

Although the city, through the recruitment of variant types to perform its diverse tasks and the accentuation of their uniqueness through competition and the premium upon eccentricity, novelty, efficient performance, and inventiveness, produces a highly differentiated population, it also exercises a leveling influence. Wherever large numbers of differently constituted individuals congregate, the process of depersonalization also enters....

NOTES

1. *Wirtschaft und Gesellschaft* (Tübingen, 1925), part I, chap. 8, p. 514.

2. "Die Grossstädte und das Geistesleben," *Die Grosstadt*, ed. Theodor Petermann (Dresden, 1903), pp. 187–206.

3. E. Durkheim, *De la division du travail social* (Paris, 1932), p. 248.

56

A Planet Under Stress

Rising to the Challenge

LESTER R. BROWN

The world's nonrenewable resources are finite—they won't last forever. Lester Brown raises questions about growing populations and reduction in resources, leaving much of the world with hunger, illiteracy, disease, water shortages, desertification—the list goes on and on. Brown outlines strategies some countries are pursuing to address environmental problems that affect the well-being of humans and the sustainability of the environment.

Think about these questions as you read this article:

1. *Which problems are intensifying as the world's population increases?*
2. *Which solutions are being tried in different parts of the world? Are you aware of any attempts to deal with problems in your community?*
3. *To what is Brown referring when he talks of "honest market," "taxing indirect costs," and "shifting subsidies"?*
4. *Which of the ideas discussed might work in your community?*

GLOSSARY **Honest market** An economic market that reflects ecological/environmental realities. **Tax shifting** Tax alternatives such as lowering income taxes while raising taxes on environmentally destructive activities. **Shifting subsidies** Changing the underwriting (paying) for environmentally destructive activities to subsidies for environmentally sound activities.

Early in this new century, the world is facing many longstanding social challenges, including hunger, illiteracy, and disease. If developing countries add nearly 3 billion people by mid-century, as projected, population growth will continue to undermine efforts to improve the human condition. The gap between the billion richest and the billion poorest will continue to widen, putting even more stress on the international political fabric.

As a species, our failure to control our numbers is taking a frightening toll. Slowing population growth is the key to eradicating poverty and its distressing symptoms, and, conversely, eradicating poverty is the key to slowing population growth. With time running out, the urgency of moving simultaneously on both fronts seems clear.

The challenge is to create quickly the social conditions that will accelerate the shift to smaller families. Among these conditions are universal

SOURCE: Originally printed in *The Futurist* Vol. 37, Issue 6 (November/December 2003), pp. 18–24.

education, good nutrition, and prevention of infectious diseases. We now have the knowledge and resources to reach these goals. In an increasingly integrated world, we also have a vested interest in doing so.

Historically, we have lived off the interest generated by the earth's natural capital assets, but now we are consuming those assets themselves. We have built an environmental bubble economy, one where economic output is artificially inflated by overconsumption of the earth's natural assets. The challenge today is to deflate the bubble before it bursts.

Keeping the bubble from bursting will require an unprecedented degree of international cooperation to stabilize population, climate, water tables, and soils—and at wartime speed. Indeed, in both scale and urgency the effort required is comparable to U.S. mobilization during World War II.

Our only hope now is rapid systemic change—change based on market signals that tell the ecological truth. This means restructuring the tax system: lowering income taxes and raising taxes on environmentally destructive activities, such as fossil fuel burning, to incorporate the ecological costs. Unless we can get the market to send signals that reflect reality, we will continue making faulty decisions as consumers, corporate planners, and government policy makers. Ill-informed economic decisions and the economic distortions they create can lead to economic decline.

Continuing with business as usual offers an unacceptable outcome—continuing environmental degradation and disruption and a bursting of the economic bubble. The warning signals are coming more frequently, whether they be collapsing fisheries, melting glaciers, or falling water tables. Thus far the wake-up calls have been local, but soon they could become global, and time is running out. Bubble economies, which by definition are artificially inflated, do not continue indefinitely. Our demands on the earth exceed its regenerative capacity by a wider margin with each passing day.

DEFLATING THE BUBBLE

Stabilizing world population at about 7.5 billion is central to avoiding economic breakdown in countries with large projected population increases that are already overconsuming their natural capital assets. Some 36 countries, all in Europe except Japan, have essentially stabilized their populations. The challenge now is to create the economic and social conditions and to adopt the priorities that will lead to population stability in all remaining countries. The keys here are extending primary education to all children, providing vaccinations and basic health care, and offering reproductive health care and family-planning services in all countries.

Shifting from a carbon-based to a hydrogen-based energy economy to stabilize climate is now technologically possible. Advances in wind turbine design and in solar cell manufacturing, the availability of hydrogen generators, and the evolution of fuel cells provide the technologies needed to build a climate-benign hydrogen economy. Moving quickly from a carbon-based to a hydrogen-based energy economy depends on getting the price right and on incorporating the indirect costs of burning fossil fuels into the market price.

Iceland is the first country to adopt a national plan to convert its carbon-based energy economy to one based on hydrogen. Denmark now gets 18% of its electricity from wind turbines and plans to increase this to 40% by 2030. Japan leads the world in electricity generation from solar cells. The Netherlands leads the industrial world in exploiting the bicycle as an alternative to the automobile. The Canadian province of Ontario is emerging as a leader in phasing out coal. It plans to replace its five coal-fired power plants with gas-fired plants, wind farms, and efficiency gains. This initiative calls for the first plant to close in 2005 and the last one in 2015. The resulting reduction in carbon emissions is equivalent to taking 4 million cars off the road.

Stabilizing water tables depends on quickly raising water productivity. It is difficult to overstate the urgency of this effort. Failure to stop the fall in water tables by systematically reducing water use will lead to the depletion of aquifers, an abrupt cutback in water supplies, and the risk of a precipitous drop in food production. By pioneering in drip irrigation technology, Israel has become the world leader in the efficient use of agricultural water. This unusually labor-intensive irrigation practice, now being used to produce high-value crops in many countries, is ideally suited where water is scarce and labor is abundant.

With soil erosion, we have no choice but to reduce the loss to the rate of new soil formation or below. The only alternative is a continuing decline in the inherent fertility of eroding soils and cropland abandonment. South Korea, with once denuded mountainsides and hills now covered with trees, has achieved a level of flood control, water storage, and hydrological stability that is a model for other countries. In the United States as well, farmers have reduced soil erosion by nearly 40% in less than two decades thanks to a combination of several programs and practices.

Thus all the things we need to do to keep the bubble from bursting are now being done in at least a few countries. If these highly successful initiatives are adopted worldwide, and quickly, we can deflate the bubble before it bursts, similar to the way U.S. mobilization helped lead Allied forces to victory in less than four years.

In retrospect, the speed of the conversion from a peacetime to a wartime economy at the beginning of World War II was stunning. One month after Pearl Harbor, President Roosevelt announced plans to produce 60,000 planes, 45,000 tanks, 20,000 antiaircraft guns, and 6 million tons of merchant shipping. The automobile industry went from producing nearly 4 million cars in 1941 to producing 24,000 tanks and 17,000 armored cars in 1942—but only 223,000 cars, and most of which were produced early in the year, before the conversion began. Essentially the auto industry was closed down from early 1942 through the end of 1944. In 1940, the United States produced some 4,000 aircraft. In 1942, it produced 48,000. By the end of the war, more than 5,000 ships were added to the 1,000 that made up the American Merchant Fleet in 1939.

Various other firms likewise converted. A sparkplug factory switched to producing machine guns; a manufacturer of stoves produced lifeboats; a merry-go-round factory made gun mounts; a toy company turned out compasses; a corset manufacturer produced grenade belts; and a pinball machine plant began to make armor-piercing shells.

This mobilization of resources within a matter of months demonstrates that a country and, indeed, the world can restructure its economy quickly if it is convinced of the need to do so.

CREATING AN HONEST MARKET

The key to restructuring the economy is the creation of an honest market, one that tells the ecological truth. The market has three fundamental weaknesses: It does not incorporate the indirect costs of providing goods or services into prices. It does not value nature's services properly. It does not respect the sustainable-yield thresholds of natural systems such as fisheries, forests, rangelands, and aquifers.

As the global economy has expanded and as technology has evolved, the indirect costs of some products have become far larger than the price fixed by the market. The price of a gallon of gasoline, for instance, includes the cost of production but not the expense of treating respiratory illnesses from breathing polluted air or the repair bill from acid rain damage. Nor does it cover the cost of rising global temperature, ice melting, more destructive storms, or the relocation of millions of refugees forced from their homes by sea-level rise.

If we have learned anything over the last few years, it is that accounting systems that do not tell the truth can be costly. Faulty corporate accounting systems that overstate income or leave costs off the books have driven some of the world's largest corporations into bankruptcy, costing millions of people their lifetime savings, retirement incomes, and jobs.

Unfortunately, we also have a faulty economic accounting system at the global level, but with potentially far more serious consequences. Economic prosperity is achieved in part by running up ecological deficits, costs that do not show up on the books, but costs that someone will eventually pay. Some of the record economic prosperity of recent decades has come from consuming the earth's productive assets and from destabilizing its climate.

No one has attempted to assess fully the worldwide costs of rising temperature and then to allocate them by gallon of gasoline or ton of coal. A summary of eight studies done during the 1990s indicates that, if the price were raised enough to make drivers pay some of the indirect costs of automobile use, a gallon of gas would cost anywhere from $3.03 to $8.64, with the variations largely due to how many indirect costs were covered. For example, some studies included the military costs of protecting petroleum supply lines and ensuring access to Middle Eastern oil, while others did not. No studies, unfortunately, incorporated all the costs of using gasoline—including the future inundation of coastal cities, island countries, and rice-growing river floodplains.

Not only are some of the looming costs associated with continued fossil fuel burning virtually incalculable, but the outcome is unacceptable. What is the cost of inundating half of Bangladesh's riceland by a one-meter rise in sea level? How much is this land worth in a country that is the size of New York state and has a population half that of the United States? And what would be the cost of relocating the 40 million Bangladeshis who would be displaced by the one-meter rise in sea level? Would they be moved to another part of the country? Or would they migrate to less densely populated countries, such as the United States, Canada, Australia, or Brazil?

Another challenge in creating an honest market is to get it to value nature's services. For example, after several weeks of flooding in the Yangtze River basin in 1998 inflicted $30 billion worth of damage, the Chinese government announced that it was banning all tree cutting in the basin. It justified the ban by saying that trees standing are worth three times as much as trees cut.

Once we calculate all the costs of a product or service, we can incorporate them into market prices by restructuring taxes. If we can get the market to tell the truth, then we can avoid being blindsided by faulty accounting systems that lead to bankruptcy.

TAXING INDIRECT COSTS

The need for tax shifting—lowering income taxes while raising taxes on environmentally destructive activities—in order to get the market to tell the truth has been widely endorsed by economists. The basic idea is to establish a tax that reflects the indirect costs to society of an economic activity. For example, a tax on coal would incorporate the increased health-care costs associated with breathing polluted air, the costs of damage from acid rain, and the costs of climate disruption.

Among the activities taxed in Europe are carbon emissions, emissions of heavy metals, and the generation of garbage (so-called landfill taxes). The Nordic countries, led by Sweden, pioneered tax shifting at the beginning of the 1990s. By 1999, a second wave of tax shifting was under way, this one including the larger economies of Germany, France, Italy, and the United Kingdom. Tax shifting does not change the level of taxes, only their composition. One of the better-known changes was a four-year plan adopted in Germany in 1999 to shift taxes from labor to energy. By 2001, this had lowered fuel use by 5%. A tax on carbon emissions adopted in Finland in 1990 lowered emissions there 7% by 1998.

There are isolated cases of environmental tax reform elsewhere. The United States, for example, imposed a stiff tax on chlorofluorocarbons to phase them out in accordance with the Montreal Protocol of 1987. At the local level, the city of Victoria, British Columbia, adopted a trash tax of $1.20 per bag of garbage, reducing its daily trash flow 18% within one year.

One of the newer taxes gaining in popularity is the so-called congestion tax. City governments are turning to a tax on vehicles entering the city, or at least the inner part of the city where traffic congestion is most serious. In London, where the average speed of an automobile was nine miles per hour—about the same as a horse-drawn carriage—a congestion tax was adopted in early 2003. The £5 ($8) charge on all motorists driving into the center city between 7 A.M. and 6:30 P.M. immediately reduced the number of vehicles by 24%, permitting traffic to flow more freely while cutting pollution and noise.

Environmental tax shifting usually brings a double dividend. In reducing taxes on income, labor becomes less costly, creating additional jobs while protecting the environment. This was the principal motivation in the German four-year shift of taxes from income to energy. The shift from fossil fuels to more energy-efficient technologies and to renewable sources of energy reduces carbon emissions and represents a shift to more labor-intensive industries. By lowering the air pollution from smokestacks and tailpipes, it also reduces respiratory illnesses, such as asthma and emphysema, and healthcare costs—a triple dividend.

When it comes to reflecting the value of nature's services, ecologists can calculate the values of services that a forest in a given location provides. Once these are determined, they can be incorporated into the price of trees as a stumpage tax of the sort that Bulgaria and Lithuania have adopted. Anyone wishing to cut a tree would have to pay a tax equal to the value of the services provided by that tree. The market would then be telling the truth. The effect of this would be to reduce tree cutting, since forest services may be worth several times as much as the timber, and to encourage wood and paper recycling.

Some 2,500 economists, including eight Nobel Prize winners in economics, have endorsed the concept of tax shifts. Former Harvard economics professor N. Gregory Mankiw, chairman of the President's Council of Economic Advisers, wrote in *Fortune* magazine: "Cutting income taxes while increasing gasoline taxes would lead to more rapid economic growth, less traffic congestion, safer roads, and reduced risk of global warming—all without jeopardizing long-term fiscal solvency. This may be the closest thing to a free lunch that economics has to offer." Mankiw could also have

added that it would reduce the military expenditures associated with ensuring access to Middle Eastern oil.

The Economist has recognized the advantage of environmental tax shifting and endorses it strongly: "On environmental grounds, never mind energy security, America taxes gasoline too lightly. Better than a one-off increase, a politically more feasible idea, and desirable in its own terms, would be a long-term plan to shift taxes from incomes to emissions of carbon." In Europe and the United States, polls indicate that at least 70% of voters support environmental tax reform once it is explained to them.

SHIFTING SUBSIDIES

Each year the world's taxpayers underwrite $700 billion of subsidies for environmentally destructive activities, such as fossil fuel burning, overpumping aquifers, clear-cutting forests, and overfishing. A 1997 Earth Council study, *Subsidizing Unsustainable Development*, observes that "there is something unbelievable about the world spending hundreds of billions of dollars annually to subsidize its own destruction."

Iran provides a classic example of extreme subsidies when it prices oil for internal use at one-tenth the world price, strongly encouraging the consumption of gasoline. The World Bank reports that if this $3.6 billion annual subsidy were phased out it would reduce Iran's carbon emissions by a staggering 49%. It would also strengthen the economy by freeing up public revenues for investment in the country's economic and social development. Iran is not alone. The Bank reports that removing energy subsidies would reduce carbon emissions in Venezuela by 26%, in Russia by 17%, in India by 14%, and in Indonesia by 11%.

Some countries are eliminating or reducing these climate-disrupting subsidies. Belgium, France, and Japan have phased out all subsidies for coal. Germany reduced its coal subsidy from $5.4 billion in 1989 to $2.8 billion in 2002, meanwhile lowering its coal use by 46%. It plans to phase them out entirely by 2010. China cut its coal subsidy from $750 million in 1993 to $240 million in 1995. More recently, it has imposed a tax on high-sulfur coals. Together these two measures helped to reduce coal use in China by 5% between 1997 and 2001 while the economy was expanding by one-third.

The environmental tax shifting described earlier reduces taxes on wages and encourages investment in such activities as wind electric generation and recycling, thus simultaneously boosting employment and lessening environmental destruction. Eliminating environmentally destructive subsidies reduces both the burden on taxpayers and the destructive activities themselves.

Subsidies are not inherently bad. Many technologies and industries were born of government subsidies. Jet aircraft were developed with military R&D expenditures, leading to modern commercial airliners. The Internet was a result of publicly funded efforts to establish links between computers in government laboratories and research institutes. And the combination of the federal tax incentive and a robust state tax incentive in California gave birth to the modern wind power industry.

But just as there is a need for tax shifting, there is also a need for subsidy shifting. A world facing the prospect of economically disruptive climate change, for example, can no longer justify subsidies to expand the burning of coal and oil. Shifting these subsidies to the development of climate-benign energy sources such as wind power, solar power, and, geothermal power is the key to stabilizing the earth's climate. Shifting subsidies from road construction to rail construction could increase mobility in many situations while reducing carbon emissions.

In a troubled world economy facing fiscal deficits at all levels of government, exploiting these tax and subsidy shifts with their double and triple dividends can help balance the books and save the environment. Tax and subsidy shifting promise both gains in economic efficiency and reductions in environmental destruction, a win-win situation.

A CALL TO GREATNESS

There is a growing sense among the more thoughtful political and opinion leaders worldwide that business as usual is no longer a viable option and that, unless we respond to the social and environmental issues that are undermining our future, we may not be able to avoid economic decline and social disintegration. The prospect of failing states is growing as mega-threats such as the HIV epidemic, water shortages, and land hunger threaten to overwhelm countries on the lower rungs of the global economic ladder. Failed states are a matter of concern not only because of the social costs to their people but also because they serve as ideal bases for international terrorist organizations.

We have the wealth to achieve these goals. What we do not yet have is the leadership. And if the past is any guide to the future, that leadership can only come from the United States. By far the wealthiest society that has ever existed, the United States has the resources to lead this effort. Economist Jeffrey Sachs sums it up well: "The tragic irony of this moment is that the rich countries are so rich and the poor so poor that a few added tenths of one percent of GNP from the rich ones ramped up over the coming decades could do what was never before possible in human history: ensure that the basic needs of health and education are met for all impoverished children in this world. How many more tragedies will we suffer in this country before we wake up to our capacity to help make the world a safer and more prosperous place not only through military might, but through the gift of life itself?"

The additional external funding needed to achieve universal primary education in the 88 developing countries that require help is conservatively estimated by the World Bank at $15 billion per year. Funding for an adult literacy program based largely on volunteers is estimated at $4 billion. Providing for the most basic health care is estimated at $21 billion by the World Health Organization. The additional funding needed to provide reproductive health and family planning services to all women in developing countries is $10 billion per year.

Closing the condom gap and providing the additional 9 billion condoms needed to control the spread of HIV in the developing world and eastern Europe requires $2.2 billion—$270 million for condoms and $1.9 billion for AIDS prevention education and condom distribution. The cost per year of extending school lunch programs to the 44 poorest countries is $6 billion per year. An additional $4 billion per year would cover the cost of assistance to preschool children and pregnant women in these countries.

In total, this comes to $62 billion. If the United States offered to cover one-third of this additional funding, the other industrial countries would almost certainly be willing to provide the remainder, and the worldwide effort to eradicate hunger, illiteracy, disease, and poverty would be under way.

This reordering of priorities means restructuring the U.S. foreign policy budget. Stephan Richter, editor of *The Globalist*, notes, "There is an emerging global standard set by industrialized countries, which spend $1 on aid for every $7 they spend on defense.... At the core, the ratio between defense spending and foreign aid signals whether a nation is guided more by charity and community—or by defensiveness." And then the punch line: "If the United States were to follow this standard, it would have to commit about $48 billion to foreign aid each year." This would be up from roughly $10 billion in 2002.

The challenge is not just to alleviate poverty, but in doing so to build an economy that is compatible with the earth's natural systems—an eco-economy, an economy that can sustain progress. This means a fundamental restructuring of the energy economy and a substantial modification of the food economy. It also means raising the productivity of energy and shifting from fossil fuels to renewables. It means raising water productivity over the next half century, much as we did land productivity over the last one.

This economic restructuring depends on tax restructuring, on getting the market to be ecologically honest. Hints of what might lie ahead came from Tokyo in early 2003 when Environment Minister Shunichi Suzuki announced that discussions were to begin on a carbon tax, scheduled for adoption in 2005. The benchmark of political leadership in all countries will be whether or not leaders succeed in restructuring the tax system.

It is easy to spend hundreds of billions in response to terrorist threats, but the reality is that the resources needed to disrupt a modern economy are small, and a Department of Homeland Security, however heavily funded, provides only minimal protection from suicidal terrorists. The challenge is not just to provide a high-tech military response to terrorism, but to build a global society that is environmentally sustainable, socially equitable, and democratically based—one where there is hope for everyone. Such an effort would more effectively undermine the spread of terrorism than a doubling of military expenditures.

We can build an economy that does not destroy its natural support systems, a global community where the basic needs of all the earth's people are satisfied, and a world that will allow us to think of ourselves as civilized. This is entirely doable. To paraphrase Franklin Roosevelt at another of those hinge points in history, let no one say it cannot be done.

The choice is ours—yours and mine. We can stay with business as usual and preside over a global bubble economy that keeps expanding until it bursts, leading to economic decline. Or we can be the generation that stabilizes population, eradicates poverty, and stabilizes climate. Historians will record the choice, but it is ours to make.

Chapter 14

Social Issues and Global Change
Society in Flux

Jeanne Ballantine

S ociologists and other social scientists bring to our attention the many issues facing the world today. Sometimes the issues seem overwhelming, but without awareness and understanding we can do little to combat problems and bring about change. In this final chapter, we turn to expert social scientists and futurists who are studying national and global issues and proposing strategies to tackle these problems.

Social change is the alteration, over time, of a basic pattern of social organization. It involves two related types of changes: changes in folkways, mores, and other cultural elements of the society; and changes in the social structure and social relations of the society. These cultural and structural changes affect all aspects of society, from international relations to our individual lifestyles. An example of a change in cultural elements is women increasingly entering the labor force; an example of structural change is the rise of secondary-group relationships with a corresponding increase in formal organization.

Social change happens through the triple impact of *diffusion* (the speed of ideas across culture), *discovery* (unpremeditated findings), and *invention* (purposeful new arrangements). Each of these actions occurs at geometrically expanding rates—that is, the period of time over which an initial invention evolves. The change may require centuries, but once it is established, further changes related to that invention are rapid. For example, it took less than fifty years to advance from the Wright brothers' 12-second flight to supersonic travel and trips to the moon. Similarly, the rates of diffusion and discovery are now greater because of advancements in communication and travel.

Change occurs in all societies, but at different rates. Consider the sources of change. Sometimes change is *planned*; leaders or planning groups develop strategic plans or five-year plans to guide the process of change in organizations or countries. However, often change is *unplanned*, as the following examples illustrate: War breaks out between rival factions in Afghanistan, Kenya, Nepal, or Sudan, leaving in its wake thousands or millions of people homeless, hungry, and struggling to survive. Natural disasters strike a region as in the tsunami that hit Indonesia, the cyclone that struck Myanmar (Burma), or the earthquake that toppled buildings in Sechuan province in China. Riots threaten governments as in Nepal, and autocratic regimes cause many people to suffer from political repression or failing economies as in North Korea or Zimbabwe. While change is rapid and radical in some societies, such as eastern European countries and the former Soviet Union, and can actually tear societies apart, as in the case of the former country of Yugoslavia, change generally occurs more slowly in stable, traditional societies.

There are times when the norms that guide societal behavior break down. At these times, *collective behavior*—behavior that is both spontaneous and organized—may predominate. These forms are major sources of change in societies. Many of us have been involved in a form of collective behavior called a social movement, a group of people attempting to resist or bring about change.

Although social movements are often planned and have leaders, the resulting change is not always predictable. Yet movements are a powerful source of change. Participants in social movements may be trying to resist change and defend a past practice, fight for a social reform, transform a society, or bring about global change in environmental and global warming issues, child labor, or animal rights. Consider the article in this section on the movement for women's literacy around the world, written by Ruth Levine. Social scientists and policy makers are tackling world problems outlined in the United Nations Millennium Goals. One of the goals calls for the education of girls in developing countries, and Levine discusses the impact this movement for change can have on improving the lives of individuals, family members, and countries.

Recent examples of social movements in the United States include the civil rights and gay rights movements, various education reform movements, fundamentalist Christian movements, the women's movement, and the environmental movement. A movement may have a short life span, or it may become an institutionalized, stable part of the ongoing society. The next time you hear about a change in the United States or some other country, think about what is causing that change. Is it a war or revolution? A social movement? Something planned and executed by leadership?

After looking at the sources of change in society, futurists predict trends that will lead us into the future. Sociologists are active in this field, contributing study methods and an understanding of social systems and the process of change. In some cases, they focus on specific areas of change.

In the first article in this chapter, James Martin lays out his view of 17 of the greatest challenges facing societies in the twenty-first century. He predicts some solutions that could reduce the problems if people—especially young people—recognize the problems and act on them now.

Anil Hira focuses on issues that affect local and global economies, where money is being spent, and how we can reach the poorest people with global aid. He proposes a plan for a global welfare system that would address many of the issues that both he and Martin lay out and that would reduce the need for expenditures to control problems caused by war, environmental degradation, control of narcotics trafficking, and other problems.

Finally, Jerome Glenn discusses goals and strategies for dealing with demographic, economic, technological, and other issues discussed in this chapter, and looks to the future solutions. Understanding worldwide trends gives us a peek into the future and helps individuals, communities, nations, and world organizations prepare for the inevitability of change in a systematic way.

As you read these selections on social change and the future, consider both the changes stimulated by major events such as epidemics, wars, natural disasters, and terrorist attacks, and those changes that are carefully planned to achieve desired results.

1. What are some current social issues, movements, and disasters that stimulate change, and how are they bringing about changes?

2. What changes do the authors propose in order to deal with issues?

3. For the predicted changes to take place, what would need to happen in institutions of societies?

4. What kind of society and world do we want, and how can we attain it?

57

The 17 Great Challenges of the Twenty-First Century

JAMES MARTIN

Futurists can be found in a number of disciplines—from social sciences to policy studies to technology and life sciences. Martin overlaps the social science and sciences to present his view of the greatest challenges for the future—challenges generally resulting from human actions. He points out what the challenges are, what needs to happen, and what is not happening and why, and he challenges what he calls the Transition Generation *to tackle the problems if our way of life is to be preserved.*

As you read this article, consider the following questions:

1. *What does Martin consider the great challenges, and why?*
2. *Why does Martin believe that solutions to the world's problems are unlikely to be implemented?*
3. *What does Martin suggest as strategies to deal with the world's problems?*
4. *When would you want to be alive, and why?*

GLOSSARY **Transition Generation** Today's teenagers, whose job it will be to get humanity through the coming instability. **Globalism** The shrinking of the planet due to communication, transportation, technology, and business. **Transhumanism** The ability to radically change humans through technology and understanding of human brains.

Young people sometimes ask me, "If you were to pick any time in history to be alive, which time would you pick?" I reply to their question, "If I could choose any time to live, I would want to be a teenager now (in a country where great education is available)." The reason I would choose to be a young person today, rather than during some earlier period in history, is that now, more than at any other time, young people will make a spectacular difference.

We are heading toward an inflection point of epic significance, when scientific advances will be beyond anything we've ever before experienced. Simultaneous to this techno-revolution, fresh water will run out in many parts of the world. Global warming will bring hurricanes far more severe than Katrina and will cause natural climate-control mechanisms to go wrong. Rising temperatures will lower crop yields in many parts of the world's poorest countries, such as those in central Africa.

Source: Martin, James. 2007. "The 17 Great Challenges of the Twenty-first Century." *The Futurist.* January–February, pp. 20–24.

The immense tensions brought about by such catastrophes will occur in a time of extremism, religious belligerence, and suicidal terrorism, and this will coincide with terrible weapons becoming much less expensive and much more widely available.

Solutions exist, or can exist, to most of the serious problems we will face in the decades ahead. The bad news is that the most powerful people today have little understanding of the solutions and little incentive to apply them. Politicians are anxious to find votes—the next election dominates their thinking. Powerful business executives are eager to achieve profits—it is their job to increase shareholder value, and shareholders will judge them by this quarter's results. So for the powerful people who control events, the desire for the short-term benefits overwhelms the desire to solve long-term problems. The job of today's teenagers, the "Transition Generation" if you will, is to get humanity through the coming instability as quickly and safely as possible.

If we are to survive, we have to know how to manage this situation. We need to put in place rules, protocols, methodologies, codes of behavior, cultural facilities, means of governance, treaties, and institutions of many types that will enable us to cooperate and thrive on planet Earth. If we can do that with whatever the twenty-first century throws at us, we'll probably be able to survive. If our twenty-first-century world falls apart at the seams, civilization will be set back many centuries.

The twenty-first century, then, brings these 17 challenges, which are all interlinked and mutually reinforcing. Together they constitute the twenty-first-century transition. As our knowledge improves, our challenges, too, will be refined and added to.

1. SAVING THE EARTH

A change in the capability to manage the Earth well is coming from the deployment of vast quantities of micro-instruments, which feed voluminous data to computer systems. Humanity is changing from being ignorant about the planet to having vast quantities of information linked to supercomputer models.

In the second half of the century, we will have learned how to live within nature's trust fund. One hopes that we will learn this by science and good teaching. If not, we will learn from catastrophe-first patterns of events. The Earth's climate will change, and we will learn to live with the changes.

2. REVERSING POVERTY

While rich nations become richer, billions of people live in extreme poverty with short, brutal lives. In his book *An End to Poverty: Economic Possibilities for Our Time* (Penguin, 2005), Jeffrey Sachs lays out nine steps for solving global poverty. They are: commit to the task, adopt a plan of action, raise the voice of the poor, redeem the United States' role in the world, rescue the International Monetary Fund and the World Bank, strengthen the United Nations, harness global science, promote sustainable development, and make a personal commitment.

3. STEADYING POPULATION GROWTH

Much of the extreme poverty on the planet relates to the population being too high. It is estimated that the Earth's population will soon be increased by 2.5 billion people, most of them in the countries least able to grow enough food. There are now non-oppressive ways to lower the birthrate. Population declines strongly in countries where almost all women can read and full women's liberation is in effect. Population also tends to decline when GDP is high. The challenge of improving lifestyles equates to the goal of lowering population.

4. ACHIEVING SUSTAINABLE LIFESTYLES

Most people (almost 9 billion) will eventually want to participate in the affluence of the planet. This cannot happen with twentieth-century lifestyles.

We need higher-quality lifestyles that are environmentally harmless. Rich, affluent, globally sustainable lifestyles, more satisfying than today's, can be achieved at the same time as healing the environment.

5. PREVENTING ALL-OUT WAR

All-out war in the twenty-first century could end everything. No economic or political benefit can justify the risk of an all-out war with nuclear and biological weapons. We absolutely must prevent war between nations with arsenals of mass destruction. There will either be no war among high-tech nations or no civilization. The existence of weapons capable of ending civilization makes this a very different century from any before.

6. DEALING EFFECTIVELY WITH GLOBALISM

Globalism is here to stay. The planet is shrinking and bandwidth is increasing, but globalism should be designed to allow local unique cultures to thrive and be protected. The right balance between what is global and what is local needs to be achieved.

Global business will continue to expand, and it needs to benefit everyone rather than bypassing some countries and leaving them destitute. Failed nations must be helped until they become developing nations.

7. PROTECTING THE BIOSPHERE

We are losing species of plants and creatures at a shocking rate. Many endangered species can be protected by identifying and preserving hot spots, those places with a high density of endangered species.

Today, 90% of the edible fish in the oceans have been caught. It is possible to create conditions in which ocean life will slowly recover. This requires well-designed marine protection areas, combined with well-managed fishing. Laws are needed for transforming the oceans from their appallingly depleted state today to a vigorous, healthy state.

Different challenges to the biosphere include reduced agricultural biodiversity to achieve high-yield farming and the increasing use of genetically modified crops.

Global management of the biosphere is essential. This requires thorough, computer-inventoried knowledge of all species.

8. DEFUSING TERRORISM

The dawn of an age of terrorism coincides with the rise of weapons of mass destruction that will become progressively less expensive. It is vital, therefore, to address the reasons why people want to become terrorists and to achieve cooperation among potentially hostile cultures.

9. CULTIVATING CREATIVITY

The technology of the near future will lead to an era of extreme creativity. Young people everywhere should participate in the excitement of this creativity. Different cultures are likely to accept one another as exciting jobs spread and rich countries help young people around the planet to be entrepreneurs. The world is becoming finely laced with supply chains of electronically connected businesses that will eventually interlink all countries and become very valuable.

10. CONQUERING DISEASE

We must thwart the rapid spread of infectious diseases that could kill many millions of people, as has happened numerous times in history. We now have sensors that can detect the existence of a dangerous virus in the air, as well as medical procedures to prevent it from spreading. We need to be ready, with all our technological resources, to stop bird flu and future pandemics that are a surprise. We are not ready today.

11. EXPANDING HUMAN POTENTIAL

A tragedy of humankind today is that most people fall outrageously short of their potential. A goal of the twenty-first century ought to be to develop the capability latent in everybody by harnessing powerful technologies that accelerate learning potential.

12. THE SINGULARITY

Decades from now, computer intelligence that is quite different from human intelligence will feed on itself, becoming more intelligent at a rapidly accelerating rate. This chain reaction of computer intelligence is referred to as the Singularity. Humanity needs to discover how to avoid being overwhelmed by accelerating change that is totally out of control and harmful. Technical controls will be needed for computing, perhaps in the form of hardware design, to ensure that, when computers become incomparably more intelligent than we are, they act in our best interests.

The main impact of the Singularity will be that the cleverest professionals will use it to achieve extraordinary results. By the time it happens, the capability for handling the Singularity will be distributed globally, particularly among appropriately educated young people. It will enable many different self-evolving technologies to become "infinite in all directions."

13. CONFRONTING EXISTENTIAL RISK

The twenty-first century is the first in which events could happen that terminate *Homo sapiens*. These are referred to as existential risks (risks to

our very existence) and include such possibilities as the unleashing of a genetically modified pathogen. In his book *Our Final Hour* (Basic Books, 2004), Martin Rees describes such risks in detail and gives humanity only a 50% chance of surviving this century.

If we do survive, our accomplishments by the end of the century will be awesome. The magnificence of what human civilizations will achieve if they continue for many centuries is beyond all imagining—so magnificent that it would be too tragic for words if humanity were terminated. To run the risk of terminating *Homo sapiens* would be the most unspeakable evil. We should regard any risk to our existence as totally unacceptable. We need to take whatever actions are necessary to bring the probability of extinction to zero.

14. EXPLORING TRANSHUMANISM

This is the first century in which we will be able to radically change human beings, and this fact alone gives very special meaning to the twenty-first century. Technology will enable us to live longer, learn more, and have interesting prostheses. Neuroscience will blossom spectacularly when we can map the brain, recording the transmission of signals among individual neurons and then emulating parts of the brain with technology millions of times faster than the brain. A new world will open up when we can connect diverse neurons in our brain to external devices. We'll connect the brain directly to nanotechnology objects on or in our skulls and to supercomputers far away. This will change human capability in extraordinary ways.

Transhumanism will be highly controversial. It will raise major ethical arguments. We might harm some of the qualities that make humanity wonderful. It will lead to fundamental advances in what humans are capable of but will create extreme differences between the haves and have-nots.

We need to understand where changes to *Homo sapiens* can be made without net-negative consequences. Transhumanism will be a prime enabler of civilizations far beyond those of today.

15. PLANNING AN ADVANCED CIVILIZATION

The twenty-first century will experience a major increase in real wealth (adjusted for inflation). Sooner or later, machines will do most of the work. What we do with our leisure will be a huge issue. A big question that we ought to be asking now is: "What could truly magnificent civilizations be like at the end of the century?" Because of transhumanism and the Singularity, the changes will be more extreme than are generally realized.

16. MODELING THE PLANET'S SYSTEMS

Because we need to make sure that we are not close to the limits beyond which runaway global warming occurs, earth system science needs to be a thorough academic discipline that comprehensively measures and models the Earth's control mechanisms. There will be uncertainties in the models, but we should not take the slightest risk of upsetting the immense forces that make our home planet livable.

Perhaps the greatest catastrophe that could befall us would be that we inadvertently push Gaia so that positive feedback causes it to become unstable or to change to a different state. The twenty-first century must put the science in place to regulate human behavior to live at peace with Gaia. This will be essential for future centuries.

17. BRIDGING THE SKILL AND WISDOM GAP

Deep wisdom about the meaning of the twenty-first century will be essential. A serious problem of our time is the gap between skill and wisdom.

Science and technology are accelerating furiously, but wisdom is not. Today, deep reflection about our future circumstances is eclipsed by a frenzy of ever more complex techniques and gadgets and preoccupation with how to increase shareholder value. The skill–wisdom gap is made greater because skills offer the ways to get wealthy. Society's best brains are saturated with immediate issues that become ever more complex, rather than reflecting on why we are doing this and what the long-term consequences will be.

THE TAPESTRY OF BIG ISSUES

The story of what is happening to humanity and its home planet can largely be put in place now. It is not sharply in focus; our knowledge in many areas is sketchy. Yet the tapestry of big issues is visible. We are damaging our future in diverse ways, but there are resolutions to these problems—numerous solutions in different disciplines. A massive transition is needed, and the agenda can be created for the generation who is going to bring about the transition. Broadly speaking, we know what needs to be done. It involves all nations. The issues are global. There is no place to hide.

There is a problem, however. Much of what needs to be done is not happening. The grand-scale transition of the twenty-first century could occur gently. There could be step-by-step replacement of carbon-based fuels, steady improvement of food-growing capability, measures to conserve water, a lockdown of sources of fissile uranium, growth of antiterrorist measures, a drive for eco-affluent lifestyles, and so on.

It's a long list, but today's computer models make clear that we are not changing our ways fast enough. We are drifting toward irreversible climate change faster than we are taking any actions to keep this from happening. Water is essential for food production, but we are taking water from aquifers at a rate that will cause many of them to run dry. In addition to water depletion and soil degradation, food production will also be seriously lowered in some countries by drought and heat waves caused by the rising quantity of greenhouse gases. As if that weren't bad enough, we are diverting huge amounts of water from farms and cities, and that trend will grow because there are massive migrations of people from the countryside to cities.

To make the twenty-first-century transition as painless as possible, we need to make positive changes before the problems become too bad. In almost all areas, however, this is not happening. Where steady transitions are possible—for example, with the change to noncarbon fuels—almost nothing is being done. The U.S. Environmental Pro-tection Agency tries to repair environmental damage after it has occurred, but seems to have almost no ability to change the economic practices that cause the damage in the first place. World summits on sustainable development have had agendas of immense importance but have taken almost no action on anything. Their follow-up can be described as studied avoidance of any changes that are controversial. The acronym UNCTAD (United Nations Conference on Trade and Development) is often said to stand for Under No Circumstances Take Any Decision.

The subsidies for fuels that damage the environment are massive, but subsidies for fuels that would help the environment are small. Ecology expert Norman Myers catalogued $2 trillion per year of "perverse subsidies"—subsidies that do more harm than good. They leave the environment or the economy worse off than if the subsidy had never been granted. If voters and taxpayers were given a listing of subsidies they pay, along with the net harm from those subsidies, they would revolt. Not surprisingly, governments tend to hide that information. The longer the transition is delayed, the more difficult it will be.

Unlike their predecessors, the Transition Generation will not sit idle and see their world go down the drain. The twenty-first-century transition could become revolution, not evolution. As often before in history, revolution will be the consequence of complacency. Craig Venter, the legendary genome mapper, commented to me that the danger to our society is not science—it's apathy.

WHEN WOULD YOU WANT TO BE ALIVE?

When I say I would rather be a young person now than during any other time in history, there's a reason. The twenty-first-century revolution is absolutely essential, and today's young people will make it happen. But there needs to be an absolute crusading determination to bring about the changes we describe. Today's young people will collectively determine whether civilization survives or not. It will be a time of revolution establishing the processes by which humankind can achieve levels of greatness never before dreamed of.

With technologies that are infinite in all directions, what can humanity become? Our future wealth will increasingly relate to knowledge in the broadest sense of the term. We might use the term "knowledge capability" to refer to the quantity of available knowledge multiplied by the power of technology to process that knowledge. The quantity of usable knowledge is rising fast (for example, precisely mapping the genome of everything biological), and the power of technology to process the knowledge capability is increasing exponentially. Combining these, knowledge capability is approximately doubling every year. It seems likely that this doubling will go on throughout the century (if there is no catastrophic disruption). That means that, during the twenty-first century, knowledge capability will increase by two to the power of 100, a thousand billion billion.

The individual is immersed in such an expanding ocean of capability to process knowledge. That makes the twenty-first century both more exciting and more perilous than any other century so far. We are heading toward an inflection point, but our leaders are not preparing to make the passage smoother for us. That will be the job of the Transition Generation.

58

Dislocations and the Global Economy: Time for a Global Welfare System?

ANIL HIRA

The world is facing a number of challenges to achieving a stable economic state as it continues to globalize. Anil Hira first discusses some of these challenges, and then suggests that the world could tackle these problems by controlling population growth and devoting a small portion of the money spent on controlling illegal immigration and narcotics trafficking, fighting wars, and dealing with environmental degradation to what he refers to as "a global welfare system." To meet the United Nations Millennium Development Goals would require a fraction of what is spent to control other problems—problems that would be alleviated if the goals were met.

As you read these articles, consider the following questions:

1. *What does Hira mean when he says that "the pressure point is population growth"?*
2. *What are the major "dislocations to the global economy"?*
3. *How could spending money on meeting the United Nations Millennium Goals help alleviate the billions spent on controlling other problems?*
4. *What does Hira mean when he says it is "time for a global welfare system"?*

GLOSSARY **Global welfare system** A global institutional system with regulatory and welfare capacity to meet a number of the world's pressing needs.

DISLOCATIONS AND THE GLOBAL ECONOMY

The standard of living in the West has improved steadily for more than 200 years, so why do we now face a wide range of seemingly intractable problems? Here is a brief overview of the dislocations wrought by the emerging global economy.

Terrorism

Terrorism is now a daily obsession to people in both the West and Middle East. Some observers

Source: Hira, Anil. 2007. "Dislocations and the Global Economy" and "Time for a Global Welfare System?" *The Futurist.* May–June, pp. 27–32.

imply that the "war" on terrorism has more to do with irreconcilable ideological differences and extremism than with poverty. But look to Ireland as an example of how this view may be wrong: It is not pure coincidence that peace has come to Ireland at the same time that the Irish economy has taken off.

We know that life for many Palestinians is appallingly insecure and difficult. Many in the Middle East suffer from the repression of autocratic governments, from the pseudo-democracies of Egypt and Jordan to the harsh repression of Syria and Saudi Arabia. The lack of opportunity and a strong sense of frustration add fuel to a latent support for terrorism in the region.

Employment Insecurity

In the last 20 to 30 years, the advanced economies have seen jobs sent to less-costly countries, as well as a steady erosion of domestic labor skills and a breakdown of employer–employee relationships.

A university education would seem more important than ever, but today, a bachelor's degree does not mean entry into a good white-collar job. The brief IT boomlet of the late 1990s and the subsequent plundering of jobs to outsourcing offshore means even sure bets, such as engineering and computer science, are now under siege.

This breakdown of steady and stable employment possibilities in the West can also be tied to changes in the world economy. The mobility of production essentially means that companies are no longer tied to local labor forces. Now, even high-tech, high-value-added complex processes for which the labor is limited are not immune from globalization's disruptions.

Financial Volatility and
Corporate Disasters

Not only do terrorists and narcotraffickers use offshore and Swiss banks, but so do most corporations. The deposits in offshore banking havens such as the Cayman Islands are designed specifically to help companies avoid paying taxes. Their existence means that we are unable to trace financial transactions for

security purposes in the fight against terrorism and drugs. It also means that billions of dollars in corporate profit each year are untaxed and thus unavailable to the community and governments from which they came. Because of the globalization of trade and finance, it is increasingly difficult to push corporations to act responsibly.

This reality undercuts individuals' long-term financial security, including our ability to save and invest and to be secure for our retirement. The 1998–1999 Asian financial crisis demonstrated that with the push of a button a flood of pension fund withdrawals could crush an entire financial system. Small levels of investment enjoy deposit insurance, but the inherent risk and volatility of the financial system reduces savings and makes our lives insecure.

Global Diseases

The billions of dollars we have spent on AIDS have not led to a cure, though treatments are remarkably improved. Yet, we are continually reminded of the ramifications of the millions of new HIV/AIDs infections.

Globalization has brought new risks, such as the possible spread and mutation of the Ebola virus from Africa, as well as mad cow disease, avian flu, and SARS from Asia. Our sense of vulnerability to a new epidemic comes from the millions of air and cargo shipments that cross borders and fill grocery stores every day.

Global Warming and Other
Environmental Disasters

There are other stories in the backburner of our minds that we need to bring forward. Every day we are put in fear about global warming, overfishing, and the extinction of species. The air is hardly breathable now in Shanghai and Mexico City—what will happen as the millions of new autos being demanded hit the road? The air in China and Mexico affects the air that everybody breathes as well as the temperature of our climate.

A generation ago, we could count on the possibility of enjoying the wild animals and environment of different areas, but those possibilities are rapidly being extinguished. As the debacle around the implementation of the Kyoto Protocol demonstrates, we seem to be unable to move from recognition of the problem to action in terms of slowing down the threat to the environment. Such inaction hurts not only our overall quality of life, but also our economic prospects.

TIME FOR A GLOBAL WELFARE SYSTEM?

We live in an era of great transformation. Citizens everywhere can feel the ground shifting from under them, from terrorism to pandemics to job insecurity, but we do not yet have a road map of the changes, let alone a blueprint for how to address them.

All of these dislocations can be tied in part to major changes in the way the global economy operates. Some of these changes have been documented by authors such as Thomas Friedman *(The World Is Flat)* and William Greider *(The Soul of Capitalism)*, both in terms of a sense of great optimism and great pessimism, respectively. Yet no one has clearly tied these things together in a way that gets us past the stage of lamentation and on to a plan to respond.

The world is not flat, but round. We are increasingly affected and interconnected from one remote corner to another, and both the positive and negative effects of this transformation are multiplying.

A golden age is well within our grasp. It is not a pipe dream. What is good for us as individuals and as nations is in many ways directly tied in with the welfare of others.

A Global Welfare System

Given the benefits of global trade, finance, and immigration, it is both self-defeating and impossible to think about reversing such trends. Yet, individual nations no longer seem to have the means of dealing with the resulting international problems.

A proposal to create a global institutional system with regulatory and welfare capacity will undoubtedly create huge nationalistic reactions. But the rise of China as the number-two economy in the world illustrates that the world order has shifted, and the architecture set up by the West will no longer be acceptable for the rest of the planet. The question then becomes not whether we need to remake global institutions, but how.

There is an opportunity now to remake those institutions, to create a system in which the global economy can prosper and offer opportunities for improving standards of living for all. The remaking of such systems does not mean a loss of sovereignty or domestic preferences, but rather the creation of institutions that are responsive on the global level to both states and citizens. A new system of global government should enable us to deal proactively with global problems, rather than relying on ad hoc and inconsistent efforts by individual or temporary groups of states.

Global environmental problems, immigration pressures, downward wage pressures, insecurity from terrorism, reliance on far-flung commodity suppliers, humanitarian crises, and global disease all bode ill for the future if individual nations are only able to deal with them in haphazard fashion. Yet, there is a key pressure point upon which all of these crises touch, which—if addressed—could turn the nightmare toward a paradise. The pressure point is population growth.

Healthy families where women have the opportunity to work have fewer children. This means slower population growth, which reduces pressures on immigration, the creation of new terrorists, and competition linked to weak labor and environmental standards. The slash-and-burn agricultural practices that are destroying the rain forests and the desperate turn to narcotrafficking or to jobs in sweatshop conditions to support oneself would all be diminishing enterprises, if we treated one of the chief causes rather than symptoms of global problems.

Reducing population pressure is a chicken-and-egg problem. Urbanization reduces family size, as do access to contraceptives, education, and health care.

Costs of a Global Welfare System

Spending on global fund to fight AIDS	$3.0 billion and climbing
Costs of illegal immigration enforcement	$28.9 billion and climbing
Costs of war in Iraq and Afghanistan, respectively	$252.0 billion and climbing
Costs of refugee crises	$1.35 billion just by the UNHCR
Costs of narcotrafficking	$12.2 billion just by U.S.
Costs of environmental degradation	incalculable
Costs [estimated] of reaching UN Millennium Development Goals related to basic improvements in health, education, gender equality, environmental improvement, and disease eradication in the developing world	$28.0 billion in 2006

SOURCE: Compiled by the author from various sources, including the United Nations and the White House.

Ensuring female access to education and health care are the surest ways to reduce population growth without controversy over abortions or contraceptive use. More-prosperous families are healthier and provide a productive contribution, rather than a drain or threat, to the global economy. However, huge populations in developing countries, even at a reduced pace, make such transformations seem glacially slow and overwhelming. The fastest-growing populations, indeed, are in the poorest regions, including South Asia and sub-Saharan Africa. A global system is needed to accelerate the change to a low-population, high-quality-of-life economy. Without such attention, the average American, Japanese, European, and Canadian will find it impossible to compete with the thousands of hungry, hard-working, and, in Asia, well-educated growing middle classes.

Such a system would ensure that no one starves, that labor has reasonable bargaining rights, that a policy that respects the environment is enforceable,

and that basic human rights are afforded to all. Indeed, the one major global welfare transfer, namely, the Marshall Plan, set the world on a growth spree for two decades, simultaneously creating new markets while lifting millions out of poverty. By recognizing that we are now living in a global economy, we simply move to regulate that economy so it can thrive. Not only do we reduce the costs of such problems as terrorism, pandemics, and environmental degradation, but we can create a whole new generation of consumers.

Let us consider some basic "back of the envelope" costs for creating a system that moves beyond aid towards basic global welfare. [See "Costs of a Global Welfare System."]

As we can see, it is clearly illogical for us to continue to pay the costs of the symptoms of global problems without addressing the causes of those problems. Addressing the causes would reduce the costs of treating the symptoms and also provide humanitarian benefits.

The main problem with the United Nations and other nonprofit organizations who have echoed this line is that they continue to rely upon a system of voluntary charity. In my book *Development Projects for a New Millennium* (Praeger, 2004), I outlined several reasons why development in general is simply ineffective, a few of which bear highlighting here. For one thing, development agencies treat world citizens as charity cases. Rather than look at education or health as an investment in a future citizen who can later pay taxes and enlarge the global economy, development agencies treat beneficiaries as helpless cases for pity.

Charity at the international level is closely tied to national interests, rarely coordinated, and arrives in haphazard and inconsistent fashion. Development projects and agencies rarely, if ever, undergo any serious independent evaluation, and there is no system for learning from projects. The main result of development aid seems to be to fund middle-class bureaucrats who have good intentions but neither the resources nor the wherewithal to make a real difference on the ground. Moreover, these organizations answer to both donor country politics and host country politics, as well as their own bureaucratic maze, and are constantly pulled in different directions. This system effectively supports the corrupt and inefficient regimes that have made the development projects necessary. By responding only to governments, development projects have come under fire from NGOs (nongovernmental organizations) and citizens who wonder why they should spend money on organizations that have no direct accountability to them.

Obviously, this system is not working. What is needed is a global system with its own revenue base and independent action, as well as direct accountability to world citizens. Let us briefly discuss how such a system could be constructed.

Building a Global Aid and Development Agency

A global welfare system does not mean giving up local decisions, or consigning individuals or nations to the whims of some distant bureaucracy. There is an infinite variety of systems by which these goals could be accomplished.

Generating Revenues for Development.   The backbone of this system is a new global taxation framework. The criteria for a global taxation authority should be that any tax should be progressive and nondiscriminatory, and should not distort trade and investment flows. For a number of years now, economists have been calling for a tax on international financial transactions to reduce the volatility of financial flows. Chile and Malaysia successfully put controls on "hot" money to reduce such volatility, and this concept could be extended to a basic tax upon all international transactions, including trade and finance. If we were able to include a global tax on corporations—which currently evade taxes by locating offshore—we would raise billions of dollars.

In the spirit of the Live Aid and Live 8 concerts, we could place a small tax on rock concerts and other entertainment, such as films and video rentals in the North, to pay for such schemes. If we added a small percentage tax on all entertainment, including the billions we spend annually on sporting events in the developed world, we would have a vast pool of resources and an entertainment industry that could make an effective difference.

We could also institute various tax sources that related problems directly to solutions. For example, we could have a small percentage tax on pharmaceutical sales and a series of sin taxes on alcohol and tobacco to raise money for health concerns. We could put a small tax on gourmet food items, such as coffee, or on luxury items to pay for improving educational access. One of the biggest revenue producers would be to create a small percentage tax on arms sales, which could then be used for peacekeeping and international security operations designed to make the world safer. We could place a small tax on energy consumption to help pay for environmental protection.

Another basic tax that would be easy to administer would be a universal tax. A global tax regime would allow us to ensure that corporations pay an appropriate share in taxes, and avoid using offshore banking or transfer pricing or a myriad of other schemes to avoid taxes.

There are a number of possible ways to administer this type of tax. One way would be to have a global income tax or a universal sales tax. A small percentage would be added onto existing tax collections to help. Both of these taxes would be progressive in the sense of taxing the larger markets and richer consumers more, as they spend more and consume the lion's share of energy and other resources. As with other sales taxes, essential items such as basic foods and transport in poorer areas could be made exempt or taxed at a lower rate.

The key point is that a few small percentage points added to each transaction would be almost unnoticeable after the initial outcry. Such taxes would be progressive by placing more of a burden upon those who consume more; however, they would also be fair in the sense that richer consumers in the developing world who have similar consumption habits would also be contributing. There would be no need for direct enforcement, as the tax would be simply added onto the already functional tax systems in the North. This would have the added benefit of taxing rich elites from the

South with assets, properties, and businesses in the North. The universal system would also end the beg-thy-neighbor type of financial havens and tax jurisdictions that have so aided dictators, terrorists, and narcotraffickers.

Collecting and Distributing Global Aid

Before answering who and how the taxes raised would be spent to deal with global problems, we have to cut through one major remaining obstacle: corruption. Direct expenditures for development now often mean dealing with corrupt international organization bureaucracies and host governments. That needs to change. International organizations such as the UN are extremely weak because they must constantly seek the support of powerful nations for global missions, and there is little scrutiny or evaluation of the results. A steady revenue stream would resolve the problem of continual coalition-building and begging for resources.

With assured resources, the proposed global agency would act in concert with powerful nations, but would have independent authority to make decisions. Such decisions would be limited to where and how much to spend on specific global welfare programs over any given five- to 10-year period. Because the mandate of the agency would be limited to these specific tasks, there would be no concern on the part of citizens or states about infringement or dictation from a world government. Many developing country governments are corrupt and ineffective. They would be given a choice of accepting world welfare as direct payments to their citizens or not receiving anything. The global welfare agency or council could thus work with host governments cooperatively without compromising their mission and direct duty to every citizen. In international finance, it would mean that local banks have to conform with a universal set of sound management and transparency principles, and greater resources would have to be provided to ensure liquidity and stability in the financial system.

The best and only way to ensure accountability would be to set up democratic systems: taxation with representation. To ensure that members of this agency were accountable to world citizens, there would need to be some agreed-upon combination of voting and representation for a deliberating council, as well as an independent auditing and evaluation arm.

A wide variety of governance systems could be envisioned. For example, one half of the council could be popularly elected and the other appointed by the 10 states with the largest contributions. Or citizens from each country could elect delegates. These delegates could elect representatives along geographic, development status, or other lines to ensure representation of the appropriate interests. The council itself would not have to become a massive bureaucracy. It could receive competing proposals for bids to complete programs or projects that it suggests.

The idea of a global welfare organization will be hard for some people to accept. But since the current policies are not working well, we need to think about new solutions. These thoughts are intended to spark greater discussion of this important issue.

59

Educating Girls, Unlocking Development

RUTH LEVINE

The United Nations Millennium Development Goals (MDG) call for universal primary education and gender parity in education in all countries by 2015. Levine talks about these goals and what difference they will make for girls, especially in developing countries. Major changes will be brought about by educating girls, from enhancing a girl's ability to improve her own health and that of her family and community, to contributing to the productivity of her nation. Levine discusses the trends, challenges, and way forward in girls' education.

As you read this article, think about the following questions:

1. *Why is it so important for girls to receive education?*
2. *What happens when girls receive education, and why?*
3. *Why is educating girls beneficial to development efforts?*
4. *What change can come about through educating girls?*

GLOSSARY **Unlocking development** The idea that to enhance development efforts in developing countries, education of girls is a key element. **Educational attainment** How much schooling children receive. **Rural/urban education gap** The difference between urban and rural educational attainment levels.

One of the most important public policy goals in the developing world is the expansion and improvement of education for girls. Vital in its own right for the realization of individual capabilities, the education of girls has the potential to transform the life chances of the girls themselves, their future families, and the societies in which they live. Girls with at least a primary school education are healthier and wealthier when they grow up and their future children have much greater opportunities than they otherwise would; even national economic outcomes appear to be positively influenced by expanded girls' education.

Unlike some development outcomes that depend on multiple factors outside the control of policy makers (either in developing countries or among donor nations), significant improvement in girls' education can be achieved through specific government actions. Expansion of basic education, making school infrastructure and curriculum more girl-friendly, and conditional cash transfers and scholarships to overcome household barriers have all been used to improve key outcomes, with demonstrable success. Lessons from regions that have made rapid advances with girls' education, and from programs that have introduced successful financing

Source: Levine, Ruth. 2006. "Educating Girls, Unlocking Development." *Current History*. March, pp. 127–131. (Citation was "Reprinted from *Current History*, March 2006, pp. 127–131. Copyright 2006 by Current History, Inc. Reprinted with permission.)

and teaching innovations, can be applied to accelerate progress.

While public policy can make the difference, policies that ignore important gender-related constraints to education at the primary and, particularly, at the postprimary educational levels can have the opposite effect, reinforcing existing patterns of gender discrimination and exclusion. Those patterns are often deepseated. Families in many societies traditionally have valued schooling less for girls than for boys. In most households, the domestic workload falls more to females than to males, leaving less time for school. If families are struggling to find income, the demand for girls' help around the house (or in wage labor) may increase. Many parents believe that the return on educational investments varies according to gender—particularly if girls, when they marry, leave their parents' households to join the husbands'.

When girls in developing countries do enroll in school, they frequently encounter gender-based discrimination and inadequate educational resources. Large numbers of girls in sub-Saharan Africa drop out, for example, when they reach puberty and the onset of menstruation simply because schools lack latrines, running water, or privacy. Parental concerns about girls' security outside the home can limit schooling where girls are vulnerable in transit and male teachers are not trusted. And in some countries, cultural aversion to the education of girls lingers. Afghanistan's Taliban insurgents, who believe that girls' education violates Islamic teachings, have succeeded in closing numerous schools, sometimes by beheading teachers. Afghanistan is an extreme case, but a reminder nonetheless of the challenges that remain on the path toward achieving the high payoffs from girls' education.

THE BENEFITS

Why is the schooling of girls so critical? Education in general is among the primary means through which societies reproduce themselves; correspondingly, changing the educational opportunities for particular groups in society—girls and minority groups—is perhaps the single most effective way to achieve lasting transformations. A considerable body of evidence has shown that the benefits of educating a girl are manifested in economic and social outcomes: her lifetime health, labor force participation, and income; her (future) children's health and nutrition; her community's and her nation's productivity. Most important, education can break the intergenerational transmission of poverty.

Female participation in the formal labor market consistently increases with educational attainment, as it does for males. In at least some settings, the returns to education of girls are superior to those for boys. Several studies have shown that primary schooling increases lifetime earnings by as much as 20 percent for girls—higher than for their brothers. If they stay in secondary school, the returns from education are 25 percent or higher.

The inverse relationship between women's education and fertility is perhaps the best studied of all health and demographic phenomena. The relationship generally holds across countries and over time, and is robust even when income is taken into account. Completion of primary school is strongly associated with later age at marriage, later age at first birth, and lower lifetime fertility. A study of eight sub-Saharan countries covering the period from 1987 to 1999 found that girls' educational attainment was the best predictor of whether they would have their first births during adolescence.

Another study examined surveys across the developing world to compare female education and fertility by region. The higher the level of female education, the lower desired family size, and the greater the success in achieving desired family size. Further, each additional year of a mother's schooling cuts the expected infant mortality rate by 5 to 10 percent.

Maternal education is a key determinant of children's attainment. Multiple studies have found that a mother's level of education has a strong positive effect on daughters' enrollment—more than on sons and significantly more than the effect of fathers' education on daughters. Studies from Egypt, Ghana, India, Kenya, Malaysia, Mexico, and Peru all find that mothers with a basic education are substantially more likely to educate their children, especially their daughters.

Children's health also is strongly associated with mothers' education. In general, this relationship holds across countries and time, although the confounding effect of household income has complicated the picture. One study, for instance, compared 17 developing countries, examining the relationship between women's education and their infants' health and nutritional status. It found the existence of an education-related health advantage in most countries, although stronger for postneonatal health than for neonatal health. (In some countries the "education advantage" did appear to be eliminated when controlling for other dimensions of socioeconomic status.)

Other studies have found clear links between women's school attainment and birth and death rates, and between women's years of schooling and infant mortality. A 1997 study for the World Bank, which focused on Morocco, found that a mother's schooling and functional literacy predicted her child's height-for-age, controlling for other socioeconomic factors.

Although the causal links are harder to establish at the macro-level, some researchers have made the attempt, with interesting results. For example, in a 100-country study, researchers showed that raising the share of women with a secondary education by 1 percent is associated with a 0.3 percent increase in annual per capita income growth. In a 63-country study, more productive farming because of increased female education accounts for 43 percent of the decline in malnutrition achieved between 1970 and 1995.

In short (and with some important nuances set aside), girls' education is a strong contributor to the achievement of multiple key development outcomes: growth of household and national income, health of women and children, and lower and wanted fertility. Compelling evidence, accumulated over the past 20 years using both quantitative and qualitative methods, has led to an almost universal recognition of the importance of focusing on girls' education as part of broader development policy.

THE TRENDS

Given the widespread understanding about the value of girls' education, the international community and national governments have established ambitious goals for increased participation in primary education and progress toward gender parity at all levels. The Millennium Development Goals (MDG), approved by all member states of the United Nations in 2000, call for universal primary education in all countries by 2015, as well as gender parity at all levels by 2015.

There is good news to report. Impressive gains have been made toward higher levels of education enrollment and completion, and girls have been catching up rapidly with their brothers. As primary schooling expands, girls tend to be the main beneficiaries because of their historically disadvantaged position.

The rate of primary school completion also has improved faster for girls than for boys, again in large part because they had more to gain at the margins. Across all developing countries, girls' primary school completion increased by 17 percent, from 65 to 76 percent, between 1990 and 2000. During the same period, boys' primary completion increased by 8 percent, from 79 to 85 percent. Global progress is not matched, however, in every region. In sub-Saharan Africa, girls did only slightly better between 1990 and 2000, with primary completion increasing from 43 to 46 percent. (The primary completion rate for boys went in the opposite direction, from 57 to 56 percent.)

The overall good news about girls' progress must be tempered by realism, and a recognition that the goal is not to have boys' and girls' educational attainment "equally bad." Today, a mere nine years from the MDG deadline, it is clear that the important improvements over the past several decades in the developing world—in many instances, unprecedented rates of increase in primary school enrollment and completion—still leave a large number of poor countries very far from the target. While girls are making up ground rapidly, in many of the poorest countries the achievements on improved gender parity must be seen in the context of overall low levels of primary school completion.

An estimated 104 million to 121 million children of primary school age across the globe are not in school, with the worst shortfalls in Africa and South Asia. Completion of schooling is a significant problem. While enrollment has been increasing, many children drop out before finishing the fifth grade. In Africa, for example, just 51 percent of children (46 percent of girls) complete primary school. In South Asia, 74 percent of children (and just 63 percent of girls) do so.

Low levels of enrollment and completion are concentrated not only in certain regions but also among certain segments of the population. In every country completion rates are lowest for children from poor households. In Western and Central Africa, the median grade completed by the bottom 40 percent of the income distribution is zero, because less than half of poor children complete even the first year of school.

The education income gap also exacerbates gender disparities. In India, for example, the gap between boys and girls from the richest households is 2.5 percent, but the difference for children from the poorest households is 24 percent.

In some countries the main reason for low educational attainment is that children do not enroll in school. In Bangladesh, Benin, Burkina Faso, Ivory Coast, India, Mali, Morocco, Niger, and Senegal, more than half of children from the bottom 40 percent of the income distribution never even enroll. Elsewhere, particularly in Latin America, enrollment may be almost universal, but high repetition and dropout rates lead to low completion rates. In both cases poor students are much more likely not to complete school.

In many countries the rural/urban education gap is a key factor explaining education differentials. In Mozambique, the rural completion rate is 12 percent, while at the national level 26 percent of children complete school. Burkina Faso, Guinea, Madagascar, Niger, and Togo all demonstrate a similar pattern. In rural areas, the gender gap in completion is pronounced in Africa: in Benin, Burkina Faso, Guinea, Madagascar, Mozambique, and Niger, a mere 15 percent of girls who start primary school make it to the end.

Policy makers increasingly are recognizing the importance of addressing the special needs and vulnerabilities of marginal populations, even in relatively well-off countries with education levels that, on average, look quite good. As my colleagues Maureen Lewis and Marlaine Lockheed at the Center for Global Development highlight in a forthcoming book, girls who are members of marginalized groups—the Roma in Eastern Europe, the indigenous populations in Central America and elsewhere, the underprivileged castes and tribes in India—suffer a double disadvantage. Low educational attainment for girls is an obvious mechanism through which historical disadvantage is perpetuated. In Laos, for example, more than 90 percent of men in the dominant Laotai group are literate, while only 30 percent of the youngest cohort of women belonging to excluded rural ethnic groups can read and write.

Beyond the primary school enrollment and completion trends, a complex problem is the quality of education. Although measurement of learning outcomes is spotty at best, analyses of internationally comparable assessments of learning achievement in mathematics, reading, and science indicate that most developing countries rank far behind the industrialized nations. This is all the more of concern because the tests are taken by the children in school who, in low-enrollment countries, are the equivalent in relative terms to the top performers in the high-enrollment developed nations. The data on national examinations is equally alarming. Student performance on national exams in South Asian and African countries shows major gaps in acquisition of knowledge and skills.

Thus, the picture of progress and gaps is a complex one: rapid improvements relative to historical trends, but far off the ideal mark in the poorest countries. Girls are catching up quickly in most countries, but the level they are catching up to is still quite low. In many nations, the "lowest hanging fruit" has already been reached; for all children, and for girls in particular, the ones now out of school come from the most economically and socially disadvantaged backgrounds, and will be the hardest to reach. Finally, even among those children in school, evidence about poor learning outcomes should be cause for alarm.

THE CHALLENGES

The central imperative for improving educational opportunities and outcomes for girls in the low enrollment countries, including in sub-Saharan Africa and parts of South Asia, is to improve overall access and the quality of primary schooling. In doing so, planners and policy makers should ensure that they are not perpetuating barriers to girls' participation.

Getting to universal primary education (either enrollment or the more ambitious goal of completion) in sub-Saharan Africa and South Asia will require large-scale expansion in physical infrastructure, the number of teachers, and teaching/learning materials. Moreover, it will require fundamental improvements in the education institutions: more attention to learning outcomes rather than enrollment numbers, greater incentives for quality teaching, and more responsiveness to parents. This is a huge agenda. The donor and international technical community can support it, but it must be grounded in the political commitment of national and subnational governments.

Secondary to the "more and better education for all" agenda, and of particular relevance in countries that have already made significant progress so that most children go to school, is the need to understand and address the needs of particular disadvantaged groups, where gender differentials are especially pronounced. Beyond the efforts to reach children from poor and rural households, public policy makers need to understand and pay attention to ethnic and linguistic minorities, reaching them with tailored approaches rather than simply an expansion of the types of educational opportunities provided to the majority population. In addressing this challenge, policy makers must accept that reaching these key populations implies higher unit costs, as well as the adoption of potentially controversial measures, such as bilingual curriculum.

Finally, success in moving close to universal primary school enrollment generates its own new challenges. As more children complete primary school, the private benefits, in higher wages, decline (though the social benefits remain high). Private rates of return—perceived and real—cease to be seen as much of a reason for sending children to primary school, unless there is access to postprimary education. In addition, both the expansion of the existing education systems in many developing countries and the "scaling-up" of other public sector functions (such as health services, water management, and general public administration) require a larger cadre of educated and trained workers, the products of postprimary education. For these reasons, attention must be given to expanded opportunities for girls at the secondary level.

While international attention and goal-setting have been directed almost exclusively at the primary level, and the donor community has been persuaded by arguments about greater economic returns from primary education and the potentially regressive effects of investments at the secondary level, a large agenda remains unattended. It is at the secondary level that many of the microeconomic, health, and fertility outcomes of girls' education are fully realized. And common sense alone suggests that the large (and growing) cohort of children moving through primary schooling will create unsustainable pressures for postprimary education opportunities. If those are severely rationed, as they are in much of sub-Saharan Africa, the negative feedback to parents who sacrificed to send their children through primary school may be profound. Sorting out the design, financing, and institutional arrangements for effective secondary schooling—that is also responsive to labor market demand—is an essential part of good policy making today.

THE WAY FORWARD

Beyond general expansion of enrollment, governments can get out-of-school children into school by crafting specific interventions to reach them, and by increasing educational opportunities (formal and informal) for girls and women. In designing these initiatives, success depends on understanding and taking into account powerful demand-side influences that may constrain girls' school participation.

Specific interventions have been shown, in some settings, to get hard-to-reach children into school. These include eliminating school fees, instituting conditional cash transfers, using school feeding programs as an incentive to attend school, and implementing school health programs to reduce absenteeism. Several interventions have proved particularly successful where girls' participation is low. These include actions that increase security and privacy for girls (for example, ensuring that sanitation facilities are girl-friendly), as well as those that reduce gender-stereotyping in curriculum and encourage girls to take an active role in their education.

While few rigorous evaluations have been undertaken, many experts suggest that literacy programs for uneducated mothers may help increase school participation by their children. Adult literacy programs may be particularly useful in settings where there are pockets of undereducated women, such as ethnic or indigenous communities.

It is tempting for policy makers to focus on specific programmatic investments. But sustained improvements in education are impossible to achieve without improving the way in which key institutions in the sector function, and without increasing parental involvement in decisions affecting their children's education. Many countries with poorly performing educational systems suffer from institutional weaknesses, including low management capacity, nontransparent resource allocation and accounting practices, and substandard human resources policies and practices. Incentive structures that fail to reward good performance create and reinforce the most deleterious characteristics of weak institutions.

Parents who are well informed of policies and resource allocations in the education sector and who are involved in decisions regarding their children's schooling exert considerable influence and help contribute solutions. Involved communities are able to articulate local school needs, hold officials accountable, and mobilize local resources to fill gaps when the government response is inadequate.

A MODEST PROPOSAL

Donor agencies have been at the leading edge of the dialogue about the importance of girls' education, often providing the financial support, research, and political stimulus that may be lacking in countries that have more than their hands full with the basics of "Education for All." There is a broad consensus in the international donor community about the value of girls' education, and innovations have been introduced through donor-funded programs under the auspices of UNICEF, the World Food Program, the U.S. Agency for International Development, and other key agencies. These have been valuable contributions, and have supported the work of champions at the national and local levels.

The donor community could come together now to accelerate progress in a very particular way. Working with both governments and nongovernmental organizations in countries where specific excluded groups—ethnic and/or linguistic minorities—have much poorer education outcomes, donors could finance the design, introduction, and rigorous evaluation of targeted programs to improve access to appropriate educational opportunities, with a particular emphasis (if warranted by the baseline research) on the needs and characteristics of girls. While different bilateral and multilateral donors could take the lead in funding specific types of programs or working in particular countries on the challenge of the "doubly disadvantaged," a shared learning agenda could be coordinated across agencies to generate much more than the spotty anecdotes and case studies on which we currently depend.

The learning agenda would include three components: first, the enduring questions to be examined—for example, determining the most effective strategies to improve learning outcomes among children who come from households where the language spoken is not the language of instruction; second, the use of methods that permit observed results to be attributed to the program; and third, the features that will ensure maximum credibility of the evaluations, such as independence, dissemination of results (whether the findings are favorable or not), and wide sharing of the data for reanalysis.

Just as education can transform individuals' lives, learning what works can transform the debates in development policy. The beneficiaries in developing countries would include not only girls who receive the education they deserve and need, but also families and communities and future generations thereby lifted over time out of poverty.

60

Scanning the Global Situation and Prospects for the Future

JEROME C. GLENN

Considering the state of the world's people, economics, environment, and other factors is the task of those who study and predict the future, and who plan policies to alleviate global problems. Glenn begins by reviewing the major trends in the world today, points out key areas that need attention and reform, discusses the role of technology, and outlines goals and strategies for the future, especially those needed to achieve the United Nations Millennium Development Goals and 15 Global Challenges.

As you read this article, think about the following questions:

1. *What is the global situation today?*
2. *How will technology and connectedness affect the future?*
3. *What are four prospects for improving humanity?*
4. *What are goals, strategies, and prospects for the future?*

GLOSSARY   **Global warming** Warming of earth due to carbon dioxide and other emissions into the earth's atmosphere.   **Transnational organized crime** Crime that crosses country borders and becomes global.

P eople around the world are becoming healthier, wealthier, better educated, more peaceful, and increasingly connected, and they are living longer. At the same time, the world is more corrupt, congested, warmer, and increasingly dangerous. Although the digital divide is beginning to close, income gaps are still expanding around the world and unemployment continues to grow.

The global economy grew at 5.4% in 2006 to $66 trillion. The population grew 1.1%, and the average world per capita income increased by 4.3%. At this rate, world poverty will be cut by more than half between 2000 and 2015,

meeting the UN Millennium Development Goal for poverty reduction except in sub-Saharan Africa.

Although the majority of the world is improving economically, income disparities are still enormous: Two percent of the world's richest people own more than 50% of the world's wealth, while the poorest 50% of people own 1%. And the income of the 225 richest people in the world is equal to those of the poorest 2.7 billion, 40% of the world.

More than half the 6.6 billion people of the world live in urban environments. The information technology industry is laying the foundations for

Source: Glenn, Jerome C. 2008. "Scanning the Global Situation and Prospects for the Future." *The Futurist*. January–February, pp. 42–46.

cities to become augmented by ubiquitous computing for collective intelligence, with just-in-time knowledge for better management. Nanosensors and transceivers in nearly everything will make it easier to manage a city as a whole—from transportation to security.

Although great human tragedies like Iraq and Darfur dominate the news, the vast majority of the world is living in peace. Conflicts actually decreased over the past decade, dialogues among differing world views are growing, intrastate conflicts are increasingly being settled by international interventions, and the number of refugees is falling. The number of African conflicts fell from a peak of 16 in 2002 to five in 2005.

The prevalence of HIV/AIDS in Africa has begun to level off and could begin to actually decrease over the next few years. Meanwhile, the disease continues to spread rapidly in eastern Europe and in Central and South Asia. AIDS is the fourth leading cause of death in the world and the leading cause of death in sub-Saharan Africa.

The world's average life expectancy is increasing from 48 years for those born in 1955 to 73 years for those who will be born in 2025, according to the World Health Organization. The global population is changing from high mortality and high fertility to low mortality and low fertility. Population may increase by another 2.8 billion by 2050 before it begins to fall, according to the UN's lower forecast, after which it could be 5.5 billion by 2100—which is 1 billion fewer people than are alive today. However, technological breakthroughs are likely to change these forecasts over the next 50 years, giving people even longer and more productive lives than most would believe possible today.

According to UNESCO, in 1970 about 37% of all people over the age of 15 were illiterate. That has fallen to less than 18% today. Between 1999 and 2004, the number of children without primary education fell by around 21 million to 77 million.

Many of these trends are extremely hopeful, or at least have a hopeful element. From our research on all of these various indicators and trends, some key themes emerge that are worthy of special consideration, as they speak to some broad challenges and opportunities that will dominate the debate about how to make the world a better place in the years ahead. The state of the global environment is rising as a unifying issue. Many obstacles remain in the way of creating a more just and free world. And information technology will increasingly provide the tools for solving the great problems of the twenty-first century.

GLOBAL WARMING RISES AS A KEY ISSUE

The increasing and overwhelming evidence for global warming, the success of Al Gore's movie *An Inconvenient Truth* with his selection for the Nobel Peace Prize, and China's passing the United States in carbon dioxide emissions have made global climate change a top issue. The Intergovernmental Panel on Climate Change (IPCC) reported that CO_2 emissions rose faster than its worst-case scenario projected during 2000–2004 and that, without new government action, greenhouse gases will rise 25%–90% over 2000 levels by 2030. Applying data from the U.S. Geological Survey and the International Energy Agency, the Netherlands Environmental Assessment Agency estimated that China passed the United States in carbon emissions in 2006 by 8%. China now consumes 2 billion tons of coal each year and may produce 4 billion tons annually by 2016. There are 28,000 coal mines in China.

The United States actually decreased its CO_2 emissions in 2006 by 1.4% from the previous year. Fossil CO_2 emissions of the EU-15 countries remained almost constant in 2006. Hence, there is some good news: The rate of increase of CO_2 emissions in 2006 from fossil fuel use was about 2.6%, while in 2005 it was 3.3%. But the positive effects of this trend in relative reduction could be short-lived, as China builds more coal plants and purchases more cars.

Approximately 800 to 1,000 conventional coal plants are in some stage of planning or construction around the world. If built, they will have expected production lives of 40 years. Reducing greenhouse gas emissions becomes less likely if the countries building these plants complete construction. One impact of continued global warming is rising sea levels that threaten more than 634 million people living in coastal areas, according to NASA.

U.S. Vice Admiral Richard H. Truly has said that global warming is a uniquely serious environmental security problem because it's not like "some hot spot we're trying to handle. It's going to happen to every country and every person in the whole world at the same time." According to the IPCC report *Climate Change 2007: Impacts, Adaptation and Vulnerability*, the most severe impacts of climate change will be experienced by people in the poorest regions who have emitted the least amount of greenhouse gases. Billionaire Richard Branson has offered $25 million for a way to remove a billion tons of carbon dioxide a year from Earth's atmosphere, and he plans to invest $3 billion in fighting global warming.

The total number of people affected by natural disasters has tripled over the past decade to 2 billion people, with the accumulated impact of natural disasters resulting in an average of 211 million people directly affected each year, according to the International Federation of the Red Cross and Red Crescent Societies. This is approximately five times the number of people thought to have been affected by conflict over the past decade.

The world should pressure the United States and China to create and lead a global strategy to create safer energy with fewer greenhouse gas emissions, which would reduce climate change and continue economic growth. Initial U.S.–China cooperation has begun on cleaner coal processing and biofuels. The alternatives to those that produce nuclear waste or CO_2 emissions are proliferating.

The options to create and update global energy strategies seem too complex and rapidly changing for decision makers to make coherent policy. Yet, the environmental and social consequences of incoherent policy are so serious that international actors might be justified in spending resources to establish a new global system for the identification, analysis, possible consequence assessment, and synthesis of energy options. Whatever body or institution takes on the task of designing such a system, it should ensure that the system can be understood and used by the general public, politicians, and non-scientists, as well as by leading scientists and engineers around the world. The Millennium Project has begun an initial feasibility study for such a system.

When humans used up natural resources in the past, they just migrated to new areas with more resources. This strategy will not work as well for the 40% of humanity who live in India and China, as their water and soil resources are depleted. By 2025, 1.8 billion people could be living in water-scarce areas; as a result, mass migrations increase in likelihood. We have to develop more freshwater supplies, not just use pricing policies to redistribute resources. Both developed and developing nations will need to initiate massive desalinization measures, as well as seawater agriculture programs along 24,000 kilometers of desert coastlines to produce biofuels, food for humans and animals, and pulp for paper industries—all of which would free up freshwater for other purposes while absorbing CO_2.

UNEVEN PROGRESS IN PROTECTING FREEDOM AND JUSTICE

The number of free countries grew from 46 to 90 over the past 30 years, accounting for 46% of the world's population. Moreover, 64% of the world's nations have become electoral democracies, according to the independent, nongovernmental organization Freedom House. Since democracies tend not to fight each other, and since humanitarian crises are far more likely under authoritarian than democratic regimes, the trend toward democracy should lead to a more peaceful future among nation states.

Unfortunately, massively destructive weapons will be more available to more individuals. Future desktop molecular and pharmaceutical manufacturing and organized crime's access to nuclear materials give single individuals the ability to make and use weapons of mass destruction—from biological weapons to low-level nuclear ("dirty") bombs. The International Atomic Energy Agency reported 149 confirmed incidents of illicit use of radioactive materials in 2006. Only 10% of the 220 million sea containers that transport 90% of the world's trade are inspected, giving criminal and terrorist groups easier supply lines.

Transnational organized crime continues to grow in the absence of a comprehensive, integrated global counter-strategy. Its total annual income could be well over $2 trillion, giving it more financial resources than all the military budgets worldwide. The 13 to 15 million AIDS orphans, with potentially another 10 million by 2010, constitute a gigantic pool of new talent for organized crime.

Progress in the area of women's rights remains gradual. The International Labor Organization reports that the proportion of Legislative, senior official, or managerial positions held by women has grown slowly from 25.6% in 1995 to 28.3% today. Although condemnation of any form of discrimination against women is almost universal, progress is mixed. About 57% of women work in the cash economy, but only 17% are national legislators. There are now 94 girls in primary school for every 100 boys, up from 92 in 1999. Of the 181 countries with 2004 data available, about two-thirds have achieved gender parity in primary education. However, only one-third of the 177 countries have achieved parity on secondary education. Some 781 million adults lack minimum literacy skills; two-thirds are women. Violence against women by men continues to cause more casualties than wars do today.

Other disturbing trends caution against too much optimism in the area of human rights. More slaves exist in the world now than during the highest point of the African slave trade. Estimates vary from 12.3 million to 27 million, with the majority being women in Asia. More than $1 trillion is paid each year in political bribes, according to World Bank estimates, of which $20 billion to $40 billion is received by public officials from developing and transition countries and $60 billion to $80 billion in more-developed countries. News media, blogs,

mobile phone cameras, ethics commissions, and organizations like Transparency International are increasingly exposing corrupt decision-making processes. However, too few media outlets pay proper attention to these essential issues of justice and freedom.

A NEW ERA OF TECHNOLOGY AND CONNECTEDNESS

If current trends in information technology continue, within 25 years, a computer could emerge possessing all the processing power of the human brain; 25 years after that, a descendant computer could have the total processing power of all human brains. Imagine every individual having computer capability equal to all the human brains on Earth!

In the meantime, more than a billion people (17.5% of the world) are connected to the Internet. The digital divide is closing. Argentina, Uruguay, Brazil, Nigeria, Libya, Pakistan, and Thailand have each put in 250,000 orders for the MIT-inspired $100 laptops (actually $178) thus far. As the integration of cell phones, video, and the Internet grows, prices will fall, accelerating globalization and allowing swarms of people to quickly form and disband, coordinate actions, and share information ranging from stock market tips to contagious new ideas.

As the world moves toward ubiquitous computing with collective intelligence for just-in-time knowledge, decisions should improve. Decision making will increasingly be augmented by the integration of ubiquitous sensors, a more intelligent Web, and institutional and personal intelligence software that helps us receive and respond to feedback for improving decisions.

The world is expected to have produced more data in 2007 than it can store. According to the firm IDC, the world produced 161 exabytes (billion gigabytes) in 2006 and had 185 exabytes of storage capacity. With the increased use of multimedia systems like YouTube, the profusion of surveillance cameras, and regulatory rules for corporate data retention, 988 exabytes (nearly 1 zettabyte) could be produced in 2010, but only 601 exabytes are expected to be available for storage by 2010.

These technological trends in data creation are having attendant consequences on global wealth and prosperity. World trade grew 15% in 2006, according to the World Trade Organization. However, higher oil and commodity prices contributed to the 30% trade growth for the least-developed countries—a world record—and their economies continued to exceed 6% growth for the third year in a row. The debt-to-GDP ratios decreased in all developing regions, partly due to debt forgiveness. Excluding South Africa, the economies of sub-Saharan Africa averaged 4.5% growth, but poverty continues to grow due to high birth rates, corruption, armed conflicts, poor governance, environmental degradation, poor health conditions, and lack of education.

Since the world has 2.4 million doctors, nurses, and midwives, according to the World Health Organization, governments, NGOs, and health organizations may increasingly turn to telemedicine, biochip sensors for self-diagnosis, and other automated systems to mitigate the effects of people living longer. The threat of SARS has been eliminated by a well-managed, coherent human response. Now the world is preparing for genetic variations that could occur in the avian flu H5N1 virus (or some mutation) that could kill 25 million people, bringing air transportation to a halt and throwing the world into a depression.

The extraordinary impacts of science and technology over the past 25 years will seem slight compared with what is likely to happen in the next 25 years. The factors that accelerated the rate of innovation are themselves changing at accelerating rates. Transistors are now smaller than light waves (65 nanometers). Intel has created the first programmable 1 Teraflop chip able to perform more than 1 trillion floating point operations per second. The brain-computer interface now lets thoughts move software, nanoparticles, and fibers stimulate neural growth, and mini-biocomputers help treat specific individual cells. Photons have been slowed and accelerated, adult stem cells have been regressed to repair damaged tissue, and microbial fuel cells have been demonstrated. China

plans to be the fourth country (after the United States, Russia, and Japan) to orbit the moon.

Some experts forecast that molecular manufacturing and 3-D printing will eventually allow people to "print" high-tech objects previously shipped around the world. If that day ever comes, then shipping bytes instead of atoms would dramatically alter industrial world trade. According to Lux Research, $12.4 billion was invested in nanotech R&D worldwide in 2006, and more than $50 billion worth of nano-enabled products were sold.

GOALS AND STRATEGIES FOR THE FUTURE

Achieving the UN Millennium Development Goals could cost $135 billion (by current estimates); by comparison, the U.S. Congress has approved $600 billion for the war in Iraq, and another $140 billion may be requested for 2008.

These numbers speak to a global need for greater governmental, corporate, academic, scientific, engineering, and medical focus on the issues most relevant to the broader global situation, achieving the eight UN Millennium Development Goals, and addressing the 15 Global Challenges described in *State of the Future*. We need transinstitutional management and more serious public education through the media.

National decision makers have not been trained in the theory and practice of decision making, and few know how advanced decision-support software could help them. Formalized ethics and decision training for decision makers could result in a significant improvement in the quality of global decisions.

In addition to policy makers needing training in how to make decisions, processes to set local, national, and international priorities need further development. We know the world is increasingly complex and that the most serious challenges are global in nature, yet we don't seem to know how to improve and deploy Internet-based management tools and concepts fast enough to get on top of the situation.

Drawing on his experience as secretary-general of the United Nations, Kofi Annan has identified four principles to improve prospects for humanity:

1. The security of everyone is the security of everyone else.

2. We are responsible for each other's welfare (global solidarity).

3. Respect for each other should be reinforced by human rights and rule of law.

4. Governments must be accountable both internally and internationally (mutual accountability); and these four principles can be achieved through multilateral institutions like the United Nations, he believes.

Increased acknowledgment of climate change and other forms of global interdependence, such as financial links and communicable diseases, demonstrates the need for global systems for resilience—the capacity to anticipate, respond to, and recover from disasters such as tsunamis, massive migrations due to water shortages, prolonged electric or Internet outages, financial crashes, and conflicts. If much of the global complexity cannot be managed efficiently by current systems, then we need new decision-making systems. The International Organization for Standardization (with more than 16,000 ISO standards) and the Internet have proved effective mediums for self-organized decision making. Hence, policy makers should look at creating self-organizing global systems for resilience.

Although many people criticize globalization's potential cultural impacts, cultural change is necessary to address global challenges. The development of genuine democracy, preventing AIDS, sustainable development, ending violence against women, and ending ethnic violence all require cultural change. We should use the tools of globalization, such as the Internet, global trade, international trade treaties, and international outsourcing, to help cultures adapt in a way that preserves their unique contributions to humanity while improving the human condition.

It has been considered ridiculous to pursue health and security for all people. Equally ridiculous today is thinking that one day an individual acting alone will not be able to create and use a weapon of mass destruction. Likewise, we cannot assume that there will not be serious pandemics as we crowd more people and animal habitats into urban concentrations, transborder travel grows easier, and biodiversity diminishes. Viewing the welfare of one being as necessary to the welfare of all could become a pragmatic long-range approach to countering terrorism, keeping airports open, and preventing destructive mass migrations and other potential threats to human security. Ridiculing idealism is shortsighted, but idealism without the rigors of pessimism is misleading. We need very hard-headed idealists who can look into the worst and best of humanity and can create and implement strategies of success.

There are many answers to many problems, but there is much extraneous information in the global media landscape. As a result, too many decision makers have difficulty identifying and concentrating on what is truly relevant. Since healthy democracies need relevant information and since democracy is becoming more global, the public will need globally relevant information to sustain this trend. We hope the annual *State of the Future* reports can help provide such information.

The insights in this eleventh year of the Millennium Project's work as reported in *2007 State of the Future* can help decision makers and educators who fight against hopeless despair, blind confidence, and ignorant indifference—attitudes that too often have blocked efforts to improve the prospects for humanity.